# Regression and Time Series Model Selection

# Regression and Time Series Model Selection

Allan D R McQuarrie
*North Dakota State University*
Chih-Ling Tsai
*University of California, Davis*

*Published by*

World Scientific Publishing Co. Pte. Ltd.

P O Box 128, Farrer Road, Singapore 912805

*USA office:* Suite 1B, 1060 Main Street, River Edge, NJ 07661

*UK office:* 57 Shelton Street, Covent Garden, London WC2H 9HE

**British Library Cataloguing-in-Publication Data**
A catalogue record for this book is available from the British Library.

First published 1998
Reprinted 1999

**REGRESSION AND TIME SERIES MODEL SELECTION**



ISBN 981-02-3242-X

Printed in Singapore.

To our parents, Donald and Carole, Liang-Chih and Chin-Lin; our wives, Penelope, Yu-Yen; our children, Antigone and Evan, Wen-Lin and Wen-Ting; our teachers; and our thesis advisers, Robert Shumway, Dennis Cook.

## Contents

Preface ..... xiii

List of Tables ..... xv

Chapter 1 Introduction 1

1.1. Background ..... 1

1.1.1. Historical Review ..... 2

1.1.2. Efficient Criteria ..... 3

1.1.3. Consistent Criteria ..... 3

1.2. Overview ..... 4

1.2.1. Distributions ..... 4

1.2.2. Model Notation ..... 5

1.2.3. Discrepancy and Distance Measures ..... 5

1.2.4. Efficiency under Kullback–Leibler and $L_2$ ..... 7

1.2.5. Overfitting and Underfitting ..... 8

1.3. Layout ..... 9

1.4. Topics Not Covered ..... 13

Chapter 2 The Univariate Regression Model 15

2.1. Model Description ..... 16

2.1.1. Model Structure and Notation ..... 16

2.1.2. Distance Measures ..... 17

2.2. Derivations of the Foundation Model Selection Criteria ..... 19

2.3. Moments of Model Selection Criteria ..... 24

2.3.1. AIC and AICc ..... 25

2.3.2. FPE and Cp ..... 27

2.3.3. SIC and HQ ..... 29

2.3.4. Adjusted $R^2$, $R^2_{adj}$ ..... 30

2.4. Signal-to-noise Corrected Variants ..... 31

2.4.1. AICu ..... 32

2.4.2. FPEu ..... 33

2.4.3. HQc ... 34
2.5. Overfitting ... 35
2.5.1. Small-sample Probabilities of Overfitting ... 36
2.5.2. Asymptotic Probabilities of Overfitting ... 40
2.5.3. Small-sample Signal-to-noise Ratios ... 43
2.5.4. Asymptotic Signal-to-noise Ratios ... 43
2.6. Small-sample Underfitting ... 45
2.6.1. Distributional Review ... 45
2.6.2. Expectations of $L_2$ and Kullback–Leibler Distance ... 48
2.6.3. Expected Values for Two Special Case Models ... 50
2.6.4. Signal-to-noise Ratios for Two Special Case Models ... 54
2.6.5. Small-sample Probabilities for Two Special Case Models ... 57
2.7. Random $X$ Regression and Monte Carlo Study ... 60
2.8. Summary ... 64
Appendix 2A. Distributional Results in the Central Case ... 66
Appendix 2B. Proofs of Theorems 2.1 to 2.6 ... 70
Appendix 2C. Small-sample and Asymptotic Properties ... 77
Appendix 2D. Moments of the Noncentral $\chi^2$ ... 87

Chapter 3 The Univariate Autoregressive Model 89
3.1. Model Description ... 89
3.1.1. Autoregressive Models ... 89
3.1.2. Distance Measures ... 91
3.2. Selected Derivations of Model Selection Criteria ... 93
3.2.1. AIC ... 93
3.2.2. AICc ... 93
3.2.3. AICu ... 94
3.2.4. FPE ... 94
3.2.5. FPEu ... 95
3.2.6. Cp ... 95
3.2.7. SIC ... 96
3.2.8. HQ ... 96
3.2.9. HQc ... 97
3.3. Small-sample Signal-to-noise Ratios ... 97
3.4. Overfitting ... 100
3.4.1. Small-sample Probabilities of Overfitting ... 100
3.4.2. Asymptotic Probabilities of Overfitting ... 106
3.4.3. Small-sample Signal-to-noise Ratios ... 108
3.4.4. Asymptotic Signal-to-noise Ratios ... 108

3.5. Underfitting for Two Special Case Models .......... 111
3.5.1. Expected Values for Two Special Case Models .......... 111
3.5.2. Signal-to-noise Ratios for Two Special Case Models .......... 114
3.5.3. Probabilities for Two Special Case Models .......... 116
3.6. Autoregressive Monte Carlo Study .......... 117
3.7. Moving Average MA(1) Misspecified as Autoregressive Models .......... 120
3.7.1. Two Special Case MA(1) Models .......... 121
3.7.2. Model and Distance Measure Definitions .......... 121
3.7.3. Expected Values for Two Special Case Models .......... 122
3.7.4. Misspecified MA(1) Monte Carlo study .......... 124
3.8. Multistep Forecasting Models .......... 126
3.8.1. Kullback–Leibler Discrepancy for Multistep .......... 127
3.8.2. AICcm, AICm, and FPEm .......... 128
3.8.3. Multistep Monte Carlo Study .......... 129
3.9. Summary .......... 130
Appendix 3A. Distributional Results in the Central Case .......... 130
Appendix 3B. Small-sample Probabilities of Overfitting .......... 132
Appendix 3C. Asymptotic Results .......... 137

Chapter 4 The Multivariate Regression Model 141
4.1. Model Description .......... 142
4.1.1. Model Structure and Notation .......... 142
4.1.2. Distance Measures .......... 144
4.2. Selected Derivations of Model Selection Criteria .......... 145
4.2.1. $L_2$-based Criteria FPE and Cp .......... 145
4.2.2. Kullback–Leibler-based Criteria AIC and AICc .......... 147
4.2.3. Consistent Criteria SIC and HQ .......... 149
4.3. Moments of Model Selection Criteria .......... 149
4.3.1. AIC and AICc .......... 150
4.3.2. SIC and HQ .......... 152
4.4. Signal-to-noise Corrected Variants .......... 154
4.4.1. AICu .......... 154
4.4.2. HQc .......... 156
4.5. Overfitting Properties .......... 157
4.5.1. Small-sample Probabilities of Overfitting .......... 157
4.5.2. Asymptotic Probabilities of Overfitting .......... 160
4.5.3. Asymptotic Signal-to-noise Ratio .......... 163
4.6. Underfitting .......... 165
4.6.1. Distributions for Underfitted Models .......... 166

4.6.2. Expected Values for Two Special Case Models ............ 168
4.6.3. Signal-to-noise Ratios for Two Special Case Models ....... 171
4.6.4. Probabilities for Two Special Case Models ................ 174
4.7. Monte Carlo Study ........................................ 175
4.8. Summary .................................................. 179
Appendix 4A. Distributional Results in the Central Case ............ 180
Appendix 4B. Proofs of Theorems 4.1 to 4.5 ........................ 183
Appendix 4C. Small-sample Probabilities of Overfitting ............. 190
Appendix 4D. Asymptotic Probabilities of Overfitting ............... 193
Appendix 4E. Asymptotic Signal-to-noise Ratios .................... 196

Chapter 5 The Vector Autoregressive Model 199
5.1. Model Description ........................................ 199
5.1.1. Vector Autoregressive Models ............................ 199
5.1.2. Distance Measures ....................................... 201
5.2. Selected Derivations of Model Selection Criteria ............ 203
5.2.1. FPE ..................................................... 203
5.2.2. AIC ..................................................... 204
5.2.3. AICc .................................................... 205
5.2.4. AICu .................................................... 205
5.2.5. SIC ..................................................... 206
5.2.6. HQ ...................................................... 206
5.2.7. HQc ..................................................... 206
5.3. Small-sample Signal-to-noise Ratios ......................... 206
5.3.1. AIC ..................................................... 207
5.3.2. AICc .................................................... 209
5.3.3. AICu .................................................... 209
5.3.4. SIC ..................................................... 210
5.3.5. HQ ...................................................... 211
5.3.6. HQc ..................................................... 212
5.4. Overfitting .............................................. 213
5.4.1. Small-sample Probabilities of Overfitting ................ 213
5.4.2. Asymptotic Probabilities of Overfitting .................. 217
5.4.3. Asymptotic Signal-to-noise Ratios ........................ 220
5.5. Underfitting in Two Special Case Models ..................... 222
5.5.1. Expected Values for Two Special Case Models ............. 223
5.5.2. Signal-to-noise Ratios for Two Special Case Models ....... 226
5.5.3. Probabilities for Two Special Case Models ................ 229
5.6. Vector Autoregressive Monte Carlo Study ..................... 230

5.7. Summary ... 234
Appendix 5A. Distributional Results in the Central Case ... 235
Appendix 5B. Small-sample Probabilities of Overfitting ... 238
Appendix 5C. Asymptotic Probabilities of Overfitting ... 244
Appendix 5D. Asymptotic Signal-to-noise Ratios ... 248

Chapter 6 Cross-validation and the Bootstrap 251
6.1. Univariate Regression Cross-validation ... 251
6.1.1. Withhold-1 Cross-validation ... 251
6.1.2. Delete-*d* Cross-validation ... 254
6.2. Univariate Autoregressive Cross-validation ... 255
6.2.1. Withhold-1 Cross-validation ... 255
6.2.2. Delete-*d* Cross-validation ... 256
6.3. Multivariate Regression Cross-validation ... 257
6.3.1. Withhold-1 Cross-validation ... 257
6.3.2. Delete-*d* Cross-validation ... 259
6.4. Vector Autoregressive Cross-validation ... 260
6.4.1. Withhold-1 Cross-validation ... 260
6.4.2. Delete-*d* Cross-validation ... 261
6.5. Univariate Regression Bootstrap ... 261
6.5.1. Overview of the Bootstrap ... 261
6.5.2. Doubly Cross-validated Bootstrap Selection Criterion ... 266
6.6. Univariate Autoregressive Bootstrap ... 268
6.7. Multivariate Regression Bootstrap ... 270
6.8. Vector Autoregressive Bootstrap ... 274
6.9. Monte Carlo Study ... 276
6.9.1. Univariate Regression ... 277
6.9.2. Univariate Autoregressive Models ... 283
6.10. Summary ... 290

Chapter 7 Robust Regression and Quasi-likelihood 293
7.1. Nonnormal Error Regression Models ... 293
7.1.1. $L_1$ Distance and Efficiency ... 294
7.2. Least Absolute Deviations Regression ... 295
7.2.1. L1AICc ... 295
7.2.2. Special Case Models ... 297
7.3. Robust Version of Cp ... 304
7.3.1. Derivation of RCp ... 304
7.4. Wald Test Version of Cp ... 306

7.5. FPE for Robust Regression ........ 307
7.6. Unification of AIC Criteria ........ 309
7.6.1. The Unification of the AIC Family ........ 310
7.6.2. Location–Scale Regression Models ........ 312
7.6.3. Monte Carlo Study ........ 315
7.7. Quasi-likelihood ........ 316
7.7.1. Selection Criteria for Extended Quasi-likelihood Models ........ 317
7.7.2. Quasi-likelihood Monte Carlo Study ........ 319
7.8. Summary ........ 326
Appendix 7A. Derivation of AICc under Quasi-likelihood ........ 327

Chapter 8 Nonparametric Regression and Wavelets 329
8.1. Model Selection in Nonparametric Regression ........ 330
8.1.1. AIC for Smoothing Parameter Selection ........ 333
8.1.2. Nonparametric Monte Carlo Study ........ 338
8.2. Semiparametric Regression Model Selection ........ 348
8.2.1. The Family of Candidate Models ........ 349
8.2.2. AICc ........ 350
8.2.3. Semiparametric Monte Carlo Study ........ 350
8.3. A Cross-validatory AIC for Hard Wavelet Thresholding ........ 351
8.3.1. Wavelet Reconstruction and Thresholding ........ 353
8.3.2. Nason's Cross-validation Method ........ 355
8.3.3. Cross-validatory AICc ........ 356
8.3.4. Properties of the AICc Selected Estimator ........ 359
8.3.5. Wavelet Monte Carlo Study ........ 362
8.4. Summary ........ 363

Chapter 9 Simulations and Examples 365
9.1. Introduction ........ 365
9.1.1. Univariate Criteria List ........ 366
9.1.2. Multivariate Criteria List ........ 367
9.1.3. Nonparametric Rank Test for Criteria Comparison ........ 368
9.2. Univariate Regression Models ........ 369
9.2.1. Model Structure ........ 369
9.2.2. Special Case Models ........ 369
9.2.3. Large-scale Small-sample Simulations ........ 371
9.2.4. Large-sample Simulations ........ 375
9.2.5. Real Data Example ........ 377
9.3. Autoregressive Models ........ 379

9.3.1. Model Structure ... 379
9.3.2. Two Special Case Models ... 380
9.3.3. Large-scale Small-sample Simulations ... 382
9.3.4. Large-sample Simulations ... 384
9.3.5. Real Data Example ... 386
9.4. Moving Average MA(1) Misspecified as Autoregressive Models ... 387
9.4.1. Model Structure ... 387
9.4.2. Two Special Case Models ... 388
9.4.3. Large-scale Small-sample Simulations ... 390
9.4.4. Large-sample Simulations ... 391
9.5. Multivariate Regression Models ... 392
9.5.1. Model Structure ... 392
9.5.2. Two Special Case Models ... 393
9.5.3. Large-scale Small-sample Simulations ... 395
9.5.4. Large-sample Simulations ... 397
9.5.5. Real Data Example ... 399
9.6. Vector Autoregressive Models ... 401
9.6.1. Model Structure ... 401
9.6.2. Two Special Case Models ... 401
9.6.3. Large-scale Small-sample Simulations ... 403
9.6.4. Large-sample Simulations ... 407
9.6.5. Real Data Example ... 409
9.7. Summary ... 410
Appendix 9A. Details of Simulation Results ... 412
Appendix 9B. Stepwise Regression ... 427

References ... 430

Author Index ... 440

Index ... 445

# Preface

Why a book on model selection? The selection of an appropriate model from a potentially large class of candidate models is an issue that is central to regression, times series modeling, and generalized linear models. The variety of model selection methods in use not only demonstrates a wide range of statistical techniques, but it also illustrates the creativity statisticians have employed to approach various problems—there are parametric procedures, data resampling and bootstrap procedures, and a full complement of nonparametric procedures as well. The object of this book is to connect many different aspects of the growing model selection field by examining the different lines of reasoning that have motivated the derivation of both classical and modern criteria, and then to examine the performance of these criteria to see how well it matches the intent of their creators. In this way we hope to bridge theory and practicality with a book that can serve both as a guide to the researcher in techniques for the application of these criteria, and also as a resource for the practicing statistician for matching appropriate selection criteria to a given problem or data set.

We begin to understand the different approaches that inspired the many criteria considered in this book by deriving some of the most commonly used selection criteria. These criteria are themselves statistics, and have their own moments and distributions. An evaluation of the properties of these moments leads us to suggest a new model selection criterion diagnostic, the *signal-to-noise ratio*. The signal-to-noise ratio and other properties such as expectations, mean and variance for differences between two models, and probabilities of selecting one model over another, can be used not only to evaluate individual criterion performance, but also to suggest modifications to improve that performance. We determine relative performance by comparing criteria against each other under a wide variety of simulated conditions. The simulation studies in this book, some of the most detailed in the literature, are a useful tool for narrowing the field of selection criteria that are applicable to a given practical scenario. We cover parametric, nonparametric, semiparametric, and wavelet regression models as well as univariate and multivariate response structures. We discuss bootstrap, cross validation, and robust methods. While we focus on Gaussian random errors, we also consider quasi-likelihood and location-scale distributions. Overall, this book collects and relates a broad range of insightful work in the field of model selection, and we hope that a diverse readership will find it accessible.

We wish to thank our families for their support, without which this book would not have been possible. We also thank Michelle Pallas for her carefully review and constructive comments, and Rhonda Boughtin for proofreading. Finally, we are grateful for the direct and indirect inspiration, assistance, suggestions, and comments made by Raj Bhansali, Peter Brockwell, Prabir Burman, Richard Davis, Clifford Hurvich, David Rocke, Elvezio Ronchetti, Ritei Shibata, Peide Shi, Jeffrey Simonoff, Robert Shumway, David Woodruff and Jeff Wu. The National Science Foundation provided partial support for Chih-Ling Tsai's research. The manuscript was typeset using TEX, and the graphs were produced using the postscript features of VTEX.

*North Dakota State University* A.D.R. McQuarrie
*University of California, Davis* C.L. Tsai

## List of Tables


Table 2.1. Probabilities of overfitting. 38
Table 2.2. Signal-to-noise ratios for overfitting. 39
Table 2.3. Asymptotic probability of overfitting by $L$ variables. 42
Table 2.4. Asymptotic signal-to-noise ratio for overfitting by $L$ variables. 45
Table 2.5. Expected values and expected efficiency for Model 1. 51
Table 2.6. Expected values and expected efficiency for Model 2. 53
Table 2.7. Signal-to-noise ratios for Model 1. 55
Table 2.8. Signal-to-noise ratios for Model 2. 56
Table 2.9. Probability of selecting a particular candidate model of order $k$ over the true order 6 for Model 1. 58
Table 2.10. Probability of selecting a particular candidate model of order $k$ over the true order 6 for Model 2. 59
Table 2.11. Simulation results for Model 1. Counts and observed efficiency. 62
Table 2.12. Simulation results for Model 2. Counts and observed efficiency. 63
Table 3.1. Probabilities of overfitting. 104
Table 3.2. Signal-to-noise ratios for overfitting. 105
Table 3.3. Asymptotic probability of overfitting by $L$ variables. 108
Table 3.4. Asymptotic signal-to-noise ratio for overfitting by $L$ variables. 110
Table 3.5. Expected values and expected efficiency for Model 3. 112
Table 3.6. Expected values and expected efficiency for Model 4. 113
Table 3.7. Signal-to-noise ratios for Model 3. 114
Table 3.8. Signal-to-noise ratios for Model 4. 115
Table 3.9. Probability of selecting order $p$ over the true order 5 for Model 3. 117
Table 3.10. Probability of selecting order $p$ over the true order 5 for Model 4. 117
Table 3.11. Simulation results for Model 3. Counts and observed efficiency. 118
Table 3.12. Simulation results for Model 4. Counts and observed efficiency. 119

Table 3.13. Expected values and expected efficiency for Model 5. ...... 122
Table 3.14. Expected values and expected efficiency for Model 6. ...... 123
Table 3.15. Simulation results for Model 5. Counts and observed efficiency. ...... 125
Table 3.16. Simulation results for Model 6. Counts and observed efficiency. ...... 126
Table 3.17. Multistep AR simulation results. ...... 129
Table 4.1. Asymptotic probability of overfitting by $L$ variables for $q = 2$. ...... 162
Table 4.2. Asymptotic probability of overfitting by $L$ variables for $q = 5$. ...... 163
Table 4.3. Asymptotic signal-to-noise ratios for overfitting by $L$ variables for $q = 2$. ...... 165
Table 4.4. Asymptotic signal-to-noise ratios for overfitting by $L$ variables for $q = 5$. ...... 165
Table 4.5. Expected values and expected efficiency for Model 7. ...... 169
Table 4.6. Expected values and expected efficiency for Model 8. ...... 170
Table 4.7. Approximate signal-to-noise ratios for Model 7. ...... 172
Table 4.8. Approximate signal-to-noise ratios for Model 8. ...... 173
Table 4.9. Probability of selecting a particular candidate model of order $k$ over the true order 5 for Model 7. ...... 174
Table 4.10. Probability of selecting a particular candidate model of order $k$ over the true order 5 for Model 8. ...... 175
Table 4.11. Simulation results for Model 7. Counts and observed efficiency. ...... 177
Table 4.12. Simulation results for Model 8. Counts and observed efficiency. ...... 178
Table 5.1. Asymptotic probability of overfitting by $L$ variables for $q = 2$. ...... 219
Table 5.2. Asymptotic probability of overfitting by $L$ variables for $q = 5$. ...... 219
Table 5.3. Asymptotic signal-to-noise ratios for overfitting by $L$ variables for $q = 2$. ...... 220
Table 5.4. Asymptotic signal-to-noise ratios for overfitting by $L$ variables for $q = 5$. ...... 220
Table 5.5. Expected values and expected efficiency for Model 9. ...... 224
Table 5.6. Expected values and expected efficiency for Model 10. ...... 225
Table 5.7. Approximate signal-to-noise ratios for Model 9. ...... 226
Table 5.8. Approximate signal-to-noise ratios for Model 10. ...... 227

Table 5.9. Probability of selecting order $k$ over the true order 4 for Model 9. .......... 229
Table 5.10. Probability of selecting order $k$ over the true order 4 for Model 10. .......... 230
Table 5.11. Simulation results for Model 9. Counts and observed efficiency. .......... 231
Table 5.12. Simulation results for Model 10. Counts and observed efficiency. .......... 232
Table 6.1. Summary of the regression models in simulation study. .......... 277
Table 6.2. Relationship between parameter structure and true order. .......... 277
Table 6.3. Bootstrap relative K-L performance. .......... 278
Table 6.4. Simulation results for Model 11. Counts and observed efficiency. .......... 280
Table 6.5. Simulation results for Model 12. Counts and observed efficiency. .......... 281
Table 6.6. Simulation results over 48 regression models—K-L observed efficiency ranks. .......... 283
Table 6.7. Simulation results over 48 regression models—$L_2$ observed efficiency ranks. .......... 283
Table 6.8. Summary of the autoregressive models. .......... 284
Table 6.9. Relationship between parameter structure and true model order. .......... 284
Table 6.10. Bootstrap relative K-L performance. .......... 285
Table 6.11. Simulation results for Model 13. Counts and observed efficiency. .......... 287
Table 6.12. Simulation results for Model 14. Counts and observed efficiency. .......... 288
Table 6.13. Simulation results over 36 autoregressive models—K-L observed efficiency ranks. .......... 289
Table 6.14. Simulation results over 36 autoregressive models—$L_2$ observed efficiency ranks. .......... 289
Table 7.1. Simulation results for Model 15. Counts and observed efficiency. .......... 299
Table 7.2. Simulation results for Model 16. Counts and observed efficiency. .......... 300
Table 7.3. Simulation results for Model 17. Counts and observed efficiency. .......... 301
Table 7.4. Simulation results for Model 18. Counts and observed efficiency. .......... 302

Table 7.5. Model selection performance of $\hat{L}(k)$. .......... 309
Table 7.6. Proportion of correct order selection. .......... 316
Table 7.7. Simulation results for Model 19. Counts and observed efficiency. .......... 321
Table 7.8. Simulation results for Model 20. Counts and observed efficiency. .......... 322
Table 7.9. Simulation results for Model 21. Counts and observed efficiency. .......... 323
Table 7.10. Simulation results for Model 22. Counts and observed efficiency. .......... 324
Table 8.1. Simulation results for the local linear estimator. .......... 341
Table 8.2. Simulation results for the local quadratic estimator. .......... 342
Table 8.3. Simulation results for the second-order convolution kernel estimator. .......... 343
Table 8.4. Simulation results for the fourth-order convolution kernel estimator. .......... 344
Table 8.5. Simulation results for the cubic smoothing spline estimator. .......... 345
Table 8.6. Simulation results for uniform random design. .......... 346
Table 8.7. Estimated probability of choosing the correct order of the true parametric component. .......... 351
Table 8.8. Average *ISE* values of hard threshold estimators. .......... 363
Table 9.1. Simulation results summary for Model 1. K-L observed efficiency ranks, $L_2$ observed efficiency ranks and counts. .......... 370
Table 9.2. Simulation results summary for Model 2. K-L observed efficiency ranks, $L_2$ observed efficiency ranks and counts. .......... 370
Table 9.3. Summary of the regression models in simulation study. .......... 371
Table 9.4. Relationship between parameter structure and true order. .......... 371
Table 9.5. Simulation results over 540 models. Summary of overall rank by K-L and $L_2$ observed efficiency. .......... 373
Table 9.6. Simulation results summary for Model A1. K-L observed efficiency ranks, $L_2$ observed efficiency ranks and counts. .......... 376
Table 9.7. Simulation results summary for Model A2. K-L observed efficiency ranks, $L_2$ observed efficiency ranks and counts. .......... 376
Table 9.8. Model choices for highway data example. .......... 378
Table 9.9. Regression statistics for model x1, x4, x8, x9, x12. .......... 378
Table 9.10. Regression statistics for model x1, x4, x9. .......... 379
Table 9.11. Simulation results summary for Model 3. K-L observed efficiency ranks, $L_2$ observed efficiency ranks and counts. .......... 381

Table 9.12. Simulation results summary for Model 4. K-L observed efficiency ranks, $L_2$ observed efficiency ranks and counts. . . 381
Table 9.13. Summary of the autoregressive models. . . . . . . . . . . . . . . . . . 382
Table 9.14. Relationship between parameter structure and true model order. . . . . . . . . . . . . . . . . . . . . . . . . . . . . . . . . . . . . . . . . . . 382
Table 9.15. Simulation results over 360 models. Summary of overall rank by K-L and $L_2$ observed efficiency. . . . . . . . . . . . . . . . . . . 383
Table 9.16. Simulation results summary for Model A3. K-L observed efficiency ranks, $L_2$ observed efficiency ranks and counts. . . 385
Table 9.17. Simulation results summary for Model A4. K-L observed efficiency ranks, $L_2$ observed efficiency ranks and counts. . . 385
Table 9.18. Model choices for Wolf sunspot data. . . . . . . . . . . . . . . . . . . . . 387
Table 9.19. AR(9) model statistics. . . . . . . . . . . . . . . . . . . . . . . . . . . . . . . . . 387
Table 9.20. Simulation results summary for Model 5. K-L observed efficiency ranks and $L_2$ observed efficiency ranks. . . . . . . . . . 389
Table 9.21. Simulation results summary for Model 6. K-L observed efficiency ranks and $L_2$ observed efficiency ranks. . . . . . . . . . 389
Table 9.22. Summary of the misspecified MA(1) models. . . . . . . . . . . . . . . 390
Table 9.23. Simulation results over 50 models. Summary of overall rank by K-L and $L_2$ observed efficiency. . . . . . . . . . . . . . . . . . . 391
Table 9.24. Simulation results summary for Model A5. K-L observed efficiency ranks and $L_2$ observed efficiency ranks. . . . . . . . . . 391
Table 9.25. Simulation results summary for Model 7. K-L observed efficiency ranks and $L_2$ observed efficiency ranks. . . . . . . . . . 394
Table 9.26. Simulation results summary for Model 8. K-L observed efficiency ranks and $L_2$ observed efficiency ranks. . . . . . . . . . 394
Table 9.27. Summary of multivariate regression models. . . . . . . . . . . . . . . 395
Table 9.28. Relationship between parameter structure and true order. . 396
Table 9.29. Simulation results over 504 Models. Summary of overall rank by K-L and $L_2$ observed efficiency. . . . . . . . . . . . . . . . . . . 397
Table 9.30. Simulation results summary for Model A6. K-L observed efficiency ranks, $L_2$ observed efficiency ranks and counts. . . 398
Table 9.31. Simulation results summary for Model A7. K-L observed efficiency ranks, $L_2$ observed efficiency ranks and counts. . . 398
Table 9.32. Multivariate real data selected models. . . . . . . . . . . . . . . . . . . 400
Table 9.33. Multivariate regression results for x1, x2, x6. . . . . . . . . . . . . . 400
Table 9.34. Multivariate regression results for x1, x2, x4, x6. . . . . . . . . . 400
Table 9.35. Simulation results summary for Model 9. K-L observed efficiency ranks, $L_2$ observed efficiency ranks and counts. . . 402

Table 9.36. Simulation results summary for Model 10. K-L observed efficiency ranks, $L_2$ observed efficiency ranks and counts. . . 403
Table 9.37. Summary of vector autoregressive (VAR) models. . . . . . . . . . 404
Table 9.38. Relationship between parameter structure and true model order. . . . . . . . . . . . . . . . . . . . . . . . . . . . . . . . . . . . . . . . 404
Table 9.39. Simulation results over 864 Models. Summary of overall rank by K-L and $L_2$ observed efficiency. . . . . . . . . . . . . . . . . . . 406
Table 9.40. Simulation results summary for Model A8. K-L observed efficiency ranks, $L_2$ observed efficiency ranks and counts. . . 407
Table 9.41. Simulation results summary for Model A9. K-L observed efficiency ranks, $L_2$ observed efficiency ranks and counts. . . 408
Table 9.42. VAR real data selected models. . . . . . . . . . . . . . . . . . . . . . . . . . 409
Table 9.43. Summary of VAR(2) model. . . . . . . . . . . . . . . . . . . . . . . . . . . . . 409
Table 9.44. Summary of VAR(11) model. . . . . . . . . . . . . . . . . . . . . . . . . . . 410
Table 9A.1. Counts and observed efficiencies for Model 1. . . . . . . . . . . . . 412
Table 9A.2. Counts and observed efficiencies for Model 2. . . . . . . . . . . . . 412
Table 9A.3. Simulation results for all 540 univariate regression models—K-L observed efficiency. . . . . . . . . . . . . . . . . . . . . . . . . . 413
Table 9A.4. Simulation results for all 540 univariate regression models—$L_2$ observed efficiency. . . . . . . . . . . . . . . . . . . . . . . . . . . 413
Table 9A.5. Counts and observed efficiencies for Model A1. . . . . . . . . . . . 414
Table 9A.6. Counts and observed efficiencies for Model A2. . . . . . . . . . . . 414
Table 9A.7. Counts and observed efficiencies for Model 3. . . . . . . . . . . . . 415
Table 9A.8. Counts and observed efficiencies for Model 4. . . . . . . . . . . . . 415
Table 9A.9. Simulation results for all 360 autoregressive models—K-L observed efficiency. . . . . . . . . . . . . . . . . . . . . . . . . . . . . . . . . . . . 416
Table 9A.10. Simulation results for all 360 autoregressive models—$L_2$ observed efficiency. . . . . . . . . . . . . . . . . . . . . . . . . . . . . . . . . . . . 416
Table 9A.11. Counts and observed efficiencies for Model A3. . . . . . . . . . . . 417
Table 9A.12. Counts and observed efficiencies for Model A4. . . . . . . . . . . . 417
Table 9A.13. Counts and observed efficiencies for Model 5. . . . . . . . . . . . . 418
Table 9A.14. Counts and observed efficiencies for Model 6. . . . . . . . . . . . . 418
Table 9A.15. Simulation results for all 50 misspecified MA(1) models—K-L observed efficiency. . . . . . . . . . . . . . . . . . . . . . . . . . 419
Table 9A.16. Simulation results for all 50 misspecified MA(1) models—$L_2$ observed efficiency. . . . . . . . . . . . . . . . . . . . . . . . . . 419
Table 9A.17. Counts and observed efficiencies for Model A5. . . . . . . . . . . . 420
Table 9A.18. Counts and observed efficiencies for Model 7. . . . . . . . . . . . . 420
Table 9A.19. Counts and observed efficiencies for Model 8. . . . . . . . . . . . . 421

Table 9A.20. Simulation results for all 504 multivariate regression models—K-L observed efficiency. .......... 421
Table 9A.21. Simulation results for all 504 multivariate regression models—$\text{tr}\{L_2\}$ observed efficiency. .......... 422
Table 9A.22. Simulation results for all 504 multivariate regression models—$\det(L_2)$ observed efficiency. .......... 422
Table 9A.23. Counts and observed efficiencies for Model A6. .......... 423
Table 9A.24. Counts and observed efficiencies for Model A7. .......... 423
Table 9A.25. Counts and observed efficiencies for Model 9. .......... 424
Table 9A.26. Counts and observed efficiencies for Model 10. .......... 424
Table 9A.27. Simulation results for all 864 VAR models—K-L observed efficiency. .......... 425
Table 9A.28. Simulation results for all 864 VAR models—$\text{tr}\{L_2\}$ observed efficiency. .......... 425
Table 9A.29. Simulation results for all 864 VAR models—$\det(L_2)$ observed efficiency. .......... 426
Table 9A.30. Counts and observed efficiencies for Model A8. .......... 426
Table 9A.31. Counts and observed efficiencies for Model A9. .......... 427
Table 9B.1. Stepwise counts and observed efficiencies for Model 1. .......... 428
Table 9B.2. Stepwise counts and observed efficiencies for Model 2. .......... 428
Table 9B.3. Stepwise results for all 540 univariate regression models—K-L observed efficiency. .......... 429
Table 9B.4. Stepwise results for all 540 univariate regression models—$L_2$ observed efficiency. .......... 429

Chapter 1
# Introduction

## 1.1. Background

A question perhaps as old as modeling is "Which variables are important?" Because the need to select a model applies to more than just variable selection in regression models, there is a rich variety of answers. For example, model selection techniques can be applied to areas such as histogram construction (see Linhart and Zucchini, 1986), to determine the number of factors in factor analysis, and to nonparametric problems such as curve smoothing and smoothing bandwidth selection. In fact, model selection criteria can be applied to any situation where one tries to balance variability with complexity.

What defines a good model? A good model certainly fits the data set under investigation well. Of course, the more variables added to the model, the better the apparent fit. One of the goals of model selection is to balance the increase in fit against the increase in model complexity. Perhaps a better defining quality of a good model is its performance on future data sets collected from the same process. A model that fits well on one of the data sets representing the process should fit well on any other data set. More importantly, a model that is too complicated but fits the current data set well may fit subsequent data sets poorly. A model that is too simple may fit none of the data sets well.

How to select a model? Once a probabilistic model has been proposed for an experiment, data can be collected, leading to a set of competing candidate models. The statistician would like to select some appropriate model from this set, where there may be more than one definition of "appropriate." Model selection criteria are one way to decide on the most appropriate model.

Model selection criteria are often compared using results from simulation studies. However, assessing subtle differences between performance results is a daunting task—no single model selection criterion will always be better than another; certain criteria perform best for specific model types. In this book we use many different models to compare performance of the criteria, sometimes narrowly focusing on only a few differences between model types and sometimes varying them very widely. Often, a count of the times that a selection criterion identifies the correct model is a useful measure of model selection performance. However, the more variety in models, the more unreliable counts can become, as we will see in some simulations throughout the book. When the true model belongs to the set of candidate models, our measure of performance is the distance between the selected model and the true model. In any set of candidate models, one of the candidates will be closest to the true model. We term the ratio that compares the distance between the closest candidate model and the selected model the *observed efficiency*, which we will discuss in more detail below. We will see that observed efficiency is a much more flexible measure of performance than comparisons of counts.

### *1.1.1. Historical Review*

Much of past model selection research has been concerned with univariate or multiple regression models. Perhaps the first model selection criterion to be widely used is the adjusted R-squared, $R^2_{adj}$, which still appears in many regression texts today. It is known that $R^2$ always increases whenever a variable is added to the model, and therefore it will always recommend additional complexity without regard to relative contribution to model fit. $R^2_{adj}$ attempts to correct for this always-increasing property. Other model selection work appeared in the late 60's and early 70's, most notably Akaike's FPE (Akaike, 1969) and Mallows's Cp (Mallows, 1973). The latter is currently one of the most commonly used model selection criteria for regression. Information theory approaches also appeared in the 1970's, with the landmark Akaike Information Criterion (Akaike, 1973, 1974), based on the Kullback–Leibler discrepancy. AIC is probably the most commonly used model selection criterion for time series data. In the late 1970's there was an explosion of work in the information theory area, when the Bayesian Information Criterion (BIC, Akaike, 1978), the Schwarz Information Criterion (SIC, Schwarz, 1978), the Hannan and Quinn Criterion (HQ, Hannan and Quinn, 1979), FPE$\alpha$ (Bhansali and Downham, 1977), and GM (Geweke and Meese, 1981) were proposed. Subsequently, in the late 1980's, Hurvich and Tsai (1989) adapted Sugiura's 1978 results to develop an improved small-sample unbiased estimator of the Kullback–Leibler discrepancy, AICc. AICc has shown itself to be one of the best model selection criteria in an increasingly crowded field.

In 1980 the notion of asymptotic efficiency appeared in the literature (Shibata, 1980) as a paradigm for selecting the most appropriate model, and SIC, HQ, and GM became associated with the notion of consistency. We briefly describe these two philosophies of model selection.

### *1.1.2. Efficient Criteria*

A common assumption in both regression and time series is that the generating or true model is of infinite dimension, or that the set of candidate models does not contain the true model. The goal is to select one model that best approximates the true model from a set of finite-dimensional candidate models. The candidate model that is closest to the true model is assumed to be the appropriate choice. Here, the term "closest" requires some well-defined distance or information measure in order to be evaluated. In large samples, a model selection criterion that chooses the model with minimum mean squared error distribution is said to be *asymptotically efficient* (Shibata, 1980). FPE, AIC, AICc, Cp are all asymptotically efficient. Researchers who believe that the system they study is infinitely complicated, or that there is no way to measure all the important variables, choose models based on efficiency. Much research has been devoted to finding small-sample improvements ("corrections") to efficient criteria. AIC is perhaps the most popular basis for correction. Perhaps the best known corrected version is AICc (Sugiura, 1978 and Hurvich and Tsai, 1989).

Sometimes the predictive ability of a candidate model is its most important attribute. An early selection criterion that modeled mean squared prediction error is PRESS (Allen, 1973). Akaike's FPE is also intended to select models that make good predictions. Both PRESS and FPE are efficient, and while we do not study predictive ability as a way to evaluate performance except with respect to bootstrapping and cross-validation methods, it is worth noting that prediction and asymptotic efficiency are related (Shibata, 1980).

### *1.1.3. Consistent Criteria*

Many researchers assume that the true model is of finite dimension, and that it is included in the set of candidate models. Under this assumption the goal of model selection is to correctly choose the true model from the list of candidate models. A model selection criterion that identifies the correct model asymptotically with probability one is said to be *consistent.* SIC, HQ, and GM are all consistent. Here the researcher believes that all variables can be measured, and furthermore, that enough is known about the physical system being studied to write the list of all important variables. These are strong assumptions to many statisticians, but they may hold in fields like physics, where there are large bodies of theory to justify assuming the existence of a true model that belongs to the set of candidate models.

Many of the classic consistent selection criteria are derived from asymptotic arguments. Less work has been focused on finding improvements to consistent criteria than efficient criteria, due in part to the fact that the consistent criteria do not estimate some distance function or discrepancy. In this book we present one corrected consistent criterion, HQc, a small-sample correction to Hannan and Quinn's criterion HQ.

Which is better, efficiency or consistency? There is little agreement. As we noted above, the choice is highly subjective and depends upon the individual researcher's assessment of the complexity and measurability of the modeling problem. To make matters more confusing, both consistency and efficiency are asymptotic properties. In small samples, the criteria can behave much differently. Because of the practical limitations on gathering and using data, small-sample performance is often more important than asymptotic properties. This issue is discussed in Chapter 2 using the signal-to-noise diagnostic.

## 1.2. Overview

### *1.2.1. Distributions*

It is important to remember that all model selection criteria are themselves random variables with their own distributions. The moments of many of the classical selection criteria have been investigated in other papers, as have their probabilities of selecting a true model, assuming that it is one of the candidate models (Nishii, 1984 and Akaike, 1969). We derive moments and probabilities for the primary criteria discussed in this book, and relate them to performance via the concept of the signal-to-noise ratio.

Differences between models are also investigated. When evaluating the relative merits of two models, the value of the selection criterion for each is compared and some decision is made. Such differences also have distributions that can be investigated, and probabilities of selecting one model over another are based on the distribution of the difference. We derive moments for these differences as well. Examination of the moments can lead to insights into the behavior of model selection criteria. These moments are used to derive the signal-to-noise ratio.

Two somewhat uncommon distributions are reviewed, the log-$\chi^2$ and log-Beta distributions. These two distributions are important to the derivations of many of the classical model selection criteria, and detailed information about them can be found in Appendix 2A to Chapter 2. They can be described as follows: If $X \sim \chi^2(m)$, then $\log(X) \sim \text{log-}\chi^2(m)$. Log-Beta is related to the usual Beta distribution. If $X \sim \text{Beta}(\alpha, \beta)$, then we take logs to give $\log(X) \sim \text{log-Beta}(\alpha, \beta)$. While the exact moments can be computed for these distributions, we will derive some useful approximations that will allow us to more easily compute small-sample signal-to-noise ratios.

Multivariate model selection criteria often make use of the generalized variance (Anderson, 1984 p. 259). In regression, the variance has either a central or noncentral Wishart distribution. Many of the classic multivariate selection criteria have moments involving the log-determinant(Wishart) distribution, and therefore exact and closed-form approximations are developed for the log-determinant(Wishart). Tests of two multivariate models are often performed via likelihood ratios or U-statistics, where U-statistics are much like a multivariate F-test for comparing the "full" model with the "reduced" model. Moments for the log-U distribution are developed so that signal-to-noise ratios can be formulated for model selection criteria in the multivariate case.

### *1.2.2. Model Notation*

Regression as well as time series autoregressive models are discussed. Since these models necessarily have different structures, different notation is used. We use $k$ to represent the model order when the model includes the intercept. If there are $p$ important variables plus the intercept, then the regression model is of order $k = p + 1$. For regression cases, all models include the intercept. The true model, if it belongs to the set of candidate models, will be of order $k_*$, where "*" denotes the true model. Our time series models do not include an intercept or constant term, and the order of the model will be equal to the number of variables, or $p$, and the true autoregressive model is denoted by $p_*$.

### *1.2.3. Discrepancy and Distance Measures*

How to measure model selection performance? If the true model belongs to the set of candidate models and consistency holds, then a natural way to measure performance is to compare the probabilities of selecting the correct model for each criterion considered. For efficiency, where the true model may not belong to the set of candidate models, selecting the closest approximation is the goal. For this some sort of distance measure is required. A distance function or metric, $d$, is a real valued function with two arguments, $u$ and $v$, which may be vectors or scalars. A distance function, $d(u, v)$, must satisfy the following three properties:

1) *Positiveness:*

$$d(u, v) > 0 \text{ for } u \neq v,$$

$$d(u, v) = 0 \text{ for } u = v.$$

2) *Symmetry:*

$$d(u, v) = d(v, u).$$

3) *Triangle Inequality:*

$$d(u, w) \leq d(v, u) + d(v, w).$$

For the purposes of selecting a model, we are interested in only the first property. By definition, a model with a better fit must have a smaller distance than a model with a poor fit. We do not need a distance function for model selection; any function satisfying Property 1 will suffice. Such a function is often referred to as a *discrepancy*, a term dating back to Haldane (1951). Other authors have continued to use the term to describe the distance between likelihoods for a variety of problems. Certainly, the set of functions satisfying Property 1 yields a large class of potential discrepancy functions, and several important ones are given in Linhart and Zucchini (1986, p. 18). The three we will use in this book are listed below.

Let $M_T$ be the true model with density $f_T$ and distribution $F_T$. Let $M_A$ denote the candidate (approximating) model with density $f_A$ and let $\Delta$ denote the discrepancy. The *Kullback–Leibler discrepancy*, (Kullback and Leibler, 1951), also called the Kullback–Leibler information number, or K-L, is based on the likelihood ratio. The Kullback–Leibler discrepancy applies to nearly all parametric models. K-L is a real valued function for univariate regression as well as multivariate regression. As such, K-L is perhaps the most important discrepancy used in model selection. In general,

$$\Delta_{\text{K-L}}(M_T, M_A) = E_{F_T}\left[log\left(\frac{f_T(x)}{f_A(x)}\right)\right].$$

The $L_2$ norm can be used as a basis for measuring distance as well. Let $\mu_{M_T}$ and $\mu_{M_A}$ denote the true and candidate model means, respectively. We can define $L_2$ as

$$\Delta_{L_2}(M_T, M_A) = \| \mu_{M_T} - \mu_{M_A} \|^2.$$

$L_2$ is a distance function and is easy to apply to univariate models. An advantage of $L_2$ is that it depends only on the means of the two distributions and not the actual densities. This means that $L_2$ can be applied when errors are not normally distributed. However, a disadvantage is that $L_2$ is a matrix in certain multivariate models.

While there are many types of discrepancy functions on which to base model choices, some are more easily applied and computed than others. The relative ease with which K-L and $L_2$ can be adapted to a variety of situations led us to choose them to measure model selection performance. Although the two measures sometimes give different indications of small-sample performance, in large samples they can be shown (via a lengthy derivation) to be equivalent. Thus, criteria that are efficient in the $L_2$ sense are also efficient in the K-L sense.

Chapter 7, "Robust Regression and Quasi-Likelihood," introduces distance using the $L_1$ norm or absolute difference norm,

$$\Delta_{L_1}(M_T, M_A) = \| \mu_{M_T} - \mu_{M_A} \| = \sum_{i=1}^{n} |\mu_{M_T i} - \mu_{M_A i}|.$$

When the error distribution is heavy-tailed and outliers are suspected, the $L_1$ norm may be more robust. The $L_2$ and $L_1$ norms are much more applicable in the robust setting because their forms do not depend on any given distribution. By contrast, when errors are nonnormal the Kullback–Leibler discrepancy must be computed for each distribution.

### *1.2.4. Efficiency under Kullback–Leibler and $L_2$*

Both K-L and $L_2$ have useful qualities—K-L is always a scalar, while $L_2$ can be applied to models with nonnormal errors. We can use these two measures to define efficiency in both the asymptotic and the small-sample (observed) sense. For $L_2$, the distance between the true and candidate models is $\| \mu_{M_T} - \mu_{M_A} \|^2$. Shibata (1980) suggested using the expected distance, $E_{F_T}[L_2] = E_{F_T}\left[\| \mu_{M_T} - \mu_{M_A} \|^2\right]$, as the distance measure. Using Shibata's measure, we assume that among the candidate models there exists model $M_c$ that is closest to the true model in terms of the expected $L_2$ distance, $E_{F_T}[L_2](M_c)$. Suppose a model selection criterion selects model $M_k$, which has an expected $L_2$ distance of $E_{F_T}[L_2](M_k)$. Of course, $E_{F_T}[L_2](M_k) \geq E_{F_T}[L_2](M_c)$. A model selection criterion is said to be *asymptotically efficient* if

$$\text{p-}\lim_{n\to\infty} \frac{E_{F_T}[L_2](M_c)}{E_{F_T}[L_2](M_k)} = 1,$$

where $n$ denotes the sample size. For small samples, we analogously define $L_2$ *observed efficiency* to be

$$L_2 \text{ observed efficiency} = \frac{L_2(M_c)}{L_2(M_k)}, \tag{1.1}$$

where $L_2 = \| \mu_{M_T} - \hat{\mu} \|^2$ and $\hat{\mu}$ is the vector of predicted values for the fitted candidate model. To define observed (small-sample) efficiency for K-L, again let $M_c$ be the candidate model that is closest to the true model, and let K-L($M_c$) denote this distance. Let $M_k$ be the candidate model, with distance K-L($M_k$), selected by some criterion. We define *Kullback–Leibler observed efficiency* as

$$\text{K-L observed efficiency} = \frac{\text{K-L}(M_c)}{\text{K-L}(M_k)}, \tag{1.2}$$

where K-L is computed using the parameters from the true model and the estimated parameters from the candidate model. The observed efficiencies given in Eq. (1.1) and Eq. (1.2) are used to assess model selection performance in simulations throughout this book. Wherever we make references to model selection performance under K-L and $L_2$, the terms K-L and $L_2$ refer to observed efficiency unless otherwise mentioned.

Chapters 2–5 include theoretical properties of model selection criteria and the $L_2$ and K-L distances. Here we use the expected values of $L_2$ and K-L when discussing theoretical distance between the candidate model and true model. As noted earlier, efficiency can be defined in terms of expected distance. For $L_2$, we define *$L_2$ expected efficiency* as

$$L_2 \text{ expected efficiency} = \frac{E_{F_T}[L_2(M_c)]}{E_{F_T}[L_2(M_k)]}, \tag{1.3}$$

where $E_{F_T}[L_2(M_c)]$ is the expected $L_2$ distance of the closest model and $E_{F_T}[L_2(M_k)]$ is the expected $L_2$ distance of the candidate model. Analogously, *K-L expected efficiency* is defined as

$$\text{K-L expected efficiency} = \frac{E_{F_T}[\text{K-L}(M_c)]}{E_{F_T}[\text{K-L}(M_k)]}, \tag{1.4}$$

where $E_{F_T}[\text{K-L}(M_c)]$ is the expected K-L distance of the closest model and $E_{F_T}[\text{K-L}(M_k)]$ is the expected K-L distance of the candidate model. In later chapters, expectation under the true model, $E_{F_T}$ is denoted by $E_*$. When the true model belongs to the set of candidate models or for general expectation, we use $E$ without subscripts.

### *1.2.5. Overfitting and Underfitting*

The terms *overfitting* and *underfitting* can be defined two ways. Under consistency, when a true model is itself a candidate model, overfitting is defined as choosing a model with extra variables, and underfitting is defined as choosing a model that either has too few variables or is incomplete. We have no term to describe choosing a model with the correct order but the wrong variables.

Using efficiency (observed or expected), overfitting can be defined as choosing a model that has more variables than the model identified as closest to the true model, thereby reducing efficiency. Underfitting is defined as choosing a model with too few variables compared to the closest model, also reducing efficiency. Both overfitting and underfitting can lead to problems with the predictive abilities of a model. An underfitted model may have poor predictive ability due to a lack of detail in the model. An overfitted model may be unstable in the sense that repeated samples from the same process can lead to widely differing predictions due to variability in the extraneous variables. A criterion that can balance the tendencies to overfit and underfit is preferable.

## 1.3. Layout

We will discuss the broad model categories of univariate models, multivariate models, data resampling techniques, and nonparametric models, and include simulation results for each category, presenting results under both K-L and $L_2$ observed efficiencies. We leave it to the practitioner to decide his or her preference. In addition, at the end of this book we devote an entire chapter of simulation studies for each model type as well as real data examples. The contents of each chapter are summarized below.

In Chapter 2 we lay the foundation for the criteria we will discuss throughout the book, and for the K-L and $L_2$ observed efficiencies. We introduce the distributions necessary to develop the concept of the signal-to-noise ratio. We begin by examining the large-sample and small-sample properties of the classical criteria AICc, AIC, FPE, and SIC for univariate regression, including their asymptotic probabilities of overfitting (the probability of preferring one overfit model to the true model) and asymptotic signal-to-noise ratios. The signal-to-noise information is analyzed in order to suggest some signal-to-noise corrected variant criteria that perform better than the parent criteria.

In this Chapter we also introduce the simulation model format we will use to illustrate criterion performance throughout the book. This includes a brief discussion of random $X$ regression, since it is used to generate the design matrices for our simulations, and also an explanation of the ranking method we will use to compare model selection criteria. Ranks for each individual simulation run are computed and averaged over all runs, and the criterion with the lowest overall average rank is considered the best; *i.e.*, the higher the observed efficiency, the lower the rank.

In general, for each model category we will begin with a simulation study of two special cases where the noncentrality parameter and true model structure is known. The expected values of AICc, AIC, FPE, and SIC are compared, and the moments of differences between the true and candidate model of these six model selection criteria are computed, as are signal-to-noise ratios. Then to measure the performance of a model selection criterion in small samples, observed efficiency is developed and a large-scale small-sample simulation is conducted.

In Chapter 3 we discuss the autoregressive model, which describes the present value $y_t$ as a linear combination of past observations $y_{t-1}, \ldots$. This linear relationship allows us to write the autoregressive model as a special case regression model similar to those in Chapter 2. Since past observations are used to model the present, we have a problem modeling the first observation $y_1$ because there are no past observations. There are several possible solutions. The one we have chosen is to begin modeling at observation $p+1$, and to lose the first $p$ observations due to conditioning on the past. Although this results in a reduced sample size, it also requires fewer model assumptions. However, this also means that the sample size for autoregressive models changes with the model, unlike the univariate regression models in Chapter 2.

Another way to model time series is with a univariate moving average model. Although we do not discuss model selection with respect to moving average models, we do address the situation where the data is truly the result of a moving average process, but is modeled using autoregressive models. Also, under certain conditions a moving average MA(1) model may be written as an infinite order AR model. This allows us to examine how criteria behave with models of infinite order where the true model does not belong to the set of candidate models. Multistep prediction AR models are discussed briefly and the performance of some multistep variants are tested via a simulation study.

In Chapter 4 we consider the multivariate regression model. Multivariate regression models are similar to univariate regression models with the important difference that the error variance is actually a covariance matrix. Since many selection criteria are functions of this matrix, a central issue is how to produce a scalar function from these matrices. Determinants and traces are common methods, but are by no means the only options. Generalized variances are popular due to the fact that their distributional properties are well-known, whereas distributions of other scalar functions of matrices, such as the trace, are not well-known. In this book we focus on the generalized variance (the determinant) so that moments and probabilities of overfitting (the probability of preferring one overfit model to the true model) can be computed. However, we also present the trace criterion results for comparison purposes. Generalizing the $L_2$ norm to a scalar is also a problem; in our simulations, the trace appears to be more useful than the determinant of $L_2$. In general, $\det(L_2)$ results are presented only when they differ substantially from those of $\mathrm{tr}\{L_2\}$.

In Chapter 5 we discuss the vector autoregressive model. Of all the models in this book, the vector autoregressive or VAR model is perhaps the most difficult to work with due to the rapid increase in parameter counts as model complexity increases. This rapid increase causes many selection criteria to perform poorly, particularly those prone to underfitting. As we did with the univariate autoregressive models, we begin modeling at $p+1$, and thus the sample size decreases as model order increases. We again condition on the past and write the vector autoregressive model as a special case multivariate regression model. This loss of sample size eliminates the need for backcasting or other assumptions about unobserved past data. Casting the VAR model into a multivariate framework allows us to compute moments of the model selection criteria as well as compute probabilities of overfitting (the probability of preferring one overfit model to the true model). Such moments allow us to better study small-sample properties of the selection criteria. Overfitting has much smaller probability of occurring in VAR models than in multivariate regression models, due in part to the rapid increase in parameters with model order and to the decrease in sample size with increasing model order. Simulation results indicate that an excessively heavy penalty function leads to decreased performance in VAR model selection.

In Chapter 6 we investigate data resampling techniques. If predictive ability is of interest for the model, then cross-validation or bootstrapping techniques can be applied. Cross-validation and bootstrapping are discussed for univariate as well as multivariate regression and time series. The PRESS statistic (Allen, 1973) is an example of cross-validation. We use the notation CV to denote both PRESS as well as cross-validation. CV is also an efficient criterion and is asymptotically equivalent to FPE. Some issues unique to bootstrapping include choosing between randomly selecting pairs $(y, x)$ or bootstrapping from the residuals. Both are considered in a simulation study. We briefly discuss the role of the number of bootstrap pseudo-samples on bootstrap model selection performance, and adjusting or "inflating" residuals to compensate for nonconstant variance of the usual residuals. Variants of bootstrapped selection criteria with penalty functions that prevent overfitting are also introduced.

In Chapter 7 we discuss robust regression and robust model selection criteria. The least squares approach does not assume normality; however, least squares can be affected by heavy-tailed distributions. We begin with least absolutes regression, or $L_1$ regression, and introduce the $L_1$ distance and observed efficiency. $L_1$ regression is equivalent to maximum likelihood if one assumes that the error distribution is the double exponential distribution. Using this assumption we will discuss the L1AICc criterion and present an $L_1$ regression simulation study.

In this Chapter we also propose a generalized Kullback-Leibler information for measuring the distance between a robust function evaluated under the true model and a fitted model. We then use this generalization to obtain robust model selection criteria that not only fit the majority of the data, but also take into account nonnormal errors. These criteria have the additional advantage of unifying most existing Akaike information criteria.

Lastly in Chapter 7 we develop criteria for quasi-likelihood models. Such models include not only regression models with normal errors, but also logistic regression models, Poisson regression models, exponential regression models, etc. The performance of these criteria are examined via simulation focusing on logistic regression.

In Chapter 8, we develop a version of AICc for use with nonparametric and semiparametric regression models. The nonparametric AICc can be used to choose smoothing parameters for any linear smoother, including local quadratic and smoothing spline estimators. It has less tendency to undersmooth and it exhibits low variability. Monte Carlo results show that the nonparametric AICc is comparable to well-behaved plug-in methods (see Ruppert, Sheather and Wand, 1995), but also performs well when the plug-in method fails or is unavailable. In semiparametric regression models, simulation studies show that AICc outperforms AIC.

We also develop a cross-validatory or cross-validation version of AICc for selecting a hard wavelet threshold (Donoho and Johnstone, 1994), and show via simulations that our method can outperform universal hard thresholding. In addition, we provide supporting theory on the rate at which our proposed method attains the optimal mean integrated squared error.

Finally, Chapter 9 is devoted almost exclusively to simulation results for each of the modeling categories in earlier chapters. Our goal is to use a wide enough range of models to compare a large enough list of selection criteria so that meaningful conclusions about performance in the "real world" can be made. Simulations include two special case models, a large-scale multi-model study, and two very large sample size models. They are presented for univariate regression and time series, and for multivariate regression and time series. Sixteen criteria are compared for the univariate models, while 18 criteria are compared for the multivariate models. While our studies are by no means comprehensive, they do illustrate the performance of a variety of selection criteria under many different modeling circumstances. Four real data examples are also analyzed for each model type. Finally, we study the performance of the stepwise procedure in the selection of variables.

## 1.4. Topics Not Covered

Unfortunately, there is much interesting work being done on topics that are outside the scope of this book, but important to the topic of model selections. Work on recent approaches for Bayesian variable selection (George and McCulloch, 1993 and 1997; Carlin and Chib, 1995; Chipman, Hamada, and Wu, 1997), asymptotic theory for linear model selection (Shibata, 1981, Nishii, 1984, Rao and Wu, 1989, Hurvich and Tsai, 1995, Zheng and Loh, 1995, and Shao, 1997), the impact of misspecification in model selection (Hurvich and Tsai, 1996), regression diagnostics in model selection (Weisberg, 1981 and Léger and Altman, 1993), the impact of model selection on inference in linear regression (Hurvich and Tsai, 1990), the use of marginal likelihood in model selection (Shi and Tsai, 1998), the impact of parameter estimation methods in autoregressive model selection (Broersen and Wensink, 1996), the application of generalized information criteria in model selection (Konishi and Kitagawa, 1996), and the identification of ARMA models (Pukkila, Koreisha and Kallinen, 1990, Brockwell and Davis, 1991, Choi, 1992, and Lai and Lee, 1997) may be of interest to the reader. In addition, there are important model categories which we do not address but are nevertheless important areas for research in variable selection. These include, but are not limited to, survival models (Lawless, 1982), regression models with ARMA errors (Tsay, 1984), measurement error models (Fuller, 1987), transformation and weighted regression models (Carroll and Ruppert, 1988), nonlinear regression models (Bates and Watts, 1988), Markov regression time series models (Zeger and Qaqish, 1988), structural time series models (Harvey, 1989), sliced inverse regression (Li, 1991), linear models with longitudinal data (Diggle, Liang and Zeger, 1994), generalized partially linear single-index models (Carroll, Fan, Gijbels and Wand, 1997) and ARCH models (Gouriéroux, 1997). Finally, a forthcoming book, Model Selection and Inference: A Practical Information Theoretic Approach, (Burnham and Anderson, 1998) covers some subjects that have not been addressed in this book. The interested reader may find it a useful reference for the study of model selection.

# Chapter 2
# The Univariate Regression Model

One of the statistician's most useful tools is the univariate multiple regression model, and as such, selection techniques for this class of models have received much attention over the last two decades. In this Chapter we give a brief history of some of the selection criteria commonly used in univariate regression modeling, and for six "foundation" criteria we will give a more detailed discussion with derivations. Two of these, the Akaike Information Criterion (AIC, Akaike, 1973) and its corrected version (AICc, Sugiura, 1978 and Hurvich and Tsai, 1989) estimate the Kullback–Leibler discrepancy (Kullback and Leibler, 1951). Two others, FPE (Akaike, 1973) and Mallows's Cp (Mallows, 1973) estimate the mean squared prediction error, similar to estimating $L_2$, where $L_2$ is the Hilbert space of sequences of real numbers with the inner product $< \cdot, \cdot >$ and the Euclidean norm $\| \cdot \|$. Finally, the Schwarz Information Criterion (SIC, Schwarz, 1978) and the Hannan and Quinn criterion (HQ, Hannan and Quinn, 1979) are derived for their asymptotic performance properties. While this list is by no means complete, these six criteria were chosen as the basis for illustrating three possible approaches to selecting a model—using efficient criteria to estimate K-L, using efficient criteria to estimate $L_2$, and using consistent criteria. With the aim of making further refinements, we will also examine the small-sample moments of three of these criteria in order to suggest improvements to their penalty functions.

We will discuss the use of the signal-to-noise ratio as a descriptive statistic for evaluating model selection criteria. Sections 2.2 and 2.3 provide examples of derivations and overfitting properties for our foundation criteria, and corresponding material for other criteria is detailed in Appendix 2C. In Section 2.4 we introduce signal-to-noise corrected variants and their asymptotic properties of overfitting. The rest of Chapter 2 examines small-sample properties, including underfitting using two special case models, and we close with a simulation study of these two models for the purposes of comparison to the expected theoretical results.

## 2.1. Model Description

### 2.1.1. Model Structure and Notation

Before we can discuss model selection in regression, we need to define the model structures with which we will work and the assumptions we will make. Here we introduce three model structures: the true model, the general model, and the fitted model.

We first define the *true regression model* to be

$$Y = \mu_* + \varepsilon_* \tag{2.1}$$

and

$$\varepsilon_* \sim N(0, \sigma_*^2 I), \tag{2.2}$$

where $Y = (y_1, \ldots, y_n)'$ is an $n \times 1$ vector of responses, $\mu_* = (\mu_{*1}, \ldots, \mu_{*n})'$ is an $n \times 1$ vector of true unknown functions, and $\varepsilon_* = (\varepsilon_{*1}, \ldots, \varepsilon_{*n})'$. In Eq. (2.2), we assume that the errors $\varepsilon_{*i}$ are independent and identically normally distributed, with constant variance $\sigma_*^2$ for $i = 1, \ldots, n$.

We next define the *general model* to be

$$Y = X\beta + \varepsilon \tag{2.3}$$

and

$$\varepsilon \sim N(0, \sigma^2 I), \tag{2.4}$$

where $X = (x_1, \ldots, x_n)'$ is a known $n \times k$ design matrix of rank $k$, $x_i$ is a $k \times 1$ vector, $\beta$ is a $k \times 1$ vector of unknown parameters, and $\varepsilon = (\varepsilon_1, \ldots, \varepsilon_n)'$. In Eq. (2.4), we assume that the errors $\varepsilon_i$ are independent and identically normally distributed with the constant variance $\sigma^2$ for $i = 1, \ldots, n$. If the constant, or y-intercept, is included in the model, the first column of $X$ will contain a column of 1's associated with the constant.

Finally we will define the *fitted model*, or the *candidate model*, with respect to the general model. In order to classify candidate model types we will partition $X$ and $\beta$ such that $X = (X_0, X_1, X_2)$ and $\beta = (\beta_0', \beta_1', \beta_2')'$, where $X_0, X_1$ and $X_2$ are $n \times k_0$, $n \times k_1$ and $n \times k_2$ matrices, and $\beta_0, \beta_1$ and $\beta_2$ are $k_0 \times 1$, $k_1 \times 1$ and $k_2 \times 1$ vectors, respectively. If $\mu_*$ is a linear combination of unknown parameters such that $\mu_* = X_*\beta_*$, then underfitting will occur when $\text{rank}(X) < \text{rank}(X_*)$, and overfitting will occur when $\text{rank}(X_*) < \text{rank}(X)$. Thus we can rewrite the model in Eq. (2.3) in the following form:

$$\begin{aligned} Y &= X_0\beta_0 + X_1\beta_1 + X_2\beta_2 + \varepsilon \\ &= X_*\beta_* + X_2\beta_2 + \varepsilon, \end{aligned}$$

where $\beta_* = (\beta_0', \beta_1')'$, $X_0$ is the design matrix for an underfitted candidate model, $X_* = (X_0, X_1)$ is the design matrix for the true model and $X = (X_0, X_1, X_2)$ is the design matrix for an overfitted model. Thus an underfitted model is written as

$$Y = X_0\beta_0 + \varepsilon, \tag{2.5}$$

and an overfitted model is written as

$$Y = X\beta + \varepsilon = X_0\beta_0 + X_1\beta_1 + X_2\beta_2 + \varepsilon. \tag{2.6}$$

Of course, the overfitted model has the same form as the general model in Eq. (2.3).

We will further assume that the method of least squares is used to fit a model to the data, and the candidate model (unless otherwise noted) will be of order $k$. When fitting a candidate model the usual ordinary least squares parameter estimate of $\beta$ is

$$\hat{\beta} = (X'X)^{-1}X'Y.$$

This is also the maximum likelihood estimate (MLE) of $\beta$, since the errors $\varepsilon$ satisfy the assumption in Eq. (2.4). The unbiased and the maximum likelihood estimates of $\sigma^2$, respectively, are given below:

$$s_k^2 = \frac{\text{SSE}_k}{n-k}, \tag{2.7}$$

and

$$\hat{\sigma}_k^2 = \frac{\text{SSE}_k}{n}, \tag{2.8}$$

where $\text{SSE}_k = \| Y - \hat{Y} \|^2$ is the usual sum of squared errors and $\hat{Y} = X\hat{\beta}$.

### 2.1.2. Distance Measures

The distance measures $L_2$ and the Kullback–Leibler discrepancy (K-L) provide a way to evaluate how well the candidate model approximates the true model given in Eq. (2.1) and Eq. (2.2) by estimating the difference between the expectations of the vector $Y$ under the true model and the candidate model. We can use the notation from Eq. (2.1) and Eq. (2.2) to define both the K-L and $L_2$ distances. Thus for $L_2$ we have

$$L_2 = \frac{1}{n}\| \mu_* - X\beta \|^2.$$

Analogously, the $L_2$ distance between the estimated candidate model and the expectation of the true model assuming $\mu_* = X_*\beta_*$ can be defined as

$$L_2 = \frac{1}{n} \| X_*\beta_* - X\hat{\beta} \|^2. \tag{2.9}$$

In order to define the Kullback–Leibler discrepancy, we must also consider the density functions of the true and candidate models. These likelihood functions will later play a key role in the derivations of K-L-based criteria AIC and AICc. Under the assumption of normality, the density of the true model $f_*$ is the joint density of $Y$, or

$$f_* = (2\pi)^{-n/2}(\sigma_*^2)^{-n/2} \exp\left(-\frac{1}{2\sigma_*^2}\sum_{i=1}^{n}(y_i - x'_{*i}\beta_*)^2\right).$$

By comparison, the likelihood function for the candidate model $f$ is

$$f = (2\pi)^{-n/2}(\sigma^2)^{-n/2} \exp\left(-\frac{1}{2\sigma^2}\sum_{i=1}^{n}(y_i - x'_i\beta)^2\right).$$

Based on these two likelihood functions we define K-L as

$$\text{K-L} = \frac{2}{n}E_*\left[\log\left(\frac{f_*}{f}\right)\right],$$

where $f_*$ and $E_*$ denote the density and the expectation under the true model. We have scaled the usual Kullback–Leibler information number by $2/n$ in order to express it as a rate or average information per observation. Taking logs,

$$\log(f_*) = -\frac{n}{2}\log(2\pi) - \frac{n}{2}\log(\sigma_*^2) - \frac{1}{2\sigma_*^2}\sum_{i=1}^{n}(y_i - x'_{*i}\beta_*)^2$$

and

$$\log(f) = -\frac{n}{2}\log(2\pi) - \frac{n}{2}\log(\sigma^2) - \frac{1}{2\sigma^2}\sum_{i=1}^{n}(y_i - x'_i\beta)^2.$$

Substituting and simplifying, we obtain

$$\text{K-L} = \log\left(\frac{\sigma^2}{\sigma_*^2}\right) + \frac{2}{n}E_*\left[-\frac{1}{2\sigma_*^2}\sum_{i=1}^{n}(y_i - x'_{*i}\beta_*)^2 + \frac{1}{2\sigma^2}\sum_{i=1}^{n}(y_i - x'_i\beta)^2\right].$$

Finally, by taking expectations with respect to the true model, we arrive at

$$\text{K-L} = \log\left(\frac{\sigma^2}{\sigma_*^2}\right) + \frac{\sigma_*^2}{\sigma^2} + \frac{\frac{1}{n}\| X_*\beta_* - X\beta \|^2}{\sigma^2} - 1.$$

In practice, the candidate model is estimated from the data. Substituting the $\hat{\sigma}^2$ in Eq. (2.8) for $\sigma^2$ and using the $L_2$ distance in Eq. (2.9), the Kullback–Leibler discrepancy between the fitted candidate model and true model is

$$\text{K-L} = \log\left(\frac{\hat{\sigma}_k^2}{\sigma_*^2}\right) + \frac{\sigma_*^2}{\hat{\sigma}_k^2} + \frac{L_2}{\hat{\sigma}_k^2} - 1. \tag{2.10}$$

## 2.2. Derivations of the Foundation Model Selection Criteria

We will begin by deriving the $L_2$-based model selection criteria FPE and Cp. We first consider FPE, which was originally derived for autoregressive time series models. A similar procedure was developed by Davisson (1965) for analyzing signal-plus-noise data; however, since Akaike published his findings in the statistical literature, FPE is usually attributed to Akaike. The derivation of FPE is straightforward for regression. Suppose we have $n$ observations from the overfitted model given by Eq. (2.6), and the resulting least squares estimate of $\beta$ is $\hat{\beta}$. Now we obtain $n$ new observations, $Y_0 = (y_{10}, \ldots, y_{n0})' = X\beta + \varepsilon_0$, from Eq. (2.6), for which the predicted value of $Y_0$ is $\hat{Y}_0 = (\hat{y}_{10}, \ldots, \hat{y}_{n0})' = X\hat{\beta}$. Hence, the mean squared prediction error is

$$\begin{aligned}\frac{1}{n}E[(Y_0 - \hat{Y}_0)'(Y_0 - \hat{Y}_0)] &= \frac{1}{n}E[(X\beta + \varepsilon_0 - X\hat{\beta})'(X\beta + \varepsilon_0 - X\hat{\beta})] \\ &= \frac{1}{n}E[(X\hat{\beta} - X\beta)'(X\hat{\beta} - X\beta)] + \frac{1}{n}E[\varepsilon_0{}'\varepsilon_0] \\ &= \sigma^2(1 + k/n).\end{aligned}$$

Conventionally, this mean squared prediction error, $\sigma^2(1 + k/n)$, is also called the final prediction error. Akaike's derivation estimated $\sigma^2$ with the unbiased estimate $s_k^2$, and this substitution yields $\widehat{\text{FPE}} = s_k^2(1 + k/n)$. Thus, $\widehat{\text{FPE}}$ is unbiased for FPE. Rewriting $\widehat{\text{FPE}}$ in terms of the maximum likelihood estimate $\hat{\sigma}_k^2$ gives us the familiar form of $\widehat{\text{FPE}}$, which we denote as FPE:

$$\text{FPE}_k = \hat{\sigma}_k^2 \frac{n + k}{n - k}, \tag{2.11}$$

or equivalently

$$\text{FPE}_k = \hat{\sigma}_k^2 \left(1 + \frac{2k}{n - k}\right).$$

It can be shown easily from Eq. (2.26) in Section 2.6 that for overfitting, $E[\text{FPE}_k]$ differs from $E[L_2]$ by $\sigma^2$. Note that $\text{FPE}_k$ balances the variance of the best linear predictor for $Y_0$ with the variance of $X\hat{\beta}$. Hence, the idea of minimizing FPE strikes a balance between these two variances.

Mallows (1973) took a different approach to obtaining an $L_2$-based model selection criterion. Consider the function

$$\begin{aligned} J_k &= \frac{1}{\sigma_*^2}\sum_{i=1}^{n}(x_i'\hat{\beta} - x_i'\beta)^2 \\ &= \frac{n}{\sigma_*^2}L_2 \\ &= \frac{1}{\sigma_*^2}(\hat{\beta} - \beta)'X'X(\hat{\beta} - \beta). \end{aligned}$$

Mallows found that $E[J_k] = V_k + B_k/\sigma_*^2$, where $V_k$ represents the variance and $B_k$ represents the bias. In regression, $V_k = k$ and $B_k =$ the noncentrality parameter. It is known that $E[\text{SSE}_k] = (n-k)\sigma_*^2 + B_k$ and

$$\begin{aligned} E[\text{SSE}_k/\sigma_*^2 - n + 2k] &= k + B_k/\sigma_*^2 \\ &= V_k + B_k/\sigma_*^2 \\ &= E[J_k]. \end{aligned}$$

Hence, $\text{SSE}_k/\sigma_*^2 - n + 2k$ is unbiased for $E[J_k]$. In Mallows' derivation, the estimate $s_K^2$ (Eq. (2.7) with $k = K$) from the largest candidate model was substituted as a potentially unbiased estimate of $\sigma_*^2$ to yield the well-known Mallows's Cp model selection criterion

$$\text{Cp} = \frac{\text{SSE}_k}{s_K^2} - n + 2k. \tag{2.12}$$

However, the quantity Cp is no longer unbiased for $E[J_k]$ since $1/s_k^2$ is not unbiased for $1/\sigma_*^2$. Recent work by Mallows (1995) indicates that any candidate model where $\text{Cp} < k$ should be carefully examined as a potential best model, and in practice this is a reasonable approach. However, in our simulations we consider the model where a criterion attains its minimum as best. Because this is not necessarily the best way to apply Cp, this may explain its disappointing performance in our simulations.

We now look at the foundation criteria that estimate the Kullback–Leibler discrepancy, AIC and AICc. AIC (Akaike, 1973) was the first of the Kullback–Leibler information based model selection criteria. It is asymptotically unbiased for K-L. In his derivation Akaike made the useful, but arguably unrealistic, assumption that the true model belongs to the set of candidate models. This assumption may be unrealistic in practice, but it allows us to compute expectations for central distributions, and it also allows us to entertain the concept of overfitting. In general,

$$\text{AIC} = -2\log(\textit{likelihood}) + 2 \times \textit{number of parameters},$$

where the *likelihood* is usually evaluated at the estimated parameters. The derivation of AIC is intended to create an estimate that is an approximation of the Kullback–Leibler discrepancy (a detailed derivation can be found in Linhart and Zucchini, 1986, p. 243). Like the Kullback–Leibler discrepancy on which it is based, AIC is readily adapted to a wide range of statistical models.

In fitting candidate models to Eqs. (2.3)–(2.4), we have

$$-2\log(\textit{likelihood}) = n\log(2\pi) + n\log(\sigma^2) + \frac{1}{\sigma^2}\sum_{i=1}^{n}(y_i - x_i'\beta)^2.$$

Using the maximum likelihood estimates under the normal error assumption,

$$-2\log(\textit{likelihood}) = n\log(2\pi) + n\log(\hat{\sigma}_k^2) + n.$$

The *number of parameters* is $k$ for the $\beta$ and 1 for $\sigma^2$. Substituting,

$$\text{AIC} = n\log(2\pi) + n\log(\hat{\sigma}_k^2) + n + 2(k+1).$$

The constants $n\log(2\pi) + n$ play no practical role in model selection and can be ignored. Now,

$$\text{AIC} = n\log(\hat{\sigma}_k^2) + 2(k+1).$$

We scale AIC by $1/n$ to express it as a rate:

$$\text{AIC}_k = \log(\hat{\sigma}_k^2) + \frac{2(k+1)}{n}. \tag{2.13}$$

Many authors have shown (Hurvich and Tsai, 1989) that the small-sample properties of AIC lead to overfitting. In response to this difficulty, Sugiura (1978) and Hurvich and Tsai (1989) derived AICc by estimating the expected Kullback–Leibler discrepancy directly in regression models. As with AIC, the candidate model is estimated via maximum likelihood. Hurvich and Tsai also adopted the assumption that the true model belongs to the set of candidate models. Under this assumption, they took expectations of Eq. (2.10):

$$E_*[\text{K-L}] = E_*\left[\log\left(\frac{\hat{\sigma}_k^2}{\sigma_*^2}\right) + \frac{\sigma_*^2}{\hat{\sigma}_k^2} + \frac{\frac{1}{n}\| X_*\beta_* - X\hat{\beta} \|^2}{\hat{\sigma}_k^2} - 1\right].$$

These expectations can be simplified due to the fact that $\| X_*\beta_* - X\hat{\beta} \|^2$ and $\hat{\sigma}_k^2$ are independent, that $E_*[\hat{\sigma}_k^2] = (n-k)\sigma_*^2/n$, and that $E_*[1/\hat{\sigma}_k^2] = n/\{(n-k-2)\sigma_*^2\}$. Substituting,

$$E_*[\text{K-L}] = E_*[\log(\hat{\sigma}_k^2)] - \log(\sigma_*^2) + \frac{n\sigma_*^2}{(n-k-2)\sigma_*^2} + \frac{k\sigma_*^2}{(n-k-2)\sigma_*^2} - 1.$$

Simplifying,

$$E_*[\text{K-L}] = E_*[\log(\hat{\sigma}_k^2)] - \log(\sigma_*^2) + \frac{n+k}{n-k-2} - 1.$$

Noticing that $\log(\hat{\sigma}_k^2)$ is unbiased for $E_*[\log(\hat{\sigma}_k^2)]$, then

$$\log(\hat{\sigma}_k^2) + \frac{n+k}{n-k-2} - \log(\sigma_*^2) - 1$$

is unbiased for $E_*[\text{K-L}]$. The constant $-\log(\sigma_*^2) - 1$ makes no contribution to model selection and can be ignored, yielding

$$\text{AICc}_k = \log(\hat{\sigma}_k^2) + \frac{n+k}{n-k-2}. \tag{2.14}$$

AICc is intended to correct the small-sample overfitting tendencies of AIC by estimating $E_*[\text{K-L}]$ directly rather than estimating an approximation to K-L. Hurvich and Tsai (1989) have shown that AICc does in fact outperform AIC in small samples, but that it is asymptotically equivalent to AIC and therefore performs just as well in large samples. Shibata (1981) showed that AIC and FPE are asymptotically efficient criteria, and other authors (*e.g.*, Nishii, 1984) have shown that AIC, FPE, and Cp are asymptotically equivalent, which implies that AICc and Cp are also asymptotically efficient.

We next consider the case where an investigator believes that the true model belongs to the set of candidate models. Here the goal is to identify the true model with an asymptotic probability of 1, the approach that resulted in the derivation of consistent model selection criteria. Two authors, Akaike (1978) and Schwarz (1978), introduced equivalent consistent model selection criteria conceived from a Bayesian perspective. Schwarz derived SIC for selecting models in the Koopman–Darmois family, whereas Akaike derived his model selection criterion BIC for the problem of selecting a model in linear regression. Although in this book we consider SIC, the reader should note that the two procedures are equivalent both in performance and by date of introduction.

Schwarz's derivation is more general than the usual linear regression. Assume that the observations come from a Koopman–Darmois family with density of the form

$$f(x,\theta) = \exp(\theta \cdot y(x) - b(\theta)),$$

where $\theta \in \Theta$, a convex subset of $\Re^K$, and $Y$ is a $K$-dimensional sufficient statistic for $\theta$. Since SIC does not depend on the prior, the exact distribution of the prior need not be known. Schwarz assumes it is of the form $\sum \alpha_j \mu_j$, where $\alpha_j$ is the prior probability for model $j$, and $\mu_j$ is the conditional prior of $\theta$ given model $j$. Finally, Schwarz assumed a fixed penalty or loss for selecting the wrong model. The Bayes solution for selecting a model is to choose the model with the largest posterior probability of being correct. In large samples, this posterior can be approximated by a Taylor expansion. Schwarz found the first term to be $n\log(\hat{\sigma}_j^2)$, the log of the MLE for the variance in model $j$. The second term was of the form $\log(n)k$ where $k$ is the dimension of the model and $n$ is the sample size. The remaining terms in the Taylor expansion were shown to be bounded and hence could be ignored in large samples. Scaling the first two terms by $n$, we have

$$\text{SIC}_k = \log(\hat{\sigma}_k^2) + \frac{\log(n)k}{n}. \tag{2.15}$$

The $2k$ term in AIC is replaced by $\log(n)k$ in SIC, resulting in a much stronger penalty for overfitting.

The other consistent criterion among our foundation criteria was proposed by Hannan and Quinn (1979). They applied the law of the iterated logarithm to derive HQ for autoregressive time series models. Although intended for use with the autoregressive model, HQ also can be applied to regression models. We postpone the derivation for HQ until Chapter 3, where we discuss autoregressive models in detail, and simply present the expression for the scaled HQ for regression with $\hat{\sigma}^2$ from Eq. (2.8):

$$\text{HQ}_k = \log(\hat{\sigma}_k^2) + \frac{2\log\log(n)k}{n}. \tag{2.16}$$

Although asymptotically consistent, many authors have pointed out that HQ behaves more like the efficient model selection criterion AIC. This can be explained by the behavior of its penalty function, which even for a sample size of 200,000 is roughly only 2.5 times larger than that of AIC (loglog(200,000) = 2.502). Thus, for most practical sample sizes, the penalty function of HQ is similar to that of AIC.

Many authors (*e.g.*, Bhansali and Downham, 1977) have examined the penalty functions of AIC and FPE, and have proposed variations that seek to modify overfitting properties by adjusting them by $\alpha$. For example, AIC$\alpha = \log(\hat{\sigma}_k^2) + \alpha k/n$ and FPE$\alpha = \hat{\sigma}_k^2(1 + \alpha k/(n-k))$. When $\alpha = 2$, we have the familiar versions of AIC and FPE. The choice of $\alpha$ follows from setting asymptotic probability of overfitting tolerances. On the basis of simulation results, Bhansali and Downham propose FPE4, although other authors have found $\alpha$ in the range of 1.5 to 5 to yield acceptable performance. Note that the penalty function of HQ falls within this $\alpha$ range for $n$ between 9 and 200,000, extremely wide limits for sample size. Our signal-to-noise ratio derivations show that in small samples, adjusting the penalty function by $\alpha$ yields much less satisfactory results than the correction proposed by AICc.
Shibata

(1981) also showed that the $\alpha = 2$ criteria AIC and FPE are asymptotically efficient, and that $\alpha = 2$ is rooted in information theory, particularly in the case of AIC. Other choices of $\alpha$ are often motivated by the resulting asymptotic probability of overfitting.

### 2.3. Moments of Model Selection Criteria

When choosing among candidate models, the standard rule is that the best model is the one for which the value of the model selection criterion used attains its minimum, and models are compared by taking the difference between the criterion values for each model. For example, suppose we have one model with $k$ variables and a second model with $L$ additional variables, and we would like to use some hypothetical model selection criterion, say MSC, to evaluate them. Model A (with $k$ variables) will be considered better than Model B (with $k + L$ variables) if $\text{MSC}_{k+L} > \text{MSC}_k$. This difference depends largely on the strength of the penalty function, and is actually a random variable for which moments can be computed for nested models. We define the *signal* as $E[\text{MSC}_{k+L} - \text{MSC}_k]$, the *noise* as the standard deviation of the difference $sd[\text{MSC}_{k+L} - \text{MSC}_k]$, and the *signal-to-noise ratio* as $E[\text{MSC}_{k+L} - \text{MSC}_k]/sd[\text{MSC}_{k+L} - \text{MSC}_k]$. We will use this definition and some convenient approximations to calculate the signal-to-noise ratios for all the criteria in this Section. While the signal depends primarily on the penalty function, the noise depends on the distribution of SSE and the distribution of differences in SSE. If the penalty function is weaker than the noise, the model selection criterion will have a weak signal, a weak signal-to-noise ratio, and tend to overfit. A large signal-to-noise ratio, which occurs when the penalty function is much stronger than the noise, will overcome this difficulty. We often use the terms strong and weak when describing the signal-to-noise ratio. In general, a strong signal-to-noise ratio refers to a large positive value (often greater than 2). A weak signal-to-noise ratio usually refers to one that is small (less than 0.5) or negative. However, if the penalty function is too large the signal-to-noise ratio becomes weak in the underfitting case, and the model selection criterion will be prone to underfitting. Because an examination of underfitting will require the use of noncentral distributions and the noncentrality parameter, $\lambda$, which we will discuss later, for now we will use the signal-to-noise ratio only to examine overfitting for our six foundation criteria. For comparison purposes we will also consider adjusted $\mathrm{R}^2$ ($\mathrm{R}^2_{adj}$), since it is still widely used.

### 2.3.1. AIC and AICc

We will first look at the K-L-based criteria AIC and AICc, which estimate the Kullback–Leibler information. For AIC, we will choose model $k$ over model $k+L$ if $\mathrm{AIC}_{k+L} > \mathrm{AIC}_k$. Let $\Delta\mathrm{AIC} = \mathrm{AIC}_{k+L} - \mathrm{AIC}_k$. Neither the expression for the signal nor the noise have closed forms, and therefore the Taylor expansions in Appendix 2A (Eq. (2A.13) and Eq. (2A.14)) will be used to derive algebraic expressions for the signal and noise of AIC. By examining this ratio we should be able to gain some insight into the behavior of AIC. Applying Eq. (2A.13), the signal is

$$E[\Delta\mathrm{AIC}] = \log\left(\frac{n-k-L}{n-k}\right) - \frac{L}{(n-k-L)(n-k)} + \frac{2L}{n},$$

and from Eq. (2A.14), the noise is

$$sd[\Delta\mathrm{AIC}] = sd[\Delta\log(\mathrm{SSE}_k)] = \frac{\sqrt{2L}}{\sqrt{(n-k-L)(n-k+2)}}.$$

The signal-to-noise ratio is

$$\frac{E[\Delta\mathrm{AIC}]}{sd[\Delta\mathrm{AIC}]} = \frac{\sqrt{(n-k-L)(n-k+2)}}{\sqrt{2L}}\left(\log\left(\frac{n-k-L}{n-k}\right) - \frac{L}{(n-k-L)(n-k)} + \frac{2L}{n}\right).$$

We will examine the behavior of the signal-to-noise ratio one term at a time. As $L$ increases to $n-k$, the first term

$$\frac{\sqrt{(n-k-L)(n-k+2)}}{\sqrt{2L}}\log\left(\frac{n-k-L}{n-k}\right) \to 0,$$

and the second term

$$-\frac{\sqrt{(n-k-L)(n-k+2)}}{\sqrt{2L}}\frac{L}{(n-k-L)(n-k)} \to -\infty.$$

The behavior of the first and second terms follows from the use of $\log(\hat{\sigma}_k^2)$ or $\log(\mathrm{SSE}/n)$. The last term,

$$\frac{\sqrt{(n-k-L)(n-k+2)}}{\sqrt{2L}}\frac{2L}{n} \to 0.$$

This third term increases for small $L$, then decreases to 0 as $L$ increases, and its behavior follows from the penalty function. The K-L-based model selection criteria therefore have two components—log(SSE) and an additive penalty function. As can be seen for AIC, the signal eventually decreases as $k$ (or $L$) increases due to the fact that its penalty function is linear in $k$ and is not strong enough to overcome log(SSE). Hence, as $L$ becomes large, AIC's signal-to-noise ratio becomes weak, and an undesirable negative signal-to-noise ratio for excessive overfitting results. Typically, the signal-to-noise ratio of AIC increases for small $L$, but as $L \to n-k$, the signal-to-noise ratio of AIC $\to -\infty$ resulting in the well-known small-sample overfitting problems of AIC. This problem will plague any criterion of the form $\log(\hat{\sigma}_k^2) + \alpha k/n$, where $\alpha$ is some constant and the penalty function is linear in $k$.

Since the noise component of all K-L-based model selection criteria is derived from log(SSE), their signal-to-noise ratios will all depend on the size of the penalty function. A small penalty function results in a weak signal-to-noise ratio, and thus will cause a criterion to be prone to overfitting. In order to overcome this difficulty, the penalty function must be *superlinear* in $k$, whereby we mean that the first derivative is positive and the penalty function is unbounded for some $k \leq n$. This ensures that the signal increases rapidly with the amount of overfitting $L$. AICc's correction term is based on just such a superlinear penalty function, as we see below.

AICc estimates small-sample properties of the expected Kullback–Leibler information. Applying Eq. (2A.13), its signal is

$$E[\Delta\mathrm{AICc}] = \log\left(\frac{n-k-L}{n-k}\right) - \frac{L}{(n-k-L)(n-k)} + \frac{2L(n-1)}{(n-k-2)(n-k-L-2)}.$$

We can see that the signal is in fact superlinear in $L$. From Eq. (2A.14), the noise is

$$sd[\Delta\text{AICc}] = sd[\Delta\log(\text{SSE}_k)] = \frac{\sqrt{2L}}{\sqrt{(n-k-L)(n-k+2)}}.$$

The signal-to-noise ratio is

$$\frac{E[\Delta\text{AICc}]}{sd[\Delta\text{AICc}]} = \frac{\sqrt{(n-k-L)(n-k+2)}}{\sqrt{2L}}\left(\log\left(\frac{n-k-L}{n-k}\right)\right.$$
$$\left. - \frac{L}{(n-k-L)(n-k)} + \frac{2L(n-1)}{(n-k-2)(n-k-L-2)}\right),$$

which increases as $L$ increases. Because the signal-to-noise ratio for AICc is large when $L$ is large, AICc should perform well from an overfitting perspective.

We end this Section with the following theorem showing that criteria with penalty functions similar to AICc, of the form $\alpha k/(n-k-2)$, have signal-to-noise ratios that increase as the amount of overfitting increases. Such criteria should overfit less than criteria with weaker (linear) penalty functions. The proof of Theorem 2.1 is given in Appendix 2B.1.

**Theorem 2.1**

Given the regression model in Eqs. (2.3)–(2.4) and the criterion of the form $\log(\hat{\sigma}_k^2) + \alpha k/(n-k-2)$, for all $n \geq 6$, $\alpha \geq 1$, $0 < L < n-k-2$, and for the overfitting case where $0 < k_* \leq k < n-3$, the signal-to-noise ratio of this criterion increases as $L$ increases.

### *2.3.2. FPE and Cp*

We will now examine model selection criteria that estimate the $L_2$ distance, or prediction error variance. Suppose that $k_*$ is the true model. We derive the moments of FPE and Cp under the more general models $k$ and $k+L$ where $k \geq k_*$ and $L > 0$, and we assume $k$ and $k+L$ form nested models for all $L$. For FPE, it follows from Eq. (2A.4) and Eq. (2A.5) that when $a = -(n+k)/n(n-k)$ and $b = -(n+k+L)/n(n-k-L)$, that the signal is

$$E[\Delta\text{FPE}] = \frac{L}{n}\sigma_*^2,$$

the noise is

$$sd[\Delta\text{FPE}] = \sqrt{\frac{(n+k)^2(n-k-L)2L\sigma_*^4 + 8n^2L^2\sigma_*^4}{n^2(n-k)^2(n-k-L)}},$$

and the signal-to-noise ratio is

$$\frac{E[\Delta\text{FPE}]}{sd[\Delta\text{FPE}]} = \frac{(n-k)\sqrt{L(n-k-L)}}{\sqrt{2(n+k)^2(n-k-L)+8n^2L}}.$$

Both the signal and the noise increase as $L$ increases. We see that the numerator of the noise decreases with $L$. The noise increases superlinearly as $L$ increases. In fact, the noise is unbounded for the saturated model where $L = n - k$. For small $L$, the signal-to-noise ratio for FPE increases as $L$ increases, and then, for large values of $L$, this ratio decreases to 0, again leading to overfitting. In general, FPE has a weak (small) signal-to-noise ratio in small samples.

To obtain the signal-to-noise ratio for Mallows's Cp, we also need to assume that there is some largest model of order $K$. Now,

$$\begin{aligned}\Delta\text{Cp} &= (n-K)\frac{\text{SSE}_{k+L}}{\text{SSE}_K} - n + 2(k+L) - (n-K)\frac{\text{SSE}_k}{\text{SSE}_K} - n + 2k \\ &= (n-K)\frac{\text{SSE}_{k+L} - \text{SSE}_k}{\text{SSE}_K} + 2L.\end{aligned}$$

In terms of distributions,

$$\Delta\text{Cp} = (n-K)\frac{-\chi^2_L}{\chi^2_{n-K}} + 2L,$$

with independent $\chi^2$ distributions for the numerator and denominator. From this, the signal is

$$\begin{aligned}E[\Delta\text{Cp}] &= (n-K)E\left[\frac{-\chi^2_L}{\chi^2_{n-K}}\right] + 2L \\ &= (n-K)\frac{-L}{n-K-2} + 2L \\ &= \left(\frac{n-K-4}{n-K-2}\right)L.\end{aligned}$$

Hence, Cp's signal increases linearly as $L$ increases. The variance is

$$\begin{aligned}var[\Delta\text{Cp}] &= (n-K)^2 var\left[\frac{-\chi^2_L}{\chi^2_{n-K}}\right] \\ &= (n-K)^2\left(\frac{2L+L^2}{(n-K-2)(n-K-4)} - \frac{L^2}{(n-K-2)^2}\right) \\ &= (n-K)^2\left(\frac{2L(n-K-2)+2L^2}{(n-K-2)^2(n-K-4)}\right),\end{aligned}$$

and the noise is

$$sd[\Delta\text{Cp}] = \frac{(n-K)\sqrt{2L(n-K-2)+2L^2}}{(n-K-2)\sqrt{n-K-4}}.$$

Note that the noise also increases approximately linearly as $L$ increases, rather than superlinearly as for FPE. Cp's noise term is therefore preferable to that of FPE. Finally, the signal-to-noise ratio for Cp is

$$\frac{E[\Delta\text{Cp}]}{sd[\Delta\text{Cp}]} = \frac{(n-K-4)^{3/2}}{n-K} \frac{L}{\sqrt{2L(n-K-2)+2L^2}},$$

which increases as $L$ increases. While the signal-to-noise ratio for Cp is often weak, which leads to some overfitting, because of its superior noise term we expect less overfitting from Cp than from FPE. Also, because the signal-to-noise ratio for Cp depends on the order $K$ of the largest candidate model, if $K$ changes then Cp must be recalculated for all candidate models.

### 2.3.3. SIC and HQ

We next consider the consistent model selection criteria SIC and HQ. For SIC, applying Eq. (2A.13), the signal is

$$E[\Delta\text{SIC}] = \log\left(\frac{n-k-L}{n-k}\right) - \frac{L}{(n-k-L)(n-k)} + \frac{\log(n)L}{n}.$$

From Eq. (2A.14), the noise is

$$sd[\Delta\text{SIC}] = sd[\Delta\log(\text{SSE}_k)] = \frac{\sqrt{2L}}{\sqrt{(n-k-L)(n-k+2)}}.$$

The signal-to-noise ratio is

$$\frac{E[\Delta\text{SIC}]}{sd[\Delta\text{SIC}]} = \frac{\sqrt{(n-k-L)(n-k+2)}}{\sqrt{2L}}\left(\log\left(\frac{n-k-L}{n-k}\right) - \frac{L}{(n-k-L)(n-k)} + \frac{\log(n)L}{n}\right).$$

A term-by-term analysis indicates that SIC suffers from the same problems in small samples that AIC does. The first two terms $\to -\infty$ as $L \to n-k$. The third term, which follows from SIC's penalty function, $\to 0$. In larger samples with $k << n$, SIC should not suffer from excessive overfitting. But although the signal-to-noise ratio of SIC increases for small $L$ to moderate $L$, as $L \to n-k$, the signal-to-noise ratio of SIC $\to -\infty$, leading to small-sample overfitting.

For HQ, applying Eq. (2A.13), the signal is

$$E[\Delta \text{HQ}] = \log\left(\frac{n-k-L}{n-k}\right) - \frac{L}{(n-k-L)(n-k)} + \frac{2\log\log(n)L}{n}.$$

From Eq. (2A.14), the noise is

$$sd[\Delta \text{HQ}] = sd[\Delta \log(\text{SSE}_k)] = \frac{\sqrt{2L}}{\sqrt{(n-k-L)(n-k+2)}},$$

and the signal-to-noise ratio is

$$\frac{E[\Delta \text{HQ}]}{sd[\Delta \text{HQ}]} = \frac{\sqrt{(n-k-L)(n-k+2)}}{\sqrt{2L}}\left(\log\left(\frac{n-k-L}{n-k}\right) - \frac{L}{(n-k-L)(n-k)} + \frac{2\log\log(n)L}{n}\right).$$

Again, the second term

$$-\frac{\sqrt{(n-k-L)(n-k+2)}}{\sqrt{2L}}\frac{L}{(n-k-L)(n-k)}$$

decreases to $-\infty$ as $L$ increases. In small samples HQ will be expected to perform similarly to AIC. Like AIC, the signal-to-noise ratio of HQ increases for small $L$, and then decreases to $-\infty$ resulting in small-sample overfitting problems.

### *2.3.4. Adjusted $R^2$, $R^2_{adj}$*

Finally we will briefly discuss the adjusted coefficient of determination. Although its performance is much less than ideal under certain circumstances it is still widely applied in many disciplines. The adjusted coefficient of determination, or adjusted R-squared statistic $\text{R}^2_{adj}$, is a standard tool in most regression and statistical software. Of possible candidate models proposed, typically the model that maximizes $\text{R}^2_{adj}$ is chosen as the best model. Consider again the regression model Eqs. (2.3)–(2.4) described in Section 2.1. In regression,

$$\text{R}^2_{adj} = 1 - \frac{\text{MSE}}{\text{MS}(Total)},$$

where MSE $= s_k^2$ (see Eq. (2.7)) and MS$(Total) = \sum_{i=1}^{n}(y_i - \bar{y})^2/(n-1)$ (Rawlings, 1988, p. 183). Since most model selection criteria choose the model where the model selection criterion attains the minimum, for consistency we will use the equivalent expression

$$1 - \mathrm{R}_{adj}^2 = \frac{s_k^2}{\mathrm{MS}(Total)}. \tag{2.17}$$

Hence the model that minimizes $1 - \mathrm{R}_{adj}^2$ is the same model that maximizes $\mathrm{R}_{adj}^2$, and is chosen as the best. Furthermore, for any regression model data, MS$(Total)$ is constant for all candidate models and $1 - \mathrm{R}_{adj}^2 \propto s_k^2$. Choosing the model $k$ that maximizes $\mathrm{R}_{adj}^2$ is the same as choosing the model $k$ that minimizes $s_k^2$.

For true model $k_*$ with $s_{k_*}^2$, we will choose model $k_*$ if $s_{k_*+L}^2 > s_{k_*}^2$ or if $\Delta s^2 = s_{k_*+L}^2 - s_{k_*}^2 > 0$. It follows from Eq. (2A.4) and Eq. (2A.5), if we let $a = -1/(n-k_*)$ and $b = -1/(n-k_*-L)$, that the signal is

$$E[\Delta s^2] = 0$$

and the noise is

$$sd[\Delta s^2] = \frac{\sqrt{2L}\sigma_*^2}{\sqrt{(n-k_*)(n-k_*-L)}}.$$

Hence, the signal-to-noise ratio is

$$\frac{E[\Delta s^2]}{sd[\Delta s^2]} = 0.$$

In fact, the signal-to-noise ratio of $\mathrm{R}_{adj}^2$ is 0 for all amounts of overfitting $L$ and sample sizes $n$. Clearly, both $s_{k_*+L}^2$ and $s_{k_*}^2$ are unbiased for $\sigma_*^2$, and choosing a model on the basis of $s_k^2$ involves the distribution of differences between $s_k^2$ for competing models. Thus, $s_k^2$ itself will not reliably choose model $k_*$ even for large $L$ or $n$. It is well-known that $s_{k_*}^2$ is unbiased for $\sigma_*^2$ when the true model is fitted or when the model is overfitted. In practice, the variance of $s_{k+L}^2 - s_k^2$ can lead to false minima in $1 - \mathrm{R}_{adj}^2$ among overfitted models, and this results in a tendency for $\mathrm{R}_{adj}^2$ to overfit.

## 2.4. Signal-to-noise Corrected Variants

The performance of model selection criteria with weak signal-to-noise ratios could be improved if their signal-to-noise ratios were strengthened. Unfortunately, there is no single appropriate correction for all criteria. We will suggest a possible approach that leads to three signal-to-noise corrected variants, FPEu, AICu, and HQc. We have chosen these three in order that a variant will be derived from each of the three broad classes of criteria we have described: K-L- and $L_2$-based efficient criteria and consistent criteria. All the corrected variants will be shown to outperform their parent criteria, and the simulation study at the end of the Chapter will show that, in particular, AICu and HQc are extremely competitive overall.

The signal-to-noise corrected variants we will introduce are AICu (McQuarrie, Shumway and Tsai, 1997), FPEu, and HQc. The "u" suffix indicates that we have used the unbiased estimate $s_k^2$ in place of the usual MLE $\hat{\sigma}_k^2$. The "c" suffix indicates that the corrected penalty function is modeled after the relationship between AIC and AICc. All three variants will be derived under the regression model Eqs. (2.3)–(2.4) described in Section 2.1, assuming general models of order $k$ and $k+L$ where $k \geq k_*$, $L > 0$, where $k_*$ is the true model. We assume $k$ and $k+L$ form nested models for all $L$.

### *2.4.1. AICu*

We have seen that AIC has a weak signal-to-noise ratio and therefore tends to overfit in small samples, and this is the problem we want to address. For AIC $= \log(\hat{\sigma}_k^2) + 2(k+1)/n$, the signal-to-noise ratio associated with $\log(\hat{\sigma}_k^2)$ $\to -\infty$ as $L$ increases. If we make two substitutions: changing $\log(\hat{\sigma}_k^2)$ to $\log(s_k^2)$, and changing the penalty function $2(k+1)/n$ to $2(k+1)/(n-k-2)$, we will get

$$\text{AICu}_k = \log(s_k^2) + \frac{2(k+1)}{n-k-2}.$$

Constants play no practical role in selecting a model, and so for convenience we can add 1 to AICu, yielding the equivalent form

$$\text{AICu}_k = \log(s_k^2) + \frac{n+k}{n-k-2}. \tag{2.18}$$

While AICu has the same penalty function as our other corrected AIC criterion, AICc, it differs by its use of the unbiased estimate $s_k^2$, Eq. (2.7), in place of $\hat{\sigma}_k^2$, Eq. (2.8). Later we will show that this substitution yields a criterion with asymptotic properties that differ from those of its counterpart using $\hat{\sigma}_k^2$. One might wonder why it is worthwhile to derive AICu when AICc itself performs well. One reason is similar to that which motivated the derivation of the $\alpha$ variants of AIC, which were intended to address the parent criterion's asymptotic properties of overfitting. In Section 2.5 we will show that the probability that an efficient model selection criterion will overfit by one particular extra variable is 0.1573, whereas consistent model selection criteria overfit with probability 0. In addition, we will see in Section 2.5 that the probability that AICu overfits by one particular extra variable is 0.0833, roughly halfway between 0 and 0.1573.

We apply our algorithm and Eq. (2A.13) to obtain the signal for AICu,

$$E[\Delta\text{AICu}] = -\frac{L}{(n-k-L)(n-k)} + \frac{2L(n-1)}{(n-k-2)(n-k-L-2)},$$

and from Eq. (2A.14) the noise is

$$sd[\Delta\text{AICu}] = \frac{\sqrt{2L}}{\sqrt{(n-k-L)(n-k+2)}},$$

and therefore the signal-to-noise ratio is

$$\frac{E[\Delta\text{AICu}]}{sd[\Delta\text{AICu}]} = \frac{\sqrt{(n-k-L)(n-k+2)}}{\sqrt{2L}} \times \left(-\frac{L}{(n-k-L)(n-k)} + \frac{2L(n-1)}{(n-k-2)(n-k-L-2)}\right). \quad (2.19)$$

The following theorem states the two main advantages of the signal-to-noise corrected variant AICu over its parent AIC. We give analogous theorems for the following two variants, and the proofs for all three can be found in Appendix 2B.

**Theorem 2.2**

Given the regression model in Eqs. (2.3)–(2.4), for all $n \geq 5$, $0 < L < n-k-2$, and for the overfitting case where $0 < k_* \leq k < n-3$, the signal-to-noise ratio of AICu increases as $L$ increases and is greater than the signal-to-noise ratio of AIC.

### *2.4.2. FPEu*

Like AIC and AICu, FPEu substitutes $s_k^2$ for $\hat{\sigma}_k^2$ in FPE:

$$\text{FPEu} = \frac{n+k}{n-k} s_k^2. \quad (2.20)$$

It follows from Eq. (2A.4) that the signal is

$$E[\Delta\text{FPEu}] = \frac{2nL}{(n-k)(n-k-L)} \sigma_*^2.$$

Using

$$a = -\frac{n+k}{(n-k)^2} \text{ and } b = -\frac{n+k+L}{(n-k-L)^2},$$

and applying Eq. (2A.5), the variance is

$$\begin{aligned} var[\Delta\text{FPEu}] =& \frac{2(n+k)^2(n-k-L)^3 L\sigma_*^4}{(n-k)^4(n-k-L)^3} \\ &+ \frac{2\left[(n+k+L)(n-k)^2 - (n+k)(n-k-L)^2\right]^2\sigma_*^4}{(n-k)^4(n-k-L)^3}. \end{aligned}$$

Therefore the signal-to-noise ratio $E[\Delta\text{FPEu}]/sd[\Delta\text{FPEu}]$ is

$$\frac{\sqrt{2}n(n-k)L\sqrt{n-k-L}}{\sqrt{(n+k)^2(n-k-L)^3L + \left[(n+k+L)(n-k)^2 - (n+k)(n-k-L)^2\right]^2}}.$$

Unlike AICu, the signal-to-noise ratio of FPEu does not strictly increase as $L$ increases. As $L \to n-k$, the numerator $\sqrt{2}n(n-k)L\sqrt{n-k-L} \to 0$ and the denominator is always greater than 0; thus the signal-to-noise ratio of FPEu $\to 0$. While this results in some overfitting by FPEu, its signal-to-noise ratio is improved over that for FPE.

**Theorem 2.3**

Given the regression model in Eqs. (2.3)–(2.4), for all $n \geq 5$, $0 < L < n-k-2$, and for the overfitting case where $0 < k_* \leq k < n-3$, the signal-to-noise ratio of FPEu is greater than or equal to the signal-to-noise ratio of FPE.

### *2.4.3. HQc*

Our modification of HQ is based on the relationship between AIC and AICc, and so we need to reexamine their respective penalty functions. In order to see more clearly how the two functions differ, we can rewrite AICc's penalty function in terms of AIC's penalty function × some scaling factor. Since adding a constant plays no role in model selection, first subtract 1 from AICc to get

$$\begin{aligned} \text{AICc} - 1 &= \log(\hat{\sigma}_k^2) + \frac{n+k}{n-k-2} - 1 \\ &= \log(\hat{\sigma}_k^2) + \frac{2(k+1)}{n-k-2}, \end{aligned}$$

which can be rewritten as

$$\text{AICc} = \log(\hat{\sigma}_k^2) + \frac{2(k+1)}{n}\frac{n}{n-k-2}.$$

Now we can easily see that the penalty function for AIC has been scaled by $n/(n-k-2)$ to yield the penalty function for AICc. If we use this same scaling to form a new penalty function for HQ, we can define HQc as

$$\mathrm{HQc} = \log(\hat{\sigma}_k^2) + \frac{2\log\log(n)k}{n-k-2}. \tag{2.21}$$

Note that HQc not only corrects the small-sample performance of HQ, but it is also an asymptotically consistent criterion. Applying Eq. (2A.13), the signal of HQc is

$$\begin{aligned} E[\Delta\mathrm{HQc}] =& \log\left(\frac{n-k-L}{n-k}\right) - \frac{L}{(n-k-L)(n-k)} \\ &+ \frac{2\log\log(n)(n-2)L}{(n-k-L-2)(n-k-2)}. \end{aligned}$$

From Eq. (2A.14), the noise is

$$sd[\Delta\mathrm{HQc}] = \frac{\sqrt{2L}}{\sqrt{(n-k-L)(n-k+2)}},$$

and therefore the signal-to-noise ratio is

$$\begin{aligned} \frac{E[\Delta\mathrm{HQc}]}{sd[\Delta\mathrm{HQc}]} =& \frac{\sqrt{(n-k-L)(n-k+2)}}{\sqrt{2L}}\left(\log\left(\frac{n-k-L}{n-k}\right)\right. \\ &\left. - \frac{L}{(n-k-L)(n-k)} + \frac{2\log\log(n)(n-2)L}{(n-k-L-2)(n-k-2)}\right). \end{aligned} \tag{2.22}$$

The signal-to-noise ratio of HQc increases as $L$ increases and is therefore stronger than the signal-to-noise ratio of its parent criterion.

**Theorem 2.4**

Given the regression model in Eqs. (2.3)–(2.4), for all $n \geq 6$, $0 < L < n-k-2$, and for the overfitting case where $0 < k_* \leq k < n-3$, the signal-to-noise ratio of HQc increases as $L$ increases and is greater than the signal-to-noise ratio of HQ.

## 2.5. Overfitting

For our six foundation criteria and the three signal-to-noise corrected variants we derived in the previous Section, we will now look at the probabilities of overfitting both in small samples and asymptotically. The probabilities for five sample sizes ($n = 15$, 25, 35, 50, 100) have been calculated for illustration purposes, and we will also obtain the small-sample and asymptotic signal-to-noise ratios for these same sample sizes in order to relate the probabilities of overfitting to their corresponding signal-to-noise ratios. Overfitting here refers to comparing one overfitted model of order $k_* + L$ to the reduced model of order $k_*$.

### *2.5.1. Small-sample Probabilities of Overfitting*

Suppose there is a true model order $k_*$ and we fit a candidate model of order $k_* + L$ where $L > 0$. Assume that only the two models $k_* + L$ and $k_*$ are compared and that they form nested models. To compute the probability of overfitting by $L$ extra variables we will compute the probability of selecting one model with $\text{rank}(X) = k_* + L$ over the true model with rank $(X_*) = k_*$. We will present detailed calculations for one criterion, AIC. Only the results of the analogous calculations for the other criteria will be shown here, but the details can be found in Appendix 2C.

**AIC**

We know that AIC overfits if $\text{AIC}_{k_*+L} < \text{AIC}_{k_*}$. For finite $n$, the probability that AIC prefers the overfitted model $k_* + L$ is

$$\begin{aligned}
&P\{\text{AIC}_{k_*+L} < \text{AIC}_{k_*}\} \\
&= P\left\{\log\left(\hat{\sigma}^2_{k_*+L}\right) + \frac{2(k_* + L + 1)}{n} < \log\left(\hat{\sigma}^2_{k_*}\right) + \frac{2(k_* + 1)}{n}\right\} \\
&= P\left\{\log\left(\text{SSE}_{k_*+L}\right) - \log(n) + \frac{2(k_* + L + 1)}{n}\right. \\
&\qquad\left. < \log\left(\text{SSE}_{k_*}\right) - \log(n) + \frac{2(k_* + 1)}{n}\right\} \\
&= P\left\{\log\left(\frac{\text{SSE}_{k_*+L}}{\text{SSE}_{k_*}}\right) < \frac{-2L}{n}\right\} \\
&= P\left\{\log\left(\frac{\text{SSE}_{k_*}}{\text{SSE}_{k_*+L}}\right) > \frac{2L}{n}\right\} \\
&= P\left\{\frac{\text{SSE}_{k_*}}{\text{SSE}_{k_*+L}} - 1 > \exp\left(\frac{2L}{n}\right) - 1\right\} \\
&= P\left\{\frac{\text{SSE}_{k_*} - \text{SSE}_{k_*+L}}{\text{SSE}_{k_*+L}} > \exp\left(\frac{2L}{n}\right) - 1\right\}.
\end{aligned}$$

Once in this form, we use independent $\chi^2$ facts from Appendix 2A and

$$\begin{aligned} &P\{\text{AIC}_{k_*+L} < \text{AIC}_{k_*}\} \\ &= P\left\{\frac{\chi^2_L}{\chi^2_{n-k_*-L}} > \exp\left(\frac{2L}{n}\right) - 1\right\} \\ &= P\left\{F_{L,n-k_*-L} > \frac{n-k_*-L}{L}\left(\exp\left(\frac{2L}{n}\right) - 1\right)\right\}. \end{aligned}$$

**AICc**

$$\begin{aligned} &P\{\text{AICc}_{k_*+L} < \text{AICc}_{k_*}\} \\ &= P\left\{F_{L,n-k_*-L} > \frac{n-k_*-L}{L} \times \right. \\ &\qquad \left.\left(\exp\left(\frac{2L(n-1)}{(n-k_*-L-2)(n-k_*-2)}\right) - 1\right)\right\}. \end{aligned}$$

**AICu**

$$\begin{aligned} &P\{\text{AICu}_{k_*+L} < \text{AICu}_{k_*}\} \\ &= P\left\{F_{L,n-k_*-L} > \frac{n-k_*-L}{L}\left(\exp\left(\frac{\log(n)L}{n}\right) - 1\right)\right\}. \end{aligned}$$

**FPE**

$$P\{\text{FPE}_{k_*+L} < \text{FPE}_{k_*}\} = P\left\{F_{L,n-k_*-L} > \frac{2n}{n+k_*}\right\}.$$

**FPEu**

$$\begin{aligned} &P\{\text{FPEu}_{k_*+L} < \text{FPEu}_{k_*}\} \\ &= P\left\{F_{L,n-k_*-L} > \frac{3n^2 - n(2k_*+L) - k_*(k_*+L)}{(n-k_*-L)(n+k_*)}\right\}. \end{aligned}$$

**Cp**

Recall that $K$ represents the number of variables in the largest model with all variables included.

$$P\{\text{Cp}_{k_*+L} < \text{Cp}_{k_*}\} = P\{F_{L,n-K} > 2\}.$$

Table 2.1. Probabilities of overfitting (one candidate versus true model).

$n=15$. True order = 6.

| $L$ | AIC | AICc | AICu | SIC | HQ | HQc | Cp | FPE | FPEu |
|---|---|---|---|---|---|---|---|---|---|
| 1 | 0.317 | 0.025 | 0.015 | 0.244 | 0.318 | 0.031 | 0.216 | 0.266 | 0.145 |
| 2 | 0.393 | 0.004 | 0.002 | 0.283 | 0.395 | 0.006 | 0.230 | 0.302 | 0.125 |
| 3 | 0.461 | 0.000 | 0.000 | 0.322 | 0.463 | 0.001 | 0.233 | 0.324 | 0.108 |
| 4 | 0.536 | 0.000 | 0.000 | 0.376 | 0.538 | 0.000 | 0.233 | 0.347 | 0.098 |

$n=25$. True order = 6.

| $L$ | AIC | AICc | AICu | SIC | HQ | HQc | Cp | FPE | FPEu |
|---|---|---|---|---|---|---|---|---|---|
| 1 | 0.237 | 0.079 | 0.044 | 0.133 | 0.201 | 0.063 | 0.191 | 0.220 | 0.118 |
| 2 | 0.257 | 0.041 | 0.016 | 0.112 | 0.204 | 0.028 | 0.191 | 0.228 | 0.089 |
| 3 | 0.266 | 0.019 | 0.005 | 0.095 | 0.202 | 0.011 | 0.185 | 0.226 | 0.066 |
| 4 | 0.277 | 0.008 | 0.002 | 0.084 | 0.202 | 0.004 | 0.178 | 0.223 | 0.051 |
| 5 | 0.291 | 0.003 | 0.000 | 0.078 | 0.207 | 0.001 | 0.173 | 0.221 | 0.040 |
| 6 | 0.310 | 0.001 | 0.000 | 0.076 | 0.216 | 0.000 | 0.168 | 0.221 | 0.032 |
| 7 | 0.334 | 0.000 | 0.000 | 0.078 | 0.231 | 0.000 | 0.164 | 0.223 | 0.026 |
| 8 | 0.364 | 0.000 | 0.000 | 0.082 | 0.251 | 0.000 | 0.161 | 0.227 | 0.022 |

$n=35$. True order = 6.

| $L$ | AIC | AICc | AICu | SIC | HQ | HQc | Cp | FPE | FPEu |
|---|---|---|---|---|---|---|---|---|---|
| 1 | 0.210 | 0.103 | 0.057 | 0.095 | 0.158 | 0.070 | 0.173 | 0.202 | 0.107 |
| 2 | 0.214 | 0.066 | 0.025 | 0.064 | 0.141 | 0.035 | 0.163 | 0.200 | 0.076 |
| 3 | 0.209 | 0.040 | 0.011 | 0.045 | 0.124 | 0.016 | 0.148 | 0.190 | 0.054 |
| 4 | 0.204 | 0.023 | 0.004 | 0.032 | 0.111 | 0.007 | 0.135 | 0.180 | 0.038 |
| 5 | 0.202 | 0.013 | 0.002 | 0.024 | 0.101 | 0.003 | 0.125 | 0.171 | 0.028 |
| 6 | 0.202 | 0.007 | 0.001 | 0.019 | 0.094 | 0.001 | 0.116 | 0.164 | 0.021 |
| 7 | 0.204 | 0.003 | 0.000 | 0.016 | 0.090 | 0.000 | 0.109 | 0.159 | 0.016 |
| 8 | 0.209 | 0.002 | 0.000 | 0.014 | 0.088 | 0.000 | 0.103 | 0.155 | 0.012 |

$n=50$. True order = 6.

| $L$ | AIC | AICc | AICu | SIC | HQ | HQc | Cp | FPE | FPEu |
|---|---|---|---|---|---|---|---|---|---|
| 1 | 0.192 | 0.120 | 0.065 | 0.068 | 0.128 | 0.072 | 0.166 | 0.188 | 0.100 |
| 2 | 0.186 | 0.086 | 0.032 | 0.037 | 0.101 | 0.038 | 0.151 | 0.180 | 0.068 |
| 3 | 0.173 | 0.059 | 0.016 | 0.021 | 0.079 | 0.019 | 0.133 | 0.165 | 0.046 |
| 4 | 0.161 | 0.039 | 0.007 | 0.012 | 0.063 | 0.009 | 0.117 | 0.151 | 0.031 |
| 5 | 0.151 | 0.026 | 0.003 | 0.007 | 0.051 | 0.004 | 0.104 | 0.139 | 0.021 |
| 6 | 0.143 | 0.017 | 0.002 | 0.005 | 0.042 | 0.002 | 0.093 | 0.128 | 0.015 |
| 7 | 0.137 | 0.010 | 0.001 | 0.003 | 0.036 | 0.001 | 0.084 | 0.120 | 0.011 |
| 8 | 0.133 | 0.006 | 0.000 | 0.002 | 0.031 | 0.000 | 0.077 | 0.112 | 0.008 |

$n=100$. True order = 6.

| $L$ | AIC | AICc | AICu | SIC | HQ | HQc | Cp | FPE | FPEu |
|---|---|---|---|---|---|---|---|---|---|
| 1 | 0.174 | 0.139 | 0.075 | 0.039 | 0.093 | 0.069 | 0.161 | 0.173 | 0.092 |
| 2 | 0.159 | 0.111 | 0.041 | 0.014 | 0.060 | 0.036 | 0.142 | 0.157 | 0.059 |
| 3 | 0.139 | 0.084 | 0.022 | 0.005 | 0.039 | 0.018 | 0.120 | 0.137 | 0.037 |
| 4 | 0.122 | 0.064 | 0.012 | 0.002 | 0.025 | 0.009 | 0.102 | 0.120 | 0.024 |
| 5 | 0.107 | 0.048 | 0.006 | 0.001 | 0.017 | 0.005 | 0.087 | 0.105 | 0.015 |
| 6 | 0.095 | 0.036 | 0.003 | 0.000 | 0.011 | 0.002 | 0.075 | 0.092 | 0.010 |
| 7 | 0.085 | 0.026 | 0.002 | 0.000 | 0.008 | 0.001 | 0.065 | 0.081 | 0.007 |
| 8 | 0.076 | 0.020 | 0.001 | 0.000 | 0.005 | 0.001 | 0.056 | 0.072 | 0.004 |

Table 2.2. Signal-to-noise ratios for overfitting.

$n = 15$. True order = 6.

| $L$ | AIC | AICc | AICu | SIC | HQ | HQc | Cp | FPE | FPEu |
|---|---|---|---|---|---|---|---|---|---|
| 1 | 0.003 | 2.850 | 3.479 | 0.255 | 0.000 | 2.583 | 0.071 | 0.271 | 0.836 |
| 2 | -0.067 | 4.600 | 5.480 | 0.264 | -0.070 | 4.180 | 0.089 | 0.341 | 1.011 |
| 3 | -0.173 | 6.627 | 7.687 | 0.198 | -0.177 | 6.039 | 0.100 | 0.369 | 1.077 |
| 4 | -0.310 | 9.455 | 10.651 | 0.074 | -0.314 | 8.642 | 0.107 | 0.374 | 1.086 |

$n = 25$. True order = 6.

| $L$ | AIC | AICc | AICu | SIC | HQ | HQc | Cp | FPE | FPEu |
|---|---|---|---|---|---|---|---|---|---|
| 1 | 0.284 | 1.479 | 2.149 | 0.888 | 0.451 | 1.742 | 0.311 | 0.405 | 1.015 |
| 2 | 0.360 | 2.200 | 3.145 | 1.189 | 0.590 | 2.585 | 0.414 | 0.536 | 1.300 |
| 3 | 0.389 | 2.842 | 3.997 | 1.372 | 0.662 | 3.332 | 0.481 | 0.615 | 1.457 |
| 4 | 0.388 | 3.474 | 4.804 | 1.485 | 0.692 | 4.062 | 0.530 | 0.666 | 1.551 |
| 5 | 0.364 | 4.128 | 5.609 | 1.546 | 0.692 | 4.814 | 0.567 | 0.698 | 1.608 |
| 6 | 0.320 | 4.828 | 6.442 | 1.564 | 0.665 | 5.616 | 0.597 | 0.716 | 1.639 |
| 7 | 0.258 | 5.599 | 7.332 | 1.545 | 0.615 | 6.496 | 0.621 | 0.723 | 1.652 |
| 8 | 0.180 | 6.470 | 8.309 | 1.492 | 0.543 | 7.486 | 0.642 | 0.721 | 1.648 |

$n = 35$. True order = 6.

| $L$ | AIC | AICc | AICu | SIC | HQ | HQc | Cp | FPE | FPEu |
|---|---|---|---|---|---|---|---|---|---|
| 1 | 0.404 | 1.177 | 1.860 | 1.269 | 0.703 | 1.613 | 0.510 | 0.476 | 1.108 |
| 2 | 0.543 | 1.720 | 2.685 | 1.743 | 0.957 | 2.349 | 0.701 | 0.641 | 1.452 |
| 3 | 0.629 | 2.178 | 3.359 | 2.070 | 1.126 | 2.965 | 0.837 | 0.749 | 1.660 |
| 4 | 0.683 | 2.604 | 3.966 | 2.314 | 1.246 | 3.533 | 0.944 | 0.826 | 1.800 |
| 5 | 0.716 | 3.018 | 4.538 | 2.500 | 1.332 | 4.081 | 1.031 | 0.882 | 1.898 |
| 6 | 0.730 | 3.431 | 5.094 | 2.642 | 1.390 | 4.624 | 1.104 | 0.923 | 1.967 |
| 7 | 0.730 | 3.853 | 5.645 | 2.747 | 1.426 | 5.174 | 1.168 | 0.953 | 2.016 |
| 8 | 0.716 | 4.288 | 6.199 | 2.821 | 1.442 | 5.740 | 1.223 | 0.974 | 2.049 |

$n = 50$. True order = 6.

| $L$ | AIC | AICc | AICu | SIC | HQ | HQc | Cp | FPE | FPEu |
|---|---|---|---|---|---|---|---|---|---|
| 1 | 0.495 | 1.004 | 1.695 | 1.645 | 0.933 | 1.579 | 0.595 | 0.536 | 1.186 |
| 2 | 0.680 | 1.450 | 2.427 | 2.286 | 1.292 | 2.274 | 0.829 | 0.732 | 1.585 |
| 3 | 0.808 | 1.815 | 3.011 | 2.751 | 1.548 | 2.837 | 1.000 | 0.867 | 1.843 |
| 4 | 0.903 | 2.143 | 3.524 | 3.119 | 1.747 | 3.339 | 1.139 | 0.968 | 2.027 |
| 5 | 0.977 | 2.452 | 3.994 | 3.422 | 1.908 | 3.807 | 1.256 | 1.047 | 2.165 |
| 6 | 1.034 | 2.749 | 4.437 | 3.676 | 2.040 | 4.255 | 1.358 | 1.110 | 2.271 |
| 7 | 1.077 | 3.040 | 4.863 | 3.892 | 2.149 | 4.691 | 1.448 | 1.161 | 2.355 |
| 8 | 1.108 | 3.330 | 5.277 | 4.076 | 2.238 | 5.121 | 1.528 | 1.202 | 2.421 |

$n = 100$. True order = 6.

| $L$ | AIC | AICc | AICu | SIC | HQ | HQc | Cp | FPE | FPEu |
|---|---|---|---|---|---|---|---|---|---|
| 1 | 0.601 | 0.840 | 1.540 | 2.306 | 1.291 | 1.632 | 0.661 | 0.615 | 1.291 |
| 2 | 0.840 | 1.200 | 2.189 | 3.237 | 1.810 | 2.326 | 0.929 | 0.854 | 1.770 |
| 3 | 1.017 | 1.485 | 2.697 | 3.936 | 2.198 | 2.872 | 1.132 | 1.027 | 2.104 |
| 4 | 1.159 | 1.733 | 3.131 | 4.511 | 2.516 | 3.343 | 1.299 | 1.165 | 2.362 |
| 5 | 1.280 | 1.958 | 3.522 | 5.007 | 2.788 | 3.769 | 1.444 | 1.280 | 2.570 |
| 6 | 1.384 | 2.167 | 3.880 | 5.443 | 3.027 | 4.163 | 1.573 | 1.378 | 2.743 |
| 7 | 1.476 | 2.366 | 4.216 | 5.835 | 3.240 | 4.534 | 1.689 | 1.463 | 2.889 |
| 8 | 1.557 | 2.557 | 4.534 | 6.190 | 3.432 | 4.889 | 1.796 | 1.537 | 3.014 |

**SIC**

$$P\{\text{SIC}_{k_*+L} < \text{SIC}_{k_*}\}$$
$$= P\left\{F_{L,n-k_*-L} > \frac{n-k_*-L}{L}\left(\exp\left(\frac{\log(n)L}{n}\right)-1\right)\right\}.$$

**HQ**

$$P\{\text{HQ}_{k_*+L} < \text{HQ}_{k_*}\}$$
$$= P\left\{F_{L,n-k_*-L} > \frac{n-k_*-L}{L}\left(\exp\left(\frac{2\log\log(n)L}{n}\right)-1\right)\right\}.$$

**HQc**

$$P\{\text{HQc}_{k_*+L} < \text{HQc}_{k_*}\}$$
$$= P\left\{F_{L,n-k_*-L} > \frac{n-k_*-L}{L}\left(\exp\left(\frac{2\log\log(n)(n-2)L}{(n-k_*-L-2)(n-k_*-2)}\right)-1\right)\right\}.$$

The probabilities calculated from the above expressions are presented in Table 2.1 for sample sizes $n$ = 15, 25, 35, 50, 100. In each case the true order is 6 and $L$ represents the number of variables by which the model is overfitted. These probabilities have close relationships to the signal-to-noise ratios in Table 2.2, which we will discuss in Section 2.5.3.

Three general patterns can be discerned from the results in Table 2.1. First, the probabilities of overfitting for model selection criteria with weak penalty functions (AIC and FPE) decrease as the sample size increases. The probabilities of overfitting for criteria with strong penalty functions (AICc and AICu) increase as the sample size increases. Finally, the probabilities of overfitting for the consistent criteria (SIC, HQ, and HQc) decrease as the sample size increases, although the probabilities for SIC decrease much faster than the probabilities for HQ.

### *2.5.2. Asymptotic Probabilities of Overfitting*

Having established some general patterns for overfitting behavior in small samples, we will now look at overfitting in large samples by obtaining asymptotic probabilities of overfitting. Again, assume that only the two models $k_* + L$ and $k_*$ are compared and that they form nested models. Detailed calculations will be given for one criterion from each class; calculations for those not shown can be found in Appendix 2C Section 2. We will make use of the following facts: first, as $n \to \infty$ and where $k_*$ and $L$ are fixed, it is known that $\chi^2_{n-k_*-L}/(n-k_*-L) \to 1$ a.s. By Slutsky's theorem, $F_{L,n-k_*-L} \to \chi^2_L/L$, and the small-sample F distribution is replaced by a $\chi^2$ distribution. The asymptotic probabilities of overfitting presented here are hence derived from the $\chi^2$ distribution. We will also use the expansion for $\exp(\cdot)$, $\exp(z) = 1 + z + \sum_{i=2}^{\infty} z^i/i!$. For small $z$, $\exp(z) = 1 + z + o(z^2)$.

**AIC**

Our example from the K-L-based criteria will be AIC. For AIC,

$$\begin{aligned}\frac{n-k_*-L}{L}\left(\exp\left(\frac{2L}{n}\right)-1\right) &= \frac{n-k_*-L}{L}\left(\frac{2L}{n}+O\left(\frac{1}{n^2}\right)\right)\\ &= \frac{n}{L}\frac{2L}{n}+O\left(\frac{1}{n}\right)\\ &\to 2.\end{aligned}$$

Given the above and that

$$F_{L,n-k_*-L} \to \frac{\chi^2_L}{L},$$

we have $P\{\text{AIC overfits by } L\} = P\{\chi^2_L > 2L\}$. The $O(1/n^2)$ will be ignored in all other derivations.

**AICc**

$P\{\text{AICc overfits by } L\} = P\{\chi^2_L > 2L\}$.

**AICu**

$P\{\text{AICu overfits by } L\} = P\{\chi^2_L > 3L\}$.

**FPE**

Our example for the $L_2$-based criteria will be FPE. For FPE,

$$F_{L,n-k_*-L} \to \frac{\chi^2_L}{L}$$

and

$$\frac{2n}{n+k_*} \to 2.$$

Therefore we have $P\{\text{FPE overfits by } L\} = P\{\chi^2_L > 2L\}$.

**FPEu**

$P\{\text{FPEu overfits by } L\} = P\{\chi^2_L > 3L\}$.

**Cp**

$P\{\text{Cp overfits by } L\} = P\{\chi^2_L > 2L\}.$

**SIC**

Our example for the consistent criteria will be SIC. For SIC, expanding

$$\begin{aligned}\frac{n-k_*-L}{L}&\left(\exp\left(\frac{\log(n)L}{n}\right)-1\right)\\ &=\frac{n-k_*-L}{L}\left(\frac{\log(n)L}{n}+O\left(\frac{1}{n^2}\right)\right)\\ &=\frac{n}{L}\frac{\log(n)L}{n}\\ &\to\infty.\end{aligned}$$

Therefore, we have $P\{\text{SIC overfits by } L\} = 0$.

**HQ** and **HQc**

As expected, for both HQ and HQc we also have $P\{\text{overfits by } L\} = 0$.

Table 2.3. Asymptotic probability of overfitting by $L$ variables. Probabilities refer to selecting one particular overfit model over the true model.

| $L$ | AIC | AICc | AICu | SIC | HQ | HQc | Cp | FPE | FPEu |
|---|---|---|---|---|---|---|---|---|---|
| 1 | 0.1573 | 0.1573 | 0.0833 | 0 | 0 | 0 | 0.1573 | 0.1573 | 0.0833 |
| 2 | 0.1353 | 0.1353 | 0.0498 | 0 | 0 | 0 | 0.1353 | 0.1353 | 0.0498 |
| 3 | 0.1116 | 0.1116 | 0.0293 | 0 | 0 | 0 | 0.1116 | 0.1116 | 0.0293 |
| 4 | 0.0916 | 0.0916 | 0.0174 | 0 | 0 | 0 | 0.0916 | 0.0916 | 0.0174 |
| 5 | 0.0752 | 0.0752 | 0.0104 | 0 | 0 | 0 | 0.0752 | 0.0752 | 0.0104 |
| 6 | 0.0620 | 0.0620 | 0.0062 | 0 | 0 | 0 | 0.0620 | 0.0620 | 0.0062 |
| 7 | 0.0512 | 0.0512 | 0.0038 | 0 | 0 | 0 | 0.0512 | 0.0512 | 0.0038 |
| 8 | 0.0424 | 0.0424 | 0.0023 | 0 | 0 | 0 | 0.0424 | 0.0424 | 0.0023 |
| 9 | 0.0352 | 0.0352 | 0.0014 | 0 | 0 | 0 | 0.0352 | 0.0352 | 0.0014 |
| 10 | 0.0293 | 0.0293 | 0.0039 | 0 | 0 | 0 | 0.0293 | 0.0293 | 0.0039 |

We have calculated the above asymptotic probabilities associated with each criterion, and the results are presented in Table 2.3. From Table 2.3 it is clear that AICc, AIC, FPE, and Cp are not consistent criteria, since their probabilities of overfitting are not 0, but that SIC, HQ, and HQc are consistent and thus do have 0 probability of overfitting. It can also be seen that AICc, AIC, FPE, and Cp are asymptotically equivalent, as are AICu and FPEu, and HQ and HQc. Both signal-to-noise corrected variants (AICu and FPEu) have much smaller asymptotic probabilities of overfitting than do AICc, AIC, FPE, and Cp. Thus, they would be expected to have superior performance if the true model belonged to the set of candidate models. They are more consistent than the efficient criteria.

The model selection criteria discussed in this Chapter can be thought of as being asymptotically equivalent to the criterion $\text{AIC}\alpha = \log(\hat{\sigma}^2) + \alpha k/n$ (see Section 2.2). Let $\alpha A$ be AIC$\alpha$ when $\alpha = A$. Then consistent criteria are asymptotically $\alpha\infty$. AICu is $\alpha3$ and the efficient criteria are $\alpha2$. Although we did not compute probabilities of overfitting for $\text{R}^2_{adj}$, it can be shown that $\text{R}^2_{adj}$ is asymptotically equivalent to $\alpha1$. The smaller the $\alpha$, the higher the probability of overfitting, thus we expect $\text{R}^2_{adj}$ to overfit more than the other criteria.

### *2.5.3. Small-sample Signal-to-noise Ratios*

Small-sample signal-to-noise ratios were derived earlier in Sections 2.3 and 2.4. They are summarized in Table 2.2 under the same conditions used for Table 2.1. The relationship between the probabilities of overfitting and the signal-to-noise ratios is clear from Tables 2.1 and 2.2: small or negative signal-to-noise ratios result in a high probability of overfitting, and large signal-to-noise ratios lead to small probabilities of overfitting.

### *2.5.4. Asymptotic Signal-to-noise Ratios*

In order to obtain asymptotic signal-to-noise ratios, we will make use of the fact that when $k_*$, $L$, and $K$ are fixed and $n \to \infty$, then $\log(1 - L/n) \doteq -L/n$ when $L << n$. We will show a sample calculation for each category of criterion, and the details for those not presented can be found in Appendix 2C.3.

**AIC**

Our example from the K-L-based criteria will be AIC. For AIC, the asymptotic signal-to-noise ratio is

$$\begin{aligned}
&\lim_{n\to\infty} \frac{\sqrt{(n-k_*-L)(n-k_*+2)}}{\sqrt{2L}} \Bigg(\log\Big(1 - \frac{L}{n-k_*}\Big) \\
&\quad - \frac{L}{(n-k_*-L)(n-k_*)} + \frac{2L}{n}\Bigg) \\
&= \lim_{n\to\infty} \frac{n}{\sqrt{2L}} \left(\log\Big(1 - \frac{L}{n}\Big) - \frac{L}{n^2} + \frac{2L}{n}\right) \\
&= \frac{-L+2L}{\sqrt{2L}} \\
&= \sqrt{\frac{L}{2}}.
\end{aligned}$$

**AICc**

The asymptotic signal-to-noise ratio for overfitting for AICc is also $\sqrt{L/2}$.

**AICu**

The asymptotic signal-to-noise ratio is $\sqrt{2L}$.

**FPE**

Our example from the $L_2$-based criteria will be FPE. For FPE, the asymptotic signal-to-noise ratio is

$$\lim_{n\to\infty} \frac{(n-k_*)\sqrt{L(n-k_*-L)}}{\sqrt{2(n+k_*)^2(n-k_*-L)+8nL}}$$
$$=\sqrt{L/2}.$$

**Cp**

The asymptotic signal-to-noise ratio for overfitting for Cp is also $\sqrt{L/2}$.

**FPEu**

As was the case for AICu, the asymptotic signal-to-noise ratio is $\sqrt{2L}$.

**SIC**

Our example from the consistent criteria will be SIC. For SIC, the asymptotic signal-to-noise ratio is

$$\lim_{n\to\infty} \frac{\sqrt{(n-k_*-L)(n-k_*+2)}}{\sqrt{2L}} \left(\log\left(1-\frac{L}{n-k_*}\right)\right.$$
$$\left. - \frac{L}{(n-k_*-L)(n-k_*)} + \frac{\log(n)L}{n}\right)$$
$$= \lim_{n\to\infty} \frac{n}{\sqrt{2L}} \left(\log\left(1-\frac{L}{n}\right) - \frac{L}{n^2} + \frac{\log(n)L}{n}\right)$$
$$=\infty.$$

**HQ** and **HQc**

Unsurprisingly, the asymptotic signal-to-noise ratios for HQ and HQc are also $\infty$.

Table 2.4. Asymptotic signal-to-noise ratio for overfitting by $L$ variables.

| $L$ | AIC | AICc | AICu | SIC | HQ | HQc | Cp | FPE | FPEu |
|---|---|---|---|---|---|---|---|---|---|
| 1 | 0.707 | 0.707 | 1.414 | $\infty$ | $\infty$ | $\infty$ | 0.707 | 0.707 | 1.414 |
| 2 | 1.000 | 1.000 | 2.000 | $\infty$ | $\infty$ | $\infty$ | 1.000 | 1.000 | 2.000 |
| 3 | 1.225 | 1.225 | 2.449 | $\infty$ | $\infty$ | $\infty$ | 1.225 | 1.225 | 2.449 |
| 4 | 1.414 | 1.414 | 2.828 | $\infty$ | $\infty$ | $\infty$ | 1.414 | 1.414 | 2.828 |
| 5 | 1.581 | 1.581 | 3.162 | $\infty$ | $\infty$ | $\infty$ | 1.581 | 1.581 | 3.162 |
| 6 | 1.732 | 1.732 | 3.464 | $\infty$ | $\infty$ | $\infty$ | 1.732 | 1.732 | 3.464 |
| 7 | 1.871 | 1.871 | 3.742 | $\infty$ | $\infty$ | $\infty$ | 1.871 | 1.871 | 3.742 |
| 8 | 2.000 | 2.000 | 4.000 | $\infty$ | $\infty$ | $\infty$ | 2.000 | 2.000 | 4.000 |
| 9 | 2.121 | 2.121 | 4.243 | $\infty$ | $\infty$ | $\infty$ | 2.121 | 2.121 | 4.243 |
| 10 | 2.236 | 2.236 | 4.472 | $\infty$ | $\infty$ | $\infty$ | 2.236 | 2.236 | 4.472 |

Table 2.4 lists asymptotic signal-to-noise ratios for selecting model $k_* + L$ over the model $k_*$. Results from Table 2.4 show what we have previously calculated—that the efficient criteria AICc, AIC, FPE, and Cp have equivalent signal-to-noise ratios, that the consistent criteria SIC, HQ, and HQc all have infinite signal-to-noise ratios, and that AICu and FPEu are asymptotically equivalent and have much larger signal-to-noise ratios (which increase with $L$) than AICc, AIC, FPE, or Cp.

## 2.6. Small-sample Underfitting

In previous sections we have been concerned only with overfitting. In this Section our study of small-sample properties will focus on underfitting as described by Eq. (2.5), but our findings in this Section have some implications for small-sample overfitting as well, which we will note. A useful reference for underfitted model selection is given by Hurvich and Tsai (1991).

The distributions of FPE, FPEu, and Cp are based on $\chi^2$, whereas the distributions of AICc, AIC, SIC, HQ, AICu, and HQc are based on log-$\chi^2$. Therefore we will need several expectations and variances for the noncentral $\chi^2$ and log-$\chi^2$ distributions, as well as the log-Beta distribution. The necessary results are listed here, and detailed derivations can be found in Appendix 2D.

Under model Eq. (2.5), $\text{SSE}_{k_0}$ has a noncentral $\chi^2$ distribution and so $\log(\text{SSE}_{k_0})$ has a noncentral log-$\chi^2$ distribution. In order to allow $s_K^2$ in the Cp criterion to have a central distribution, we assume that the largest candidate model includes all the variables in the true model.

### 2.6.1. Distributional Review

We now briefly review some useful facts regarding noncentral $\chi^2$, log-$\chi^2$, noncentral Beta, and log-Beta distributions. Central Beta and log-Beta facts are also presented. We focus on means and variances and some approximations to these moments. Details can be found in Appendices 2A.1 and 2A.2.

**Noncentral $\chi^2$**

We will first review the noncentral $\chi^2$ distribution and the noncentrality parameter for underfitted regression models. Let $Z \sim \chi^2(m, \lambda)$ with $m$ degrees of freedom and noncentrality parameter $\lambda$. For regression models $\lambda = E[Y'](I-X(X'X)^{-1}X')E[Y]/\sigma_*^2$ (Rao, 1973, p. 190). Let $H_j$ be the projection matrix (or hat matrix) corresponding to $X_j$ (j=0,1 and 2, see Section 2.1.1),

$$H_j = X_j(X_j'X_j)^{-1}X_j'.$$

Under model Eq. (2.5), the noncentrality parameter for the underfitted model $Y = X_0\beta + \varepsilon$ is

$$\begin{aligned}\lambda_0 &= E[Y'](I - H_0)E[Y]/\sigma_*^2 \\ &= (X_*\beta_*)'(I - H_0)(X_*\beta_*)/\sigma_*^2 \\ &= (X_0\beta_0 + X_1\beta_1)'(I - H_0)(X_0\beta_0 + X_1\beta_1)/\sigma_*^2 \\ &= (X_1\beta_1)'(I - H_0)(X_1\beta_1)/\sigma_*^2.\end{aligned}$$

The density function for the noncentral $\chi^2$ distribution is

$$f_Z(z) = \sum_{r=0}^{\infty} e^{-\lambda/2}\frac{(\lambda/2)^r}{r!}\chi^2(m + 2r|z),$$

where $\chi^2(m+2r|z)$ represents the density of a central $\chi^2$ with $m+2r$ degrees of freedom. The following moments of the noncentral $Z \sim \chi^2(m, \lambda)$ will be needed to compute the signal-to-noise ratios for underfitting. Note that for finite $\lambda$ the moments of the noncentral $\chi^2$ are also finite. The first two moments and variance are

$$\begin{aligned}E[Z] &= m + \lambda, \\ E[Z^2] &= 2m + m^2 + 2(m + 1)\lambda + 2\lambda + \lambda^2, \\ var[Z] &= 2(m + 2\lambda).\end{aligned}$$

For $\text{SSE}_{k_0} \sim \sigma_*^2\chi^2(n - k_0, \lambda_0)$,

$$E[\text{SSE}_{k_0}] = (n - k_0 + \lambda_0)\sigma_*^2 \tag{2.23}$$

and

$$var[\text{SSE}_{k_0}] = 2(n - k_0 + 2\lambda_0)\sigma_*^4.$$

For $Z \sim \chi^2_m$,

$$E\left[\frac{1}{Z}\right] = \frac{1}{m-2}.$$

Hence, the expected value of $1/\text{SSE}_{k_0}$ is

$$E\left[\frac{1}{\text{SSE}_{k_0}}\right] = \frac{1}{\sigma_*^2}\sum_{r=0}^{\infty} e^{-\lambda_0/2}\frac{(\lambda_0/2)^r}{r!}\frac{1}{n-k_0-2+2r}.$$

**Noncentral Log-$\chi^2$**

The noncentral log-$\chi^2$ distribution with noncentrality parameter $\lambda$ for $W = \log(Z)$ has density

$$f_{\log(Z)}(w) = \sum_{r=0}^{\infty} e^{-\lambda/2}\frac{(\lambda/2)^r}{r!}\text{log-}\chi^2(m+2r|w),$$

where log-$\chi^2(m+2r|w)$ denotes the density of a central log-$\chi^2$. The expected value of $\log(\text{SSE}_{k_0})$ follows from Eq. (2A.6):

$$E[\log(\text{SSE}_{k_0})] = \log(\sigma_*^2) + \log(2) + \sum_{r=0}^{\infty} e^{-\lambda_0/2}\frac{(\lambda_0/2)^r}{r!}\psi\left(\frac{n-k_0}{2}+r\right). \quad (2.24)$$

**Log-Beta and Noncentral Log-Beta**

Small-sample signal-to-noise ratios for log(SSE)-based model selection criteria involve the log-Beta distribution and noncentral log-Beta distribution, as we see when we consider comparing the reduced model Eq. (2.5) of order $k_0$ with the true model of order $k_* > k_0$. We have $\text{SSE}_{k_*} \sim \chi^2_{n-k_*}$ and $\text{SSE}_{k_0} \sim \chi^2(n-k_0, \lambda_0)$ with $\lambda_0 = (X_*\beta_*)'(I-H_0)(X_*\beta_*)/\sigma_*^2$. Furthermore, the distribution of

$$\frac{\text{SSE}_{k_*}}{\text{SSE}_{k_0}} = \frac{\chi^2_{n-k_*}}{\chi^2_{n-k_*} + \chi^2(k_*-k_0, \lambda_0)}.$$

If we let $T = \frac{\text{SSE}_{k_*}}{\text{SSE}_{k_0}}$, we have densities

$$f_T(t) = \sum_{r=0}^{\infty} e^{-\lambda_0/2}\frac{(\lambda_0/2)^r}{r!}\text{Beta}\left(\frac{n-k_*}{2}, \frac{k_*-k_0+2r}{2}\,\middle|\, t\right),$$

where $\text{Beta}(\alpha, \beta|t)$ denotes the density of a central Beta. In addition,

$$f_{\log(T)}(u) = \sum_{r=0}^{\infty} e^{-\lambda_0/2}\frac{(\lambda_0/2)^r}{r!}\text{log-Beta}\left(\frac{n-k_*}{2}, \frac{k_*-k_0+2r}{2}\,\middle|\, u\right),$$

where log-Beta$(\alpha, \beta|u)$ denotes the density of a central log-Beta for $U = \log(T)$. The variance

$$var\left[\log\left(\frac{\text{SSE}_{k_*}}{\text{SSE}_{k_0}}\right)\right] = \sum_{r=0}^{\infty} e^{-\lambda_0/2}\frac{(\lambda_0/2)^r}{r!}\left(\psi'\left(\frac{n-k_*}{2}\right) - \psi'\left(\frac{n-k_0}{2}+r\right)\right) \tag{2.25}$$

follows from Eq. (2A.10).

### 2.6.2. Expectations of $L_2$ and Kullback–Leibler Distance

We can now use the preceding distributional information to obtain the expected values for the $L_2$ distance and K-L distance measures we introduced in Section 2.1. From Eq. (2.9), the $L_2$ distance between the candidate model given by Eq. (2.5) and the true model is

$$L_2 = \frac{1}{n}(X_0\beta_0 + X_1\beta_1 - X_0\hat{\beta}_0)'(X_0\beta_0 + X_1\beta_1 - X_0\hat{\beta}_0),$$

where $X_0\beta_0$ is known from the true model and $\hat{\beta}_0$ is estimated from the candidate model using $X_0$ as the design matrix. Furthermore, let $X_0\hat{\beta}_0 = H_0Y$, and rewrite the above equation as

$$\begin{aligned}
nL_2 =& (X_0\beta_0 + X_1\beta_1 - H_0Y)'(X_0\beta_0 + X_1\beta_1 - H_0Y) \\
=& \Big((Y - H_0Y) - (Y - X_0\beta_0 - X_1\beta_1 - H_0Y)\Big)' \\
& \Big((Y - H_0Y) - (Y - X_0\beta_0 - X_1\beta_1 - H_0Y)\Big) \\
=& \Big((I - H_0)Y - \varepsilon)\Big)'\Big((I - H_0)Y - \varepsilon)\Big) \\
=& Y'(I - H_0)Y - Y'(I - H_0)\varepsilon - \varepsilon'(I - H_0)Y + \varepsilon'\varepsilon.
\end{aligned}$$

Substituting for $Y = X_0\beta_0 + X_1\beta_1 + \varepsilon$, we have

$$\begin{aligned}
nL_2 =& (X_0\beta_0 + X_1\beta_1 + \varepsilon)'(I - H_0)(X_0\beta_0 + X_1\beta_1 + \varepsilon) \\
& - (X_0\beta_0 + X_1\beta_1 + \varepsilon)'(I - H_0)\varepsilon \\
& - \varepsilon'(I - H_0)(X_0\beta_0 + X_1\beta_1 + \varepsilon) + \varepsilon'\varepsilon \\
=& (X_1\beta_1)'(I - H_0)(X_1\beta_1) + \varepsilon'H_0\varepsilon \\
=& \lambda_0\sigma_*^2 + \varepsilon'H_0\varepsilon.
\end{aligned}$$

Hence,

$$E[L_2] = \frac{\sigma_*^2}{n}(k_0 + \lambda_0).$$

For the general model of order $k$,

$$E[L_2] = \frac{\sigma_*^2}{n}(k + \lambda_k), \tag{2.26}$$

where $\lambda_k = 0$ in the overfitting case. Using Eq. (2.26) we can compute $E[L_2]$ for all candidate models under consideration to find the model $\tilde{k}$ for which $E[L_2]$ attains a minimum. A good selection criterion should have an expected minimum value at or near model order $\tilde{k}$.

Next we will find the expected value for the K-L discrepancy in Eq. (2.10). For underfitted models using $X_0$ as the design matrix, Eq. (2.10) becomes

$$\text{K-L} = \log\left(\frac{\hat{\sigma}_0^2}{\sigma_*^2}\right) + \frac{\sigma_*^2}{\hat{\sigma}_0^2} + \frac{L_2}{\hat{\sigma}_0^2} - 1,$$

where $\hat{\sigma}_0^2$ is the MLE estimated from the candidate model. Looking at the terms individually and using our previously determined facts about the noncentral $\chi^2$ distribution, we have

$$\begin{aligned} E\left[\log\left(\frac{\hat{\sigma}_0^2}{\sigma_*^2}\right)\right] &= E\left[\log\left(\frac{\text{SSE}_{k_0}}{\sigma_*^2}\right) - \log(n)\right] \\ &= \log(2) - \log(n) + \sum_{r=0}^{\infty} e^{-\lambda_0/2}\frac{(\lambda_0/2)^r}{r!}\psi\left(\frac{n-k_0}{2} + r\right), \end{aligned}$$

$$\begin{aligned} E\left[\frac{\sigma_*^2}{\hat{\sigma}_0^2}\right] &= E\left[\frac{n}{\text{SSE}_{k_0}/\sigma_*^2}\right] \\ &= n\sum_{r=0}^{\infty} e^{-\lambda_0/2}\frac{(\lambda_0/2)^r}{r!}\frac{1}{n-k_0-2+2r}. \end{aligned}$$

From the fact that $L_2$ and SSE are independent we also know that

$$\begin{aligned} E\left[\frac{L_2}{\hat{\sigma}_0^2}\right] &= E[L_2]E\left[\frac{n}{\text{SSE}_{k_0}}\right] \\ &= \frac{\sigma_*^2}{n}(k_0 + \lambda_0)\frac{n}{\sigma_*^2}\sum_{r=0}^{\infty} e^{-\lambda_0/2}\frac{(\lambda_0/2)^r}{r!}\frac{1}{n-k_0-2+2r}, \end{aligned}$$

and substituting yields the

$$\begin{aligned} E[\text{K-L}] = &\log(2) - \log(n) - 1 \\ &+ \sum_{r=0}^{\infty} e^{-\lambda_0/2}\frac{(\lambda_0/2)^r}{r!}\left(\psi\left(\frac{n-k_0}{2} + r\right) + \frac{n+k_0+\lambda_0}{n-k_0-2+2r}\right). \end{aligned} \tag{2.27}$$

In contrast to underfitted models, note that for the overfitted candidate model $Y = X\beta + \varepsilon$ with $\text{rank}(X) = k > k_*$, the distributions will be central. Using Eq. (2.27) we can compute $E$[K-L] for all candidate models under consideration and find the order $\hat{k}$ where $E$[K-L] attains a minimum. A good model selection criterion should have an expected minimum value at or near $\hat{k}$. Model selection criteria with stronger penalty functions tend to make model choices similar to those of K-L, while good model selection criteria with weaker penalty functions tend to make choices similar to those of $L_2$. The expected K-L distance between an underfitted model and the true model tends to be less than the corresponding expected $L_2$ distance when compared to their respective distances between the closest model and the true model. This is particularly true in models with rapidly decaying $\beta$ parameters, as we will see in Section 2.7.

### *2.6.3. Expected Values for Two Special Case Models*

The moments of most model selection criteria are dependent on the model structure. In particular, as we have seen, underfitting involves the noncentrality parameter $\lambda$. Here we will illustrate the behavior of model selection criteria and distance measures using two example regression models. For both models,

$$n = 25, \sigma_*^2 = 1, k_* = 6, X'X = nI, \text{ and the intercept } \beta_0 = 1. \tag{2.28}$$

Model 1:

$$\beta_1 = \beta_2 = \beta_3 = \beta_4 = \beta_5 = 1. \tag{2.29}$$

Model 2 is identical to Model 1 with the exception that the $\beta$ parameters decay rapidly. Model 2: $\beta_j = 1/j$

$$\beta_1 = 1,\ \beta_2 = 1/2,\ \beta_3 = 1/3,\ \beta_4 = 1/4,\ \beta_5 = 1/5. \tag{2.30}$$

The constant is always included in the candidate models, which are nested in increasing order (where order $k = 1$ represents the model with the constant only).

Expected values of the six foundation and three variant criteria are presented in Tables 2.5 and 2.6. For FPE and Cp, orders $k = 1$ through 5 involve noncentral $\chi^2$ with expected values following from Eq. (2A.4) and Eq. (2.23). Orders $k \geq 6$ involve central $\chi^2$ with expected values following from Eq. (2A.4). Expectations of AICc, AIC, SIC, and HQ involve log-$\chi^2$ and noncentral log-$\chi^2$ expectations. For these criteria the expected values of orders $k = 1$ through 5 follow from Eq. (2.24), where the noncentral moments presented were computed by evaluating the sums out to 1000 terms. Expected values of orders

$k \geq 6$ follow from Eq. (2A.6). The $\psi$ appearing in all expected values of log-$\chi^2$ were computed using the recursions in Eqs. (2A.7)–(2A.8).

While Tables 2.5 and 2.6 show expected values for the selection criteria, the values for $L_2$ and K-L are expected efficiencies, which are presented to illustrate the differing behavior of the two distance functions. $L_2$ tends to

Table 2.5. Expected values and expected efficiency for Model 1, $\beta_j = 1$.

| $k$ | AIC | AICc | AICu | SIC | HQ | HQc | Cp | FPE | FPEu | $L_2$ | K-L |
|---|---|---|---|---|---|---|---|---|---|---|---|
| 1 | 1.933 | 2.954 | 2.995 | 1.901 | 1.866 | 1.879 | 166.0 | 6.457 | 6.726 | 0.048 | 0.273 |
| 2 | 1.818 | 2.864 | 2.947 | 1.836 | 1.765 | 1.801 | 134.6 | 5.776 | 6.278 | 0.059 | 0.302 |
| 3 | 1.657 | 2.737 | 2.865 | 1.724 | 1.618 | 1.688 | 103.1 | 4.938 | 5.612 | 0.077 | 0.344 |
| 4 | 1.419 | 2.546 | 2.720 | 1.534 | 1.393 | 1.512 | 71.7 | 3.922 | 4.669 | 0.111 | 0.417 |
| 5 | 1.032 | 2.219 | 2.442 | 1.196 | 1.020 | 1.202 | 40.3 | 2.700 | 3.375 | 0.200 | 0.571 |
| 6 | 0.232 | **1.496** | **1.770** | **0.445** | **0.233** | **0.497** | **8.9** | **1.240** | **1.632** | 1.000 | 1.000 |
| 7 | 0.255 | 1.615 | 1.943 | 0.516 | 0.270 | 0.638 | 9.6 | 1.280 | 1.778 | 0.857 | 0.806 |
| 8 | 0.274 | 1.754 | 2.140 | 0.584 | 0.303 | 0.801 | 10.3 | 1.320 | 1.941 | 0.750 | 0.657 |
| 9 | 0.290 | 1.918 | 2.365 | 0.649 | 0.332 | 0.993 | 11.0 | 1.360 | 2.125 | 0.667 | 0.540 |
| 10 | 0.301 | 2.113 | 2.624 | 0.709 | 0.356 | 1.220 | 11.7 | 1.400 | 2.333 | 0.600 | 0.445 |
| 11 | 0.307 | 2.347 | 2.927 | 0.763 | 0.376 | 1.490 | 12.4 | 1.440 | 2.571 | 0.545 | 0.368 |
| 12 | 0.307 | 2.631 | 3.285 | 0.812 | 0.389 | 1.818 | 13.1 | 1.480 | 2.846 | 0.500 | 0.304 |
| 13 | 0.300 | 2.980 | 3.714 | 0.854 | 0.396 | 2.220 | 13.9 | 1.520 | 3.167 | 0.462 | 0.250 |
| 14 | 0.285 | 3.419 | 4.240 | 0.888 | 0.395 | 2.722 | 14.6 | 1.560 | 3.545 | 0.429 | 0.205 |
| 15 | 0.260 | 3.980 | 4.897 | 0.912 | 0.383 | 3.364 | 15.3 | 1.600 | 4.000 | 0.400 | 0.166 |
| 16 | **0.223** | 4.720 | 5.742 | 0.923 | 0.360 | 4.207 | 16.0 | 1.640 | 4.556 | 0.375 | 0.133 |

Boldface type indicates the minimum expectation.

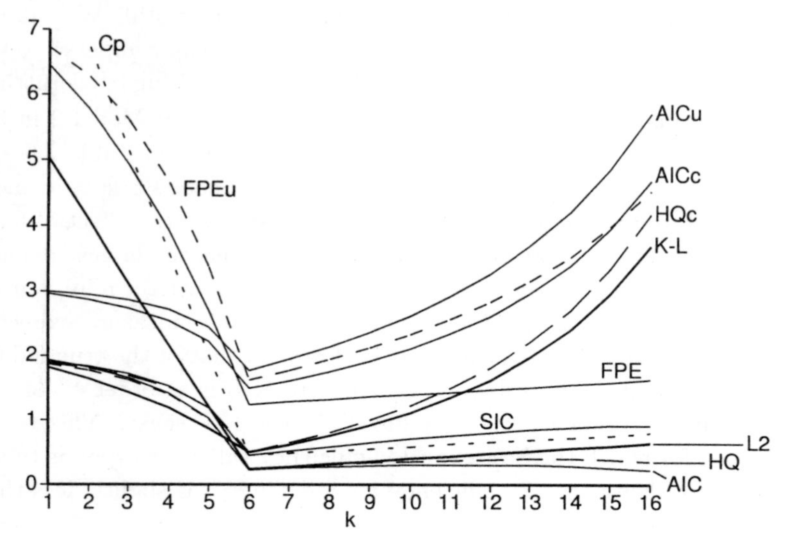


Figure 2.1. Expected values and expected distance for Model 1.

penalize, loss of efficiency, underfitting more than overfitting, while K-L penalizes, loss of efficiency, overfitting more than underfitting. We calculate the distance measure expected efficiencies as follows: first, $E[L_2]$ is computed for each candidate model order, where expectations for orders $k \leq 5$ follow from Eq. (2.26) and expectations of orders $k \geq 6$ follow with $\lambda = 0$. Then we define the $L_2$ expected efficiency for selecting order $k$ by using Eq. (2.26) in Eq. (1.3). By analogy, we arrive at the K-L expected efficiency using Eq. (2.27) and Eq. (1.4). Thus the candidate order closest to the true model has an expected efficiency of 1, and efficiency for selecting the wrong candidate order is less than 1. Results from Table 2.5 are also presented in graphical form in Figure 2.1. However, Figure 2.1 plots actual expected values for $L_2$ and K-L distance rather than the expected efficiencies given in Table 2.5. Note that for presentation purposes, Cp in Figure 2.1 has been scaled by a factor of 20.

All of the selection criteria, except AIC, have a well-defined minimum for Model 1 at the true order 6. AIC has a local minimum at the true order 6. The expected criterion values for overfitted models increase rapidly for AICc, AICu, and HQc due to their strong penalty functions, which are superlinear in $k$. On the other hand, Cp and FPE have weaker penalty functions which increase linearly with $k$, and the expected criterion values for overfitted models increase much less quickly. The penalty functions for AIC and HQ are also linear in $k$, and due to this and the $\log(\hat{\sigma}^2)$ term their expected values first increase, then decrease with increasing $k$. AIC has a global minimum at $k = 16$. A similar pattern would appear for SIC if $k$ were allowed to be larger still. We will see in the next Section that these penalty functions are too weak compared to the variation in the differences (the noise) in small samples, resulting in overfitting.

We have evaluated the criteria and distance measures for Model 2 in the same manner as for Model 1, and the results are presented in Table 2.6 and Figure 2.2. In this case the rapidly decaying $\beta_j$ parameters result in more variation in performance across criteria. We can clearly see the effect of large penalty functions on small-sample underfitting. AICu has the largest penalty function for small samples, and underfits most severely with a minimum at order $k = 3$. Unlike Model 1, for Model 2 the distance measure expected efficiencies do not agree on the order that is closest to that of the true model. The best expected efficiency value with respect to K-L is at order 4. In this case the true model order of 6 is "overfitted" in the K-L sense. This is not the case for $L_2$, for which the best expected efficiency value corresponds to the true model order. Note that $L_2$ efficiency for order $k = 5$ is slightly less than 1, but appears to be 1 due to rounding.

The criteria that more closely parallel K-L—AIC, SIC, HQc, Cp, and

FPEu—not surprisingly all have expected minima at order 4. All of these model selection criteria underfit in the $L_2$ sense. However, HQ and FPE, which more closely parallel the $L_2$, have minima at the true order 6. As was the case for Model 1, AIC has an artificial global minimum at order 16. The AIC and HQ minima at the correct order are only local; their penalty functions are too

Table 2.6. Expected values and expected efficiency for Model 2, $\beta_j = 1/j$.

| $k$ | AIC | AICc | AICu | SIC | HQ | HQc | Cp | FPE | FPEu | $L_2$ | K-L |
|---|---|---|---|---|---|---|---|---|---|---|---|
| 1 | 1.018 | 2.040 | 2.081 | 0.987 | 0.952 | 0.965 | 52.33 | 2.626 | 2.735 | 0.160 | 0.391 |
| 2 | 0.525 | 1.571 | 1.654 | 0.543 | 0.472 | 0.508 | 20.90 | 1.624 | 1.765 | 0.441 | 0.751 |
| 3 | 0.365 | 1.445 | **1.573** | 0.431 | 0.326 | 0.396 | 13.58 | 1.392 | 1.582 | 0.719 | 0.974 |
| 4 | 0.293 | **1.419** | 1.594 | **0.408** | 0.267 | **0.385** | 10.72 | 1.302 | **1.549** | 0.914 | 1.000 |
| 5 | 0.255 | 1.442 | 1.665 | 0.419 | 0.243 | 0.424 | 9.43 | 1.260 | 1.575 | 1.000 | 0.897 |
| 6 | 0.232 | 1.496 | 1.770 | 0.445 | **0.233** | 0.497 | **8.86** | **1.240** | 1.632 | 1.000 | 0.750 |
| 7 | 0.255 | 1.615 | 1.943 | 0.516 | 0.270 | 0.638 | 9.57 | 1.280 | 1.778 | 0.857 | 0.605 |
| 8 | 0.274 | 1.754 | 2.140 | 0.584 | 0.303 | 0.801 | 10.29 | 1.320 | 1.941 | 0.750 | 0.493 |
| 9 | 0.290 | 1.918 | 2.365 | 0.649 | 0.332 | 0.993 | 11.00 | 1.360 | 2.125 | 0.667 | 0.405 |
| 10 | 0.301 | 2.113 | 2.624 | 0.709 | 0.356 | 1.220 | 11.71 | 1.400 | 2.333 | 0.600 | 0.334 |
| 11 | 0.307 | 2.347 | 2.927 | 0.763 | 0.376 | 1.490 | 12.43 | 1.440 | 2.571 | 0.545 | 0.276 |
| 12 | 0.307 | 2.631 | 3.285 | 0.812 | 0.389 | 1.818 | 13.14 | 1.480 | 2.846 | 0.500 | 0.228 |
| 13 | 0.300 | 2.980 | 3.714 | 0.854 | 0.396 | 2.220 | 13.86 | 1.520 | 3.167 | 0.462 | 0.188 |
| 14 | 0.285 | 3.419 | 4.240 | 0.888 | 0.395 | 2.722 | 14.57 | 1.560 | 3.545 | 0.429 | 0.154 |
| 15 | 0.260 | 3.980 | 4.897 | 0.912 | 0.383 | 3.364 | 15.29 | 1.600 | 4.000 | 0.400 | 0.125 |
| 16 | **0.223** | 4.720 | 5.742 | 0.923 | 0.360 | 4.207 | 16.00 | 1.640 | 4.556 | 0.375 | 0.100 |

Boldface type indicates the minimum expectation.

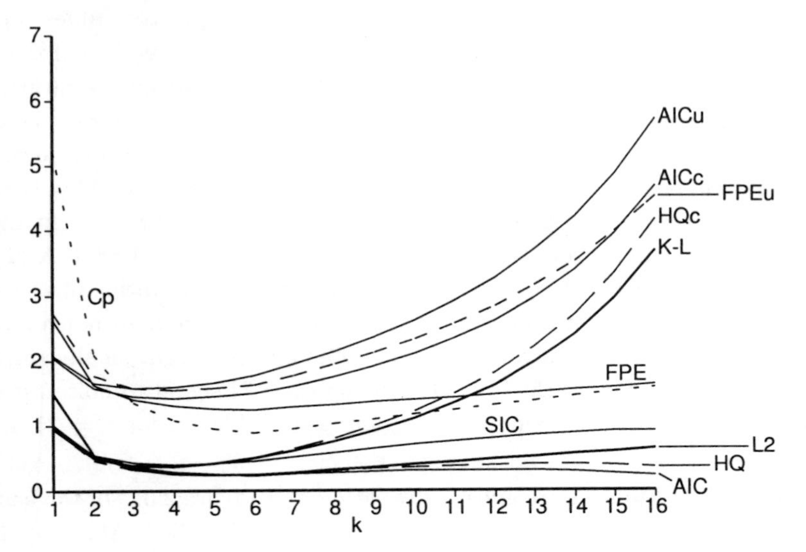


Figure 2.2. Expected values and expected distance for Model 2.

weak in small samples. This can result in decreasing expectations and artificial minima for very large amounts of overfitting.

The expected efficiencies for the two distance measures also differ when an incorrect model is selected. K-L favors model selection criteria prone to some underfitting (*i.e.*, those criteria with strong penalty functions, such as AICc and AICu). For overfitting, the efficiencies of K-L are smaller than the efficiencies of $L_2$; in other words, K-L penalizes overfitting more than $L_2$ by a larger loss in efficiency. By contrast, $L_2$ favors selection criteria which overfit slightly, and has much smaller efficiency than K-L for underfitted model orders.

Figure 2.2 plots actual expected values for $L_2$ and K-L distance rather than the expected efficiencies found in Table 2.6. Note that for presentation purposes, Cp in Figure 2.2 has been scaled by a factor of 10.

### *2.6.4. Signal-to-noise Ratios for Two Special Case Models*

We next examine the behavior of the six foundation and three variant criteria by looking at their signal-to-noise ratios for the two special case models. As before, the true model order is $k_* = 6$, and we will compare the true order to candidate models of order $k$. The true model is also the model where $E[L_2]$ attains a minimum, and consequently attains maximum efficiency. For convenience the signal-to-noise ratio at order $k_*$ is given as 0, since the signal-to-noise ratio is undefined when comparing a model to itself.

The expected differences, or signals, follow from the expected values computed in Section 2.6.1. The standard deviations, or noise, of the log(SSE) terms are computed using the log-Beta distribution. Signal-to-noise ratios of AICc, AIC, SIC, and HQ for orders $k \leq 5$ involve differences of their expected values, and variances of noncentral log-Beta Eq. (2.25). All sums were evaluated out to 1000 terms. Orders $k \geq 7$ involve the variance of central log-Beta, Eq. (2A.10), and represent small-sample signal-to-noise ratios for overfitting. All variances of log-Beta were computed using the recursions in Eqs. (2A.11)–(2A.12) for $\psi'$. For FPE, Cp, and FPEu the signal-to-noise ratios of orders $k \leq 5$ involve differences of their expected values and variances of noncentral $\chi^2$. Orders $k \geq 7$ involve central $\chi^2$ distributions and represent small-sample signal-to-noise ratios for overfitting. Overfitting ($k \geq 7$) signal-to-noise ratios for Models 1 and 2 are the same as those found in Table 2.2 for $n = 25$ and $L = k - 6$.

The results for Model 1 are given in Table 2.7 and Figure 2.3. On the basis of the signal-to-noise ratios, Model 1 should be easily identified. Many of the criteria have large signal-to-noise ratios even though they slightly underfit or overfit. The signal-to-noise ratio trends reflect those for the expected criterion values in Table 2.5. The criteria with strong superlinear penalty functions, AICc, AICu, and HQc, have large signal-to-noise ratios, particularly against overfitting. The signal-to-noise ratios and expected values for overfitted models increase with $k$ for both Cp and FPE, but are still relatively weak when compared to the noise. Hence we expect that some overfitting will occur with

Table 2.7. Signal-to-noise ratios for Model 1, $\beta_j = 1$.

| $k$ | AIC | AICc | AICu | SIC | HQ | HQc | Cp | FPE | FPEu |
|---|---|---|---|---|---|---|---|---|---|
| 1 | 5.451 | 4.676 | 3.927 | 4.670 | 5.234 | 4.429 | 1.594 | 5.281 | 4.852 |
| 2 | 5.164 | 4.455 | 3.833 | 4.529 | 4.988 | 4.244 | 1.575 | 4.749 | 4.412 |
| 3 | 4.756 | 4.143 | 3.654 | 4.268 | 4.620 | 3.973 | 1.545 | 4.132 | 3.877 |
| 4 | 4.151 | 3.671 | 3.321 | 3.810 | 4.056 | 3.546 | 1.490 | 3.389 | 3.206 |
| 5 | 3.155 | 2.852 | 2.650 | 2.963 | 3.102 | 2.778 | 1.356 | 2.406 | 2.293 |
| 6 | 0 | 0 | 0 | 0 | 0 | 0 | 0 | 0 | 0 |
| 7 | 0.284 | 1.479 | 2.149 | 0.888 | 0.451 | 1.742 | 0.311 | 0.405 | 1.015 |
| 8 | 0.360 | 2.200 | 3.145 | 1.189 | 0.590 | 2.585 | 0.414 | 0.536 | 1.300 |
| 9 | 0.389 | 2.842 | 3.997 | 1.372 | 0.662 | 3.332 | 0.481 | 0.615 | 1.457 |
| 10 | 0.388 | 3.474 | 4.804 | 1.485 | 0.692 | 4.062 | 0.530 | 0.666 | 1.551 |
| 11 | 0.364 | 4.128 | 5.609 | 1.546 | 0.692 | 4.814 | 0.567 | 0.698 | 1.608 |
| 12 | 0.320 | 4.828 | 6.442 | 1.564 | 0.665 | 5.616 | 0.597 | 0.716 | 1.639 |
| 13 | 0.258 | 5.599 | 7.332 | 1.545 | 0.615 | 6.496 | 0.621 | 0.723 | 1.652 |
| 14 | 0.180 | 6.470 | 8.309 | 1.492 | 0.543 | 7.486 | 0.642 | 0.721 | 1.648 |
| 15 | 0.085 | 7.481 | 9.413 | 1.407 | 0.452 | 8.632 | 0.659 | 0.711 | 1.632 |
| 16 | -0.024 | 8.689 | 10.703 | 1.290 | 0.340 | 9.997 | 0.674 | 0.695 | 1.603 |

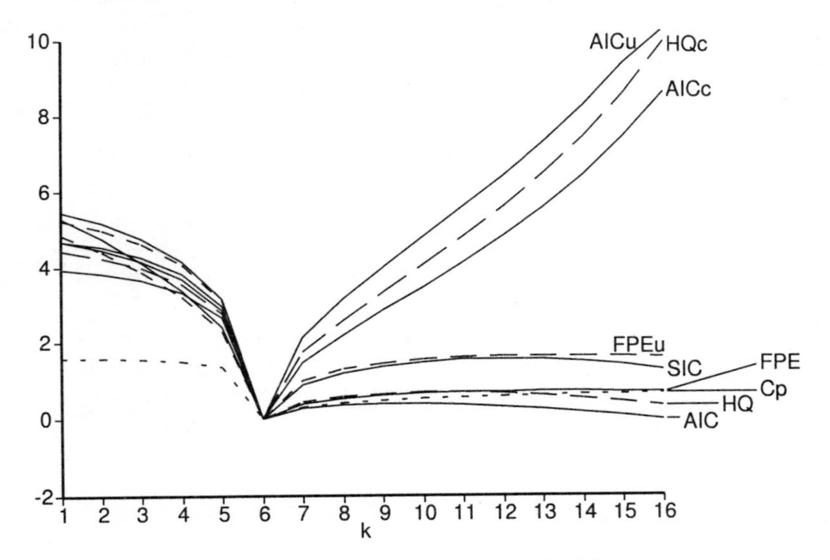


Figure 2.3. Signal-to-noise for Model 1.

both Cp and FPE. Although FPEu is a signal-to-noise corrected variant, its signal-to-noise ratio for overfitting is still much weaker than those of AICc, AICu, and HQc. Once again, AIC's overfitting problem can be seen in the negative signal-to-noise at $k = 16$. As we see from the strong signal-to-noise ratios for orders 1–5 there is little problem with underfitting for any of the criteria due to the strongly identifiable $\beta$ parameters in Model 1. Penalty functions

Table 2.8. Signal-to-noise ratios for Model 2, $\beta_j = 1/j$.

| $k$ | AIC | AICc | AICu | SIC | HQ | HQc | Cp | FPE | FPEu |
|---|---|---|---|---|---|---|---|---|---|
| 1 | 2.839 | 1.966 | 1.123 | 1.958 | 2.594 | 1.687 | 1.636 | 2.481 | 1.786 |
| 2 | 1.309 | 0.337 | -0.515 | 0.439 | 1.067 | 0.048 | 1.657 | 1.055 | 0.309 |
| 3 | 0.721 | -0.274 | -1.070 | -0.072 | 0.501 | -0.551 | 1.599 | 0.541 | -0.143 |
| 4 | 0.415 | -0.519 | -1.201 | -0.249 | 0.231 | -0.763 | 1.443 | 0.283 | -0.296 |
| 5 | 0.222 | -0.523 | -1.021 | -0.251 | 0.091 | -0.707 | 1.124 | 0.133 | -0.289 |
| 6 | 0 | 0 | 0 | 0 | 0 | 0 | 0 | 0 | 0 |
| 7 | 0.284 | 1.479 | 2.149 | 0.888 | 0.451 | 1.742 | 0.311 | 0.405 | 1.015 |
| 8 | 0.360 | 2.200 | 3.145 | 1.189 | 0.590 | 2.585 | 0.414 | 0.536 | 1.300 |
| 9 | 0.389 | 2.842 | 3.997 | 1.372 | 0.662 | 3.332 | 0.481 | 0.615 | 1.457 |
| 10 | 0.388 | 3.474 | 4.804 | 1.485 | 0.692 | 4.062 | 0.530 | 0.666 | 1.551 |
| 11 | 0.364 | 4.128 | 5.609 | 1.546 | 0.692 | 4.814 | 0.567 | 0.698 | 1.608 |
| 12 | 0.320 | 4.828 | 6.442 | 1.564 | 0.665 | 5.616 | 0.597 | 0.716 | 1.639 |
| 13 | 0.258 | 5.599 | 7.332 | 1.545 | 0.615 | 6.496 | 0.621 | 0.723 | 1.652 |
| 14 | 0.180 | 6.470 | 8.309 | 1.492 | 0.543 | 7.486 | 0.642 | 0.721 | 1.648 |
| 15 | 0.085 | 7.481 | 9.413 | 1.407 | 0.452 | 8.632 | 0.659 | 0.711 | 1.632 |
| 16 | -0.024 | 8.689 | 10.703 | 1.290 | 0.340 | 9.997 | 0.674 | 0.695 | 1.603 |

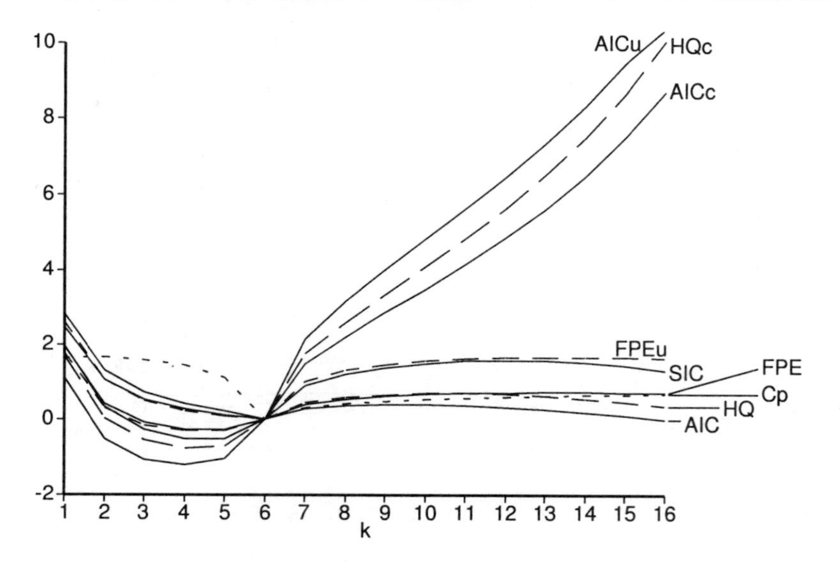


Figure 2.4. Signal-to-noise for Model 2.

strengthened against overfitting lead to improved signal-to noise ratios for overfitting, but also tend to result in smaller signal-to-noise ratios for underfitting. For example, AICu has one of the largest penalty functions and has the largest signal-to-noise ratio for overfitting, but also has the smallest signal-to-noise ratio for underfitting. However, Model 1 is sufficiently strongly identifiable that all underfitting signal-to-noise ratios are large.

On the other hand, Model 2 is weakly identifiable in that all underfitting signal-to-noise ratios tend to be weak (small or negative). Model 2 nicely illustrates the tradeoff between overfitting and underfitting that results from stronger penalty functions (Table 2.8 and Figure 2.4). The overfitting signal-to-noise ratios are the same for Model 2 as for Model 1, but Model 2 has weaker $\beta$ parameters that make detecting the correct order very difficult, and make the model more susceptible to underfitting. Here we see negative signal-to-noise ratios appearing for underfitting, where AICu has the most negative signal-to-noise ratio, indicating that it is the most likely to underfit. AICc, SIC, HQc, and FPEu also have negative signal-to-noise for orders 3, 4 and 5 indicating that some underfitting will be present. On the other hand, AIC, Cp, and FPE have positive signal-to-noise ratios against underfitting. This is not surprising; since AIC and FPE are both known to overfit, we do not expect much underfitting from them.

### *2.6.5. Small-sample Probabilities for Two Special Case Models*

We once again use the two special case models defined in Eqs. (2.28)–(2.30) to illustrate small-sample probabilities of selecting various model orders. We will compute the probability of selecting a particular candidate model of order $k$ over the true order $k_*$, where for underfitting $k < k_*$, and for overfitting $k > k_*$. Model 1, with its strong $\beta$ parameters, is expected to yield low probabilities for both underfitting and overfitting for all criteria. Model 2, with its quickly decaying $\beta$ parameters, is expected to have somewhat higher probabilities of underfitting and the same small probabilities of overfitting. All such probabilities are easily written as Beta probabilities, which can be numerically evaluated by the incomplete beta function (Press, *et al.*, 1986, pp. 166–168).

Orders $k = 1$ through 5 involve the noncentral Beta distributions for reduced models of order $k$ and true model of order 6. The sums were evaluated out to 1000 terms. Orders $k = 7$ and higher involve the central Beta distribution, and represent probabilities of overfitting by $L = k - 6$ additional variables (see Table 2.1, $n = 25$).

For Model 1, Table 2.9 shows the probability of choosing the candidate model of order $k$ over the true model of order 6 for all nine selection criteria. The probability of choosing order 6 over itself is left undefined and is denoted by an asterisk.

Table 2.9. Probability of selecting a particular candidate model of order $k$ over the true order 6 for Model 1, $\beta_j = 1$.

| $k$ | AIC | AICc | AICu | SIC | HQ | HQc | Cp | FPE | FPEu |
|---|---|---|---|---|---|---|---|---|---|
| 1 | 0.000 | 0.000 | 0.000 | 0.000 | 0.000 | 0.000 | 0.000 | 0.000 | 0.000 |
| 2 | 0.000 | 0.000 | 0.000 | 0.000 | 0.000 | 0.000 | 0.000 | 0.000 | 0.000 |
| 3 | 0.000 | 0.000 | 0.000 | 0.000 | 0.000 | 0.000 | 0.000 | 0.000 | 0.000 |
| 4 | 0.000 | 0.000 | 0.000 | 0.000 | 0.000 | 0.000 | 0.000 | 0.000 | 0.000 |
| 5 | 0.000 | 0.001 | 0.003 | 0.000 | 0.000 | 0.001 | 0.000 | 0.000 | 0.001 |
| 6 | * | * | * | * | * | * | * | * | * |
| 7 | 0.237 | 0.079 | 0.044 | 0.133 | 0.201 | 0.063 | 0.191 | 0.220 | 0.118 |
| 8 | 0.257 | 0.041 | 0.016 | 0.112 | 0.204 | 0.028 | 0.191 | 0.228 | 0.089 |
| 9 | 0.266 | 0.019 | 0.005 | 0.095 | 0.202 | 0.011 | 0.185 | 0.226 | 0.066 |
| 10 | 0.277 | 0.008 | 0.002 | 0.084 | 0.202 | 0.004 | 0.178 | 0.223 | 0.051 |
| 11 | 0.291 | 0.003 | 0.000 | 0.078 | 0.207 | 0.001 | 0.173 | 0.221 | 0.040 |
| 12 | 0.310 | 0.001 | 0.000 | 0.076 | 0.216 | 0.000 | 0.168 | 0.221 | 0.032 |
| 13 | 0.334 | 0.000 | 0.000 | 0.078 | 0.231 | 0.000 | 0.164 | 0.223 | 0.026 |
| 14 | 0.364 | 0.000 | 0.000 | 0.082 | 0.251 | 0.000 | 0.161 | 0.227 | 0.022 |
| 15 | 0.402 | 0.000 | 0.000 | 0.092 | 0.279 | 0.000 | 0.158 | 0.233 | 0.020 |
| 16 | 0.446 | 0.000 | 0.000 | 0.106 | 0.315 | 0.000 | 0.156 | 0.242 | 0.018 |

We can see that the probability of underfitting Model 1 is approximately 0 for all criteria, as we would expect. However, the model selection criteria differ in their probabilities of overfitting. AICc has the smallest probability of overfitting. For AIC and FPE the probabilities of overfitting increase with $k$, which leads to severe overfitting. SIC initially shows small probabilities of overfitting, but these increase for somewhat larger models even though, asymptotically, SIC has 0 probability of overfitting. In small samples, however, SIC can overfit to a greater degree than many efficient model selection criteria.

The implications of these probabilities are in accordance with the signal-to-noise ratios found in Table 2.7. Large signal-to-noise ratios for selecting candidate order $k$ over the true order correspond to small probabilities of selecting candidate order $k$. Small signal-to-noise ratios correspond to moderately high probabilities of selecting the wrong model. The highest probabilities of misselection correspond to negative signal-to-noise ratios, and hence we expect AICu, which has the most strongly positive signal-to-noise ratios, to perform well in Model 1. In fact, AICu has the largest signal-to-noise ratio against overfitting and also has the smallest probabilities of overfitting.

We next look at the results for Model 2, given in Table 2.10. Once again

Model 2 illustrates the trade-off between strong and weak penalty functions with respect to the same as those for Model 1 in Table 2.9, since the $\beta_j$ parameters in Model 2 decrease rapidly, we observe higher probabilities of underfitting. SIC is the most likely to underfit due to its large penalty function. Selection criteria with weak penalty functions have smaller probabilities of underfitting than selection criteria with strong penalty functions, and thus AIC, HQ, and FPE underfit with smaller probability and AICu has the highest chance of underfitting.

Table 2.10. Probability of selecting a particular candidate model of order $k$ over the true order 6 for Model 2, $\beta_j = 1/j$.

| $k$ | AIC | AICc | AICu | SIC | HQ | HQc | Cp | FPE | FPEu |
|---|---|---|---|---|---|---|---|---|---|
| 1 | 0.001 | 0.026 | 0.158 | 0.027 | 0.003 | 0.053 | 0.004 | 0.001 | 0.027 |
| 2 | 0.124 | 0.428 | 0.699 | 0.393 | 0.188 | 0.527 | 0.173 | 0.134 | 0.401 |
| 3 | 0.311 | 0.646 | 0.825 | 0.586 | 0.391 | 0.719 | 0.373 | 0.328 | 0.601 |
| 4 | 0.450 | 0.730 | 0.851 | 0.663 | 0.517 | 0.780 | 0.511 | 0.467 | 0.680 |
| 5 | 0.579 | 0.767 | 0.843 | 0.711 | 0.622 | 0.798 | 0.627 | 0.593 | 0.726 |
| 6 | * | * | * | * | * | * | * | * | * |
| 7 | 0.237 | 0.079 | 0.044 | 0.133 | 0.201 | 0.063 | 0.191 | 0.220 | 0.118 |
| 8 | 0.257 | 0.041 | 0.016 | 0.112 | 0.204 | 0.028 | 0.191 | 0.228 | 0.089 |
| 9 | 0.266 | 0.019 | 0.005 | 0.095 | 0.202 | 0.011 | 0.185 | 0.226 | 0.066 |
| 10 | 0.277 | 0.008 | 0.002 | 0.084 | 0.202 | 0.004 | 0.178 | 0.223 | 0.051 |
| 11 | 0.291 | 0.003 | 0.000 | 0.078 | 0.207 | 0.001 | 0.173 | 0.221 | 0.040 |
| 12 | 0.310 | 0.001 | 0.000 | 0.076 | 0.216 | 0.000 | 0.168 | 0.221 | 0.032 |
| 13 | 0.334 | 0.000 | 0.000 | 0.078 | 0.231 | 0.000 | 0.164 | 0.223 | 0.026 |
| 14 | 0.364 | 0.000 | 0.000 | 0.082 | 0.251 | 0.000 | 0.161 | 0.227 | 0.022 |
| 15 | 0.402 | 0.000 | 0.000 | 0.092 | 0.279 | 0.000 | 0.158 | 0.233 | 0.020 |
| 16 | 0.446 | 0.000 | 0.000 | 0.106 | 0.315 | 0.000 | 0.156 | 0.242 | 0.018 |

When we examine a weakly identifiable model we realize that underfitting and overfitting can be defined from two perspectives—first, if one assumes a true order $k_*$, then underfitting is selecting an order $k < k_*$, and overfitting is selecting an order $k > k_*$. We used this idea in the signal-to-noise examples in Tables 2.7 and 2.8. However, for Model 2 we see that that knowing the true order is not necessarily sufficient to allow us to define overfitting and underfitting. Here we have the additional complication that the distance measures or their efficiencies themselves can be used to define over/underfitting with respect to their identification of the model *closest* to the true model. In cases where the true model does not belong to the set of candidate models, we can suppose that the true model is of infinite dimension but we only have a finite sample size. In this case underfitting can be defined as loss of expected (or observed) efficiency due to selecting a model with too few variables, and overfitting can be defined as loss of expected (or observed) efficiency due to selecting a model with too many variables. In Chapter 3 we will consider more closely the case where the true model does not belong to the set of candidate models.

## 2.7. Random $X$ Regression and Monte Carlo Study

After examining the derivations and theoretical properties of the nine model selection criteria in this Chapter, we now explore their behavior via a simulation study using our two univariate special case models. We will make two notable modifications to these models for this study: first, $X'X$ is generated by a random design matrix rather than assumed to be $nI$. Randomly generating $X$ for each replication reduces the effect of an unusual $X$ matrix that may favor a particular selection criterion, and it is also computationally simple. Second, because in practice a researcher does not know the order of importance of the variables, all subsets for models of eleven variables (ten plus the intercept) are considered rather than the ordered models previously studied. This results in 1024 different subsets, each of which includes the intercept (or constant). For comparison we have also computed observed values for the distance measures K-L and $L_2$. Measures of model selection performance are a count of the number of times the correct model is chosen out of 10,000 realizations, and K-L and $L_2$ observed efficiencies as described in Chapter 1.

We generate $X$ with $x_{ij}$ as *i.i.d.* $N(0,1)$, $i = 1, \dots, n$ and $j = 1, \dots, 10$, and $x_{i,0} = 1$, $i = 1, \dots, n$. $X_i$ *i.i.d.* $N_k(0, I)$ is a special case of $X_i$ *i.i.d.* $N_k(0, \Sigma)$ (ignoring $x_0$). One consequence of the random $X$ structure is that the noncentrality parameter, $\lambda$, is also random and $\text{SSE}_{k_0}$ no longer has a noncentral $\chi^2$ distribution in the underfitting case. For the purposes of this simulation we have derived the $\lambda$ and the unconditional $\text{SSE}_{k_0}$ distributions, and the results are stated in the following Theorems. The derivations themselves can be found in Appendix 2B.

**Theorem 2.5**

Suppose $Y = X_*\beta_* + \varepsilon_*$ is the true model, where $X_* = (X_0, X_1) = (1, \tilde{X}_0, X_1)$, and $\beta_* = (\beta_0', \beta_1')'$. Ignoring the constant, the rows of $X_*$ are independent with the following multivariate normal distribution:

$$\tilde{X}_{*i} = \begin{pmatrix} \tilde{X}_{0i} \\ X_{1i} \end{pmatrix} \sim N\left( \begin{pmatrix} 0 \\ 0 \end{pmatrix}, \begin{pmatrix} \Sigma_{00} & \Sigma_{10} \\ \Sigma_{01} & \Sigma_{11} \end{pmatrix} \right).$$

If the model $Y = X_0\beta_0$ is fit to the data, then the noncentrality parameter, $\lambda$, is random, and

$$\lambda \sim \left(\beta_1' \Sigma_{11\cdot 0} \beta_1 / \sigma_*^2\right) \chi^2_{n-k_0},$$

where $k_0 = \text{rank}(X_0)$, $\Sigma_{11\cdot 0} = \Sigma_{11} - \Sigma_{01}'\Sigma_{00}^{-1}\Sigma_{01}$ and $\Sigma_{11} = var[\tilde{X}_{1i}]$, $\Sigma_{00} = var[X_{0i}]$, and $\Sigma_{01} = cov[\tilde{X}_{0i}, X_{1i}]$.

**Theorem 2.6**

Suppose $Y = X_*\beta_* + \varepsilon_*$ is the true model, where $X_* = (X_0, X_1) = (1, \tilde{X}_0, X_1)$, and $\beta_* = (\beta_0', \beta_1')'$. Ignoring the constant, the rows of $X_*$ are independent with the following multivariate normal distribution:

$$\tilde{X}_{*i} = \begin{pmatrix} \tilde{X}_{0i} \\ X_{1i} \end{pmatrix} \sim N\left( \begin{pmatrix} 0 \\ 0 \end{pmatrix}, \begin{pmatrix} \Sigma_{00} & \Sigma_{10} \\ \Sigma_{01} & \Sigma_{11} \end{pmatrix} \right).$$

If the model $Y = X_0\beta_0$ is fit to the data, then

$$\frac{\text{SSE}_{k_0}}{\sigma_*^2} \sim \left(1 + \frac{\beta_1'\Sigma_{11\cdot 0}\beta_1}{\sigma_*^2}\right)\chi^2_{n-k_0},$$

where $k_0 = \text{rank}(X_0)$, $\Sigma_{11\cdot 0} = \Sigma_{11} - \Sigma_{01}'\Sigma_{00}^{-1}\Sigma_{01}$ and $\Sigma_{11} = var[\tilde{X}_{1i}]$, $\Sigma_{00} = var[X_{0i}]$, and $\Sigma_{01} = cov[\tilde{X}_{0i}, X_{1i}]$.

For our special case $\Sigma = I$, Theorem 2.5 simplifies to $\lambda \sim \beta_1'\beta_1/\sigma_*^2\chi^2_{n-k_0}$ and $E[\lambda] = (n - k_0)\beta_1'\beta_1/\sigma_*^2$. Theorem 2.6 simplifies to $\text{SSE}_{k_0}/\sigma_*^2 \sim \big(1 + \beta_1'\beta_1/\sigma_*^2\big)\chi^2_{n-k_0}$. Although $E[X'X] = nI$, the random $X$ matrix models have similar but smaller noncentrality parameters than the models under the $X'X = nI$ assumption in earlier sections.

We can now specifically describe our two simulation models. Model 1 is given by Eq. (2.28) with $\beta$ parameter structure defined by Eq. (2.29). Model 2 is given by Eq. (2.28) with $\beta$ parameter structure defined by Eq. (2.30). In both models the $X'X = nI$ assumption is replaced by a random design matrix as described above. The method of least squares is used to calculate the regression parameters, and all 1024 subsets are considered as potential candidate models. Since there are too many subsets to consider individually, subsets are summarized by the rank $k$ of their design matrix. For each of the 10,000 realizations, a new $X$ matrix and $\varepsilon$ vector were generated. Although Table 2.11 gives counts for the selection of model order $k$, and as before the true model order is 6, models of order $k = 6$ may include some with the correct number of variables but which are not the true model. The "true" row lists the number of times the correct model was chosen out of the total count for order $k_* = 6$.

Observed efficiency is also computed for each realization. K-L observed efficiency is computed using Eq. (1.2) and Eq. (2.10). $L_2$ observed efficiency is computed similarly using Eq. (1.1) and Eq. (2.9). For each of the 10,000 realizations the criteria select a model and the observed efficiency of this selection is recorded. Averages, medians and standard deviations are also computed for the 10,000 K-L and $L_2$ observed efficiencies. Performance is based on average observed efficiencies, where higher observed efficiency denotes better performance. The criterion with the highest observed efficiency is given rank 1 (best) while the criterion with the lowest observed efficiency is given rank 9 (lowest of the 9 criteria considered here).

We know that Model 1 represents the case where a strong, easily identified true model is also a candidate model. The count results are one way to measure consistency, and we might therefore expect the consistent model selection criteria to have the highest counts. However, this is not always true in small samples. In fact, we see from Table 2.11 that AICu has the highest true model count, identifying the correct model from the 1024 candidate models more than 65% of the time. Of the consistent criteria HQc has the best performance, but it still overfits more than AICu. In general, the performance differences are not greatest between efficient and consistent criteria, but rather between model selection criteria with strong penalty functions and those with weak penalty

Table 2.11. Simulation results for Model 1, $\beta_j = 1$. Counts and observed efficiency.

| | | | | | counts | | | | | | |
|---|---|---|---|---|---|---|---|---|---|---|---|
| $k$ | AIC | AICc | AICu | SIC | HQ | HQc | Cp | FPE | FPEu | $L_2$ | K-L |
| 1 | 0 | 0 | 3 | 0 | 0 | 0 | 0 | 0 | 0 | 0 | 0 |
| 2 | 0 | 1 | 11 | 1 | 0 | 2 | 0 | 0 | 1 | 0 | 0 |
| 3 | 0 | 12 | 71 | 7 | 1 | 24 | 2 | 1 | 7 | 0 | 0 |
| 4 | 7 | 84 | 254 | 50 | 12 | 140 | 22 | 8 | 56 | 0 | 4 |
| 5 | 76 | 498 | 978 | 270 | 119 | 664 | 200 | 83 | 314 | 1 | 171 |
| 6 | 2463 | 6240 | 6888 | 4548 | 3040 | 6635 | 4154 | 2723 | 5020 | 9979 | 7818 |
| 7 | 3284 | 2533 | 1541 | 3209 | 3417 | 2106 | 3418 | 3470 | 3107 | 12 | 941 |
| 8 | 2542 | 568 | 228 | 1371 | 2231 | 383 | 1561 | 2435 | 1165 | 7 | 747 |
| 9 | 1179 | 59 | 25 | 444 | 880 | 43 | 520 | 971 | 283 | 1 | 274 |
| 10 | 397 | 4 | 1 | 89 | 257 | 3 | 111 | 268 | 43 | 0 | 45 |
| 11 | 52 | 1 | 0 | 11 | 43 | 0 | 12 | 41 | 4 | 0 | 0 |
| true | 2338 | 5875 | 6509 | 4307 | 2881 | 6243 | 3925 | 2585 | 4748 | 9970 | 7169 |
| | | | | K-L observed efficiency | | | | | | | |
| | AIC | AICc | AICu | SIC | HQ | HQc | Cp | FPE | FPEu | $L_2$ | K-L |
| ave | 0.498 | 0.679 | 0.714 | 0.595 | 0.525 | 0.699 | 0.577 | 0.511 | 0.619 | 0.916 | 1.000 |
| med | 0.430 | 0.743 | 0.884 | 0.526 | 0.456 | 0.838 | 0.506 | 0.445 | 0.557 | 1.000 | 1.000 |
| sd | 0.288 | 0.324 | 0.317 | 0.323 | 0.301 | 0.322 | 0.318 | 0.293 | 0.325 | 0.170 | 0.000 |
| rank | 9 | 3 | 1 | 5 | 7 | 2 | 6 | 8 | 4 | | |
| | | | | $L_2$ observed efficiency | | | | | | | |
| | AIC | AICc | AICu | SIC | HQ | HQc | Cp | FPE | FPEu | $L_2$ | K-L |
| ave | 0.652 | 0.789 | 0.802 | 0.727 | 0.673 | 0.800 | 0.714 | 0.663 | 0.745 | 1.000 | 0.832 |
| med | 0.642 | 1.000 | 1.000 | 0.744 | 0.662 | 1.000 | 0.718 | 0.653 | 0.809 | 1.000 | 1.000 |
| sd | 0.255 | 0.277 | 0.289 | 0.275 | 0.262 | 0.280 | 0.272 | 0.258 | 0.277 | 0.000 | 0.281 |
| rank | 9 | 3 | 1 | 5 | 7 | 2 | 6 | 8 | 4 | | |

functions. Or, equivalently, the model selection criteria with larger signal-to-noise ratios perform better than the model selection criteria with small signal-to-noise ratios.

Table 2.11 also shows that for Model 1, the observed efficiencies parallel the counts. AICu and HQc, the criteria with the highest counts, also have the highest Kullback–Leibler (K-L) and $L_2$ observed efficiencies. As expected due to its strong $\beta$ parameters, little underfitting is evident for Model 1; both distance measures attain their minima either at the true model or at an overfitted model with extra variables. $L_2$ penalizes overfitting less than K-L, and since all the criteria overfit to some degree, their $L_2$ observed efficiencies are higher than their K-L observed efficiencies. Model 2 represents the case where the true model is only weakly identifiable. Table 2.12 shows the results for this case.

In the situation where there is a weak true model, observed efficiency is a more useful measure of model selection performance than consistency. Since

Table 2.12. Simulation results for Model 2, $\beta_j = 1/j$. Counts and observed efficiency.

| | | | | | counts | | | | | | |
|---|---|---|---|---|---|---|---|---|---|---|---|
| $k$ | AIC | AICc | AICu | SIC | HQ | HQc | Cp | FPE | FPEu | $L_2$ | K-L |
| 1 | 2 | 9 | 30 | 16 | 4 | 14 | 4 | 2 | 14 | 0 | 1 |
| 2 | 84 | 405 | 1229 | 561 | 165 | 620 | 234 | 86 | 531 | 2 | 19 |
| 3 | 598 | 2133 | 3414 | 1973 | 932 | 2642 | 1374 | 644 | 2006 | 101 | 310 |
| 4 | 1631 | 3427 | 3274 | 2738 | 2034 | 3480 | 2515 | 1763 | 2906 | 746 | 1668 |
| 5 | 2351 | 2610 | 1543 | 2348 | 2537 | 2246 | 2700 | 2470 | 2423 | 2998 | 3752 |
| 6 | 2395 | 1066 | 410 | 1398 | 2160 | 775 | 1840 | 2448 | 1355 | 5402 | 3501 |
| 7 | 1659 | 292 | 84 | 667 | 1308 | 188 | 899 | 1552 | 562 | 647 | 612 |
| 8 | 830 | 49 | 14 | 223 | 580 | 32 | 327 | 713 | 164 | 100 | 118 |
| 9 | 353 | 9 | 2 | 61 | 224 | 3 | 91 | 262 | 33 | 4 | 16 |
| 10 | 89 | 0 | 0 | 14 | 53 | 0 | 15 | 57 | 5 | 0 | 3 |
| 11 | 8 | 0 | 0 | 1 | 3 | 0 | 1 | 3 | 1 | 0 | 0 |
| true | 201 | 120 | 48 | 156 | 195 | 90 | 180 | 207 | 149 | 3459 | 1459 |

| | | | | K-L observed efficiency | | | | | | | |
|---|---|---|---|---|---|---|---|---|---|---|---|
| | AIC | AICc | AICu | SIC | HQ | HQc | Cp | FPE | FPEu | $L_2$ | K-L |
| ave | 0.332 | 0.411 | 0.449 | 0.391 | 0.349 | 0.425 | 0.371 | 0.337 | 0.397 | 0.887 | 1.000 |
| med | 0.282 | 0.374 | 0.421 | 0.350 | 0.299 | 0.393 | 0.323 | 0.288 | 0.356 | 0.998 | 1.000 |
| sd | 0.208 | 0.225 | 0.227 | 0.225 | 0.215 | 0.226 | 0.220 | 0.209 | 0.225 | 0.176 | 0.000 |
| rank | 9 | 3 | 1 | 5 | 7 | 2 | 6 | 8 | 4 | | |

| | | | | $L_2$ observed efficiency | | | | | | | |
|---|---|---|---|---|---|---|---|---|---|---|---|
| | AIC | AICc | AICu | SIC | HQ | HQc | Cp | FPE | FPEu | $L_2$ | K-L |
| ave | 0.474 | 0.492 | 0.486 | 0.484 | 0.479 | 0.492 | 0.484 | 0.476 | 0.486 | 1.000 | 0.884 |
| med | 0.449 | 0.469 | 0.464 | 0.458 | 0.453 | 0.468 | 0.458 | 0.451 | 0.462 | 1.000 | 0.998 |
| sd | 0.207 | 0.214 | 0.214 | 0.212 | 0.209 | 0.214 | 0.212 | 0.207 | 0.213 | 0.000 | 0.185 |
| rank | 9 | 1 | 3 | 5 | 7 | 1 | 5 | 8 | 3 | | |

all the selection criteria overfit to some degree, and K-L penalizes overfitting more than $L_2$, the K-L observed efficiencies are lower than those for $L_2$. As was the case for Model 1 AICu has the highest K-L observed efficiency, and although it underfits in terms of counts, the models selected are close to the correct model as identified by K-L. However, AICu does not perform as well in the $L_2$ sense, where its strong penalty function causes it to underfit with respect to the closest $L_2$ model. AICc and HQc also have a strong penalty functions, although smaller than that of AICu, and they perform best in the $L_2$ sense.

In comparing the two models, we observe that the overall observed efficiencies for Model 2 are lower than for Model 1 due to the increased difficulties with model identification. Also, in contrast to Model 1, underfitting is a problem for Model 2, and this results in a loss of observed efficiency. Counts of correct model choice are also drastically lower than for Model 1 due to the difficulty in identifying the true model or even the closest model: none of the selection criteria correctly identify the correct model more than 2% of the time. For weakly identifiable models counts are unlikely to be a meaningful measure of performance.

Overall, over both models, AICc, AICu and HQc perform best. With so many candidate models, all subsets regression presents many opportunities for overfitting and therefore a strong penalty function is important. For both Models 1 and 2 the true model contains the intercept as well as the first 5 variables of $X$, leaving five irrelevant variables. This means that there are five individual cases in which we must compare the correct model to a candidate model that overfits by one variable. When we computed the probabilities of overfitting by one variable in Section 2.5 we were considering only one possible overfitted model. In Chapter 9, we will look at the issue of irrelevant variables in more detail, using both special case models with a limited number of extra variables, and a large-scale simulation study with 540 models.

## 2.8. Summary

In this Chapter we have considered many different theoretical properties of our foundation and variant criteria. In each case we have found that the behavior we expect based on the small-sample and asymptotic behavior of these properties correlates very well with the actual performance of the criteria as shown by counts, probabilities, and observed efficiencies. The general trends we have observed in this Chapter for K-L, $L_2$, and efficient criterion classes as well as for the strong-penalty/weak-penalty classes will be seen repeatedly in subsequent chapters. We will find that these trends are remarkably consistent across the different categories of models discussed in this book.

The consequences of criterion expectations and standard deviations are brought together conceptually with the idea of the signal-to-noise ratio. Criteria with weak penalty functions are unable to overcome increasing variability as the number of variables considered increases, and tend to overfit. On the other hand, criteria with penalty functions that excessively resist overfitting may be prone to underfitting, particularly when the true model is weak. The relationship of the penalty function to $L$ (amount of overfitting) is the key. Criteria with penalty functions linearly related to $L$ have signal-to-noise ratios that are relatively strong when $L$ is small relative to $n - k$, but weaken as $L$ approaches $n - k$ and give rise to small-sample overfitting. Criteria with penalty functions superlinearly related to $L$ can overcome this difficulty by allowing the signal-to-noise ratio to grow stronger as $L$ increases.

We found that signal-to-noise ratio values are good predictors for most of the performance measures we considered. High probabilities of selecting the correct model order correspond to high signal-to-noise ratios for the model type in question. High signal-to-noise ratios for overfitted or underfitted models correspond to low probabilities of overfitting and underfitting. We can also predict relative distance measure observed efficiencies, provided we bear in mind that in small samples, $L_2$ observed efficiencies behave differently than K-L observed efficiencies. In particular, $L_2$ penalizes underfitting more than overfitting—$L_2$ distance increases slowly when extra variables are included. K-L penalizes overfitting more than underfitting—the K-L distance increases quickly when extra variables are included. Thus model selection criteria with strong penalty functions and strong signal-to-noise ratios tend to choose models similar to the K-L, and those with weaker penalty functions perform better under $L_2$.

Finally, we found that the identifiability of the true model (strength of the true parameters) causes us to view the concept of a "true" model in a somewhat different light. When a true model is only weakly identifiable, for the purposes of evaluating criterion performance the true model is defined by the distance measures considered. Given what we know about the differences in performance under K-L and $L_2$, this leads us to recommend that criteria that perform well under both distance measures should be used to select a model in practice.

## Chapter 2 Appendices

### Appendix 2A. Distributional Results in the Central Case

*2A.1. Distributions*

Assume all distributions are central. We will take advantage of the well-known property of hierarchical (or nested) models that for $k \geq k_*$ and $L > 0$,

$$\mathrm{SSE}_k - \mathrm{SSE}_{k+L} \sim \sigma_*^2 \chi_L^2, \tag{2A.1}$$

$$\mathrm{SSE}_k \sim \sigma_*^2 \chi_{n-k}^2 \tag{2A.2}$$

and

$$\mathrm{SSE}_k - \mathrm{SSE}_{k+L} \text{ is independent of } \mathrm{SSE}_{k+L}. \tag{2A.3}$$

We will also need the distribution of linear combinations of $\mathrm{SSE}_k$ and $\mathrm{SSE}_{k+L}$. Consider the linear combination $a\mathrm{SSE}_k - b\mathrm{SSE}_{k+L}$ where a and b are scalars. It follows from Eq. (2A.2) that

$$E[a\mathrm{SSE}_k - b\mathrm{SSE}_{k+L}] = a(n-k)\sigma_*^2 - b(n-k-L)\sigma_*^2 \tag{2A.4}$$

and

$$\begin{aligned} var[a\mathrm{SSE}_k - b\mathrm{SSE}_{k+L}] &= var[a\mathrm{SSE}_k - a\mathrm{SSE}_{k+L} + a\mathrm{SSE}_{k+L} - b\mathrm{SSE}_{k+L}] \\ &= var[a(\mathrm{SSE}_k - \mathrm{SSE}_{k+L}) + (a-b)\mathrm{SSE}_{k+L}]. \end{aligned}$$

Applying Eqs. (2A.1)–(2A.3), we have

$$var[a\mathrm{SSE}_k - b\mathrm{SSE}_{k+L}] = 2a^2 L\sigma_*^4 + 2(a-b)^2(n-k-L)\sigma_*^4. \tag{2A.5}$$

Since many model selection criteria (such as AIC) use some function of $\log(\mathrm{SSE}_k)$, we will also introduce some useful distributional results involving $\log(\mathrm{SSE}_k)$. It can be shown (Gradshteyn, 1965, p. 576) that

$$\int_0^\infty z^{\nu-1} e^{-\mu z} \log(z) dz = \frac{1}{\mu^\nu}\Gamma(\nu)\left[\psi(\nu) - \log(\mu)\right],\ \mu > 0, \nu > 0,$$

where

$$\psi(\nu) = -C - \sum_{j=0}^{\infty}\left(\frac{1}{j+\nu} - \frac{1}{j+1}\right),$$

C = 0.577 215 664 901 is Euler's constant, and $\psi$ is Euler's psi function. For $Z \sim \chi^2_m$ with $m$ degrees of freedom,

$$E\big[\log(Z)\big] = \int_0^\infty \log(z) \frac{1}{2^{m/2}\Gamma(m/2)} z^{m/2-1} e^{-z/2} dz$$
$$= \log(2) + \psi\left(\frac{m}{2}\right),$$

which has no closed-form solution. For $\mathrm{SSE}_k \sim \sigma_*^2 \chi^2_{n-k}$,

$$E\big[\log(\mathrm{SSE}_k)\big] = \log(\sigma_*^2) + \log(2) + \psi\left(\frac{n-k}{2}\right). \tag{2A.6}$$

Although $\psi$ has no closed-form solution, a simple recursion exists which is useful for computing exact expectations in small samples (Gradshteyn, 1965 pp. 943–945):

$$\psi(v+1) = \psi(v) + \frac{1}{v},\; v > 0, \tag{2A.7}$$

where

$$\psi\left(\tfrac{1}{2}\right) = -C - 2\log(2) \text{ and } \psi(1) = -C. \tag{2A.8}$$

This recursion will be used to check the accuracy of the Taylor expansion derived in Section 2.3 as well as in studying small-sample properties in Section 2.6.

The distribution of differences between $\log(\mathrm{SSE}_k)$ and $\log(\mathrm{SSE}_{k+L})$ is more involved. Since we have differences of logs,

$$\log(\mathrm{SSE}_{k+L}) - \log(\mathrm{SSE}_k) = \log\left(\frac{\mathrm{SSE}_{k+L}}{\mathrm{SSE}_k}\right).$$

Let $Q = \mathrm{SSE}_{k+L}/\mathrm{SSE}_k$. Assuming nested models and applying Eqs. (2A.1)–(2A.3), we have

$$Q \sim \frac{\chi^2_{n-k-L}}{\chi^2_{n-k-L} + \chi^2_L}.$$

In nested models these two $\chi^2$ are independent, and $Q$ has the Beta distribution

$$Q \sim \mathrm{Beta}\left(\frac{n-k-L}{2}, \frac{L}{2}\right).$$

Since $Q = \mathrm{SSE}_{k+L}/\mathrm{SSE}_k$, the log-distribution is

$$\log\left(\frac{\mathrm{SSE}_{k+L}}{\mathrm{SSE}_k}\right) \sim \text{log-Beta}\left(\frac{n-k-L}{2}, \frac{L}{2}\right). \tag{2A.9}$$

It can be shown (Gradshteyn, 1965 p. 538 and p. 541) that

$$\int_0^1 t^{\mu-1}(1-t^r)^{\nu-1}\log(t)dt = \frac{1}{r^2}B\left(\frac{\mu}{r}+\nu,\nu\right)\left[\psi\left(\frac{\mu}{r}\right)-\psi\left(\frac{\mu}{r}+\nu\right)\right]$$

and

$$\begin{aligned}\int_0^1 t^{\mu-1}(1-t^r)^{\nu-1}\log^2(t)dt =& B\left(\frac{\mu}{r}+\nu,\nu\right)\left[\left(\psi\left(\frac{\mu}{r}\right)-\psi\left(\frac{\mu}{r}+\nu\right)\right)^2\right.\\ &\left.+\psi'(\mu)-\psi'(\mu+\nu)\right],\end{aligned}$$

where

$$\psi'(v) = \sum_{j=0}^{\infty}\frac{1}{(j+v)^2}$$

and $\mu > 0, \nu > 0$. For $Q \sim Beta(\frac{m}{2}, \frac{L}{2})$,

$$\begin{aligned}E\left[\log(Q)\right] &= \int_0^1 \log(t)B^{-1}\left(\frac{m}{2},\frac{L}{2}\right)t^{\frac{m}{2}-1}(1-t)^{\frac{L}{2}-1}dt\\ &= \psi\left(\frac{m}{2}\right)-\psi\left(\frac{m}{2}+\frac{L}{2}\right)\end{aligned}$$

and

$$\begin{aligned}E\left[\log^2(Q)\right] &= \int_0^1 \log^2(t)B^{-1}\left(\frac{m}{2},\frac{L}{2}\right)t^{\frac{m}{2}-1}(1-t)^{\frac{L}{2}-1}dt\\ &= \left(\psi\left(\frac{m}{2}\right)-\psi\left(\frac{m}{2}+\frac{L}{2}\right)\right)^2+\psi'\left(\frac{m}{2}\right)-\psi'\left(\frac{m}{2}+\frac{L}{2}\right).\end{aligned}$$

Hence,

$$var\left[\log(Q)\right] = \psi'\left(\frac{m}{2}\right)-\psi'\left(\frac{m}{2}+\frac{L}{2}\right)$$

and

$$var\left[\log\left(\frac{\text{SSE}_{k+L}}{\text{SSE}_k}\right)\right] = \psi'\left(\frac{n-k-L}{2}\right)-\psi'\left(\frac{n-k}{2}\right). \tag{2A.10}$$

Eq. (2A.10) also has no closed-form, but again, convenient recursions for $\psi'$ exist which are useful in small samples (Gradshteyn, 1965 pp. 945–946):

$$\psi'\left(\frac{m}{2}\right) = \psi'\left(\frac{m}{2}-1\right)-\frac{4}{(m-2)^2}, \tag{2A.11}$$

where

$$\psi'\left(\frac{1}{2}\right) = \frac{\pi^2}{2} \text{ and } \psi'(1) = \frac{\pi^2}{6}. \qquad (2A.12)$$

### *2A.2. Approximate Expected Value of log(SSE$_k$)*

AIC, AICc, SIC, and HQ are all functions of $\log(\mathrm{SSE}_k)$, and therefore computing moments for these model selection criteria involves the moments of $\log(\mathrm{SSE}_k)$, which have no closed form. It is often more useful to find an approximating function that can be written in closed form to use instead. Beginning with $E[\log(\mathrm{SSE}_k)]$, we will derive some useful approximations using Taylor expansions.

Suppose $Z \sim \chi^2_m$. Expanding $\log(Z)$ about $E[Z] = m$, we have

$$\log(Z) \doteq \log(m) + \frac{1}{m}(x - m) - \frac{1}{2m^2}(x - m)^2$$

and

$$E\big[\log(Z)\big] \doteq \log(m) - \frac{1}{m}.$$

Numerically evaluating Eq. (2A.6) and recursions Eqs. (2A.7)–(2A.8) and comparing the results to $\log(m) - 1/m$, we find that this approximation is quite good for $E\big[\log(Z)\big]$. For $m > 5$, $E[\log(Z)] \doteq \log(m) - 1/m$ yields an approximation with only 0.5% error which improves as $m$ increases. This approximation is used to compute moments of $\log(\mathrm{SSE}_k)$-based model selection criteria in Section 2.3. However, exact moments are computed for all tables involving moments of $\log(\mathrm{SSE}_k)$. For $\mathrm{SSE}_k \sim \sigma_*^2\chi^2_{n-k}$,

$$E\big[\log(\mathrm{SSE}_k)\big] \doteq \log(\sigma_*^2) + \log(n-k) - \frac{1}{n-k}. \qquad (2A.13)$$

### *2A.3. Approximate Variance of log(SSE$_{k+L}$/SSE$_k$)*

Since $Q = \mathrm{SSE}_{k+L}/\mathrm{SSE}_k$, Eq. (2A.9) can be rewritten as

$$\log(Q) \sim \text{log-Beta}\left(\frac{n-k-L}{2}, \frac{L}{2}\right).$$

However, there is no closed form to compute the variance of $\log(\mathrm{SSE}_{k+L}/\mathrm{SSE}_k)$ $(= \log(Q))$ in nested models. Therefore, expanding $\log(Q)$ about $E[Q] = (n-k-L)/(n-k)$, we have

$$\log(Q) \doteq \log\left(\frac{n-k-L}{n-k}\right) + \frac{n-k}{n-k-L}\left(Q - \frac{n-k-L}{n-k}\right)$$

and

$$\begin{aligned}
var\left[\log\left(\frac{\mathrm{SSE}_{k+L}}{\mathrm{SSE}_k}\right)\right] &= var[\log(Q)] \\
&\doteq var\left[\log\left(\frac{n-k-L}{n-k}\right) + \frac{n-k}{n-k-L}\left(Q - \frac{n-k-L}{n-k}\right)\right] \\
&\doteq \frac{(n-k)^2}{(n-k-L)^2} var[Q] \\
&= \frac{(n-k)^2}{(n-k-L)^2} \frac{L(n-k-L)/4}{((n-k)/2)^2((n-k)/2+1)} \\
&= \frac{(n-k)^2}{(n-k-L)^2} \frac{2L(n-k-L)}{(n-k)^2(n-k+2)} \\
&= \frac{2L}{(n-k-L)(n-k+2)},
\end{aligned}$$

and thus the standard deviation is

$$sd[\Delta \log(\mathrm{SSE}_k)] \doteq \frac{\sqrt{2L}}{\sqrt{(n-k-L)(n-k+2)}}. \qquad (2A.14)$$

The expansion used for the variance is shorter than the expansion used for the first moment; the longer expansion for the variance yields a much messier expression. The simpler expansion is preferable, and performs adequately. Numerically evaluating Eq. (2A.10) and recursions Eqs. (2A.11)–(2A.12) and comparing the results to Eq. (2A.14), we find the approximation has less than 10% error for $n - k - L \geq 18$, and the percentage error improves as $n - k - L$ increases.

## Appendix 2B. Proofs of Theorems 2.1 to 2.6

### *2B.1. Theorem 2.1*

Given the regression model in Eqs. (2.3)–(2.4) and the criterion of the form $\log(\hat{\sigma}_k^2) + \alpha k/(n-k-2)$, for all $n \geq 6$, $\alpha \geq 1$, $0 < L < n - k - 2$, and for the overfitting case where $0 < k_* \leq k < n - 3$, the signal-to-noise ratio of this criterion increases as $L$ increases.

**Proof:**

The signal-to-noise ratio of this criterion can be written as a function of the amount of overfitting, $L$, as follows:

$$\begin{aligned}\mathrm{STN}(L) =& \frac{\sqrt{(n-k-L)(n-k+2)}}{\sqrt{2L}}\left(\log\left(\frac{n-k-L}{n-k}\right)\right.\\ &\left. - \frac{L}{(n-k-L)(n-k)} + \frac{\alpha(n-2)L}{(n-k-L-2)(n-k-2)}\right).\end{aligned}$$

Since $\mathrm{STN}(L)$ is continuous in $L$, it is enough to show that its derivative is positive:

$$\begin{aligned}\frac{d}{dL}\mathrm{STN}(L) =& -\frac{1}{2}\frac{\sqrt{n-k+2}}{\sqrt{2L(n-k-L)}}\left(\log\left(\frac{n-k-L}{n-k}\right)\right.\\ &\left. - \frac{L}{(n-k-L)(n-k)} + \frac{\alpha(n-2)L}{(n-k-L-2)(n-k-2)}\right)\\ &- \frac{1}{2}\frac{\sqrt{(n-k-L)(n-k+2)}}{\sqrt{2L^3}}\left(\log\left(\frac{n-k-L}{n-k}\right)\right.\\ &\left. - \frac{L}{(n-k-L)(n-k)} + \frac{\alpha(n-2)L}{(n-k-L-2)(n-k-2)}\right)\\ &+ \frac{\sqrt{(n-k-L)(n-k+2)}}{\sqrt{2L}}\times\\ &\left(-\frac{1}{n-k-L} - \frac{1}{(n-k-L)(n-k)} - \frac{L}{(n-k-L)^2(n-k)}\right.\\ &\left. + \frac{\alpha(n-2)}{(n-k-L-2)(n-k-2)} + \frac{\alpha(n-2)L}{(n-k-L-2)^2(n-k-2)}\right),\end{aligned}$$

which can be written as $A(L)B(L)$, where

$$A(L) = -\frac{1}{2}\frac{\sqrt{(n-k-L)(n-k+2)}}{\sqrt{2L}} < 0$$

and

$$
\begin{aligned}
B(L) =& \frac{1}{(n-k-L)} \log\left(\frac{n-k-L}{n-k}\right) - \frac{L}{(n-k-L)^2(n-k)} \\
&+ \frac{\alpha(n-2)L}{(n-k-L)(n-k-L-2)(n-k-2)} \\
&+ \frac{1}{L}\log\left(\frac{n-k-L}{n-k}\right) - \frac{1}{(n-k-L)(n-k)} \\
&+ \frac{\alpha(n-2)}{(n-k-L-2)(n-k-2)} + \frac{2}{n-k-L} \\
&+ \frac{2}{(n-k-L)(n-k)} + \frac{2L}{(n-k-L)^2(n-k)} \\
&- \frac{2\alpha(n-2)}{(n-k-L-2)(n-k-2)} - \frac{2\alpha(n-2)L}{(n-k-L-2)^2(n-k-2)}.
\end{aligned}
$$

Furthermore,

$$
\begin{aligned}
B(L) =& \frac{n-k}{L(n-k-L)} \log\left(\frac{n-k-L}{n-k}\right) + \frac{2}{n-k-L} + \frac{1}{(n-k-L)(n-k)} \\
&+ \frac{L}{(n-k-L)^2(n-k)} - \frac{\alpha(n-2)}{(n-k-L-2)(n-k-2)} \\
&+ \frac{\alpha(n-2)}{(n-k-L-2)(n-k-2)} - \frac{2\alpha(n-2)L}{(n-k-L-2)^2(n-k-2)} \\
<& \frac{n-k}{L(n-k-L)} \log\left(\frac{n-k-L}{n-k}\right) + \frac{2}{n-k-L} + \frac{1}{(n-k-L)^2} \\
&- \frac{\alpha(n-2)}{(n-k-L-2)(n-k-2)} \\
&- \frac{\alpha(n-2)L}{(n-k-L)(n-k-L-2)(n-k-2).}
\end{aligned}
$$

Using the fact that

$$
\begin{aligned}
&\frac{n-k}{L(n-k-L)} \log\left(\frac{n-k-L}{n-k}\right) + \frac{2}{n-k-L} - \frac{\alpha(n-2)}{(n-k-L-2)(n-k-2)} \\
&= -\frac{n-k}{L(n-k-L)} \sum_{j=2}^{\infty} \frac{1}{j}\left(\frac{L}{n-k}\right)^j + \frac{1}{n-k-L} \\
&\quad - \frac{\alpha(n-2)}{(n-k-L-2)(n-k-2)} \\
&< -\frac{n-k}{L(n-k-L)} \sum_{j=2}^{\infty} \frac{1}{j}\left(\frac{L}{n-k}\right)^j + \frac{1}{n-k-L}\left(1 - \alpha\frac{n-2}{n-k-2}\right) \\
&< 0
\end{aligned}
$$

and that

$$\begin{aligned}
&\frac{1}{(n-k-L)^2} - \frac{\alpha(n-2)L}{(n-k-L)(n-k-L-2)(n-k-2)} \\
&\quad < \frac{1}{(n-k-L)^2}\left(1 - \alpha L\frac{n-2}{n-k-2}\right) \\
&\quad < 0,
\end{aligned}$$

we have $B(L) < 0$. Hence the signal-to-noise ratio increases as $L$ increases.

### *2B.2. Theorem 2.2*

Given the regression model in Eqs. (2.3)–(2.4), for all $n \geq 5$, $0 < L < n-k-2$, and for the overfitting case where $0 < k_* \leq k < n-3$, the signal-to-noise ratio of AICu increases as $L$ increases and is greater than the signal-to-noise ratio of AIC.

**Proof:**

Consider the signal-to-noise ratio of AICu, Eq. (2.19). As $L$ increases,

$$\begin{aligned}
-\frac{L}{(n-k-L)(n-k)}+&\frac{2L(n-1)}{(n-k-2)(n-k-L-2)} \\
&> -\frac{L}{(n-k-L)(n-k)} + \frac{2L(n-1)}{(n-k)(n-k-L)} \\
&= \frac{2Ln-3L)}{(n-k)(n-k-L)} \\
&= \frac{2L(n-2)}{(n-k)(n-k-L)}.
\end{aligned}$$

The signal-to-noise ratio of AICu is greater than

$$\frac{\sqrt{(n-k-L)(n-k+2)}}{\sqrt{2L}}\left(\frac{2L(n-2)}{(n-k)(n-k-L)}\right),$$

which can be written as

$$\frac{(n-2)\sqrt{n-k+2}}{\sqrt{n-k}}\frac{\sqrt{2L}}{\sqrt{n-k-L}}.$$

This increases as $L$ increases. Hence, the signal-to-noise ratio of AICu increases as $L$ increases. Since AIC and AICu have the same noise, comparing signal-to-noise ratios is equivalent to comparing signals.

$$\begin{aligned}
\text{signal of AIC} &= \log\left(\frac{n-k-L}{n-k}\right) - \frac{L}{(n-k-L)(n-k)} + \frac{2L}{n} \\
&< -\frac{L}{(n-k-L)(n-k)} + \frac{2L}{n} \\
&< -\frac{L}{(n-k-L)(n-k)} + \frac{2L(n-1)}{(n-k-2)(n-k-L-2)} \\
&= \text{signal of AICu},
\end{aligned}$$

which completes the proof.

*2B.3. Theorem 2.3*

Given the regression model in Eqs. (2.3)–(2.4), for all $n \geq 5$, $0 < L < n-k-2$, and for the overfitting case where $0 < k_* \leq k < n-3$, the signal-to-noise ratio of FPEu is greater than or equal to the signal-to-noise ratio of FPE.

**Proof:**

Rewrite the signal-to-noise ratio of FPE as

$$\frac{n\sqrt{2L}}{n\sqrt{2L}} \frac{(n-k)\sqrt{L(n-k-L)}}{\sqrt{2(n+k)^2(n-k-L)+8n^2L}},$$

so that the signal-to-noise ratios of FPE and FPEu have common numerators. For all $n$, $k < n$, $L < n-k$, if we compare the denominators we see that

$$(n+k)^2(n-k-L)^3L \leq 4Ln^2(n+k)^2(n-k-L)$$

and

$$\begin{aligned}
&\left[(n+k+L)(n-k)^2 - (n+k)(n-k-L)^2\right]^2 \\
&\qquad = \left[3n^2 - 2nk - k^2 - Ln - kL\right]^2 \\
&\qquad < 16n^4L^2.
\end{aligned}$$

Hence,

$$\begin{aligned}
(n+k)^2(n-k-L)^3L &+ \left[(n+k+L)(n-k)^2 - (n+k)(n-k-L)^2\right]^2 \\
&< 4Ln^2(n+k)^2(n-k-L) + 16n^4L^2,
\end{aligned}$$

which is equivalent to

$$(n+k)^2(n-k-L)^3L + \left[(n+k+L)(n-k)^2 - (n+k)(n-k-L)^2\right]^2 < n^2 2L(2(n+k)^2(n-k-L) + 8n^2L).$$

Taking square roots of both sides, we have

$$\sqrt{(n+k)^2(n-k-L)^3L + \left[(n+k+L)(n-k)^2 - (n+k)(n-k-L)^2\right]^2} < n\sqrt{2L}\sqrt{2(n+k)^2(n-k-L) + 8n^2L}.$$

Taking inverses and multiplying both sides by $\sqrt{2}n(n-k)L\sqrt{n-k-L}$, we have

$$\frac{\sqrt{2}n(n-k)L\sqrt{n-k-L}}{\sqrt{(n+k)^2(n-k-L)^3L + \left[(n+k+L)(n-k)^2 - (n+k)(n-k-L)^2\right]^2}} \geq \frac{n\sqrt{2L}}{n\sqrt{2L}} \frac{(n-k)\sqrt{L(n-k-L)}}{\sqrt{2(n+k)^2(n-k-L) + 8n^2L}},$$

and thus the signal-to-noise of FPEu $\geq$ signal-to-noise of FPE. This completes the proof.

### *2B.4. Theorem 2.4*

Given the regression model in Eqs. (2.3)–(2.4), for all $n \geq 6$, $0 < L < n-k-2$, and for the overfitting case where $0 < k_* \leq k < n-3$, the signal-to-noise ratio of HQc increases as $L$ increases and is greater than the signal-to-noise ratio of HQ.

**Proof:**

Let $\alpha = 2\log\log(n)$ in Theorem 2.1. Then, by applying Theorem 2.1, we obtain that the signal-to-noise ratio of HQc increases as $L$ increases.

HQ and HQc have the same noise, and therefore a comparison of their signals is equivalent to a comparison of their signal-to-noise ratios:

$$\begin{aligned}\text{signal of HQ} &= \log\left(\frac{n-k-L}{n-k}\right) - \frac{L}{(n-k-L)(n-k)} + \frac{2\log\log(n)L}{n}. \\ &< \log\left(\frac{n-k-L}{n-k}\right) - \frac{L}{(n-k-L)(n-k)} + \frac{2\log\log(n)(n-2)L}{(n-k-L-2)(n-k-2)} \\ &= \text{signal of HQc}.\end{aligned}$$

Thus the signal-to-noise ratio of HQc > the signal-to-noise ratio of HQ, which completes the proof.

*2B.5. Theorem 2.5*

Suppose $Y = X_*\beta_* + \varepsilon_*$ is the true model, where $X_* = (X_0, X_1) = (1, \tilde{X}_0, X_1)$, and $\beta_* = (\beta_0', \beta_1')'$. Ignoring the constant, the rows of $X_*$ are independent with the following multivariate normal distribution:

$$\tilde{X}_{*i} = \begin{pmatrix} \tilde{X}_{0i} \\ X_{1i} \end{pmatrix} \sim N\left(\begin{pmatrix} 0 \\ 0 \end{pmatrix}, \begin{pmatrix} \Sigma_{00} & \Sigma_{10} \\ \Sigma_{01} & \Sigma_{11} \end{pmatrix}\right).$$

If the model $Y = X_0\beta_0$ is fit to the data, then the noncentrality parameter, $\lambda$, is random, and

$$\lambda \sim \left(\beta_1'\Sigma_{11\cdot 0}\beta_1/\sigma_*^2\right)\chi^2_{n-k_0},$$

where $k_0 = \text{rank}(X_0)$, $\Sigma_{11\cdot 0} = \Sigma_{11} - \Sigma_{01}'\Sigma_{00}^{-1}\Sigma_{01}$ and $\Sigma_{11} = var[\tilde{X}_{1i}]$, $\Sigma_{00} = var[X_{0i}]$, and $\Sigma_{01} = cov[\tilde{X}_{0i}, X_{1i}]$.

**Proof:**

An underfitted model results in the noncentrality parameter

$$\lambda = E_*[Y'](I - H_0)E_*[Y]/\sigma_*^2,$$

where

$$H_0 = X_0(X_0'X_0)^{-1}X_0'.$$

Conditioning on $X_*$, we have

$$\lambda|X_* = \beta_1'X_1'(I - H_0)X_1\beta_1/\sigma_*^2.$$

Under the above partition of $X_{*i}$, $X_{1i}|\tilde{X}_{0i} \sim N(0, \Sigma_{11\cdot 0})$. $X_1\beta_1|X_0$ has the same distribution as $X_1\beta_1|\tilde{X}_0$. Since the rows of $\tilde{X}_*$ are independent, $X_1\beta_1|X_0 \sim N(0, \beta_1'\Sigma_{11\cdot 0}\beta_1 I)$. It follows that

$$\lambda|X_0 = \beta_1'X_1'(I - H_0)X_1\beta_1/\sigma_*^2|X_0 \sim \frac{\beta_1'\Sigma_{11\cdot 0}\beta_1}{\sigma_*^2}\chi^2_{n-k_0},$$

which does not depend on $X_0$. Unconditionally,

$$\lambda = \beta_1'X_1'(I - H_0)X_1\beta_1/\sigma_*^2 \sim \frac{\beta_1'\Sigma_{11\cdot 0}\beta_1}{\sigma_*^2}\chi^2_{n-k_0},$$

which completes the proof.

*2B.6. Theorem 2.6*

Suppose $Y = X_*\beta_* + \varepsilon_*$ is the true model, where $X_* = (X_0, X_1) = (1, \tilde{X}_0, X_1)$, and $\beta_* = (\beta_0', \beta_1')'$. Ignoring the constant, the rows of $X_*$ are independent with the following multivariate normal distribution:

$$\tilde{X}_{*i} = \begin{pmatrix} \tilde{X}_{0i} \\ X_{1i} \end{pmatrix} \sim N\left(\begin{pmatrix} 0 \\ 0 \end{pmatrix}, \begin{pmatrix} \Sigma_{00} & \Sigma_{10} \\ \Sigma_{01} & \Sigma_{11} \end{pmatrix}\right).$$

If the model $Y = X_0\beta_0$ is fit to the data, then

$$\frac{\text{SSE}_{k_0}}{\sigma_*^2} \sim \left(1 + \frac{\beta_1'\Sigma_{11\cdot0}\beta_1}{\sigma_*^2}\right)\chi^2_{n-k_0},$$

where $k_0 = \text{rank}(X_0)$, $\Sigma_{11\cdot0} = \Sigma_{11} - \Sigma_{01}'\Sigma_{00}^{-1}\Sigma_{01}$ and $\Sigma_{11} = var[\tilde{X}_{1i}]$, $\Sigma_{00} = var[X_{0i}]$, and $\Sigma_{01} = cov[\tilde{X}_{0i}, X_{1i}]$.

**Proof:**

The true model can be written as

$$\begin{aligned} Y &= X_0\beta_0 + X_1\beta_1 + \varepsilon_* \\ &= X_0\beta_0 + \tilde{\varepsilon}_*. \end{aligned}$$

If we condition on $X_0$, then $\varepsilon_* \sim N(0, \sigma_*^2 I)$ and $X_1\beta_1|X_0 \sim N(0, \beta_1'\Sigma_{11\cdot0}\beta_1 I)$, and both are independent. Therefore, $\tilde{\varepsilon}_* \sim N(0, (\sigma_*^2 + \beta_1'\Sigma_{11\cdot0}\beta_1)I)$. Hence,

$$\text{SSE}_{k_0} \sim (\sigma_*^2 + \beta_1'\Sigma_{11\cdot0}\beta_1)\chi^2_{n-k_0}$$

and

$$\frac{\text{SSE}_{k_0}}{\sigma_*^2} \sim \left(1 + \frac{\beta_1'\Sigma_{11\cdot0}\beta_1}{\sigma_*^2}\right)\chi^2_{n-k_0},$$

which completes the proof.

## Appendix 2C. Small-sample and Asymptotic Properties

### *2C.1. Small-sample Probabilities of Overfitting*

#### *2C.1.1. AICc*

AICc overfits if $\text{AICc}_{k_*+L} < \text{AICc}_{k_*}$. For finite $n$, the probability that AIC

prefers the overfitted model $k_* + L$ is

$$
\begin{aligned}
&P\{\text{AICc}_{k_*+L} < \text{AICc}_{k_*}\} \\
&= P\left\{\log\left(\hat{\sigma}^2_{k_*+L}\right) + \frac{n+k_*+L}{n-k_*-L-2} < \log\left(\hat{\sigma}^2_{k_*}\right) + \frac{n+k_*}{n-k_*-2}\right\} \\
&= P\left\{\log\left(\frac{\text{SSE}_{k_*+L}}{\text{SSE}_{k_*}}\right) < \frac{n+k_*}{n-k_*-2} - \frac{n+k_*+L}{n-k_*-L-2}\right\} \\
&= P\left\{\frac{\chi^2_L}{\chi^2_{n-k_*-L}} > \exp\left(\frac{n+k_*+L}{n-k_*-L-2} - \frac{n+k_*}{n-k_*-2}\right) - 1\right\} \\
&= P\left\{F_{L,n-k_*-L} > \frac{n-k_*-L}{L} \times \right. \\
&\qquad \left.\left(\exp\left(\frac{2L(n-1)}{(n-k_*-L-2)(n-k_*-2)}\right) - 1\right)\right\}.
\end{aligned}
$$

*2C.1.2. AICu*

AICu overfits if $\text{AICu}_{k_*+L} < \text{AICu}_{k_*}$. For finite $n$, the probability that AICu prefers the overfitted model $k_* + L$ is

$$
\begin{aligned}
&P\{\text{AICu}_{k_*+L} < \text{AICu}_{k_*}\} \\
&= P\left\{\log\left(s^2_{k_*+L}\right) + \frac{n+k_*+L}{n-k_*-L-2} < \log\left(s^2_{k_*}\right) + \frac{n+k_*}{n-k_*-2}\right\} \\
&= P\left\{\log\left(\frac{\text{SSE}_{k_*+L}}{\text{SSE}_{k_*}}\right) < -\log\left(\frac{n-k_*}{n-k_*-L}\right)\right. \\
&\qquad \left. + \frac{n+k_*}{n-k_*-2} - \frac{n+k_*+L}{n-k_*-L-2}\right\} \\
&= P\left\{\frac{\chi^2_L}{\chi^2_{n-k_*-L}} > \exp\left(\log\left(\frac{n-k_*}{n-k_*-L}\right)\right.\right. \\
&\qquad \left.\left. + \frac{n+k_*+L}{n-k_*-L-2} - \frac{n+k_*}{n-k_*-2}\right) - 1\right\} \\
&= P\left\{F_{L,n-k_*-L} > \frac{n-k_*-L}{L} \times \right. \\
&\qquad \left.\left(\frac{n-k_*}{n-k_*-L}\exp\left(\frac{2L(n-1)}{(n-k_*-L-2)(n-k_*-2)}\right) - 1\right)\right\}.
\end{aligned}
$$

### *2C.1.3. FPE*

FPE overfits if $\text{FPE}_{k_*+L} < \text{FPE}_{k_*}$. For finite $n$, the probability that FPE prefers the overfitted model $k_* + L$ is

$$\begin{aligned}
&P\{\text{FPE}_{k_*+L} < \text{FPE}_{k_*}\} \\
&= P\left\{\frac{n+k_*+L}{n-k_*-L}\hat{\sigma}^2_{k_*+L} < \frac{n+k_*}{n-k_*}\hat{\sigma}^2_{k_*}\right\} \\
&= P\left\{\frac{\text{SSE}_{k_*}}{\text{SSE}_{k_*+L}} > \left(\frac{n-k_*}{n+k_*}\right)\left(\frac{n+k_*+L}{n-k_*-L}\right)\right\} \\
&= P\left\{\frac{\chi^2_L}{\chi^2_{n-k_*-L}} > \left(\frac{n-k_*}{n+k_*}\right)\left(\frac{n+k_*+L}{n-k_*-L}\right) - 1\right\} \\
&= P\left\{F_{L,n-k_*-L} > \left(\frac{n-k_*-L}{L}\right)\left(\frac{2Ln}{(n-k_*-L)(n+k*)}\right)\right\} \\
&= P\left\{F_{L,n-k_*-L} > \frac{2n}{n+k_*}\right\}.
\end{aligned}$$

### *2C.1.4. FPEu*

FPEu overfits if $\text{FPEu}_{k_*+L} < \text{FPEu}_{k_*}$. For finite $n$, the probability that FPEu prefers the overfitted model $k_* + L$ is

$$\begin{aligned}
&P\{\text{FPEu}_{k_*+L} < \text{FPEu}_{k_*}\} \\
&= P\left\{\frac{n+k_*+L}{n-k_*-L}s^2_{k_*+L} < \frac{n+k_*}{n-k_*}s^2_{k_*}\right\} \\
&= P\left\{\frac{n+k_*+L}{n-k_*-L}\left(\frac{\text{SSE}_{k_*+L}}{n-k_*-L}\right) < \frac{n+k_*}{n-k_*}\left(\frac{\text{SSE}_{k_*}}{n-k_*}\right)\right\} \\
&= P\left\{\frac{\text{SSE}_{k_*}}{\text{SSE}_{k_*+L}} > \left(\frac{(n-k_*)^2}{n+k_*}\right)\left(\frac{n+k_*+L}{(n-k_*-L)^2}\right)\right\} \\
&= P\left\{\frac{\chi^2_L}{\chi^2_{n-k_*-L}} > \left(\frac{(n-k_*)^2}{n+k_*}\right)\left(\frac{n+k_*+L}{(n-k_*-L)^2}\right) - 1\right\} \\
&= P\left\{F_{L,n-k_*-L} > \frac{3n^2 - n(2k_*+L) - k_*(k_*+L)}{(n-k_*-L)(n+k_*)}\right\}.
\end{aligned}$$

### *2C.1.5. Cp*

Cp overfits if $\text{Cp}_{k_*+L} < \text{Cp}_{k_*}$. Recall that $K$ represents the number of variables in the largest model with all variables included. For finite $n$, the probability that Cp prefers the overfitted model $k_* + L$ is

$$
\begin{aligned}
&P\{\text{Cp}_{k_*+L} < \text{Cp}_{k_*}\}\\
&= P\left\{(n-K)\frac{\text{SSE}_{k_*+L}}{\text{SSE}_K} - n + 2(k_* + L) < (n-K)\frac{\text{SSE}_{k_*}}{\text{SSE}_K} - n + 2k_*\right\}\\
&= P\left\{(n-K)\frac{\text{SSE}_{k_*+L} - \text{SSE}_{k_*}}{\text{SSE}_K} < -2L\right\}\\
&= P\left\{(n-K)\frac{\text{SSE}_{k_*} - \text{SSE}_{k_*+L}}{\text{SSE}_K} > 2L\right\}\\
&= P\left\{(n-K)\frac{\chi^2_L}{\chi^2_K} > 2L\right\}\\
&= P\left\{F_{L,n-K} > 2\right\}.
\end{aligned}
$$

### 2C.1.6. SIC

SIC overfits if $\text{SIC}_{k_*+L} < \text{SIC}_{k_*}$. For finite $n$, the probability that SIC prefers the overfitted model $k_* + L$ is

$$
\begin{aligned}
&P\{\text{SIC}_{k_*+L} < \text{SIC}_{k_*}\}\\
&= P\left\{\log\left(\hat{\sigma}^2_{k_*+L}\right) + \frac{\log(n)(k_* + L)}{n} < \log\left(\hat{\sigma}^2_{k_*}\right) + \frac{\log(n)k_*}{n}\right\}\\
&= P\left\{\log\left(\frac{\text{SSE}_{k_*+L}}{\text{SSE}_{k_*}}\right) < \frac{\log(n)\,k_*}{n} - \frac{\log(n)\,(k_* + L)}{n}\right\}\\
&= P\left\{\frac{\chi^2_L}{\chi^2_{n-k_*-L}} > \exp\left(\frac{\log(n)L}{n}\right) - 1\right\}\\
&= P\left\{F_{L,n-k_*-L} > \frac{n - k_* - L}{L}\left(\exp\left(\frac{\log(n)L}{n}\right) - 1\right)\right\}.
\end{aligned}
$$

### 2C.1.7. HQ

HQ overfits if $\text{HQ}_{k_*+L} < \text{HQ}_{k_*}$. For finite $n$, the probability that HQ

prefers the overfitted model $k_* + L$ is

$$
\begin{aligned}
&P\{\mathrm{HQ}_{k_*+L} < \mathrm{HQ}_{k_*}\}\\
&= P\left\{\log\left(\hat{\sigma}^2_{k_*+L}\right) + \frac{2\log\log(n)(k_*+L)}{n} < \log\left(\hat{\sigma}^2_{k_*}\right) + \frac{2\log\log(n)k_*}{n}\right\}\\
&= P\left\{\log\left(\frac{\mathrm{SSE}_{k_*+L}}{\mathrm{SSE}_{k_*}}\right) < \frac{2\log\log(n)\,k_*}{n} - \frac{2\log\log(n)\,(k_*+L)}{n}\right\}\\
&= P\left\{\frac{\chi^2_L}{\chi^2_{n-k_*-L}} > \exp\left(\frac{2\log\log(n)L}{n}\right) - 1\right\}\\
&= P\left\{F_{L,n-k_*-L} > \frac{n-k_*-L}{L}\left(\exp\left(\frac{2\log\log(n)L}{n}\right) - 1\right)\right\}.
\end{aligned}
$$

### *2C.1.8. HQc*

HQc overfits if $\mathrm{HQc}_{k_*+L} < \mathrm{HQc}_{k_*}$. For finite $n$, the probability that HQc prefers the overfitted model $k_* + L$ is

$$
\begin{aligned}
&P\{\mathrm{HQc}_{k_*+L} < \mathrm{HQc}_{k_*}\}\\
&= P\left\{\log\left(\hat{\sigma}^2_{k_*+L}\right) + \frac{2\log\log(n)(k_*+L)}{n-k_*-L-2} < \log\left(\hat{\sigma}^2_{k_*}\right) + \frac{2\log\log(n)k_*}{n-k_*-2}\right\}\\
&= P\left\{\log\left(\frac{\mathrm{SSE}_{k_*+L}}{\mathrm{SSE}_{k_*}}\right) < \frac{2\log\log(n)k_*}{n-k_*-2} - \frac{2\log\log(n)\,(k_*+L)}{n-k_*-L-2}\right\}\\
&= P\left\{\frac{\chi^2_L}{\chi^2_{n-k_*-L}} > \exp\left(\frac{2\log\log(n)(n-2)L}{(n-k_*-L-2)(n-k_*-2)}\right) - 1\right\}\\
&= P\left\{F_{L,n-k_*-L} > \frac{n-k_*-L}{L}\times\right.\\
&\qquad\left.\left(\exp\left(\frac{2\log\log(n)(n-2)L}{(n-k_*-L-2)(n-k_*-2)}\right) - 1\right)\right\}.
\end{aligned}
$$

### *2C.1.9. General Case*

Consider a model selection criterion, say $\mathrm{MSC}_k$, of the form $\log(\mathrm{SSE}) + \alpha(n,k)$ where $\alpha(n,k)$ is the penalty function of $\mathrm{MSC}_k$. MSC overfits if $\mathrm{MSC}_{k_*+L} < \mathrm{MSC}_{k_*}$. For finite $n$, the probability that MSC prefers the overfitted model

$k_* + L$ is

$$
\begin{aligned}
&P\{\mathrm{MSC}_{k_*+L} < \mathrm{MSC}_{k_*}\} \\
&= P\Big\{\log(\mathrm{SSE}_{k_*+L}) + \alpha(n, k_* + L) < \log(\mathrm{SSE}_{k_*+L}) + \alpha(n, k_* + L)\Big\} \\
&\Rightarrow P\Big\{\log\left(\frac{\mathrm{SSE}_{k_*+L}}{\mathrm{SSE}_{k_*}}\right) < \alpha(n, k_*) - \alpha(n, k_* + L)\Big\} \\
&= P\Big\{\frac{\chi^2_L}{\chi^2_{n-k_*-L}} > \exp\big(\alpha(n, k_* + L) - \alpha(n, k_*)\big) - 1\Big\} \\
&= P\Big\{F_{L,n-k_*-L} > \frac{n - k_* - L}{L}\Big(\exp\Big(\alpha(n, k_* + L) - \alpha(n, k_*)\Big) - 1\Big)\Big\}.
\end{aligned}
$$

*2C.2. Asymptotic Probabilities of Overfitting*

Recall that for $n$, and $\lim_{n\to\infty} f_n \to f$ then

$$\lim_{n\to\infty} P\{F_{L,n-k_*-L} > f_n\} \to P\{\chi^2_L > fL\}.$$

*2C.2.1. AICc*

$$
\begin{aligned}
&\frac{n - k_* - L}{L}\left(\exp\left(\frac{2L(n-1)}{(n - k_* - L - 2)(n - k_* - 2)}\right) - 1\right) \\
&\qquad = \frac{n - k_* - L}{L}\left(\frac{2L\,(n-1)}{(n - k_* - L - 2)\,(n - k_* - 2)} + O\left(\frac{1}{n^2}\right)\right) \\
&\qquad = \frac{n}{L}\frac{2L}{n} \\
&\qquad \to 2.
\end{aligned}
$$

Therefore the asymptotic probability that AICc prefers the overfitted model $k_* + L$ is $P\{\chi^2_L > 2L\}$.

*2C.2.2. AICu*

$$\begin{aligned}
&\frac{n-k_*-L}{L}\left(\frac{n-k_*}{n-k_*-L}\exp\left(\frac{2L(n-1)}{(n-k_*-L-2)(n-k_*-2)}\right)-1\right)\\
&=\frac{n-k_*-L}{L}\left(\frac{n-k_*}{n-k_*-L}\times\right.\\
&\left.\left(1+\frac{2L(n-1)}{(n-k_*-L-2)(n-k_*-2)}+O\left(\frac{1}{n^2}\right)\right)-1\right)\\
&=\frac{n-k_*-L}{L}\left(\frac{n-k_*}{n-k_*-L}-1+\frac{n-k_*}{n-k_*-L}\times\right.\\
&\left.\left(\frac{2L(n-1)}{(n-k_*-L-2)(n-k_*-2)}+O\left(\frac{1}{n^2}\right)\right)\right)\\
&=\frac{n-k_*-L}{L}\left(\frac{L}{n-k_*-L}+\frac{n-k_*}{n-k_*-L}\times\right.\\
&\left.\left(\frac{2L(n-1)}{(n-k_*-L-2)(n-k_*-2)}+O\left(\frac{1}{n^2}\right)\right)\right)\\
&=\frac{n}{L}\left(\frac{L}{n}+\frac{2L}{n}\right)\\
&\rightarrow 3.
\end{aligned}$$

Therefore the asymptotic probability that AICu prefers the overfitted model $k_*+L$ is $P\{\chi^2_L > 3L\}$.

*2C.2.3. FPEu*

$$\begin{aligned}
\frac{3n^2-n(2k_*+L)-k_*(k_*+L)}{(n+k_*)(n-k_*-L)} &= \frac{3n^2}{(n+k_*)(n-k_*-L)}+O\left(\frac{1}{n}\right)\\
&\rightarrow 3.
\end{aligned}$$

Therefore the asymptotic probability that FPEu prefers the overfitted model $k_*+L$ is $P\{\chi^2_L > 3L\}$.

*2C.2.4. Cp*

$$F_{L,n-K}\rightarrow\frac{\chi^2_L}{L}.$$

Therefore the asymptotic probability that Cp prefers the overfitted model $k_*+L$ is $P\{\chi^2_L > 2L\}$.

*2C.2.5. HQ*

Expanding

$$\begin{aligned}
&\frac{n-k_*-L}{L}\left(\exp\left(\frac{2\log\log(n)L)}{n}\right)-1\right)\\
&\qquad=\frac{n-k_*-L}{L}\left(\frac{2\log\log(n)L)}{n}+O\left(\frac{1}{n^2}\right)\right)\\
&\qquad=\frac{n}{L}\frac{2\log\log(n)L}{n}\\
&\qquad\to\infty.
\end{aligned}$$

Therefore the asymptotic probability that HQ prefers the overfitted model $k_*+L$ is 0.

*2C.2.6. HQc*

Expanding

$$\begin{aligned}
&\frac{n-k_*-L}{L}\left(\exp\left(\frac{2\log\log(n)(n-2)L}{(n-k_*-L-2)(n-k_*-2)}\right)-1\right)\\
&\qquad=\frac{n-k_*-L}{L}\left(\frac{2\log\log(n)(n-2)L}{(n-k_*-L-2)\,(n-k_*-2)}+O\left(\frac{1}{n^2}\right)\right)\\
&\qquad=\frac{n}{L}\frac{2\log\log(n)L}{n}\\
&\qquad\to\infty.
\end{aligned}$$

Therefore the asymptotic probability that HQc prefers the overfitted model $k_*+L$ is 0.

## 2C.3. Asymptotic Signal-to-noise Ratios

### 2C.3.1. AICc

The asymptotic signal-to-noise ratio is

$$\lim_{n\to\infty} \frac{\sqrt{(n-k_*-L)(n-k_*+2)}}{\sqrt{2L}} \left( \log\left(1-\frac{L}{n-k_*}\right) - \frac{L}{(n-k_*-L)(n-k_*)} + \frac{2L(n-1)}{(n-k_*-L-2)(n-k_*-2)} \right)$$
$$= \lim_{n\to\infty} \frac{n}{\sqrt{2L}} \left( \log\left(1-\frac{L}{n}\right) - \frac{L}{n^2} + \frac{2L}{n} \right)$$
$$= \sqrt{\frac{L}{2}}.$$

### 2C.3.2. AICu

The asymptotic signal-to-noise ratio is

$$\lim_{n\to\infty} \frac{\sqrt{(n-k_*-L)(n-k_*+2)}}{\sqrt{2L}} \left( -\frac{L}{(n-k_*-L)(n-k_*)} + \frac{2L(n-1)}{(n-k_*-L-2)(n-k_*-2)} \right)$$
$$= \lim_{n\to\infty} \frac{n}{\sqrt{2L}} \left( -\frac{L}{n^2} + \frac{2L}{n} \right)$$
$$= \frac{2L}{\sqrt{2L}}$$
$$= \sqrt{2L}.$$

### 2C.3.3. FPEu

The asymptotic signal-to-noise ratio is

$$\lim_{n\to\infty} \frac{\sqrt{2}n(n-k_*)L\sqrt{n-k_*-L}}{\sqrt{(n+k_*)^2(n-k_*-L)^3L + \left((n+k_*)(n-k_*-L)^2 - (n+k_*+L)(n-k_*)^2\right)^2}}$$
$$= \lim_{n\to\infty} \frac{\sqrt{n^5}\sqrt{2}L}{\sqrt{n^5 L}}$$
$$= \sqrt{2L}.$$

### *2C.3.4. Cp*

The asymptotic signal-to-noise ratio is

$$\lim_{n\to\infty} \frac{(n-K-4)^{3/2}}{n-K} \frac{L}{\sqrt{2L(n-K-2)+2L^2}}$$
$$= \lim_{n\to\infty} \frac{n^{3/2}}{n} \frac{L}{\sqrt{2Ln}}$$
$$= \sqrt{\frac{L}{2}}.$$

### *2C.3.5. HQ*

The asymptotic signal-to-noise ratio is

$$\lim_{n\to\infty} \frac{\sqrt{(n-k_*-L)(n-k_*+2)}}{\sqrt{2L}} \left(\log\left(1-\frac{L}{n-k_*}\right) - \frac{L}{(n-k_*-L)(n-k_*)} + \frac{2\log\log(n)L}{n}\right)$$
$$= \lim_{n\to\infty} \frac{n}{\sqrt{2L}} \left(\log\left(1-\frac{L}{n}\right) - \frac{L}{n^2} + \frac{2\log\log(n)L}{n}\right)$$
$$= \infty.$$

### *2C.3.6. HQc*

The asymptotic signal-to-noise ratio is

$$\lim_{n\to\infty} \frac{\sqrt{(n-k_*-L)(n-k_*+2)}}{\sqrt{2L}} \left(\log\left(1-\frac{L}{n-k_*}\right) - \frac{L}{(n-k_*-L)(n-k_*)} + \frac{2\log\log(n)(n-2)L}{(n-k_*-L-2)(n-k_*-2)}\right)$$
$$= \lim_{n\to\infty} \frac{n}{\sqrt{2L}} \left(\log\left(1-\frac{L}{n}\right) - \frac{L}{n^2} + \frac{2\log\log(n)L}{n}\right)$$
$$= \infty.$$

## Appendix 2D. Moments of the Noncentral $\chi^2$

Let $Z \sim \chi^2(m, \lambda)$.

$$\begin{aligned}
E[Z] &= \sum_{r=0}^{\infty} e^{-\lambda/2} \frac{(\lambda/2)^r}{r!} E\left[\chi^2(m+2r|z)\right] \\
&= \sum_{r=0}^{\infty} e^{-\lambda/2} \frac{(\lambda/2)^r}{r!} E\left[m+2r\right] \\
&= m + 2E\left[\text{Poisson}(\lambda/2)\right] \\
&= m + \lambda.
\end{aligned}$$

$$\begin{aligned}
E[Z^2] &= \sum_{r=0}^{\infty} e^{-\lambda/2} \frac{(\lambda/2)^r}{r!} E\left[(\chi^2(m+2r|z))^2\right] \\
&= \sum_{r=0}^{\infty} e^{-\lambda/2} \frac{(\lambda/2)^r}{r!} E\left[2(m+2r) + (m+2r)^2\right] \\
&= \sum_{r=0}^{\infty} e^{-\lambda/2} \frac{(\lambda/2)^r}{r!} E\left[2m + 4r + m^2 + 4mr + 4r^2\right] \\
&= 2m + m^2 + 4(m+1)E\left[\text{Poisson}(\lambda/2)\right] + 4E\left[\text{Poisson}^2(\lambda/2)\right] \\
&= 2m + m^2 + 4(m+1)(\lambda/2) + 4(\lambda/2 + (\lambda/2)^2) \\
&= 2m + m^2 + 2(m+1)\lambda + 2\lambda + \lambda^2.
\end{aligned}$$

$$\begin{aligned}
var[Z] &= 2m + m^2 + 2(m+1)\lambda + 2\lambda + \lambda^2 - (m+\lambda^2)^2 \\
&= 2(m + 2\lambda).
\end{aligned}$$

$$E\left[\frac{1}{Z}\right] = \sum_{r=0}^{\infty} e^{-\lambda/2} \frac{(\lambda/2)^r}{r!} \frac{1}{m-2+2r}.$$

$$E[\log(Z)] = \log(2) + \sum_{r=0}^{\infty} e^{-\lambda/2} \frac{(\lambda/2)^r}{r!} \psi\left(\frac{m}{2} + r\right).$$

Chapter 3

# The Univariate Autoregressive Model

The autoregressive model is one of the most popular for time series data. In this Chapter we describe the univariate autoregressive model, its notation, and its relationship to multiple regression, and we review the derivations of the criteria from Chapter 2 with respect to the autoregressive model. Two of the criteria, FPE (Akaike, 1973) and the Hannan and Quinn criterion (HQ, Hannan and Quinn, 1979) were originally derived for the autoregressive model. We present an overview of their derivations. We then derive small-sample and asymptotic signal-to-noise ratios and relate them to overfitting. Finally, we explore the behavior of the selection criteria via simulation studies using two special case models under both an autoregressive and moving average framework. For the moving average MA(1) case we are interested to find out how a misspecification of the model (truly MA(1) but fitted under AR) affects the performance of the criteria. Also, because the MA(1) model can be written as an AR model with an infinite number of autoregressive parameters, it provides us with the opportunity to study models of truly infinite order, and to evaluate the performance of the selection criteria when the true model does not belong to the set of candidate models. Finally, multistep prediction AR models are discussed and a simulation study illustrating performance in terms of mean squared prediction error is presented.

## 3.1. Model Description

### 3.1.1. Autoregressive Models

We first define the *general autoregressive model* of order $p$, denoted AR($p$), as

$$y_t = \phi_1 y_{t-1} + \cdots + \phi_p y_{t-p} + w_t, \quad t = p+1, \ldots, n, \tag{3.1}$$

where

$$w_t \text{ are } i.i.d.\ N(0, \sigma^2), \tag{3.2}$$

and $y_1, \ldots, y_n$ are an observed series of data. However, although we have $n$

observations, because we use $y_{t-p}$ to model $y_t$ the effective series length is $T = n - p$. No intercept is included in the model.

Reducing the number of observations by $p$ is just one of the possible methods for AR($p$) modeling. Alternatively, Priestley (1981) suggests replacing past observations $y_0, y_{-1}, \ldots$ with 0 when $\{y_t\}$ is a zero mean time series. Box and Jenkins suggest backcasting to estimate the past observations. Each of these methods has its own merits; however, we prefer losing the first $p$ observations primarily because these other methods require extra assumptions, such as 0 mean, stationarity, or that the backcasts are good estimates for $y_0, y_{-1}, \ldots$. Another advantage of the least squares method is that it can be applied to some nonstationary AR models as well.

The assumption in Eq. (3.2) is identical to that made for the multiple regression model in Chapter 2, Eq. (2.2). For now, assume that the candidate model is an AR($p$) model. If there is a finite true order, $p_*$, then we further assume the true AR($p_*$) model belongs to the set of candidate models of orders 1 to $P$, where $P \geq p*$. Later, in the simulation studies in Sections 3.5 and 3.6, we will examine both the case where we assume such a true model exists and the case where an infinite order true model is assumed. In order to fit a candidate AR($p$) model we first define the observation vector $Y$ as $Y = (y_{p+1}, \cdots, y_n)'$, and then obtain a regression model from Eq. (3.1) and Eq. (3.2) by conditioning on the past and forming the design matrix, $X$, with elements

$$(X)_{t,j} = x_{t,j} = y_{t-j} \text{ for } j = 1, \ldots, p \text{ and } t = p+1, \ldots, n.$$

Since $X$ is formed by conditioning on past values of $Y$, least squares estimation of parameters in this context is often referred to as *conditional least squares.* By conditioning on the past we are assuming $X$ is known, and has dimension $(n-p) \times p$. Furthermore, we assume $X$ is of full rank $p$.

The conditional least squares parameter estimate of $\phi = (\phi_1, \ldots, \phi_p)'$ is

$$\hat{\phi} = (X'X)^{-1}X'Y.$$

This is also the conditional maximum likelihood estimate of $\phi$ (see Priestley, 1981, p. 348). The unbiased and the maximum likelihood estimates of $\sigma^2$ are given below:

$$s_p^2 = \frac{\text{SSE}_p}{T-p} \tag{3.3}$$

and

$$\hat{\sigma}_p^2 = \frac{\text{SSE}_p}{T}, \tag{3.4}$$

where $\text{SSE}_p = \| Y - \hat{Y} \|^2$, and $\hat{Y} = X\hat{\phi}$. We will refer to a candidate model by its order $p$, and whenever possible, we use $T$ to refer to the effective sample size $n - p$. However, when comparing two models, we will use $n - p$ for the reduced model and $n - p - L$ for the full model.

We next define the *true autoregressive model* to be

$$y_t = \mu_{*t} + w_{*t}, \quad t = 1, \ldots, n \tag{3.5}$$

with

$$w_{*t} \ i.i.d. \ N(0, \sigma_*^2). \tag{3.6}$$

If the true model is an autoregressive model, then the true model becomes $y_t = \phi_1 y_{t-1} + \cdots + \phi_{p_*} y_{t-p_*} + w_{*t}$ with true order $p_*$. Using this model, we can also define an overfitted and an underfitted model. Underfitting occurs when an AR($p$) candidate model is fitted with $p < p_*$, and overfitting occurs when $p > p_*$. If the true model does not belong to the set of candidate models, then the definitions of underfitting and overfitting depend on the discrepancy or distance used. For example, we define $\tilde{p}$ such that the AR($\tilde{p}$) model is closest to the true model. Underfitting in the $L_2$ sense can now be stated as choosing the AR($p$) model where $p < \tilde{p}$, and overfitting in the $L_2$ sense is stated as choosing the AR($p$) model where $p > \tilde{p}$. These definitions can be obtained analogously for the Kullback–Leibler distance. In the next Section we consider the use of K-L and $L_2$ as distance measures under autoregressive models.

### 3.1.2. Distance Measures

Regardless of the mechanism that generated the data, we can compute the $L_2$ or the Kullback–Leibler distance between each AR($p$) model and the true model. The $L_2$ distance between the true model and the candidate model AR($p$) is defined as

$$L_2 = \frac{1}{T} \sum_{t=p+1}^{n} (\hat{y}_t - \mu_{*t})^2, \tag{3.7}$$

where $\hat{y}_t = \sum_{j=1}^{p} \hat{\phi}_j y_{t-j}$, and it is the predicted value for $y_t$ computed from the candidate model. Notice that $L_2$ is scaled by the usable sample size, expressing it as a rate or average distance and allowing comparisons to be made between candidate models that have different effective sample sizes.

For the Kullback–Leibler information, when comparing the true model Eqs. (3.5)–(3.6) to the AR($p$) candidate model we start with the assumption that $\mu_{*1}, \ldots, \mu_{*n}$ are known and that the usable sample size is determined by the candidate model. Under the normality assumption, the density $f_*$ of the true model is

$$f_* = (2\pi)^{-T/2}(\sigma_*^2)^{-T/2} \exp\left(-\frac{1}{2\sigma_*^2}\sum_{t=p+1}^{n}(y_t - \mu_{*t})^2\right),$$

and the likelihood function $f_p$ of the candidate model is

$$f_p = (2\pi)^{-T/2}(\sigma_p^2)^{-T/2} \exp\left(-\frac{1}{2\sigma_p^2}\sum_{t=p+1}^{n}(y_t - \phi_1 y_{t-1} - \cdots - \phi_p y_{t-p})^2\right).$$

The Kullback–Leibler discrepancy, which compares these two density functions, is defined as

$$\text{K-L} = \frac{2}{T}E_*\left[\log\left(\frac{f_*}{f_p}\right)\right],$$

where $E_*$ denotes expectation under the true model. The K-L is also scaled by $2/T$ to express it as a rate. Next, taking logs we have

$$\log(f_*) = -\frac{T}{2}\log(2\pi) - \frac{T}{2}\log(\sigma_*^2) - \frac{1}{2\sigma_*^2}\sum_{t=p+1}^{n}(y_t - \mu_{*t})^2$$

and

$$\log(f_p) = -\frac{T}{2}\log(2\pi) - \frac{T}{2}\log(\sigma^2) - \frac{1}{2\sigma^2}\sum_{t=p+1}^{n}(y_t - \phi_1 y_{t-1} - \cdots - \phi_p y_{t-p})^2.$$

Substituting and simplifying, we obtain

$$\text{K-L} = \log\left(\frac{\sigma^2}{\sigma_*^2}\right) + \frac{2}{T}E_*\left[-\frac{1}{2\sigma_*^2}\sum_{t=p+1}^{n}(y_t - \mu_{*t})^2 - \frac{1}{2\sigma^2}\sum_{t=p+1}^{n}(y_t - \phi_1 y_{t-1} - \cdots - \phi_p y_{t-p})^2\right].$$

Taking expectations with respect to the true model yields

$$\text{K-L} = \log\left(\frac{\sigma^2}{\sigma_*^2}\right) + \frac{\sigma_*^2}{\sigma^2} + \frac{\frac{1}{T}\sum_{t=p+1}^{n}(\mu_{*t} - \phi_1 y_{t-1} - \cdots - \phi_p y_{t-p})^2}{\sigma^2} - 1.$$

Finally, if we replace $\phi$ with $\hat{\phi}$ and $\sigma^2$ with $\hat{\sigma}_p^2$, and use $L_2$ in Eq. (3.7) we obtain the Kullback–Leibler discrepancy for autoregressive models,

$$\text{K-L} = \log\left(\frac{\hat{\sigma}_p^2}{\sigma_*^2}\right) + \frac{\sigma_*^2}{\hat{\sigma}_p^2} + \frac{L_2}{\hat{\sigma}_p^2} - 1. \tag{3.8}$$

## 3.2. Selected Derivations of Model Selection Criteria

Derivations for autoregressive model selection criteria usually parallel those for multiple regression models with the exception that for autoregressive models the sample size is a function of the order of the candidate model. This is the case for the signal-to-noise corrected variants AICu, HQc, and FPEu, which will be presented without proof since the proofs are very similar to those given in Chapter 2. More detail will be given for the rest of the criteria, in particular FPE and HQ. The motivation for their derivations is of special interest because they were originally proposed for use in the autoregressive setting.

### *3.2.1. AIC*

AIC can be adapted quite simply to the AR model using the $\hat{\sigma}_p^2$ in Eq. (3.4) as follows:

$$\text{AIC} = \log(\hat{\sigma}_p^2) + \frac{2(p+1)}{T}. \tag{3.9}$$

### *3.2.2. AICc*

We recall that Hurvich and Tsai (1989) derived AICc by estimating the expected Kullback–Leibler discrepancy in small-sample autoregressive models of the form Eqs. (3.1)–(3.2), assuming that the true model AR($p_*$) is a member of the set of candidate models. Taking the expectation of K-L by assuming that $p > p_*$, where $p$ is the order of the candidate model, we have

$$\begin{aligned} E_*[\text{K-L}] &= E_*\left[\log\left(\frac{\hat{\sigma}_p^2}{\sigma_*^2}\right) + \frac{\sigma_*^2}{\hat{\sigma}_p^2} + \frac{L_2}{\hat{\sigma}_p^2} - 1\right] \\ &= E_*[\log(\hat{\sigma}_p^2)] - \log(\sigma_*^2) + \frac{T+p}{T-p-2} - 1, \end{aligned}$$

where $E_*$ denotes expectations under the true model. Note that $\log(\hat{\sigma}_p^2)$ is unbiased for $E_*[\log(\hat{\sigma}_p^2)]$. This leads to

$$\log(\hat{\sigma}_p^2) + \frac{T+p}{T-p-2} - \log(\sigma_*^2) - 1,$$

which is unbiased for $E_*[\text{K-L}]$. Because the constants $-\log(\sigma_*^2) - 1$ play no role in model selection we will ignore them, yielding

$$\text{AICc} = \log(\hat{\sigma}_p^2) + \frac{T+p}{T-p-2}. \tag{3.10}$$

### 3.2.3. AICu

AICu, Eq. (2.18), can be adapted to the autoregressive framework using $s_p^2$ in Eq. (3.3), as follows:

$$\text{AICu} = \log(s_p^2) + \frac{T+p}{T-p-2}. \tag{3.11}$$

### 3.2.4. FPE

The derivation of FPE (Akaike, 1969) begins by considering the observed series $y_1, \ldots, y_n$ from the AR($p$) model given by Eqs. (3.1)–(3.2). Let $\{x_t\}$ be an observed series from another AR($p$) autoregressive model that is independent of $\{y_t\}$ but where $\{x_t\}$ and $\{y_t\}$ have the same model structure. Thus the model is

$$x_t = \phi_1 x_{t-1} + \cdots + \phi_p x_{t-p} + u_t, \quad t = p+1, \ldots, n,$$

where the $u_t$ are *i.i.d.* $N(0, \sigma^2)$. Note that $u_t$ and $w_t$ have the same distribution but are independent of each other. Akaike estimated the mean squared prediction error for predicting the next $\{x_t\}$ observation, $x_{n+1}$, by estimating the parameters from the $\{y_t\}$ data and using these estimated parameters to make the prediction for $x_{n+1}$ using $\{x_t\}$. The prediction is

$$\hat{x}_{n+1} = \hat{\phi}_1 x_n + \cdots + \hat{\phi}_p x_{n-p+1}.$$

Akaike showed that the mean squared prediction error for large $T$ is

$$E\left[(\hat{x}_{n+1} - x_{n+1})^2\right] \doteq \sigma^2\left(1 + \frac{p}{T}\right).$$

Assuming that the true model is AR($p_*$) and $p > p_*$, the expectation of the MLE in Eq. (3.4) is

$$E[\hat{\sigma}^2] = \left(1 - \frac{p}{T}\right)\sigma^2.$$

Hence, $\hat{\sigma}^2/(1 - p/T)$ is unbiased for $\sigma^2$. Substituting this unbiased estimate for $\sigma^2$ leads to an unbiased estimate of the mean squared prediction error, or what Akaike defined as FPE,

$$\text{FPE} = \hat{\sigma}_p^2 \left( \frac{T+p}{T-p} \right). \tag{3.12}$$

The model minimizing FPE will have the minimum mean squared prediction error, or in other words, it should be the best model in terms of predicting future observations.

### 3.2.5. FPEu

FPEu, Eq. (2.20), adapted for autoregressive models using $s_p^2$ in Eq. (3.3) is defined as

$$\text{FPEu} = \frac{T+p}{T-p} s_p^2. \tag{3.13}$$

### 3.2.6. Cp

Mallows's Cp can be adapted to autoregressive models as follows. First, define $J_p$ as

$$J_p = \frac{1}{\sigma_*^2} \sum_{t=p+1}^{n} (X_t\hat{\phi} - X_t\phi)^2.$$

As was the case under regression, Mallows found $E[J_p] = V_p + B_p/\sigma_*^2$, where $V_p$ represents the variance and $B_p$ represents the bias. For autoregressive models, $V_p = p$ and $B_p =$ the noncentrality parameter. It is known that $E[\text{SSE}_p] = (T-p)\sigma_*^2 + B_p$ and that

$$\begin{aligned} E[\text{SSE}_p/\sigma_*^2 - T + 2p] &= p + B_p/\sigma_*^2 \\ &= V_p + B_p/\sigma_*^2 \\ &= E[J_p]. \end{aligned}$$

Hence, $\text{SSE}_p/\sigma_*^2 - T + 2p$ is unbiased for $E[J_p]$. Substituting $\sigma_*^2$ with $s_P^2$, calculated from the largest candidate model under consideration, yields Cp for autoregressive models:

$$\text{Cp} = \frac{\text{SSE}_p}{s_P^2} - T + 2p. \tag{3.14}$$

### 3.2.7. SIC

Schwarz's (1978) SIC criterion is approximately equivalent to Akaike's (1978) BIC criterion for autoregressive models (see Priestley, 1981, p. 376). We continue using SIC, defined here as

$$\text{SIC} = \log(\hat{\sigma}_p^2) + \frac{\log(T)p}{T}. \tag{3.15}$$

### 3.2.8. HQ

Hannan and Quinn (1979) derived HQ under asymptotic conditions for autoregressive models, assuming that

$$y_t = \phi_1 y_{t-1} + \cdots + \phi_p y_{t-p} + v_t, \quad t = p+1, \ldots, n$$

for large $n$. They also made three assumptions about $v_t$:

$$E[v_n|\mathcal{F}_{n-1}] = 0, \; E[v_n^2|\mathcal{F}_{n-1}] = \sigma^2, \text{ and } E[v_n^4|\mathcal{F}_{n-1}] < \infty,$$

where $\mathcal{F}_n$ is the $\sigma$-algebra generated by $\{v_n, v_{n-1}, \cdots, v_1\}$. If the true order is $p_*$ and a candidate model of order $p > p_*$ is fit, then Hannan and Quinn showed that the last $\phi$ parameter, $\phi_p$, is bounded as per the law of the iterated logarithm. Indeed,

$$\hat{\phi}_p = b_p(T)\left(\frac{2\log\log(T)}{T}\right)^{1/2},$$

where the sequence $b_p(T)$ has limit points in the interval [-1,1]. Hannan and Quinn wanted a model selection criterion of a form similar to AIC yet still strongly consistent for the order $p_*$. AIC has a penalty function of the form $\alpha p/T$, so they proposed a criterion of the form $\log(\hat{\sigma}_p^2) + pC_T$, where $C_T$ decreased as fast as possible. In AIC, $C_T = 2/T$. In SIC, $C_T = \log(T)/T$. Hannan and Quinn showed that consistency can be maintained with a $C_T$ smaller than $\log(T)/T$. In fact, consistency can be maintained for $C_T = 2c\log\log(T)/T$ where $c > 1$. This led to the proposed criterion

$$\Psi(p) = \log(\hat{\sigma}_p^2) + 2c\frac{\log\log(T)p}{T}, \quad c > 1.$$

The loglog($T$) penalty function results in a strongly consistent model selection criterion. Hannan (1980, p. 1072) has commented that strong consistency may hold for $c = 1$, and in practice, $c = 1$ is a common choice. Using $c = 1$, we obtain Hannan and Quinn's model selection criterion

$$\mathrm{HQ} = \log(\hat{\sigma}_p^2) + \frac{2 \log \log(T) p}{T}. \tag{3.16}$$

### *3.2.9. HQc*

HQc, Eq. (2.21), for autoregressive models is defined as

$$\mathrm{HQc} = \log(\hat{\sigma}_p^2) + \frac{2 \log \log(T) p}{T - p - 2}. \tag{3.17}$$

## 3.3. Small-sample Signal-to-noise Ratios

In this Section we will derive small-sample signal-to-noise ratios for the criteria under the autoregressive model. We will use the approximation in Eq. (3A.6), from Appendix 3A, to obtain the following expectations:

$$E\left[\log(\hat{\sigma}^2)\right] = \log(\sigma_*^2) + \log(n - 2p) - \frac{1}{n - 2p} - \log(n - p)$$

and

$$E\left[\log(s^2)\right] = \log(\sigma_*^2) - \frac{1}{n - 2p}.$$

The signal-to-noise ratios for AIC, AICc, AICu, SIC, HQ, and HQc can be obtained using Eq. (3A.6) and the noise term, Eq. (3A.7). Because they all are obtained in the same fashion, of these six only the calculations for AIC will be presented in detail. Signal-to-noise ratios for FPE, FPEu, and Cp are obtained differently, and the details for these calculations are also presented.

### AIC

From Eq. (3A.6) the signal for AIC is

$$\log\left(1 - \frac{2L}{n - 2p_*}\right) - \log\left(1 - \frac{L}{n - p_*}\right) - \frac{2L}{(n - 2p_* - 2L)(n - 2p_*)} + \frac{2L(n + 1)}{(n - p_* - L)(n - p_*)}.$$

The noise term from Eq. (3A.7) is

$$\frac{\sqrt{4L}}{\sqrt{(n - 2p_* - 2L)(n - 2p_* + 2)}},$$

and thus the signal-to-noise ratio for AIC overfitting is

$$\frac{\sqrt{(n-2p_*-2L)(n-2p_*+2)}}{\sqrt{4L}}\left(\log\left(1-\frac{2L}{n-2p_*}\right)-\log\left(1-\frac{L}{n-p_*}\right)\right.$$
$$\left.-\frac{2L}{(n-2p_*-2L)(n-2p_*)}+\frac{2L(n+1)}{(n-p_*-L)(n-p_*)}\right).$$

**AICc**

The signal-to-noise ratio for AICc overfitting is

$$\frac{\sqrt{(n-2p_*-2L)(n-2p_*+2)}}{\sqrt{4L}}\left(\log\left(1-\frac{2L}{n-2p_*}\right)-\log\left(1-\frac{L}{n-p_*}\right)\right.$$
$$\left.-\frac{2L}{(n-2p_*-2L)(n-2p_*)}+\frac{2Ln}{(n-2p_*-2L-2)(n-2p_*-2)}\right).$$

**AICu**

The signal-to-noise ratio for AICu overfitting is

$$\frac{\sqrt{(n-2p_*-2L)(n-2p_*+2)}}{\sqrt{4L}}\left(-\frac{2L}{(n-2p_*-2L)(n-2p_*)}\right.$$
$$\left.+\frac{2Ln}{(n-2p_*-2L-2)(n-2p_*-2)}\right).$$

**FPE**

For FPE, the signal is (applying Eq. (3A.4))

$$\sigma_*^2\frac{Ln}{(n-p_*-L)(n-p_*)},$$

and the noise is (applying Eq. (3A.5))

$$\sigma_*^2\sqrt{\frac{4Ln^2(n-2p_*-2L)(n-p_*-L)^2+2L^2(3n^2-4np_*+2L^2)}{(n-2p_*-2L)(n-p_*-L)^2(n-2p_*)^2(n-p_*)^2}}.$$

Thus the signal-to-noise ratio for FPE for overfitting is

$$\frac{n(n-2p_*)L\sqrt{n-2p_*-2L}}{\sqrt{4Ln^2(n-2p_*-2L)(n-p_*-L)^2+2L^2(3n^2-4np_*+2L^2)}}.$$

**FPEu**

For FPEu, the signal follows from Eq. (3A.4) and the noise follows from Eq. (3A.5). The signal is

$$\sigma_*^2 \frac{2Ln}{(n-2p_*-2L)(n-2p_*)},$$

and the noise is

$$\sigma_*^2 \sqrt{\frac{4Ln^2(n-2p_*-2L)^3 + 32n^2L^2(n-2p_*-L)^2}{(n-2p_*-2L)^3(n-2p_*)^4}}.$$

Thus the signal-to-noise ratio for FPEu for overfitting is

$$\frac{(n-2p_*)\sqrt{L(n-2p_*-2L)}}{\sqrt{(n-2p_*-2L)^3 + 8L(n-2p_*-L)^2}}.$$

**Cp**

Both the signal and the noise of Cp follow from Eqs. (3A.1)–(3A.5). The signal is

$$(n-2P)\left(\frac{-2L}{n-2P-2}\right) + 3L,$$

and the noise is

$$(n-2P)\sqrt{\frac{4L+4L^2}{(n-2P-2)(n-2P-4)} - \frac{4L^2}{(n-2P-2)^2}}.$$

Thus the signal-to-noise ratio for Cp for overfitting is

$$\frac{(n-2P)\left(\frac{-2L}{n-2P-2}\right) + 3L}{(n-2P)\sqrt{\frac{4L+4L^2}{(n-2P-2)(n-2P-4)} - \frac{4L^2}{(n-2P-2)^2}}}.$$

**SIC**

The signal-to-noise ratio for SIC for overfitting is

$$\frac{\sqrt{(n-2p_*-2L)(n-2p_*+2)}}{\sqrt{4L}}\left(\log\left(1-\frac{2L}{n-2p_*}\right) - \log\left(1-\frac{L}{n-p_*}\right)\right.$$
$$\left. - \frac{2L}{(n-2p_*-2L)(n-2p_*)} + \frac{\log(n-p_*-L)(p_*+L)}{n-p_*-L} - \frac{\log(n-p_*)p_*}{n-p_*}\right).$$

**HQ**

The signal-to-noise ratio for HQ overfitting is

$$\frac{\sqrt{(n-2p_*-2L)(n-2p_*+2)}}{\sqrt{4L}}\left(\log\left(1-\frac{2L}{n-2p_*}\right)\right.$$
$$-\log\left(1-\frac{L}{n-p_*}\right)-\frac{2L}{(n-2p_*-2L)(n-2p_*)}$$
$$\left.+\frac{2\log\log(n-p_*-L)(p_*+L)}{n-p_*-L}-\frac{2\log\log(n-p_*)p_*}{n-p_*}\right).$$

**HQc**

The signal-to-noise ratio for HQc for overfitting is

$$\frac{\sqrt{(n-2p_*-2L)(n-2p_*+2)}}{\sqrt{4L}}\left(\log\left(1-\frac{2L}{n-2p_*}\right)\right.$$
$$-\log\left(1-\frac{L}{n-p_*}\right)-\frac{2L}{(n-2p_*-2L)(n-2p_*)}$$
$$\left.+\frac{2\log\log(n-p_*-L)(p_*+L)}{n-2p_*-2L-2}-\frac{2\log\log(n-p_*)p_*}{n-2p_*-2}\right).$$

In the next Section, we will see how these signal-to-noise ratios relate to probabilities of overfitting.

## 3.4. Overfitting

In this Section we derive small-sample and asymptotic overfitting properties for our nine model selection criteria. We present the small-sample and asymptotic signal-to-noise ratios for each criterion, and calculate the small-sample probabilities of overfitting for sample sizes $n = 15$, 25, 35, 50, and 100. We expect that small or negative signal-to-noise ratios will lead to a high probability of overfitting, while large signal-to-noise ratios will lead to small probabilities of overfitting.

### *3.4.1. Small-sample Probabilities of Overfitting*

Suppose there is a true AR model of order $p_*$, and we fit a candidate AR model of order $p_* + L$, where $L > 0$. To calculate the asymptotic probability of overfitting by $L$ extra variables, we will compute the probability of selecting the overfitted AR($p_* + L$) model over the true AR($p_*$) model. However, we must first derive finite sample probabilities of overfitting. We know that AIC, AICc, FPE, and Cp are efficient model selection criteria and have nonzero probabilities of overfitting, and that SIC and HQ are consistent model selection criteria and therefore by definition will have zero asymptotic probability of overfitting. Detailed calculations will be presented for one criterion, AIC. Only the results of the analogous calculations for the other criteria will be shown here, but the details can be found in Appendix 3B. In each case we assume for finite $n$ that the model selection criterion MSC overfits if $\text{MSC}_{p_*+L} < \text{MSC}_{p_*}$.

### AIC

We know that AIC overfits if $\text{AIC}_{p_*+L} < \text{AIC}_{p_*}$. In terms of the original sample size $n$, $\text{AIC} = \log(\hat{\sigma}_p^2) + 2(p+1)/(n-p)$. For finite $n$, the probability that AIC prefers the overfitted model $p_* + L$ is

$$
\begin{aligned}
&P\{\text{AIC}_{p_*+L} < \text{AIC}_{p_*}\} \\
&= P\left\{\log\left(\hat{\sigma}^2_{p_*+L}\right) + \frac{2(p_*+L+1)}{n-p_*-L} < \log\left(\hat{\sigma}^2_{p_*}\right) + \frac{2(p_*+1)}{n-p_*}\right\} \\
&= P\left\{\log\left(\text{SSE}_{p_*+L}\right) - \log(n-p_*-L) + \frac{2(p_*+L+1)}{n-p_*-L}\right. \\
&\qquad \left. < \log\left(\text{SSE}_{p_*}\right) - \log(n-p_*) + \frac{2(p_*+1)}{n-p_*}\right\} \\
&= P\left\{\log\left(\frac{\text{SSE}_{p_*+L}}{\text{SSE}_{p_*}}\right) < \log\left(\frac{n-p_*-L}{n-p_*}\right) + \frac{2(p_*+1)}{n-p_*} - \frac{2(p_*+L+1)}{n-p_*-L}\right\} \\
&= P\left\{\log\left(\frac{\text{SSE}_{p_*}}{\text{SSE}_{p_*+L}}\right) > \log\left(\frac{n-p_*}{n-p_*-L}\right) + \frac{2L(n+1)}{(n-p_*-L)(n-p_*)}\right\} \\
&= P\left\{\frac{\text{SSE}_{p_*}}{\text{SSE}_{p_*+L}} - 1 > \frac{n-p_*}{n-p_*-L}\exp\left(\frac{2L(n+1)}{(n-p_*-L)(n-p_*)}\right) - 1\right\} \\
&= P\left\{\frac{\text{SSE}_{p_*} - \text{SSE}_{p_*+L}}{\text{SSE}_{p_*+L}} > \frac{n-p_*}{n-p_*-L}\exp\left(\frac{2L(n+1)}{(n-p_*-L)(n-p_*)}\right) - 1\right\} \\
&= P\left\{\frac{\chi^2_{2L}}{\chi^2_{n-2p_*-2L}} > \frac{n-p_*}{n-p_*-L}\exp\left(\frac{2L(n+1)}{(n-p_*-L)(n-p_*)}\right) - 1\right\} \\
&= P\left\{F_{2L,n-2p_*-2L} > \frac{n-2p_*-2L}{2L}\times\right. \\
&\qquad \left.\left(\frac{n-p_*}{n-p_*-L}\exp\left(\frac{2L(n+1)}{(n-p_*-L)(n-p_*)}\right) - 1\right)\right\}.
\end{aligned}
$$

**AICc**

$$
\begin{aligned}
&P\{\text{AICc}_{p_*+L} < \text{AICc}_{p_*}\} \\
&= P\Bigg\{ F_{2L,n-2p_*-2L} > \frac{n-2p_*-2L}{2L} \times \\
&\qquad \left( \frac{n-p_*}{n-p_*-L} \exp\left( \frac{2Ln}{(n-2p_*-2L-2)(n-2p_*-2)} \right) - 1 \right) \Bigg\}.
\end{aligned}
$$

**AICu**

$$
\begin{aligned}
&P\{\text{AICu}_{p_*+L} < \text{AICu}_{p_*}\} \\
&= P\Bigg\{ F_{2L,n-2p_*-2L} > \frac{n-2p_*-2L}{2L} \times \\
&\qquad \left( \frac{n-2p_*}{n-2p_*-2L} \exp\left( \frac{2Ln}{(n-2p_*-2L-2)(n-2p_*-2)} \right) - 1 \right) \Bigg\}.
\end{aligned}
$$

**FPE**

$$
P\{\text{FPE}_{p_*+L} < \text{FPE}_{p_*}\} = P\left\{ F_{2L,n-2p_*-2L} > 1 + \frac{n-2p_*}{2(n-p_*-L)} \right\}.
$$

**FPEu**

$$
P\{\text{FPEu}_{p_*+L} < \text{FPEu}_{p_*}\} = P\left\{ F_{2L,n-2p_*-2L} > 1 + \frac{n-2p_*}{n-2p_*-2L} \right\}.
$$

**Cp**

Recall that $P$ represents the order of the largest AR($p$) model considered.

$$
P\{\text{Cp}_{p_*+L} < \text{Cp}_{p_*}\} = P\left\{ F_{2L,n-2P} > 1.5 \right\}
$$

**SIC**

$$
\begin{aligned}
&P\{\mathrm{SIC}_{p_*+L} < \mathrm{SIC}_{p_*}\} \\
&= P\left\{ F_{2L,n-2p_*-2L} > \frac{n-2p_*-2L}{2L} \times \right. \\
&\left. \left( \frac{n-p_*}{n-p_*-L} \exp\left( \frac{\log(n-p_*-L)(p_*+L)}{n-p_*-L} - \frac{\log(n-p_*)p_*}{n-p_*} \right) - 1 \right) \right\}.
\end{aligned}
$$

**HQ**

$$
\begin{aligned}
&P\{\mathrm{HQ}_{p_*+L} < \mathrm{HQ}_{p_*}\} \\
&= P\left\{ F_{2L,n-2p_*-2L} > \frac{n-2p_*-2L}{2L} \left( \frac{n-p_*}{n-p_*-L} \times \right. \right. \\
&\left. \left. \exp\left( \frac{2\log\log(n-p_*-L)(p_*+L)}{n-p_*-L} - \frac{2\log\log(n-p_*)p_*}{n-p_*} \right) - 1 \right) \right\}.
\end{aligned}
$$

**HQc**

$$
\begin{aligned}
&P\{\mathrm{HQc}_{p_*+L} < \mathrm{HQc}_{p_*}\} \\
&= P\left\{ F_{2L,n-2p_*-2L} > \frac{n-2p_*-2L}{2L} \left( \frac{n-p_*}{n-p_*-L} \times \right. \right. \\
&\left. \left. \exp\left( \frac{2\log\log(n-p_*-L)(p_*+L)}{n-2p_*-2L-2} - \frac{2\log\log(n-p_*)p_*}{n-2p_*-2} \right) - 1 \right) \right\}.
\end{aligned}
$$

The calculated probabilities for sample sizes $n = 15$, 25, 35, 50, 100 are given in Table 3.1. The true order is $p_* = 5$, except for $n = 15$, where the true order is 2. $P = \min(15, n/2 - 2)$ is the maximum order used for the Cp criterion. These probabilities have close relationships with the signal-to-noise ratios in Table 3.2, which we will discuss in Section 3.4.3.

Three general patterns emerge from the results in Table 3.1. First, the probabilities of overfitting for model selection criteria with weak penalty functions (AIC and FPE) decrease with increasing sample size, and second, probabilities for criteria with strong penalty functions (AICc and AICu) increase with increasing sample size. Third, while the probabilities of overfitting for the consistent model selection criteria decrease as the sample size increases, the probability of overfitting for SIC decreases much faster with increasing $n$

Table 3.1. Probabilities of overfitting (one candidate versus true model).

$n = 15$. True order = 2.

| $L$ | AIC | AICc | AICu | SIC | HQ | HQc | Cp | FPE | FPEu |
|---|---|---|---|---|---|---|---|---|---|
| 1 | 0.277 | 0.082 | 0.048 | 0.252 | 0.332 | 0.137 | 0.309 | 0.283 | 0.164 |
| 2 | 0.303 | 0.019 | 0.008 | 0.279 | 0.399 | 0.061 | 0.329 | 0.300 | 0.130 |
| 3 | 0.355 | 0.001 | 0.000 | 0.343 | 0.496 | 0.010 | 0.337 | 0.324 | 0.111 |

$n = 25$. True order = 5.

| $L$ | AIC | AICc | AICu | SIC | HQ | HQc | Cp | FPE | FPEu |
|---|---|---|---|---|---|---|---|---|---|
| 1 | 0.294 | 0.074 | 0.041 | 0.221 | 0.301 | 0.082 | 0.309 | 0.283 | 0.156 |
| 2 | 0.319 | 0.022 | 0.008 | 0.221 | 0.336 | 0.028 | 0.329 | 0.292 | 0.117 |
| 3 | 0.350 | 0.004 | 0.001 | 0.236 | 0.380 | 0.006 | 0.337 | 0.298 | 0.090 |
| 4 | 0.403 | 0.000 | 0.000 | 0.279 | 0.449 | 0.001 | 0.340 | 0.313 | 0.075 |
| 5 | 0.494 | 0.000 | 0.000 | 0.368 | 0.558 | 0.000 | 0.342 | 0.342 | 0.070 |

$n = 35$. True order = 5.

| $L$ | AIC | AICc | AICu | SIC | HQ | HQc | Cp | FPE | FPEu |
|---|---|---|---|---|---|---|---|---|---|
| 1 | 0.261 | 0.128 | 0.072 | 0.152 | 0.229 | 0.106 | 0.309 | 0.260 | 0.147 |
| 2 | 0.260 | 0.074 | 0.028 | 0.118 | 0.217 | 0.054 | 0.329 | 0.254 | 0.105 |
| 3 | 0.256 | 0.037 | 0.010 | 0.097 | 0.208 | 0.025 | 0.337 | 0.243 | 0.076 |
| 4 | 0.257 | 0.016 | 0.003 | 0.085 | 0.207 | 0.010 | 0.340 | 0.236 | 0.056 |
| 5 | 0.266 | 0.006 | 0.001 | 0.081 | 0.215 | 0.003 | 0.342 | 0.231 | 0.042 |
| 6 | 0.284 | 0.001 | 0.000 | 0.085 | 0.234 | 0.001 | 0.344 | 0.232 | 0.033 |
| 7 | 0.316 | 0.000 | 0.000 | 0.097 | 0.267 | 0.000 | 0.345 | 0.237 | 0.027 |
| 8 | 0.366 | 0.000 | 0.000 | 0.124 | 0.322 | 0.000 | 0.346 | 0.250 | 0.024 |

$n = 50$. True order = 5.

| $L$ | AIC | AICc | AICu | SIC | HQ | HQc | Cp | FPE | FPEu |
|---|---|---|---|---|---|---|---|---|---|
| 1 | 0.245 | 0.163 | 0.094 | 0.111 | 0.187 | 0.114 | 0.247 | 0.246 | 0.142 |
| 2 | 0.233 | 0.116 | 0.047 | 0.070 | 0.156 | 0.065 | 0.240 | 0.233 | 0.100 |
| 3 | 0.217 | 0.079 | 0.023 | 0.045 | 0.131 | 0.036 | 0.229 | 0.216 | 0.070 |
| 4 | 0.204 | 0.051 | 0.010 | 0.031 | 0.113 | 0.018 | 0.219 | 0.201 | 0.049 |
| 5 | 0.195 | 0.031 | 0.005 | 0.023 | 0.101 | 0.009 | 0.211 | 0.188 | 0.036 |
| 6 | 0.189 | 0.018 | 0.002 | 0.017 | 0.092 | 0.004 | 0.204 | 0.178 | 0.026 |
| 7 | 0.187 | 0.010 | 0.001 | 0.014 | 0.087 | 0.002 | 0.199 | 0.170 | 0.019 |
| 8 | 0.188 | 0.005 | 0.000 | 0.012 | 0.086 | 0.001 | 0.194 | 0.165 | 0.015 |

$n = 100$. True order = 5.

| $L$ | AIC | AICc | AICu | SIC | HQ | HQc | Cp | FPE | FPEu |
|---|---|---|---|---|---|---|---|---|---|
| 1 | 0.232 | 0.196 | 0.116 | 0.069 | 0.143 | 0.113 | 0.230 | 0.234 | 0.138 |
| 2 | 0.212 | 0.161 | 0.070 | 0.030 | 0.101 | 0.067 | 0.212 | 0.214 | 0.095 |
| 3 | 0.190 | 0.129 | 0.042 | 0.014 | 0.072 | 0.039 | 0.191 | 0.192 | 0.065 |
| 4 | 0.170 | 0.102 | 0.025 | 0.007 | 0.052 | 0.023 | 0.173 | 0.172 | 0.045 |
| 5 | 0.154 | 0.080 | 0.015 | 0.003 | 0.038 | 0.013 | 0.158 | 0.155 | 0.032 |
| 6 | 0.139 | 0.062 | 0.009 | 0.002 | 0.028 | 0.007 | 0.145 | 0.140 | 0.022 |
| 7 | 0.127 | 0.048 | 0.005 | 0.001 | 0.021 | 0.004 | 0.134 | 0.127 | 0.016 |
| 8 | 0.117 | 0.037 | 0.003 | 0.000 | 0.017 | 0.002 | 0.125 | 0.117 | 0.011 |

Table 3.2. Signal-to-noise ratios for overfitting.

$n = 15$. True order = 2.

| $L$ | AIC | AICc | AICu | SIC | HQ | HQc | Cp | FPE | FPEu |
|---|---|---|---|---|---|---|---|---|---|
| 1 | 0.283 | 1.503 | 2.046 | 0.380 | 0.104 | 0.987 | 0.063 | 0.378 | 0.844 |
| 2 | 0.295 | 2.742 | 3.529 | 0.377 | 0.015 | 1.768 | 0.111 | 0.468 | 1.017 |
| 3 | 0.172 | 4.982 | 5.958 | 0.206 | -0.188 | 3.181 | 0.114 | 0.483 | 1.045 |

$n = 25$. True order = 5.

| $L$ | AIC | AICc | AICu | SIC | HQ | HQc | Cp | FPE | FPEu |
|---|---|---|---|---|---|---|---|---|---|
| 1 | 0.223 | 1.606 | 2.203 | 0.509 | 0.201 | 1.505 | 0.229 | 0.346 | 0.881 |
| 2 | 0.246 | 2.622 | 3.481 | 0.608 | 0.196 | 2.422 | 0.269 | 0.448 | 1.108 |
| 3 | 0.195 | 3.847 | 4.916 | 0.580 | 0.111 | 3.503 | 0.286 | 0.495 | 1.205 |
| 4 | 0.068 | 5.669 | 6.913 | 0.439 | -0.050 | 5.093 | 0.283 | 0.504 | 1.223 |
| 5 | -0.151 | 9.249 | 10.625 | 0.170 | -0.299 | 8.210 | 0.277 | 0.477 | 1.168 |

$n = 35$. True order = 5.

| $L$ | AIC | AICc | AICu | SIC | HQ | HQc | Cp | FPE | FPEu |
|---|---|---|---|---|---|---|---|---|---|
| 1 | 0.342 | 1.057 | 1.626 | 0.883 | 0.475 | 1.248 | 0.253 | 0.397 | 0.926 |
| 2 | 0.451 | 1.606 | 2.423 | 1.186 | 0.625 | 1.872 | 0.312 | 0.534 | 1.217 |
| 3 | 0.508 | 2.132 | 3.148 | 1.367 | 0.701 | 2.452 | 0.322 | 0.619 | 1.389 |
| 4 | 0.526 | 2.696 | 3.886 | 1.466 | 0.724 | 3.057 | 0.339 | 0.675 | 1.495 |
| 5 | 0.509 | 3.344 | 4.692 | 1.497 | 0.702 | 3.737 | 0.349 | 0.707 | 1.555 |
| 6 | 0.454 | 4.134 | 5.629 | 1.460 | 0.634 | 4.551 | 0.361 | 0.721 | 1.580 |
| 7 | 0.358 | 5.163 | 6.793 | 1.352 | 0.517 | 5.597 | 0.352 | 0.716 | 1.572 |
| 8 | 0.213 | 6.619 | 8.368 | 1.164 | 0.343 | 7.064 | 0.355 | 0.693 | 1.530 |

$n = 50$. True order = 5.

| $L$ | AIC | AICc | AICu | SIC | HQ | HQc | Cp | FPE | FPEu |
|---|---|---|---|---|---|---|---|---|---|
| 1 | 0.406 | 0.816 | 1.363 | 1.195 | 0.679 | 1.172 | 0.313 | 0.431 | 0.952 |
| 2 | 0.558 | 1.203 | 1.986 | 1.654 | 0.935 | 1.712 | 0.422 | 0.591 | 1.285 |
| 3 | 0.662 | 1.540 | 2.509 | 1.978 | 1.111 | 2.170 | 0.495 | 0.701 | 1.504 |
| 4 | 0.737 | 1.865 | 2.996 | 2.224 | 1.239 | 2.600 | 0.549 | 0.783 | 1.661 |
| 5 | 0.789 | 2.193 | 3.472 | 2.414 | 1.333 | 3.026 | 0.592 | 0.845 | 1.777 |
| 6 | 0.823 | 2.537 | 3.952 | 2.557 | 1.398 | 3.462 | 0.626 | 0.893 | 1.863 |
| 7 | 0.839 | 2.905 | 4.451 | 2.659 | 1.435 | 3.921 | 0.655 | 0.927 | 1.925 |
| 8 | 0.837 | 3.310 | 4.979 | 2.721 | 1.447 | 4.416 | 0.679 | 0.951 | 1.967 |

$n = 100$. True order = 5.

| $L$ | AIC | AICc | AICu | SIC | HQ | HQc | Cp | FPE | FPEu |
|---|---|---|---|---|---|---|---|---|---|
| 1 | 0.461 | 0.628 | 1.152 | 1.680 | 0.946 | 1.179 | 0.444 | 0.467 | 0.978 |
| 2 | 0.646 | 0.904 | 1.648 | 2.361 | 1.329 | 1.689 | 0.619 | 0.652 | 1.354 |
| 3 | 0.785 | 1.128 | 2.043 | 2.874 | 1.616 | 2.095 | 0.748 | 0.787 | 1.624 |
| 4 | 0.898 | 1.326 | 2.389 | 3.296 | 1.852 | 2.452 | 0.852 | 0.896 | 1.836 |
| 5 | 0.994 | 1.511 | 2.706 | 3.660 | 2.054 | 2.780 | 0.940 | 0.988 | 2.011 |
| 6 | 1.078 | 1.687 | 3.003 | 3.980 | 2.231 | 3.089 | 1.017 | 1.066 | 2.159 |
| 7 | 1.151 | 1.859 | 3.288 | 4.266 | 2.388 | 3.386 | 1.085 | 1.135 | 2.286 |
| 8 | 1.216 | 2.028 | 3.564 | 4.524 | 2.529 | 3.675 | 1.146 | 1.195 | 2.396 |

than HQ. This is due to the $\log(n-p)$ penalty in SIC as compared to the much smaller $2\log\log(n-p)$ penalty in HQ. However, at very small $n$ ($n=15$), SIC and HQ both overfit with especially high probability. We also see that the signal-to-noise correction intended to improve small-sample performance for HQ has in fact worked; HQc outperforms both of the other consistent criteria in the smaller sample sizes ($n = 15, 25$ and $35$).

### *3.4.2. Asymptotic Probabilities of Overfitting*

Table 3.1 establishes some general patterns for overfitting behavior in small samples, and we next want to see what will happen in large samples by deriving asymptotic probabilities of overfitting. Detailed derivations will be given for one criterion from each class; calculations for those not presented here can be found in Appendix 3C. We will make use of the following facts: first, as $n \to \infty$, and where $p_*$ and $L$ are fixed, it is known that $\chi^2_{n-2p_*-2L}/(n-2p_*-2L) \to 1$ a.s. By Slutsky's theorem, $F_{2L,n-2p_*-2L} \to \chi^2_{2L}/(2L)$ (*i.e.*, the small-sample F distribution is replaced by a $\chi^2$ distribution.) Hence the asymptotic probabilities of overfitting presented here are derived from the $\chi^2$ distribution. We will also use the expansion for $\exp(\cdot)$, $\exp(z) = 1 + z + \sum_{i=2}^{\infty} z^i/i!$. For small $z$, $\exp(z) = 1 + z + o(z^2)$.

#### *3.4.2.1. K-L Criteria*

Our example for the K-L criteria will be AIC. As $n \to \infty$,

$$
\begin{aligned}
&\frac{n-2p_*-2L}{2L}\left(\frac{n-p_*}{n-p_*-L}\exp\left(\frac{2L(n+1)}{(n-p_*-L)(n-p_*)}\right)-1\right)\\
&\quad=\frac{n-2p_*-2L}{2L}\left(\frac{n-p_*}{n-p_*-L}\left(1+\frac{2L(n+1)}{(n-p_*-L)(n-p_*)}\right.\right.\\
&\qquad\left.\left.+O\Big(\frac{1}{n^2}\Big)\right)-1\right)\\
&\quad=\frac{n-2p_*-2L}{2L}\left(\frac{n-p_*}{n-p_*-L}+\frac{2L(n+1)}{(n-p_*-L)^2}-1+O\Big(\frac{1}{n^2}\Big)\right)\\
&\quad=\frac{n-2p_*-2L}{2L}\left(\frac{L}{n-p_*-L}+\frac{2L(n+1)}{(n-p_*-L)^2}+O\Big(\frac{1}{n^2}\Big)\right)\\
&\quad\to\frac{n}{2L}\left(\frac{L}{n}+\frac{2L}{n}\right)\\
&\quad\to 1.5.
\end{aligned}
$$

Using the above fact,

$$F_{2L,n-2p_*-2L} \to \frac{\chi^2_{2L}}{2L},$$

and thus we have $P\{\text{AIC overfits by } L\} = P\{\chi^2_{2L} > 3L\}$. We will ignore the $O(1/n^2)$ term in the derivations for the other criteria here and in Appendix 3C. The probability that AICc overfits by $L$ is $P\{\chi^2_{2L} > 3L\}$, and for AICu it is $P\{\chi^2_{2L} > 4L\}$.

*3.4.2.2. $L_2$ Criteria*

Our example for the $L_2$ criteria will be FPE. As $n \to \infty$,

$$1 + \frac{n - 2p_*}{2(n - p_* - L)} \to 1.5.$$

Thus we have $P\{\text{FPE overfits by } L\} = P\{\chi^2_{2L} > 3L\}$. For FPEu we have the probability of choosing a model of order $p_* + L$ over the true order $p_*$ is $P\{\chi^2_{2L} > 4L\}$, and for Cp we have $P\{\text{Cp overfits by } L\} = P\{\chi^2_{2L} > 3L\}$. Thus FPEu should overfit less in large samples than either FPE or Cp.

*3.4.2.3. Consistent Criteria*

Our example for the consistent criteria will be SIC. As $n \to \infty$,

$$\begin{aligned}
&\frac{n-2p_*-2L}{2L}\left(\frac{n-p_*}{n-p_*-L}\times\right.\\
&\qquad\left.\exp\left(\frac{\log(n-p_*-L)(p_*+L)}{n-p_*-L}-\frac{\log(n-p_*)p_*}{n-p_*}\right)-1\right)\\
&\qquad=\frac{n-2p_*-2L}{2L}\left(\frac{n-p_*}{n-p_*-L}\times\right.\\
&\qquad\left.\left(1+\frac{\log(n-p_*-L)(p_*+L)}{n-p_*-L}-\frac{\log(n-p_*)p_*}{n-p_*}\right)-1\right)\\
&\qquad\to\frac{n}{2L}\left(\frac{L}{n}+\frac{\log(n)L}{n}\right)\\
&\qquad\to\infty.
\end{aligned}$$

Thus we have $P\{\text{SIC overfits by } L\} = 0$. The probabilities of overfitting for HQ and HQc are also 0. The calculated values for the above asymptotic probabilities of overfitting by $L$ variables are presented in Table 3.3. From the results in Table 3.3 we see that the efficient criteria AICc, AIC, FPE, and Cp are all asymptotically equivalent, and that the consistent criteria SIC,

HQ, and HQc have 0 probability of overfitting, as expected. The signal-to-noise corrected variants AICu and FPEu are asymptotically equivalent, and have an asymptotic probability of overfitting that lies between the efficient and consistent model selection criteria.

Table 3.3. Asymptotic probability of overfitting by $L$ variables.

| $L$ | AIC | AICc | AICu | SIC | HQ | HQc | Cp | FPE | FPEu |
|---|---|---|---|---|---|---|---|---|---|
| 1 | 0.2231 | 0.2231 | 0.1353 | 0 | 0 | 0 | 0.2231 | 0.2231 | 0.1353 |
| 2 | 0.1991 | 0.1991 | 0.0916 | 0 | 0 | 0 | 0.1991 | 0.1991 | 0.0916 |
| 3 | 0.1736 | 0.1736 | 0.0620 | 0 | 0 | 0 | 0.1736 | 0.1736 | 0.0620 |
| 4 | 0.1512 | 0.1512 | 0.0424 | 0 | 0 | 0 | 0.1512 | 0.1512 | 0.0424 |
| 5 | 0.1321 | 0.1321 | 0.0293 | 0 | 0 | 0 | 0.1321 | 0.1321 | 0.0293 |
| 6 | 0.1157 | 0.1157 | 0.0203 | 0 | 0 | 0 | 0.1157 | 0.1157 | 0.0203 |
| 7 | 0.1016 | 0.1016 | 0.0142 | 0 | 0 | 0 | 0.1016 | 0.1016 | 0.0142 |
| 8 | 0.0895 | 0.0895 | 0.0100 | 0 | 0 | 0 | 0.0895 | 0.0895 | 0.0100 |
| 9 | 0.0790 | 0.0790 | 0.0071 | 0 | 0 | 0 | 0.0790 | 0.0790 | 0.0071 |
| 10 | 0.0699 | 0.0699 | 0.0050 | 0 | 0 | 0 | 0.0699 | 0.0699 | 0.0050 |

### *3.4.3. Small-sample Signal-to-noise Ratios*

In Section 3.3 we derived the small-sample signal-to-noise ratios for our nine criteria. To characterize the small-sample properties of these ratios, Table 3.2 summarizes the signal-to-noise ratios for overfitting for sample sizes $n$ = 15, 25, 35, 50, and 100. For $n = 15$, the true order is $p_* = 2$; all the rest have true order $p_* = 5$. $P = \min(15, n/2 - 2)$ is the maximum order used for the Cp criterion.

We see that the relationships we expected between probabilities of overfitting (Table 3.1) and signal-to-noise ratios (Table 3.2) are evident: small or negative signal-to-noise ratios correspond to high probabilities of overfitting, while large signal-to-noise ratios correspond to small probabilities of overfitting.

### *3.4.4. Asymptotic Signal-to-noise Ratios*

Now that we have examined the small-sample case, in order to determine asymptotic signal-to-noise ratios for our six criteria we will make use of the fact that $\log(1 - L/n) \doteq -L/n$ and $\log(1 - 2L/n) \doteq -2L/n$ when $L << n$. This holds assuming $p_*, L$, and $P$ are fixed and that $n \to \infty$. We will show a sample calculation for each category of criterion. The details for those not presented can be found in Appendix 3C.

#### *3.4.4.1. K-L Criteria*

Our detailed example for the K-L criteria will be AIC. Using the above facts, we find the asymptotic signal-to-noise ratio for AIC to be

$$\begin{aligned}
&\lim_{n\to\infty} \frac{\sqrt{(n-2p_*-2L)(n-2p_*+2)}}{\sqrt{4L}} \left( \log\left(1-\frac{2L}{n-2p_*}\right) - \log\left(1-\frac{L}{n-p_*}\right) \right. \\
&\qquad \left. - \frac{2L}{(n-2p_*-2L)(n-2p_*)} + \frac{2L(n+1)}{(n-p_*-L)(n-p_*)} \right) \\
&= \lim_{n\to\infty} \frac{n}{2\sqrt{L}} \left( -\frac{2L}{n} + \frac{L}{n} - \frac{2L}{n^2} + \frac{2L}{n} \right) \\
&= \frac{-2L+L+2L}{2\sqrt{L}} \\
&= \frac{\sqrt{L}}{2}.
\end{aligned}$$

The asymptotic signal-to-noise ratio for AICc is also $\sqrt{L}/2$, and the asymptotic signal-to-noise ratio for AICu is $\sqrt{L}$.

#### 3.4.4.2. $L_2$ Criteria

Our example from the $L_2$-based criteria will be FPE. The asymptotic signal-to-noise ratio for FPE overfitting is

$$\begin{aligned}
&\lim_{n\to\infty} \frac{n(n-2p_*)L\sqrt{n-2p_*-2L}}{\sqrt{4Ln^2(n-2p_*-2L)(n-p_*-L)^2 + 2L^2(3n^2-4np_*+2L^2)}} \\
&= \lim_{n\to\infty} \frac{L\sqrt{n^5}}{\sqrt{4Ln^5}} \\
&= \frac{\sqrt{L}}{2}.
\end{aligned}$$

The asymptotic signal-to-noise ratio for Cp is also $\sqrt{L}/2$, and for FPEu (as for AICu), the asymptotic signal-to-noise ratio is $\sqrt{L}$.

#### 3.4.4.3. Consistent Criteria

Our example from the consistent criteria will be SIC. We find the asymptotic signal-to-noise ratio for SIC to be

$$
\begin{aligned}
&\lim_{n\to\infty} \frac{\sqrt{(n-2p_*-2L)(n-2p_*+2)}}{\sqrt{4L}} \left( \log\left(1-\frac{2L}{n-2p_*}\right) - \log\left(1-\frac{L}{n-p_*}\right) \right. \\
&\quad \left. - \frac{2L}{(n-2p_*-2L)(n-2p_*)} + \frac{\log(n-p_*-L)(p_*+L)}{n-p_*-L} - \frac{\log(n-p_*)p_*}{n-p_*} \right) \\
&= \lim_{n\to\infty} \frac{n}{2\sqrt{L}} \left( -\frac{2L}{n} + \frac{L}{n} - \frac{2L}{n^2} + \frac{\log(n)L}{n} \right) \\
&= \lim_{n\to\infty} \log(n) \frac{1}{2\sqrt{L}} \\
&= \infty.
\end{aligned}
$$

Unsurprisingly, the asymptotic signal-to-noise ratios for HQ and HQc are also $\infty$.

Table 3.4 gives the calculated values of asymptotic signal-to-noise ratios for model $p_*$ versus model $p_* + L$. Results from Table 3.4 show what we have previously calculated—that the efficient criteria AICc, AIC, FPE, and Cp have equivalent signal-to-noise ratios, that the consistent criteria SIC, HQ, and HQc all have infinite signal-to-noise ratios, and that AICu and FPEu are asymptotically equivalent and have twice the signal-to-noise ratios of AICc, AIC, FPE, and Cp.

For the asymptotic case we will not discuss underfitting, since the asymptotic noncentrality parameter is infinite for fixed $\phi$, assuming a true AR model of finite dimension exists. If $\phi$ is allowed to vary with the sample size, we can derive models with finite asymptotic noncentrality parameters.

Table 3.4. Asymptotic signal-to-noise ratio for overfitting by $L$ variables.

| $L$ | AIC | AICc | AICu | SIC | HQ | HQc | Cp | FPE | FPEu |
|---|---|---|---|---|---|---|---|---|---|
| 1 | 0.500 | 0.500 | 1.000 | $\infty$ | $\infty$ | $\infty$ | 0.500 | 0.500 | 1.000 |
| 2 | 0.707 | 0.707 | 1.414 | $\infty$ | $\infty$ | $\infty$ | 0.707 | 0.707 | 1.414 |
| 3 | 0.866 | 0.866 | 1.732 | $\infty$ | $\infty$ | $\infty$ | 0.866 | 0.866 | 1.732 |
| 4 | 1.000 | 1.000 | 2.000 | $\infty$ | $\infty$ | $\infty$ | 1.000 | 1.000 | 2.000 |
| 5 | 1.118 | 1.118 | 2.236 | $\infty$ | $\infty$ | $\infty$ | 1.118 | 1.118 | 2.236 |
| 6 | 1.225 | 1.225 | 2.450 | $\infty$ | $\infty$ | $\infty$ | 1.225 | 1.225 | 2.450 |
| 7 | 1.323 | 1.323 | 2.646 | $\infty$ | $\infty$ | $\infty$ | 1.323 | 1.323 | 2.646 |
| 8 | 1.414 | 1.414 | 2.828 | $\infty$ | $\infty$ | $\infty$ | 1.414 | 1.414 | 2.828 |
| 9 | 1.500 | 1.500 | 3.000 | $\infty$ | $\infty$ | $\infty$ | 1.500 | 1.500 | 3.000 |
| 10 | 1.581 | 1.581 | 3.162 | $\infty$ | $\infty$ | $\infty$ | 1.581 | 1.581 | 3.162 |

To test whether our nine criteria behave in reality as we expect on the basis of our theoretical calculations, in the next Section we will examine their performance with respect to two special case models.

## 3.5. Underfitting for Two Special Case Models

In this Section we will examine small-sample performance ($n = 35$) with respect to underfitting in autoregressive models. Two special case models with different degrees of identifiability will be used to test our nine criteria. For each model we will calculate expected values for the criteria and the K-L and $L_2$ distances, signal-to-noise ratios, and probabilities of underfitting.

For both models,

$$n = 35, \quad p_* = 5, \quad \sigma_*^2 = 1. \tag{3.18}$$

Model 3 is an AR model with strongly identifiable parameters:

$$y_t = y_{t-5} + w_{*t}. \tag{3.19}$$

This model is in fact a seasonal random walk with season = 5. By contrast, Model 4 is an AR model with weaker parameters that are much more difficult to identify at the true order 5:

$$y_t = 0.434y_{t-1} + 0.217y_{t-2} + 0.145y_{t-3} + 0.108y_{t-4} + 0.087y_{t-5} + w_{*t}. \tag{3.20}$$

The coefficient of $y_{t-j}$ in Model 4 is proportional to $1/j$ for $j = 1, ..., 5$. Each time series $Y$ was generated starting at $y_{-50}$ with $y_t = 0$ for all $t < -50$, but only observations $y_1, \ldots, y_{35}$ were kept. For Cp, the maximum order is $P = 10$.

We expect to see that the relationships between signal-to-noise ratio and probabilities established for overfitting to hold for underfitting as well: the larger the signal to noise ratio for the underfitted order, the less likely underfitting is to occur. However, we also recall from the results in Chapter 2 for these same models that strong penalty functions intended to prevent overfitting can in fact lead to some underfitting in small samples, and that K-L and $L_2$ differ in the degree to which they punish underfitting. K-L is less harsh than $L_2$, and favors criteria that underfit slightly (those with stronger penalty functions).

### *3.5.1. Expected Values for Two Special Case Models*

The underfitting expectations in Tables 3.5 and 3.6 are approximated on the basis of 100,000 replications. Values for $L_2$ and K-L are expected efficiencies. $L_2$ expected efficiency is computed using the expectation of Eq. (3.7) in Eq. (1.3). K-L expected efficiency is computed using the expectation of Eq. (3.8) in Eq. (1.4). The results from Tables 3.5 and 3.6 are also presented graphically in Figures 3.1 and 3.2. Figure 3.1 plots actual expected values for $L_2$ and K-L rather than the expected efficiencies found in Table 3.5. Note that for presentation purposes, Cp in Figure 3.1 has been scaled by a factor of 40.

Table 3.5 summarizes the expected values for Model 3. We see that all the selection criteria have a minimum expected value at the correct order of 5. Likewise, the $L_2$ and K-L distances attain a maximum efficiency of 1 at the correct order. We know that K-L should favor criteria that underfit slightly, that $L_2$ should favor criteria that overfit slightly, and that both penalize excessive overfitting. This is in fact what we see from the relative expected efficiencies for K-L and $L_2$: for model orders greater than five, K-L efficiencies are lower than those for $L_2$. On the other hand, for underfitting (model orders less than 5) $L_2$ efficiencies are lower than those for K-L.

Table 3.5. Expected values and expected efficiency for Model 3, $\phi_5 = 1$.

| $p$ | AIC | AICc | AICu | SIC | HQ | HQc | Cp | FPE | FPEu | $L_2$ | K-L |
|---|---|---|---|---|---|---|---|---|---|---|---|
| 1 | 2.393 | 3.404 | 3.434 | 2.379 | 2.349 | 2.356 | 404.18 | 12.348 | 12.722 | 0.016 | 0.129 |
| 2 | 2.305 | 3.330 | 3.393 | 2.335 | 2.275 | 2.296 | 337.48 | 11.417 | 12.154 | 0.018 | 0.136 |
| 3 | 2.213 | 3.259 | 3.358 | 2.288 | 2.196 | 2.239 | 284.34 | 10.694 | 11.801 | 0.020 | 0.143 |
| 4 | 2.071 | 3.148 | 3.286 | 2.191 | 2.066 | 2.143 | 229.19 | 9.691 | 11.127 | 0.024 | 0.154 |
| 5 | **0.177** | **1.298** | **1.481** | **0.344** | **0.185** | **0.309** | **5.72** | **1.166** | **1.399** | 1.000 | 1.000 |
| 6 | 0.209 | 1.393 | 1.625 | 0.423 | 0.228 | 0.420 | 6.55 | 1.209 | 1.524 | 0.815 | 0.776 |
| 7 | 0.242 | 1.512 | 1.800 | 0.503 | 0.272 | 0.557 | 7.41 | 1.257 | 1.676 | 0.683 | 0.610 |
| 8 | 0.275 | 1.668 | 2.019 | 0.585 | 0.315 | 0.731 | 8.32 | 1.312 | 1.865 | 0.584 | 0.483 |
| 9 | 0.308 | 1.872 | 2.297 | 0.667 | 0.357 | 0.956 | 9.24 | 1.375 | 2.104 | 0.506 | 0.382 |
| 10 | 0.328 | 2.140 | 2.651 | 0.735 | 0.383 | 1.246 | 10.00 | 1.434 | 2.390 | 0.436 | 0.292 |

Boldface type indicates the minimum expectation.

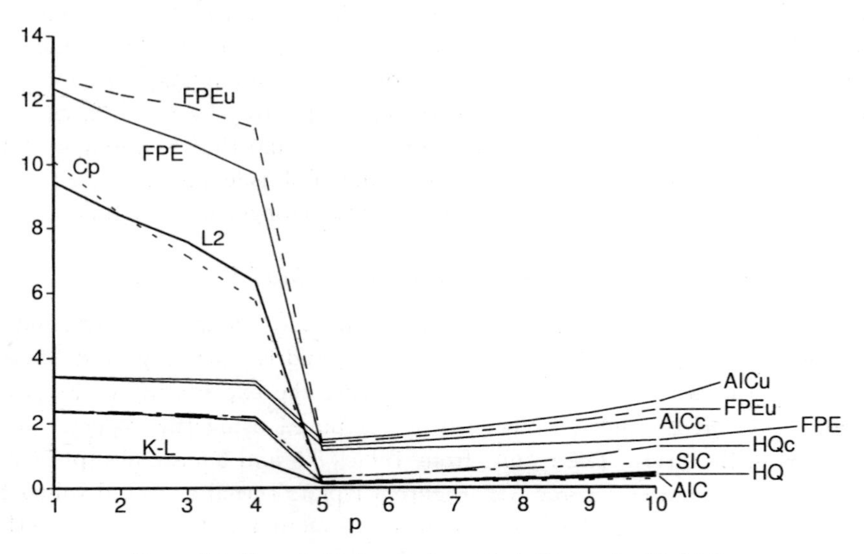


Figure 3.1. Expected values and expected distance for Model 3.

Table 3.6 summarizes the expected values for Model 4. Because it has much weaker parameters than Model 3 the correct model should be more difficult to detect, especially in a sample size of only 35. Indeed, we see that none of the selection criteria have a well-defined minimum at the correct order 5. Both K-L and $L_2$ attain a minimum at order 3, indicating a tendency toward underfitting for both distance measures.

Figure 3.2 plots actual expected values for $L_2$ and K-L rather than the expected efficiencies found in Table 3.6. Note that for presentation purposes, Cp in Figure 3.2 has been scaled by a factor of 40.

Table 3.6. Expected values and expected efficiency for Model 4, $\phi_j \propto 1/j$.

| $p$ | AIC | AICc | AICu | SIC | HQ | HQc | Cp | FPE | FPEu | $L_2$ | K-L |
|---|---|---|---|---|---|---|---|---|---|---|---|
| 1 | 0.269 | 1.280 | 1.310 | 0.255 | 0.225 | 0.233 | 8.860 | 1.285 | 1.324 | 0.407 | 0.618 |
| 2 | 0.166 | **1.191** | **1.254** | **0.197** | 0.136 | **0.157** | 4.281 | 1.153 | **1.227** | 0.839 | 0.996 |
| 3 | **0.146** | 1.193 | 1.291 | 0.221 | **0.129** | 0.173 | 3.396 | **1.128** | 1.245 | 1.000 | 1.000 |
| 4 | 0.164 | 1.242 | 1.380 | 0.285 | 0.160 | 0.237 | **3.863** | 1.149 | 1.319 | 0.910 | 0.838 |
| 5 | 0.177 | 1.298 | 1.481 | 0.344 | 0.185 | 0.309 | 5.716 | 1.166 | 1.399 | 0.757 | 0.657 |
| 6 | 0.209 | 1.393 | 1.625 | 0.423 | 0.228 | 0.420 | 6.547 | 1.209 | 1.524 | 0.622 | 0.519 |
| 7 | 0.242 | 1.512 | 1.800 | 0.503 | 0.272 | 0.557 | 7.410 | 1.257 | 1.676 | 0.521 | 0.412 |
| 8 | 0.275 | 1.668 | 2.019 | 0.585 | 0.315 | 0.731 | 8.315 | 1.312 | 1.865 | 0.447 | 0.331 |
| 9 | 0.308 | 1.872 | 2.297 | 0.667 | 0.357 | 0.956 | 9.240 | 1.375 | 2.104 | 0.368 | 0.263 |
| 10 | 0.328 | 2.140 | 2.651 | 0.735 | 0.383 | 1.246 | 10.000 | 1.434 | 2.390 | 0.339 | 0.209 |

Boldface type indicates the minimum expectation.

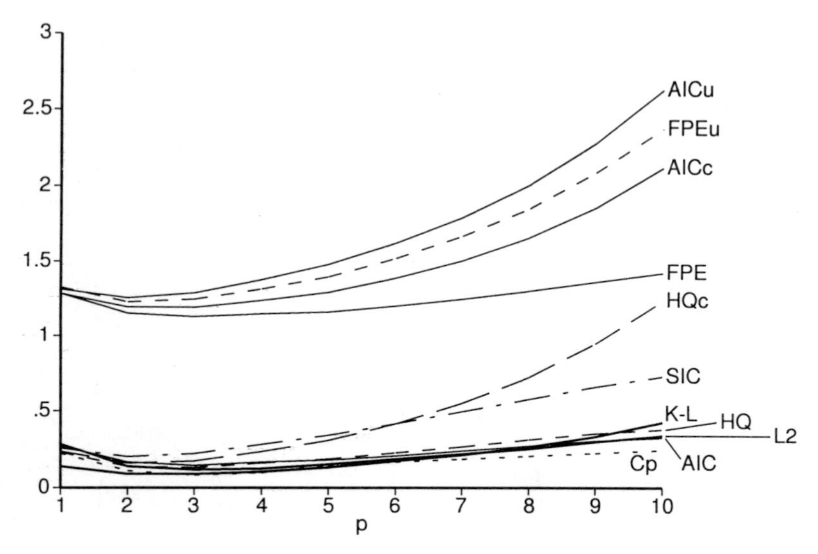


Figure 3.2. Expected values and expected distance for Model 4.

### 3.5.2. Signal-to-noise Ratios for Two Special Case Models

Next we will examine the signal-to-noise ratios and probabilities for over- and underfitting for the two special case models. We want to determine whether the results we expect on the basis of the signal-to-noise ratios calculated in Tables 3.7 and 3.8 and plotted in Figures 3.3 and 3.4 are reflected in the next Section by the probabilities in Tables 3.9 and 3.10. In Tables 3.7 and 3.8 the correct order 5 is compared to the other candidate orders $p$, and the signal-to-noise ratio is defined to be 0 when comparing the correct model to itself. Note that underfitted model signal-to-noise ratios were simulated on the basis of 100,000 runs.

Table 3.7. Signal-to-noise ratios for Model 3, $\phi_5 = 1$.

| $p$ | AIC | AICc | AICu | SIC | HQ | HQc | Cp | FPE | FPEu |
|---|---|---|---|---|---|---|---|---|---|
| 1 | 3.447 | 3.275 | 3.038 | 3.166 | 3.367 | 3.185 | 1.220 | 1.499 | 1.474 |
| 2 | 3.277 | 3.128 | 2.943 | 3.066 | 3.218 | 3.059 | 1.176 | 1.429 | 1.409 |
| 3 | 2.979 | 2.869 | 2.746 | 2.845 | 2.942 | 2.824 | 1.099 | 1.314 | 1.300 |
| 4 | 2.595 | 2.534 | 2.473 | 2.531 | 2.578 | 2.513 | 0.985 | 1.153 | 1.147 |
| 5 | 0 | 0 | 0 | 0 | 0 | 0 | 0 | 0 | 0 |
| 6 | 0.342 | 1.057 | 1.626 | 0.883 | 0.475 | 1.248 | 0.253 | 0.397 | 0.926 |
| 7 | 0.451 | 1.606 | 2.423 | 1.186 | 0.625 | 1.872 | 0.312 | 0.534 | 1.217 |
| 8 | 0.508 | 2.132 | 3.148 | 1.367 | 0.701 | 2.452 | 0.322 | 0.619 | 1.389 |
| 9 | 0.526 | 2.696 | 3.886 | 1.466 | 0.724 | 3.057 | 0.339 | 0.675 | 1.495 |
| 10 | 0.509 | 3.344 | 4.692 | 1.497 | 0.702 | 3.737 | 0.349 | 0.707 | 1.555 |

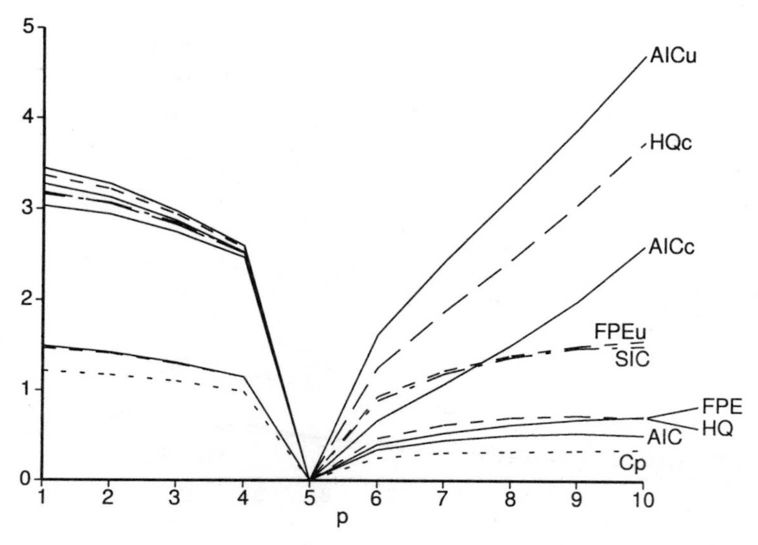


Figure 3.3. Signal-to-noise for Model 3.

The signal-to-noise ratio plot shows that Model 3 should be easy to identify due to the large, positive signal-to-noise ratios for both underfitting as well as overfitting. The larger the signal-to-noise ratio, the smaller the chance for selecting the wrong model. The strongly identifiable parameters in Model 3 are reflected in the large signal-to-noise ratios for underfitting for all the criteria. The signal-to-noise corrected variants AICu, AICc, and HQc have strong penalty functions and thus large signal-to-noise ratios for overfitting also, and should perform well overall for Model 3.

The signal-to-noise ratio plot for Model 4 has negative signal-to-noise ratios for underfitting. Underfitting is expected from all the criteria, making the true

Table 3.8. Signal-to-noise ratios for Model 4, $\phi_j \propto 1/j$.

| $p$ | AIC | AICc | AICu | SIC | HQ | HQc | Cp | FPE | FPEu |
|---|---|---|---|---|---|---|---|---|---|
| 1 | 0.403 | -0.151 | -0.915 | -0.504 | 0.144 | -0.442 | 0.433 | 0.427 | -0.332 |
| 2 | -0.145 | -0.773 | -1.551 | -1.033 | -0.392 | -1.063 | -0.045 | -0.148 | -0.889 |
| 3 | -0.359 | -0.999 | -1.711 | -1.139 | -0.571 | -1.259 | -0.285 | -0.375 | -1.026 |
| 4 | -0.289 | -0.816 | -1.340 | -0.840 | -0.435 | -1.002 | -0.265 | -0.314 | -0.814 |
| 5 | 0 | 0 | 0 | 0 | 0 | 0 | 0 | 0 | 0 |
| 6 | 0.342 | 1.057 | 1.626 | 0.883 | 0.475 | 1.248 | 0.253 | 0.397 | 0.926 |
| 7 | 0.451 | 1.606 | 2.423 | 1.186 | 0.625 | 1.872 | 0.312 | 0.534 | 1.217 |
| 8 | 0.508 | 2.132 | 3.148 | 1.367 | 0.701 | 2.452 | 0.322 | 0.619 | 1.389 |
| 9 | 0.526 | 2.696 | 3.886 | 1.466 | 0.724 | 3.057 | 0.339 | 0.675 | 1.495 |
| 10 | 0.509 | 3.344 | 4.692 | 1.497 | 0.702 | 3.737 | 0.349 | 0.707 | 1.555 |

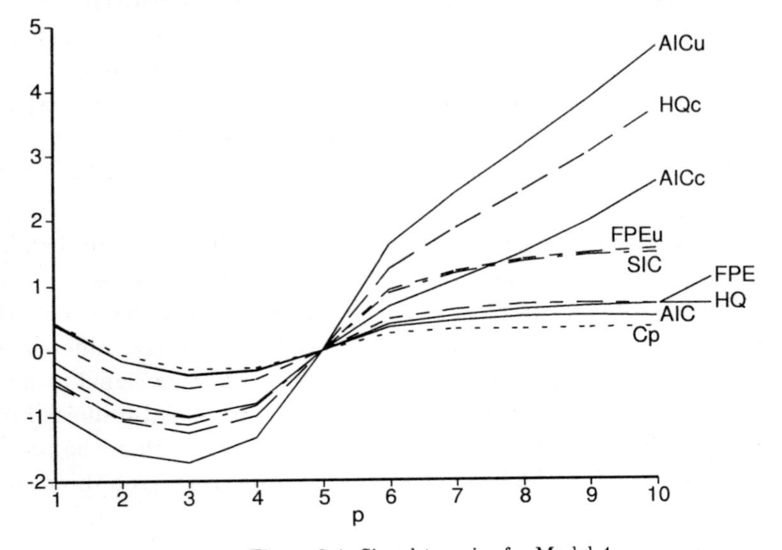


Figure 3.4. Signal-to-noise for Model 4.

order difficult to detect. AICu has by far the largest negative signal-to-noise ratio for orders $p < p_*$, indicating that AICu will tend to underfit more than the other selection criteria for this model. This is a result of AICu's strong penalty function, which prevents overfitting at the expense of some underfitting when the parameters are weak. AIC also has a negative underfitting signal-to-noise ratio as well as a weak overfitting signal-to-noise ratio, indicating that, in general, AIC will tend to underfit, and occasionally overfit excessively.

### 3.5.3. Probabilities for Two Special Case Models

Tables 3.9 and 3.10 summarize approximate probabilities for selecting the candidate order $p$ over the correct order 5. Underfitting probabilities are approximated from the same 100,000 replications used to estimate the underfitting signal-to-noise ratios. Once again, the probability of selecting the correct order 5 over itself is undefined and is noted with an asterisk. Overfitting probabilities are computed using the F-statistic.

All the selection criteria have small probabilities of underfitting for Model 3, and the only significant differences lie in probabilities of overfitting. The signal-to-noise corrected variants AICc, AICu, and HQc have the strongest penalty functions, the largest signal-to-noise ratios with respect to overfitting, and consequently, the smallest probabilities of overfitting. In comparison, their parent criteria AIC, HQ, and FPE have large probabilities of overfitting, and we expect them to perform much more poorly when unnecessarily large orders are included in the study. Overall the probabilities in Tables 3.9 and 3.10 are directly related to the signal-to-noise ratios: large signal-to-noise ratios correspond to small probabilities, and weak signal-to-noise ratios correspond to moderate probabilities of selecting the incorrect order. Strongly negative signal-to-noise ratios result in probabilities of almost 1 for selecting the wrong model.

In general, the larger the signal-to-noise ratio, the smaller the probability of misselection. Small, particularly negative, signal-to-noise ratios are associated with larger probabilities of misselection. Not all the criteria have the same distribution for overfitting. For instance, AIC and Cp have different F-distributions for overfitting (see Section 3.4.1). The overfitting signal-to-noise ratio for AIC in Table 3.7 is greater than the overfitting signal-to-noise ratio for Cp, the probability that AIC overfits in Table 3.10 is less than the probability that Cp overfits. For the log(SSE)-based criteria, there is a direct and comparable relationship between signal-to-noise ratios and probabilities.

We next look at the probabilities for Model 4, Table 3.10. Here we see large probabilities for selecting underfitted candidate orders. Here AICu's strengthened penalty function is a handicap, causing it to underfit most severely. AIC, with its relatively weak penalty function, does best with respect to limiting underfitting.

Table 3.9. Probability of selecting order $p$ over the true order 5 for Model 3, $\phi_5 = 1$.

| $p$ | AIC | AICc | AICu | SIC | HQ | HQc | Cp | FPE | FPEu |
|---|---|---|---|---|---|---|---|---|---|
| 1 | 0.000 | 0.000 | 0.001 | 0.000 | 0.000 | 0.000 | 0.000 | 0.000 | 0.000 |
| 2 | 0.000 | 0.000 | 0.000 | 0.000 | 0.000 | 0.000 | 0.000 | 0.000 | 0.000 |
| 3 | 0.000 | 0.000 | 0.001 | 0.000 | 0.000 | 0.000 | 0.000 | 0.000 | 0.000 |
| 4 | 0.000 | 0.000 | 0.001 | 0.000 | 0.000 | 0.000 | 0.000 | 0.000 | 0.000 |
| 5 | * | * | * | * | * | * | * | * | * |
| 6 | 0.261 | 0.128 | 0.072 | 0.152 | 0.229 | 0.106 | 0.309 | 0.260 | 0.147 |
| 7 | 0.260 | 0.074 | 0.028 | 0.118 | 0.217 | 0.054 | 0.329 | 0.254 | 0.105 |
| 8 | 0.256 | 0.037 | 0.010 | 0.097 | 0.208 | 0.025 | 0.337 | 0.243 | 0.076 |
| 9 | 0.257 | 0.016 | 0.003 | 0.085 | 0.207 | 0.010 | 0.340 | 0.236 | 0.056 |
| 10 | 0.266 | 0.006 | 0.001 | 0.081 | 0.215 | 0.003 | 0.342 | 0.231 | 0.042 |

Table 3.10. Probability of selecting order $p$ over the true order 5 for Model 4, $\phi_j \propto 1/j$.

| $p$ | AIC | AICc | AICu | SIC | HQ | HQc | Cp | FPE | FPEu |
|---|---|---|---|---|---|---|---|---|---|
| 1 | 0.378 | 0.599 | 0.825 | 0.719 | 0.484 | 0.700 | 0.390 | 0.368 | 0.665 |
| 2 | 0.619 | 0.800 | 0.923 | 0.853 | 0.700 | 0.858 | 0.629 | 0.612 | 0.828 |
| 3 | 0.709 | 0.853 | 0.935 | 0.874 | 0.766 | 0.890 | 0.724 | 0.705 | 0.862 |
| 4 | 0.723 | 0.835 | 0.902 | 0.839 | 0.760 | 0.863 | 0.745 | 0.721 | 0.834 |
| 5 | * | * | * | * | * | * | * | * | * |
| 6 | 0.261 | 0.128 | 0.072 | 0.152 | 0.229 | 0.106 | 0.309 | 0.260 | 0.147 |
| 7 | 0.260 | 0.074 | 0.028 | 0.118 | 0.217 | 0.054 | 0.329 | 0.254 | 0.105 |
| 8 | 0.256 | 0.037 | 0.010 | 0.097 | 0.208 | 0.025 | 0.337 | 0.243 | 0.076 |
| 9 | 0.257 | 0.016 | 0.003 | 0.085 | 0.207 | 0.010 | 0.340 | 0.236 | 0.056 |
| 10 | 0.266 | 0.006 | 0.001 | 0.081 | 0.215 | 0.003 | 0.342 | 0.231 | 0.042 |

In the next two Sections we will use Monte Carlo results for Model 3 and Model 4 to see whether our theoretical results will correspond to actual performance in simulations.

## 3.6. Autoregressive Monte Carlo Study

Here we will use Model 3, given by Eqs. (3.18)–(3.19), and Model 4, given by Eq. (3.18) and Eq. (3.20), in a simulation study to investigate observed efficiency and model order selection for our nine criteria. Ten thousand realizations are generated for each order $p$, and for each realization the criteria select a model. $L_2$ and K-L distances are also computed for each candidate model, and the observed efficiency of the selected model is then computed using Eq. (3.7) and Eq. (1.1) for $L_2$ and Eq. (3.8) and Eq. (1.2) for K-L, respectively. Averages, medians and standard deviations are computed for the 10,000 observed efficiencies, and the criteria are ranked according to their average observed efficiency. The criterion with the highest observed efficiency is given rank 1 (best), the criterion with the lowest observed efficiency is given rank 9 (worst), and ties get the best rank within the tied group. Results for Model 3 are summarized in Table 3.11, and for Model 4 in Table 3.12.

Recall from Chapter 2 that AICc, AICu, and HQc all have superlinear penalty functions that prevent excessive overfitting, but that AIC, SIC, and HQ have penalty functions that can result in excessive overfitting in small samples. Cp and FPE also overfit. Therefore we are not surprised that AICu, HQc, and AICc identify the true order most frequently. Of the consistent criteria HQc performs the best, correctly identifying the true model 87% of the time. Although the variant FPEu performs better than FPE, it still allows some excessive overfitting. We note that K-L does occasionally choose a model order less than the true order 5, illustrating its slight tendency to favor underfitting.

Table 3.11. Simulation results for Model 3, $\phi_5 = 1$. Counts and observed efficiency.

| | | | | | counts | | | | | | |
|---|---|---|---|---|---|---|---|---|---|---|---|
| $p$ | AIC | AICc | AICu | SIC | HQ | HQc | Cp | FPE | FPEu | $L_2$ | K-L |
| 1 | 0 | 1 | 11 | 1 | 0 | 1 | 0 | 0 | 1 | 0 | 1 |
| 2 | 0 | 0 | 1 | 1 | 0 | 1 | 0 | 0 | 0 | 0 | 5 |
| 3 | 1 | 1 | 2 | 1 | 1 | 2 | 1 | 1 | 1 | 0 | 8 |
| 4 | 4 | 7 | 9 | 7 | 6 | 8 | 6 | 4 | 7 | 0 | 22 |
| 5 | 5213 | 8430 | 9188 | 7562 | 5912 | 8739 | 6044 | 5436 | 7869 | 7868 | 7726 |
| 6 | 1342 | 1037 | 615 | 1074 | 1269 | 894 | 1247 | 1394 | 1104 | 1226 | 1471 |
| 7 | 894 | 349 | 130 | 487 | 792 | 255 | 769 | 917 | 463 | 496 | 471 |
| 8 | 779 | 134 | 38 | 334 | 633 | 77 | 605 | 757 | 265 | 222 | 191 |
| 9 | 660 | 29 | 5 | 231 | 519 | 18 | 501 | 587 | 148 | 126 | 76 |
| 10 | 1107 | 12 | 1 | 302 | 868 | 5 | 827 | 904 | 142 | 62 | 29 |
| | | | | | K-L observed efficiency | | | | | | |
| | AIC | AICc | AICu | SIC | HQ | HQc | Cp | FPE | FPEu | $L_2$ | K-L |
| ave | 0.663 | 0.884 | 0.927 | 0.817 | 0.707 | 0.902 | 0.715 | 0.681 | 0.843 | 0.984 | 1.000 |
| med | 0.845 | 1.000 | 1.000 | 1.000 | 0.940 | 1.000 | 0.949 | 0.887 | 1.000 | 1.000 | 1.000 |
| sd | 0.356 | 0.234 | 0.184 | 0.300 | 0.349 | 0.215 | 0.347 | 0.351 | 0.277 | 0.063 | 0.000 |
| rank | 9 | 3 | 1 | 5 | 7 | 2 | 6 | 8 | 4 | | |
| | | | | | $L_2$ observed efficiency | | | | | | |
| | AIC | AICc | AICu | SIC | HQ | HQc | Cp | FPE | FPEu | $L_2$ | K-L |
| ave | 0.731 | 0.902 | 0.933 | 0.851 | 0.766 | 0.915 | 0.772 | 0.745 | 0.871 | 1.000 | 0.984 |
| med | 0.867 | 1.000 | 1.000 | 1.000 | 0.959 | 1.000 | 0.970 | 0.905 | 1.000 | 1.000 | 1.000 |
| sd | 0.299 | 0.195 | 0.158 | 0.247 | 0.290 | 0.180 | 0.288 | 0.294 | 0.228 | 0.000 | 0.070 |
| rank | 9 | 3 | 1 | 5 | 7 | 2 | 6 | 8 | 4 | | |

The ranked observed efficiencies in Table 3.11 exactly parallel the counts. The higher the frequency of selecting the true order, the higher the observed efficiency. Both observed efficiency measures yield similar results in terms of relative performance among the criteria. Since underfitting is almost nonexistent, only overfitting is a concern here. $L_2$ penalizes overfitting less than K-L resulting in higher $L_2$ observed efficiencies than K-L observed efficiencies in Table 3.11.

In contrast to the first model, Model 4 (Table 3.12) is an example of a time series with weakly identifiable and decaying parameters. In this case we expect underfitting tendencies will have as much of an impact on correct model selection as overfitting tendencies.

Since the true model is difficult to detect, the correct counts for Model 4 are of course fewer. We see that the criteria with the strongest penalty functions, AICc, AICu, and HQc, underfit the model most dramatically. SIC and FPEu also underfit more than they overfit. While AIC, HQ, Cp and FPE underfit less severely, they also have a larger problem with overfitting.

Table 3.12. Simulation results for Model 4, $\phi_j \propto 1/j$. Counts and observed efficiency.

| | counts | | | | | | | | | | |
|---|---|---|---|---|---|---|---|---|---|---|---|
| $p$ | AIC | AICc | AICu | SIC | HQ | HQc | Cp | FPE | FPEu | $L_2$ | K-L |
| 1 | 1149 | 1901 | 3124 | 2829 | 1609 | 2393 | 1518 | 1111 | 2398 | 111 | 522 |
| 2 | 2434 | 3634 | 4098 | 3672 | 2905 | 3878 | 2775 | 2437 | 3671 | 1538 | 2421 |
| 3 | 2142 | 2511 | 1949 | 1945 | 2130 | 2313 | 2066 | 2186 | 2177 | 3447 | 3311 |
| 4 | 1216 | 1108 | 570 | 734 | 1089 | 865 | 1121 | 1263 | 884 | 2961 | 2399 |
| 5 | 822 | 530 | 191 | 363 | 681 | 366 | 696 | 858 | 446 | 1314 | 953 |
| 6 | 514 | 174 | 46 | 148 | 391 | 112 | 414 | 541 | 172 | 343 | 259 |
| 7 | 421 | 82 | 15 | 98 | 296 | 47 | 319 | 428 | 105 | 126 | 82 |
| 8 | 384 | 35 | 4 | 72 | 265 | 15 | 326 | 378 | 63 | 93 | 35 |
| 9 | 381 | 22 | 3 | 73 | 264 | 10 | 339 | 373 | 50 | 38 | 13 |
| 10 | 537 | 3 | 0 | 66 | 370 | 1 | 426 | 425 | 34 | 29 | 5 |
| | K-L observed efficiency | | | | | | | | | | |
| | AIC | AICc | AICu | SIC | HQ | HQc | Cp | FPE | FPEu | $L_2$ | K-L |
| ave | 0.520 | 0.610 | 0.606 | 0.582 | 0.543 | 0.611 | 0.532 | 0.526 | 0.592 | 0.946 | 1.000 |
| med | 0.534 | 0.622 | 0.610 | 0.591 | 0.563 | 0.620 | 0.547 | 0.541 | 0.604 | 1.000 | 1.000 |
| sd | 0.299 | 0.253 | 0.246 | 0.260 | 0.289 | 0.249 | 0.293 | 0.297 | 0.258 | 0.128 | 0.000 |
| rank | 9 | 2 | 3 | 5 | 6 | 1 | 7 | 8 | 4 | | |
| | $L_2$ observed efficiency | | | | | | | | | | |
| | AIC | AICc | AICu | SIC | HQ | HQc | Cp | FPE | FPEu | $L_2$ | K-L |
| ave | 0.564 | 0.609 | 0.572 | 0.565 | 0.569 | 0.596 | 0.561 | 0.570 | 0.584 | 1.000 | 0.938 |
| med | 0.555 | 0.605 | 0.558 | 0.550 | 0.561 | 0.587 | 0.550 | 0.563 | 0.573 | 1.000 | 1.000 |
| sd | 0.276 | 0.256 | 0.259 | 0.263 | 0.271 | 0.257 | 0.273 | 0.275 | 0.261 | 0.000 | 0.146 |
| rank | 7 | 1 | 4 | 6 | 5 | 2 | 8 | 5 | 3 | | |

For the observed efficiencies, we see from Table 3.12 that the criteria with the strongest penalty functions and the most likely to underfit also have the highest K-L observed efficiency. Since K-L favors criteria with a slight tendency to underfit, this is not surprising. We also see from Table 3.12 that both AICc and HQc strike a good balance between overfitting and underfitting. AIC, HQ, Cp, and FPE all overfit and consequently have much lower observed efficiencies than the best performers.

We know that $L_2$ penalizes underfitting more than K-L, and we see from Table 3.12 that although AICc and HQc are still in the top ranks, AICu, with the largest penalty function, drops to fourth place with respect to $L_2$ observed efficiency. For both distance measures excessive underfitting from large penalty functions seems to be less of a problem than excessive overfitting from weak penalty functions. As we have noted before, model selection criteria that overfit excessively have lower observed efficiency in small samples than those that underfit. Ideally, in practice we would like to use a criterion that balances underfitting and overfitting, such as the signal-to-noise corrected variants that perform well under both distance measures. Simulations in Chapter 9 over many different models and for large sample sizes will further examine the ability of these variants to balance overfitting and underfitting in practice.

In the beginning of this Chapter we observed that autoregressive models are of finite order; however, time series models do allow us a convenient way to generate models of infinite order by using moving average models. In the next Section we will examine the behavior of selection criteria when applied to such a model, once again by using Monte Carlo simulation.

## 3.7. Moving Average MA(1) Misspecified as Autoregressive Models

In this Section the two special case models we will consider are MA(1) models, but the candidate models will be AR models. MA(1) models have the form $y_t = \theta_1 w_{*t-1} + w_{*t}$, where the $w_{*t}$ are *i.i.d.* $N(0, \sigma_*^2)$. This means that the true model does not belong to the set of candidate models, and thus by definition efficient model selection criteria should outperform consistent criteria. We will rely on observed efficiencies to measure performance since there is no true model in the set of candidate models for which to count the number of correct selections. We take this approach because fitting an MA true model with AR candidates makes it possible to perform a simple simulation study where the true model does not belong to the set of candidate models, and the goal is to select the model that is closest to the true model.

### 3.7.1. Two Special Case MA(1) Models

Both Models 5 and 6 share the following features:

$$n = 35, \sigma_*^2 = 1 \text{ and } y_t = \theta_1 w_{*t-1} + w_{*t}. \tag{3.21}$$

Our two special case MA(1) true models are described as follows. Model 5 has

$$\theta_1 = 0.5 \tag{3.22}$$

and Model 6 has

$$\theta_1 = 0.9. \tag{3.23}$$

Both are stationary and can be written in terms of an infinite order AR model. The $\phi_j = \theta_1^j$ parameters decay much more quickly for Model 5 than for Model 6, and therefore in small samples, Model 5 may be approximated by a finite order AR model. Model 6 has AR parameters that decay much more slowly, and in small samples, no good approximation may exist.

### 3.7.2. Model and Distance Measure Definitions

For MA(1), $y_t = \theta_1 w_{*t-1} + w_{*t}$ where the $w_{*t}$ are *i.i.d.* $N(0, \sigma_*^2)$. An MA(1) model can be expressed as the infinite order autoregressive model $y_t = \sum_{j=0}^{\infty} \theta_1^j y_{t-j}$. For each finite order candidate AR($p$) model and the true model MA(1), $L_2$ can be computed as follows:

$$L_2 = \frac{1}{n-p} \sum_{t=p+1}^{n} (\hat{y}_t - \mu_{*t})^2, \tag{3.24}$$

where

$$\mu_{*t} = \theta_1 w_{*t-1}, \text{ and } \hat{y}_t = \sum_{j=1}^{p} \hat{\phi}_j y_{t-j}$$

is computed from the AR($p$) candidate model.

Analogously, we can compute K-L distance based on the finite order candidate model AR($p$) and the true MA(1) model:

$$\text{K-L} = \log\left(\frac{\hat{\sigma}_p^2}{\sigma_*^2}\right) + \frac{\sigma_*^2}{\hat{\sigma}_p^2} + \frac{L_2}{\hat{\sigma}_p^2} - 1, \tag{3.25}$$

where $L_2$ is given by Eq. (3.24) and $\hat{\sigma}_p^2$ is given by Eq. (3.4). K-L observed efficiency then can be defined using Eq. (1.2) and the K-L distance given in Eq. (3.25).

### *3.7.3. Expected Values for Two Special Case Models*

Each time series $Y$ was generated starting at $y_{-50} = 0$ and $y_t = 0$ for all $t < -50$. Only the observations $y_1, \ldots, y_{35}$ were kept, and 100,000 realizations were simulated. Candidate AR models of order 1 through 15 were fit to the data. Model 5 is defined by Eq. (3.21) and Eq. (3.22). Model 6 is

Table 3.13. Expected values and expected efficiency, for Model 5, $\theta_1 = 0.5$.

| $p$ | AIC | AICc | AICu | SIC | HQ | HQc | Cp | FPE | FPEu | $L_2$ | K-L |
|---|---|---|---|---|---|---|---|---|---|---|---|
| 1 | 0.112 | **1.123** | **1.153** | **0.098** | **0.068** | **0.075** | 2.687 | 1.089 | **1.122** | 0.887 | 1.000 |
| 2 | **0.101** | 1.126 | 1.189 | 0.131 | 0.071 | 0.092 | **2.098** | **1.077** | 1.146 | 1.000 | 0.958 |
| 3 | 0.129 | 1.175 | 1.273 | 0.203 | 0.112 | 0.155 | 2.843 | 1.106 | 1.221 | 0.763 | 0.731 |
| 4 | 0.160 | 1.237 | 1.375 | 0.280 | 0.155 | 0.232 | 3.710 | 1.143 | 1.312 | 0.576 | 0.548 |
| 5 | 0.197 | 1.318 | 1.501 | 0.363 | 0.205 | 0.329 | 4.737 | 1.188 | 1.426 | 0.455 | 0.424 |
| 6 | 0.234 | 1.418 | 1.650 | 0.448 | 0.254 | 0.445 | 5.751 | 1.238 | 1.561 | 0.370 | 0.332 |
| 7 | 0.275 | 1.546 | 1.834 | 0.537 | 0.306 | 0.591 | 6.850 | 1.298 | 1.730 | 0.311 | 0.265 |
| 8 | 0.316 | 1.708 | 2.059 | 0.626 | 0.356 | 0.772 | 7.921 | 1.363 | 1.937 | 0.265 | 0.212 |
| 9 | 0.358 | 1.922 | 2.347 | 0.717 | 0.407 | 1.006 | 9.027 | 1.442 | 2.205 | 0.230 | 0.170 |
| 10 | 0.398 | 2.211 | 2.722 | 0.806 | 0.454 | 1.317 | 10.123 | 1.532 | 2.553 | 0.200 | 0.135 |
| 11 | 0.437 | 2.619 | 3.232 | 0.893 | 0.497 | 1.749 | 11.233 | 1.642 | 3.031 | 0.176 | 0.106 |
| 12 | 0.465 | 3.224 | 3.961 | 0.971 | 0.527 | 2.382 | 12.271 | 1.772 | 3.705 | 0.155 | 0.081 |
| 13 | 0.481 | 4.208 | 5.102 | 1.035 | 0.542 | 3.400 | 13.290 | 1.938 | 4.737 | 0.137 | 0.059 |
| 14 | 0.465 | 6.037 | 7.135 | 1.067 | 0.521 | 5.272 | 14.215 | 2.152 | 6.456 | 0.121 | 0.040 |
| 15 | 0.387 | 10.454 | 11.840 | 1.034 | 0.433 | 9.759 | 15.000 | 2.456 | 9.823 | 0.106 | 0.024 |

Boldface type indicates the minimum expectation.

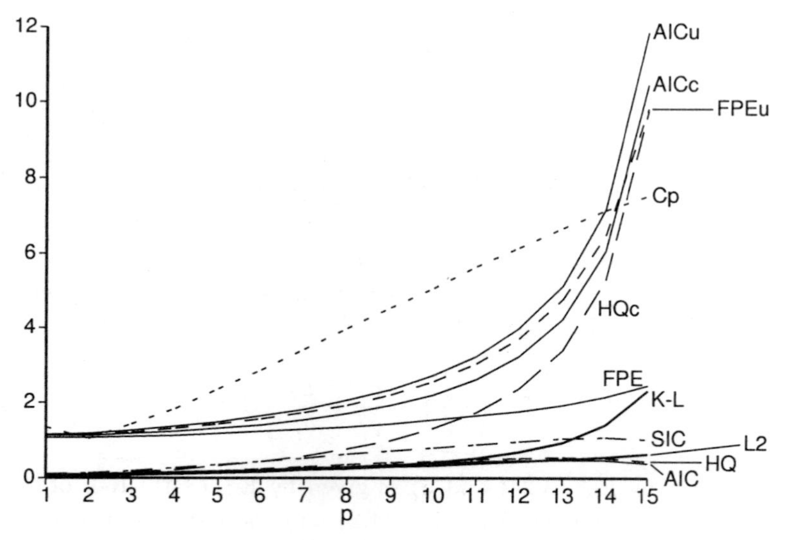


Figure 3.5. Expected values and expected distance for Model 5.

similarly defined by Eq. (3.21) and Eq. (3.23). Tables 3.13 and 3.14 and Figures 3.5 and 3.6 summarize the approximate expected values for the selection criteria computed by averaging over all realizations. $L_2$ expected distance is computed by averaging the $L_2$ distance Eq. (3.24) over the 100,000 realizations. These estimated expected distances are then used in Eq. (1.3) to compute $L_2$ expected efficiency. Similarly, K-L expected distance is computed by averaging

Table 3.14. Expected values and expected efficiency for Model 6, $\theta_1 = 0.9$.

| $p$ | AIC | AICc | AICu | SIC | HQ | HQc | Cp | FPE | FPEu | $L_2$ | K-L |
|---|---|---|---|---|---|---|---|---|---|---|---|
| 1 | 0.371 | 1.383 | 1.413 | 0.357 | 0.328 | 0.335 | 13.964 | 1.420 | 1.463 | 0.555 | 0.741 |
| 2 | 0.279 | **1.304** | **1.367** | **0.309** | 0.249 | **0.270** | 8.977 | 1.292 | **1.375** | 0.804 | 0.940 |
| 3 | 0.259 | 1.306 | 1.404 | 0.334 | **0.242** | 0.286 | 7.599 | 1.266 | 1.397 | 0.951 | 1.000 |
| 4 | **0.251** | 1.329 | 1.467 | 0.372 | 0.247 | 0.324 | **6.875** | **1.256** | 1.443 | 1.000 | 0.941 |
| 5 | 0.269 | 1.391 | 1.573 | 0.436 | 0.277 | 0.401 | 7.135 | 1.281 | 1.537 | 0.973 | 0.838 |
| 6 | 0.287 | 1.470 | 1.702 | 0.501 | 0.306 | 0.498 | 7.451 | 1.307 | 1.648 | 0.908 | 0.709 |
| 7 | 0.318 | 1.589 | 1.876 | 0.580 | 0.348 | 0.634 | 8.182 | 1.356 | 1.808 | 0.829 | 0.595 |
| 8 | 0.347 | 1.739 | 2.090 | 0.656 | 0.387 | 0.802 | 8.868 | 1.408 | 2.000 | 0.748 | 0.487 |
| 9 | 0.384 | 1.948 | 2.373 | 0.742 | 0.432 | 1.032 | 9.800 | 1.480 | 2.264 | 0.670 | 0.397 |
| 10 | 0.416 | 2.228 | 2.739 | 0.823 | 0.471 | 1.334 | 10.668 | 1.561 | 2.601 | 0.599 | 0.317 |
| 11 | 0.451 | 2.633 | 3.246 | 0.908 | 0.511 | 1.764 | 11.655 | 1.668 | 3.079 | 0.534 | 0.250 |
| 12 | 0.475 | 3.234 | 3.971 | 0.981 | 0.537 | 2.392 | 12.571 | 1.791 | 3.744 | 0.476 | 0.191 |
| 13 | 0.488 | 4.216 | 5.109 | 1.042 | 0.549 | 3.407 | 13.511 | 1.953 | 4.775 | 0.423 | 0.140 |
| 14 | 0.464 | 6.036 | 7.134 | 1.065 | 0.520 | 5.270 | 14.296 | 2.153 | 6.459 | 0.374 | 0.094 |
| 15 | 0.376 | 10.442 | 11.829 | 1.023 | 0.422 | 9.748 | 15.000 | 2.437 | 9.749 | 0.329 | 0.054 |

Boldface type indicates the minimum expectation.

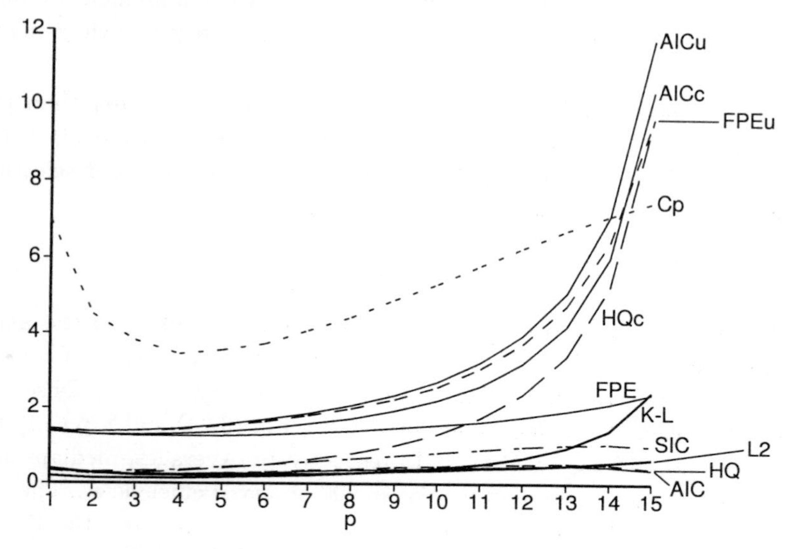


Figure 3.6. Expected values and expected distance for Model 6.

Eq. (3.25) and K-L expected efficiency uses these averaged values in Eq. (1.4).

Because Model 5 has a small MA(1) parameter and the AR parameters decay very quickly, it should be best approximated by a short AR model. In fact the selection criteria all attain minima at orders 1 or 2. In terms of K-L, the AR(1) model is closest to the true MA(1). Under $L_2$ the AR(2) model is closest. Here underfitting may be defined in terms of selecting an AR model of higher order than the order of the closest AR model. In this sense, some of the selection criteria still overfit.

Figure 3.5 plots actual expected values for $L_2$ and K-L rather than the expected efficiencies found in Table 3.13. Note that for presentation purposes, Cp in Figure 3.5 has been scaled by 1/2. We next look at the results for Model 6.

For Model 6 the closest AR model in terms of K-L is the AR(3) model, whereas for $L_2$ the closest model is the AR(4) model. AICu, SIC, HQc, and FPEu all have a minimum at order 2. These criteria underfit in the sense that order 2 has reduced K-L expected efficiency and reduced $L_2$ expected efficiency. HQ has a slightly larger penalty function than AIC ($2.5p$ versus $2p$), and here this makes the difference between HQ's minimum at order 2 and AIC's minimum at order 4. AICc and HQ have a minimum at order 3. The criteria with the weakest penalty functions in small samples, AIC, Cp, and FPE, have their minimum at order 4. We see that in a small sample size of $n = 35$, the criteria do not split along the consistency/efficiency line. Instead their small-sample performance depends on the strength of the penalty function.

Figure 3.6 plots actual expected values for $L_2$ and K-L rather than the expected efficiencies in Table 3.14. Note that for presentation purposes, Cp in Figure 3.6 has been scaled by 1/2. Next we will look at the order selection counts and criterion efficiencies for the simulation.

### *3.7.4. Misspecified MA(1) Monte Carlo study*

Ten thousand realizations are generated for Models 5 and 6 in the manner described in the previous Section. AR($p$) candidate models are fitted to the data and the criteria then select a model for each realization. Observed efficiency of the selected model is computed using Eq. (3.24) and Eq. (1.1) for $L_2$, and Eq. (3.25) and Eq. (1.2) for K-L, respectively. Averages, medians and standard deviations are computed for the 10,000 observed efficiencies. The criteria are ranked according to their average observed efficiency over the 10,000 realizations as described in Section 3.6. Results for Model 5 are summarized in Table 3.15 and for Model 6 in Table 3.16.

Overall, the performance of each of the model selection criteria is similar for Models 5 and 6. However, Model 6 has a much larger $\theta_1$ parameter which results in an higher order AR model choice. This can be seen from the K-L and $L_2$ columns in Table 3.16, where the closest model order tends to be in the range of 2 to 7. Since the true model does not belong to the set of candidate models, it is difficult to discuss overfitting patterns. However, we can define overfitting in terms of the distance measure. Let $\tilde{p}$ be the order of the model that is closest to the true model with respect to the distance measure (*e.g.*, $L_2$). Overfitting is defined as selecting a model of order greater than $\tilde{p}$ that causes a loss in observed efficiency. Underfitting is defined as selecting an order less than $\tilde{p}$ that causes a loss in observed efficiency. In other words, any order that is different from $\tilde{p}$ causes a loss in observed efficiency. With this in mind,

Table 3.15. Simulation results for Model 5, $\theta_1 = 0.5$. Counts and observed efficiency.

| | | | | | counts | | | | | | |
|---|---|---|---|---|---|---|---|---|---|---|---|
| $p$ | AIC | AICc | AICu | SIC | HQ | HQc | Cp | FPE | FPEu | $L_2$ | K-L |
| 1 | 3012 | 5241 | 6894 | 6173 | 3820 | 6004 | 5311 | 3347 | 5993 | 2395 | 3215 |
| 2 | 1817 | 2761 | 2271 | 2028 | 1917 | 2595 | 1717 | 2089 | 2398 | 4541 | 4115 |
| 3 | 831 | 1067 | 573 | 587 | 777 | 821 | 605 | 976 | 768 | 2187 | 1895 |
| 4 | 544 | 490 | 179 | 217 | 435 | 332 | 342 | 646 | 336 | 568 | 513 |
| 5 | 322 | 232 | 57 | 130 | 233 | 152 | 208 | 412 | 183 | 181 | 177 |
| 6 | 237 | 111 | 14 | 67 | 176 | 54 | 165 | 303 | 92 | 68 | 49 |
| 7 | 196 | 66 | 10 | 53 | 145 | 33 | 118 | 254 | 73 | 31 | 24 |
| 8 | 178 | 27 | 2 | 41 | 112 | 9 | 129 | 225 | 35 | 17 | 7 |
| 9 | 145 | 5 | 0 | 26 | 101 | 0 | 105 | 192 | 24 | 2 | 1 |
| 10 | 129 | 0 | 0 | 21 | 73 | 0 | 101 | 165 | 19 | 4 | 3 |
| 11 | 122 | 0 | 0 | 17 | 80 | 0 | 91 | 145 | 10 | 3 | 0 |
| 12 | 161 | 0 | 0 | 29 | 118 | 0 | 114 | 161 | 6 | 3 | 1 |
| 13 | 258 | 0 | 0 | 44 | 202 | 0 | 140 | 220 | 13 | 0 | 0 |
| 14 | 478 | 0 | 0 | 108 | 396 | 0 | 252 | 287 | 16 | 0 | 0 |
| 15 | 1570 | 0 | 0 | 459 | 1415 | 0 | 602 | 578 | 34 | 0 | 0 |
| | | | | K-L observed efficiency | | | | | | | |
| | AIC | AICc | AICu | SIC | HQ | HQc | Cp | FPE | FPEu | $L_2$ | K-L |
| ave | 0.412 | 0.626 | 0.664 | 0.602 | 0.460 | 0.644 | 0.540 | 0.468 | 0.628 | 0.961 | 1.000 |
| med | 0.360 | 0.655 | 0.687 | 0.641 | 0.466 | 0.670 | 0.582 | 0.457 | 0.659 | 1.000 | 1.000 |
| sd | 0.362 | 0.285 | 0.260 | 0.309 | 0.361 | 0.273 | 0.344 | 0.351 | 0.285 | 0.106 | 0.000 |
| rank | 9 | 4 | 1 | 5 | 8 | 2 | 6 | 7 | 3 | | |
| | | | | $L_2$ observed efficiency | | | | | | | |
| | AIC | AICc | AICu | SIC | HQ | HQc | Cp | FPE | FPEu | $L_2$ | K-L |
| ave | 0.424 | 0.598 | 0.610 | 0.561 | 0.459 | 0.604 | 0.513 | 0.472 | 0.589 | 1.000 | 0.950 |
| med | 0.342 | 0.584 | 0.599 | 0.554 | 0.410 | 0.592 | 0.494 | 0.419 | 0.577 | 1.000 | 1.000 |
| sd | 0.339 | 0.289 | 0.270 | 0.298 | 0.337 | 0.280 | 0.320 | 0.332 | 0.285 | 0.000 | 0.131 |
| rank | 9 | 3 | 1 | 5 | 8 | 2 | 6 | 7 | 4 | | |

we see that the same overfitting problems that are evident when the true model belongs to the set of candidate models are present here as well. Overfitting and underfitting problems exist regardless of whether or not the true model belongs to the set of candidate models. More extensive MA(1) simulation studies will be presented in Chapter 9.

Table 3.16. Simulation results for Model 6, $\theta_1 = 0.9$. Counts and observed efficiency.

| | | | | | counts | | | | | | |
|---|---|---|---|---|---|---|---|---|---|---|---|
| $p$ | AIC | AICc | AICu | SIC | HQ | HQc | Cp | FPE | FPEu | $L_2$ | K-L |
| 1 | 565 | 1592 | 3136 | 2438 | 931 | 2215 | 1925 | 649 | 2156 | 48 | 663 |
| 2 | 1404 | 3236 | 3766 | 2941 | 1820 | 3502 | 2427 | 1631 | 3198 | 601 | 1403 |
| 3 | 1227 | 2268 | 1788 | 1583 | 1313 | 2105 | 1398 | 1488 | 1930 | 1635 | 2224 |
| 4 | 1089 | 1595 | 888 | 957 | 1073 | 1320 | 981 | 1326 | 1276 | 2120 | 2041 |
| 5 | 625 | 689 | 282 | 404 | 582 | 476 | 510 | 817 | 562 | 2078 | 1676 |
| 6 | 595 | 391 | 98 | 264 | 491 | 251 | 399 | 728 | 358 | 1434 | 971 |
| 7 | 348 | 150 | 34 | 134 | 273 | 90 | 266 | 461 | 182 | 916 | 551 |
| 8 | 313 | 65 | 8 | 88 | 229 | 35 | 221 | 408 | 101 | 513 | 277 |
| 9 | 218 | 12 | 0 | 55 | 157 | 6 | 156 | 269 | 53 | 326 | 127 |
| 10 | 212 | 2 | 0 | 36 | 148 | 0 | 159 | 268 | 35 | 144 | 41 |
| 11 | 222 | 0 | 0 | 46 | 162 | 0 | 150 | 258 | 31 | 84 | 18 |
| 12 | 228 | 0 | 0 | 40 | 163 | 0 | 130 | 205 | 17 | 50 | 7 |
| 13 | 335 | 0 | 0 | 92 | 262 | 0 | 172 | 269 | 30 | 26 | 1 |
| 14 | 660 | 0 | 0 | 191 | 553 | 0 | 348 | 418 | 22 | 18 | 0 |
| 15 | 1959 | 0 | 0 | 731 | 1843 | 0 | 758 | 805 | 49 | 7 | 0 |
| | | | | | K-L observed efficiency | | | | | | |
| | AIC | AICc | AICu | SIC | HQ | HQc | Cp | FPE | FPEu | $L_2$ | K-L |
| ave | 0.404 | 0.657 | 0.656 | 0.566 | 0.436 | 0.657 | 0.521 | 0.480 | 0.629 | 0.914 | 1.000 |
| med | 0.395 | 0.680 | 0.668 | 0.616 | 0.474 | 0.676 | 0.572 | 0.521 | 0.657 | 1.000 | 1.000 |
| sd | 0.333 | 0.216 | 0.203 | 0.283 | 0.332 | 0.210 | 0.303 | 0.317 | 0.233 | 0.160 | 0.000 |
| rank | 9 | 1 | 3 | 5 | 8 | 1 | 6 | 7 | 4 | | |
| | | | | | $L_2$ observed efficiency | | | | | | |
| | AIC | AICc | AICu | SIC | HQ | HQc | Cp | FPE | FPEu | $L_2$ | K-L |
| ave | 0.533 | 0.671 | 0.625 | 0.587 | 0.543 | 0.652 | 0.575 | 0.592 | 0.640 | 1.000 | 0.926 |
| med | 0.527 | 0.682 | 0.625 | 0.593 | 0.545 | 0.659 | 0.580 | 0.610 | 0.649 | 1.000 | 1.000 |
| sd | 0.262 | 0.199 | 0.207 | 0.231 | 0.257 | 0.202 | 0.241 | 0.248 | 0.209 | 0.000 | 0.136 |
| rank | 9 | 1 | 4 | 6 | 8 | 2 | 7 | 5 | 3 | | |

## 3.8. Multistep Forecasting Models

All time series models in this Chapter thus far have focused on the usual one step prediction error. Suppose the goal is to make longer range, $h$-step forecasts? In this case the goal is to predict $y_{n+h}$ from the time series $y_1, \ldots, y_n$. There are several methods for making such predictions. The first is to use some model of the form Eq. (3.1) and then generate forecasts $y_{n+1}, \ldots, y_{n+h}$.

We will focus on models of the form suggested by Bhansali (1997):

$$y_{t+h} = \phi_1 y_t + \cdots + \phi_p y_{t-p+1} + w_t, \quad t = p+1, \ldots, n,$$

which makes the $h$-step prediction directly from the observed data.

We first need to derive the optimal predictors for multistep linear prediction. Let $\{y_t\}$ be a weakly stationary time series with zero mean and autocovariance function $\{r_j\}$. The best linear predictor in terms of minimizing the mean squared error for predicting $y_{t+h}$ using $y_t, \ldots, y_{t-p+1}$ is

$$\hat{y}_{t+h} = -\sum_{j=h}^{h+p-1} a_j(h,p) y_{t+h-j},$$

where the coefficients $-a_j(h,p)$ satisfy

$$\sum_{j=h}^{h+p-1} r_{|i-j|} a_j(h,p) = -r_i, \quad i = h, \ldots, h+p-1.$$

It is simpler to consider these coefficients in vector form, and so we let $a(h,p) = [a_0(h,p), \ldots, a_{h+p-1}(h,p)]'$ be a vector with $a_0(h,p) = 1$ for all $(h,p)$ and $a_1(h,p) = \cdots = a_{h-1}(h,p) = 0$ if $h > 1$. The quantity $a(h,p)$ is referred to as the optimal prediction error filter. The $h$-step prediction error is $y_{t+h} - \hat{y}_{t+h} = a(h,p)' y(h,p)$ where $y(h,p) = [y_{t+h}, \ldots, y_{t-p+1}]'$. The $h$-step mean squared prediction error of $\hat{y}_{t+h}$ is $\sigma^2(h,p) = E[(y_{t+h} - \hat{y}_{t+h})^2] = a(h,p)' R_{h+p} a(h,p)$, where $R_{h+p}$ is the $(h+p) \times (h+p)$ covariance matrix of $y_1, \ldots, y_{h+p}$.

In practice, the autocovariances $r_j$ are unknown. If estimates of $r_j$ are available, Hurvich and Tsai (1997) suggest obtaining $\hat{a}_j(h,p)$ by solving

$$\sum_{j=h}^{h+p-1} \hat{r}_{|i-j|} \hat{a}_j(h,p) = -\hat{r}_i, \quad i = h, \ldots, h+p-1,$$

where the Burg Method (Burg 1978, Hainz 1995) is used to estimate the $r_j$.

### 3.8.1. Kullback–Leibler Discrepancy for Multistep

The Kullback–Leibler discrepancy can be derived for use with multistep prediction. Consider a candidate linear predictor or model of the form

$$\tilde{y}_{t+h} = -\sum_{j=h}^{h+p-1} b_j(h,p) y_{t+h-j},$$

where the $b_j(h,p)$ are arbitrary real constants. Considering these coefficients in vector form, let $b(h,p) = [b_0(h,p), \ldots, b_{h+p-1}(h,p)]'$ be a vector with $b_0(h,p) = 1$ for all $(h,p)$ and $b_1(h,p) = \cdots = b_{h-1}(h,p) = 0$ if $h > 1$. The structure of $b(h,p)$ is similar to that of $a(h,p)$. The mean squared prediction error of $\tilde{y}_{t+h}$ is

$$b(h,p)'R_{h+p}b(h,p) = \sigma^2(h,p) + [b(h,p) - a(h,p)]'R_{h+p}[b(h,p) - a(h,p)].$$

Now we can characterize the linear prediction of $y_{t+h}$ based on the $p$ past values $y_t, \ldots, y_{t-p+1}$ by using the optimal prediction error filter $a(h,p)$ together with the optimal prediction error variance $\sigma^2(h,p)$. However, $a(h,p)$ and $\sigma^2(h,p)$ are unknown parameters. If we let $\tau^2(h,p) > 0$ be some candidate for the prediction error variance, then we can use Hurvich and Tsai's proposed generalization of the Kullback–Leibler discrepancy between the true parameters $a(h,p)$, $\sigma^2(h,p)$ and candidate parameters $b(h,p)$, $\tau^2(h,p)$:

$$d_{h,p,a,\sigma^2}(b,\tau^2) = \log \tau^2(h,p) + b(h,p)'R_{h+p}b(h,p)/\tau^2(h,p).$$

$d_{h,p,a,\sigma^2}(b,\tau^2)$ is minimized when $b(h,p) = a(h,p)$ and $\tau^2(h,p) = \sigma^2(h,p)$.

### *3.8.2. AICcm, AICm, and FPEm*

We will obtain multistep versions for three of the criteria that we have considered in this Chapter, AICc, AIC, and FPE. The multistep variants are denoted with an "m." We will go through the detailed derivation for AICcm, but present only the results for AICm and FPEm since the procedures used to obtain them are similar.

So far we have compared the true prediction filter with an arbitrary candidate filter, where both filters use $p$ observations. For real data, one may be interested in estimating $a(h,p)$ and $\sigma^2(h,p)$. The candidate model will be the one estimated from the data with discrepancy

$$d_{h,p,a,\sigma^2}(\hat{a},\hat{\sigma}^2) = \log \hat{\sigma}^2(h,p) + \hat{a}(h,p)'R_{h+p}\hat{a}(h,p)/\hat{\sigma}^2(h,p).$$

The criterion AICcm models $E[d_{h,p,a,\sigma^2}(\hat{a},\hat{\sigma}^2)]$, and under the strong assumption that certain asymptotic distributions hold for the finite sample, Hurvich and Tsai (1997) showed that AICcm has the familiar form

$$\text{AICcm} = \log(\hat{\sigma}^2(h,p)) + \frac{n+p}{n-p-2}.$$

The analogous results for AICm and FPEm are

$$\text{AICm} = \log(\hat{\sigma}^2(h,p)) + \frac{2(p+1)}{n}$$

and

$$\text{FPEm} = \hat{\sigma}^2(h,p)\frac{n+p}{n-p}.$$

Recently, Liu (1996) obtained multistep versions of BIC, FIC, and Cp, which are not discussed here. The interested reader may want to compare their performance with that of AICcm, AICm and FPEm.

### *3.8.3. Multistep Monte Carlo Study*

To illustrate $h$-step forecasting performance, observed mean squared prediction errors are compared. For a time series with known autocovariance function, the optimal filter can be found as well as the optimal order $p$. Let this order be $p^*$. Of all the orders $p$ and optimal filters $a(h,p)$, $a(h,p^*)$ has the smallest MSE for $h$-step prediction. We can compute $\hat{\sigma}^2(h,p)$ for each order $p$, and compare the results for AICcm, AICm, and FPEm.

The AR(4) true model is given by

$$y_t = 2.7607y_{t-1} - 3.8106y_{t-2} + 2.6535y_{t-3} - 0.9238y_{t-4} + w_{*t},$$

with $w_{*t}$ *i.i.d.* $N(0,1)$. Simulations were conducted for sample sizes of $n = 30$, 50, 75 and steps $h = 1$, 2, 5 for candidate orders $p = 0, \ldots, 20$. One hundred realizations were used to compute the average MSE for the $h$-step prediction. Table 3.17 (Table 2 from Hurvich and Tsai, 1997 used with permission from Statistica Sinica) summarizes the results.

Table 3.17 Multistep AR simulation results.

| | $n=30$ | | | $n=50$ | | | $n=75$ | | |
|---|---|---|---|---|---|---|---|---|---|
| Ave MSE | $h=1$ | $h=2$ | $h=5$ | $h=1$ | $h=2$ | $h=5$ | $h=1$ | $h=2$ | $h=5$ |
| AICcm | 1.63 | 15.72 | 61.73 | 1.20 | 11.00 | 42.78 | 1.13 | 10.20 | 39.64 |
| AICm | 3.39 | 33.95 | 146.84 | 1.29 | 11.90 | 47.13 | 1.14 | 10.39 | 40.17 |
| FPEm | 2.52 | 26.22 | 124.27 | 1.27 | 11.80 | 45.81 | 1.14 | 10.36 | 40.00 |
| $p^*$ | 1.62 | 15.46 | 59.27 | 1.18 | 10.76 | 41.43 | 1.13 | 10.14 | 38.30 |

AICcm performs well in terms of prediction MSE; at its worst, its error relative to the optimal $p^*$ is only 4.15%. By contrast, AICm has up to 147.75% relative error while FPEm has up to 109.67% relative error. As the sample size increases, the performance differences diminish, and all three criteria perform about the same for $n = 75$.

## 3.9. Summary

The main concepts that we identified in Chapter 2 for criterion behavior with univariate regression models carry over to univariate autoregressive models as well. We see that for autoregressive models the structure of the penalty function is the most important feature in determining small-sample performance while asymptotic properties play less of a role. This is illustrated by the performance results for the AIC and HQ criteria—although AIC is asymptotically efficient and HQ is asymptotically consistent, both have similar penalty function structure, and overfit excessively in small samples. AICc, AICu, and HQc have penalty functions that increase unboundedly as $p$ increases, or superlinearly. These superlinear penalty functions prevent small-sample overfitting and yield model selection criteria with good small-sample performance. For example, we have seen that HQc, although asymptotically consistent, performs well both when the true model belongs to the set of candidate models and also the true model does not belong to the set of candidate models.

Signal-to-noise ratio values remain good predictors for most of the performance measures we considered. High probabilities of selecting the correct model order correspond to high signal-to-noise ratios for the model type in question. High signal-to-noise ratios for overfitted or underfitted models correspond to low probabilities of overfitting and underfitting.

Finally, we once again found that when the true model is only weakly identifiable (or nonexistent) and is therefore defined by the distance measure considered, it is particularly important that criteria that perform well under both K-L and $L_2$. These trends will be revisited in Chapters 4 and 5, and addressed in much greater detail in the large-scale simulation studies in Chapter 9.

## Chapter 3 Appendices

### Appendix 3A. Distributional Results in the Central Case

As with regression models in Chapter 2, derivations of some model selection criteria and their signal-to-noise corrected variants under autoregressive models depend on the distribution of $\text{SSE}_p$, as well as the distributions of linear combinations of $\text{SSE}_p$ and $\text{SSE}_{p+L}$.

Assume all distributions below are central. In conditional AR($p$) models, the loss of $p$ observations for conditioning on the past results in the loss of 2 degrees of freedom for estimating the $p$ unknown $\phi$ parameters whenever $p$

increases by one. Recall that $T = n - p$. Thus,

$$\text{SSE}_p - \text{SSE}_{p+L} \sim \sigma_*^2 \chi_{2L}^2, \tag{3A.1}$$

$$\text{SSE}_p \sim \sigma_*^2 \chi_{T-p}^2 \tag{3A.2}$$

and

$$\text{SSE}_p - \text{SSE}_{p+L} \text{ independent of } \text{SSE}_p. \tag{3A.3}$$

Next we consider the linear combination $a\text{SSE}_p - b\text{SSE}_{p+L}$ where a and b are scalars. It follows from Eq. (3A.3) that

$$\begin{aligned} E[a\text{SSE}_p - b\text{SSE}_{p+L}] &= a(T-p)\sigma_*^2 - b(T-p-2L)\sigma_*^2 \\ &= a(n-2p)\sigma_*^2 - b(n-2p-2L)\sigma_*^2. \end{aligned} \tag{3A.4}$$

In addition,

$$\begin{aligned} var[a\text{SSE}_p - b\text{SSE}_{p+L}] &= var[a\text{SSE}_p - a\text{SSE}_{p+L} + a\text{SSE}_{p+L} - b\text{SSE}_{p+L}] \\ &= var[a(\text{SSE}_p - \text{SSE}_{p+L}) + (a-b)\text{SSE}_{p+L}]. \end{aligned}$$

Applying Eqs. (3A.1)–(3A.3), we have

$$\begin{aligned} var[a\text{SSE}_p - b\text{SSE}_{p+L}] &= 4a^2 L\sigma_*^4 + 2(a-b)^2(T-p-2L)\sigma_*^4 \\ &= 4a^2 L\sigma_*^4 + 2(a-b)^2(n-2p-2L)\sigma_*^4. \end{aligned} \tag{3A.5}$$

Lastly, we review some useful distributions involving $\log(\text{SSE}_p)$. It follows from (2A.13) that

$$E\big[\log(\text{SSE}_p)\big] = \log(\sigma_*^2) + \log(2) + \psi\left(\frac{T-p}{2}\right).$$

Although we know that $\psi$ has no closed form, we can expand $\log[(\text{SSE}_p)/\sigma_*^2]$ around $E(\text{SSE}_p/\sigma_*^2) = T-p$ as a Taylor expansion, yielding the approximation

$$\begin{aligned} E\big[\log(\text{SSE}_p)\big] &= \log(\sigma_*^2) + \log(T-p) - \frac{1}{T-p} \\ &= \log(\sigma_*^2) + \log(n-2p) - \frac{1}{n-2p}. \end{aligned} \tag{3A.6}$$

Furthermore, applying Eqs. (3A.1)–(3A.3), we have

$$\frac{\text{SSE}_{p+L}}{\text{SSE}_p} \sim \frac{\chi_{T-p-2L}^2}{\chi_{T-p-2L}^2 + \chi_{2L}^2},$$

$$\frac{\mathrm{SSE}_{p+L}}{\mathrm{SSE}_p} \sim \mathrm{Beta}\left(\frac{T-p-2L}{2}, \frac{2L}{2}\right)$$

and

$$\log\left(\frac{\mathrm{SSE}_{p+L}}{\mathrm{SSE}_p}\right) \sim \text{log-Beta}\left(\frac{T-p-2L}{2}, \frac{2L}{2}\right),$$

with variance

$$var\left[\log\left(\frac{\mathrm{SSE}_{p+L}}{\mathrm{SSE}_p}\right)\right] = \psi'\left(\frac{T-p-2L}{2}\right) - \psi'\left(\frac{T-p}{2}\right),$$

which also has no closed form. Again we take a Taylor expansion (see Chapter 2, Appendix 2A) to find the approximation

$$\begin{aligned} var\left[\log\left(\frac{\mathrm{SSE}_{p+L}}{\mathrm{SSE}_p}\right)\right] &\doteq \frac{4L}{(T-p-2L)(T-p+2)} \\ &\doteq \frac{4L}{(n-2p-2L)(n-2p+2)}. \end{aligned} \tag{3A.7}$$

This approximation is used for computing the signal-to-noise ratios of model selection criteria.

## Appendix 3B. Small-sample Probabilities of Overfitting

### *3B.1. AICc*

In terms of the original sample size $n$, AICc $= \log(\hat{\sigma}_p^2) + n/(n-2p-2)$. AICc overfits if $\mathrm{AICc}_{p_*+L} < \mathrm{AICc}_{p_*}$. For finite $n$, $P\{\text{overfit}\}$ is

$$\begin{aligned} &P\{\mathrm{AICc}_{p_*+L} < \mathrm{AICc}_{p_*}\} \\ &= P\left\{\log\left(\hat{\sigma}^2_{p_*+L}\right) + \frac{n}{n-2p_*-2L-2} < \log\left(\hat{\sigma}^2_{p_*}\right) + \frac{n}{n-2p_*-2}\right\} \\ &= P\left\{\log\left(\frac{\mathrm{SSE}_{p_*+L}}{\mathrm{SSE}_{p_*}}\right) < \log\left(\frac{n-p_*-L}{n-p_*}\right) + \frac{n}{n-2p_*-2} - \frac{n}{n-2p_*-2L-2}\right\} \\ &= P\left\{\frac{\chi^2_{2L}}{\chi^2_{n-2p_*-2L}} > \frac{n-p_*}{n-p_*-L}\exp\left(\frac{2Ln}{(n-2p_*-2L-2)(n-2p_*-2)}\right) - 1\right\} \\ &= P\left\{F_{2L,n-2p_*-2L} > \frac{n-2p_*-2L}{2L} \times \right. \\ &\qquad \left.\left(\frac{n-p_*}{n-p_*-L}\exp\left(\frac{2Ln}{(n-2p_*-2L-2)(n-2p_*-2)}\right) - 1\right)\right\}. \end{aligned}$$

### *3B.2. AICu*

In terms of the original sample size $n$, AICu $= \log(s_p^2) + n/(n-2p-2)$. AICu overfits if $\text{AICu}_{p_*+L} < \text{AICu}_{p_*}$. For finite $n$, $P\{\text{overfit}\}$ is

$$
\begin{aligned}
&P\{\text{AICu}_{p_*+L} < \text{AICu}_{p_*}\}\\
&= P\left\{\log\left(s^2_{p_*+L}\right) + \frac{n}{n-2p_*-2L-2} < \log\left(s^2_{p_*}\right) + \frac{n}{n-2p_*-2}\right\}\\
&= P\left\{\log\left(\frac{\text{SSE}_{p_*+L}}{\text{SSE}_{p_*}}\right) < \log\left(\frac{n-2p_*-2L}{n-2p_*}\right)\right.\\
&\qquad\left. + \frac{n}{n-2p_*-2} - \frac{n}{n-2p_*-2L-2}\right\}\\
&= P\left\{\frac{\chi^2_{2L}}{\chi^2_{n-2p_*-2L}} > \frac{n-2p_*}{n-2p_*-2L}\times\right.\\
&\qquad\left.\exp\left(\frac{2Ln}{(n-2p_*-2L-2)(n-2p_*-2)}\right) - 1\right\}\\
&= P\left\{F_{2L,n-2p_*-2L} > \frac{n-2p_*-2L}{2L}\times\right.\\
&\qquad\left.\left(\frac{n-2p_*}{n-2p_*-2L}\exp\left(\frac{2Ln}{(n-2p_*-2L-2)(n-2p_*-2)}\right) - 1\right)\right\}.
\end{aligned}
$$

### *3B.3. FPE*

Rewriting FPE in terms of the original sample size $n$, we have FPE $= \hat{\sigma}_p^2 n/(n-2p)$. FPE overfits if $\text{FPE}_{p_*+L} < \text{FPE}_{p_*}$. For finite $n$, $P\{\text{overfit}\}$ is

$$
\begin{aligned}
&P\{\text{FPE}_{p_*+L} < \text{FPE}_{p_*}\}\\
&= P\left\{\frac{n}{n-2p_*-2L}\hat{\sigma}^2_{p_*+L} < \frac{n}{n-2p_*}\hat{\sigma}^2_{p_*}\right\}\\
&= P\left\{\frac{n}{(n-2p_*-2L)(n-p_*-L)}\text{SSE}_{p_*+L} < \frac{n}{(n-2p_*)(n-p_*)}\text{SSE}_{p_*}\right\}\\
&= P\left\{\frac{\chi^2_{2L}}{\chi^2_{n-2p_*-2L}} > \frac{(n-2p_*)(n-p_*)}{(n-2p_*-2L)(n-p_*-L)} - 1\right\}\\
&= P\left\{F_{2L,n-2p_*-2L} > \frac{n-2p_*-2L}{2L}\left(\frac{(n-2p_*)(n-p_*)}{(n-2p_*-2L)(n-p_*-L)} - 1\right)\right\}\\
&= P\left\{F_{2L,n-2p_*-2L} > 1 + \frac{n-2p_*}{2(n-p_*-L)}\right\}.
\end{aligned}
$$

*3B.4. FPEu*

Rewriting FPEu in terms of the original sample size $n$, we have FPEu $= s_p^2 n/(n-2p)$. FPEu overfits if $\text{FPEu}_{p_*+L} < \text{FPEu}_{p_*}$. For finite $n$, $P\{\text{overfit}\}$ is

$$\begin{aligned}
&P\{\text{FPEu}_{p_*+L} < \text{FPEu}_{p_*}\} \\
&= P\left\{\frac{n}{n-2p_*-2L}s^2_{p_*+L} < \frac{n}{n-2p_*}s^2_{p_*}\right\} \\
&= P\left\{\frac{n}{(n-2p_*-2L)^2}\text{SSE}_{p_*+L} < \frac{n}{(n-2p_*)^2}\text{SSE}_{p_*}\right\} \\
&= P\left\{\frac{\chi^2_{2L}}{\chi^2_{n-2p_*-2L}} > \frac{(n-2p_*)^2}{(n-2p_*-2L)^2} - 1\right\} \\
&= P\left\{F_{2L,n-2p_*-2L} > \frac{n-2p_*-2L}{2L}\left(\frac{(n-2p_*)^2}{(n-2p_*-2L)^2} - 1\right)\right\} \\
&= P\left\{F_{2L,n-2p_*-2L} > 1 + \frac{n-2p_*}{n-2p_*-2L}\right\}.
\end{aligned}$$

*3B.5. Cp*

In terms of the original sample size $n$, Cp $= \text{SSE}_p/s_P^2 - n + 3p$. Cp overfits if $\text{Cp}_{p_*+L} < \text{Cp}_{p_*}$. Recall that $P$ represents the order of the largest AR($p$) model considered. For finite $n$, $P\{\text{overfit}\}$ is

$$\begin{aligned}
&P\{\text{Cp}_{p_*+L} < \text{Cp}_{p_*}\} \\
&= P\left\{(n-2P)\frac{\text{SSE}_{p_*+L}}{\text{SSE}_P} - n + 3(p_*+L) < (n-2P)\frac{\text{SSE}_{p_*}}{\text{SSE}_P} - n + 3p_*\right\} \\
&= P\left\{(n-2P)\frac{\text{SSE}_{p_*} - \text{SSE}_{p_*+L}}{\text{SSE}_P} > 3L\right\} \\
&= P\left\{(n-2P)\frac{\chi^2_{2L}}{\chi^2_{n-2P}} > 3L\right\} \\
&= P\left\{F_{2L,n-2P} > 1.5\right\}.
\end{aligned}$$

*3B.6. SIC*

Rewriting SIC in terms of the original sample size $n$, we have SIC $= \log(\hat{\sigma}_p^2) + \log(n-p)p/(n-p)$. SIC overfits if $\text{SIC}_{p_*+L} < \text{SIC}_{p_*}$. For finite $n$,

$P\{\text{overfit}\}$ is

$$
\begin{aligned}
&P\{\mathrm{SIC}_{p_*+L} < \mathrm{SIC}_{p_*}\} \\
&= P\left\{\log\left(\hat{\sigma}^2_{p_*+L}\right) + \frac{\log(n-p_*-L)(p_*+L)}{n-p_*-L} < \log\left(\hat{\sigma}^2_{p_*}\right) + \frac{\log(n-p_*)p_*}{n-p_*}\right\} \\
&= P\left\{\log\left(\frac{\mathrm{SSE}_{p_*+L}}{\mathrm{SSE}_{p_*}}\right) < \log\left(\frac{n-p_*-L}{n-p_*}\right)\right. \\
&\qquad \left. + \frac{\log(n-p_*)p_*}{n-p_*} - \frac{\log(n-p_*-L)(p_*+L)}{n-p_*-L}\right\} \\
&= P\left\{F_{2L,n-2p_*-2L} > \frac{n-2p_*-2L}{2L}\left(\frac{n-p_*}{n-p_*-L}\times\right.\right. \\
&\qquad \left.\left.\exp\left(\frac{\log(n-p_*-L)(p_*+L)}{n-p_*-L} - \frac{\log(n-p_*)p_*}{n-p_*}\right) - 1\right)\right\}.
\end{aligned}
$$

## *3B.7. HQ*

Rewriting HQ in terms of the original sample size $n$, we have $\mathrm{HQ} = \log(\hat{\sigma}^2_p) + 2\log\log(n-p)p/(n-p)$. HQ overfits if $\mathrm{HQ}_{p_*+L} < \mathrm{HQ}_{p_*}$. For finite $n$, $P\{\text{overfit}\}$ is

$$
\begin{aligned}
&P\{\mathrm{HQ}_{p_*+L} < \mathrm{HQ}_{p_*}\} \\
&= P\left\{\log\left(\hat{\sigma}^2_{p_*+L}\right) + \frac{2\log\log(n-p_*-L)(p_*+L)}{n-p_*-L}\right. \\
&\qquad \left. < \log\left(\hat{\sigma}^2_{p_*}\right) + \frac{2\log\log(n-p_*)p_*}{n-p_*}\right\} \\
&= P\left\{\log\left(\frac{\mathrm{SSE}_{p_*+L}}{\mathrm{SSE}_{p_*}}\right) < \log\left(\frac{n-p_*-L}{n-p_*}\right)\right. \\
&\qquad \left. + \frac{2\log\log(n-p_*)p_*}{n-p_*} - \frac{2\log\log(n-p_*-L)(p_*+L)}{n-p_*-L}\right\} \\
&= P\left\{F_{2L,n-2p_*-2L} > \frac{n-2p_*-2L}{2L}\left(\frac{n-p_*}{n-p_*-L}\times\right.\right. \\
&\qquad \left.\left.\exp\left(\frac{2\log\log(n-p_*-L)(p_*+L)}{n-p_*-L} - \frac{2\log\log(n-p_*)p_*}{n-p_*}\right) - 1\right)\right\}.
\end{aligned}
$$

## *3B.8. HQc*

Rewriting HQc in terms of the original sample size $n$, we have $\mathrm{HQc} = \log(\hat{\sigma}^2_p) + 2\log\log(n-p)p/(n-2p-2)$. HQc overfits if $\mathrm{HQc}_{p_*+L} < \mathrm{HQc}_{p_*}$. For finite $n$, $P\{\text{overfit}\}$ is

$$
\begin{aligned}
&P\{\text{HQc}_{p_*+L} < \text{HQc}_{p_*}\} \\
&= P\bigg\{\log\left(\hat{\sigma}^2_{p_*+L}\right) + \frac{2\log\log(n-p_*-L)(p_*+L)}{n-2p_*-2L-2} \\
&\qquad < \log\left(\hat{\sigma}^2_{p_*}\right) + \frac{2\log\log(n-p_*)p_*}{n-2p_*-2}\bigg\} \\
&= P\bigg\{\log\left(\frac{\text{SSE}_{p_*+L}}{\text{SSE}_{p_*}}\right) < \log\left(\frac{n-p_*-L}{n-p_*}\right) \\
&\qquad + \frac{2\log\log(n-p_*)p_*}{n-2p_*-2} - \frac{2\log\log(n-p_*-L)(p_*+L)}{n-2p_*-2L-2}\bigg\} \\
&= P\bigg\{F_{2L,n-2p_*-2L} > \frac{n-2p_*-2L}{2L}\bigg(\frac{n-p_*}{n-p_*-L}\times \\
&\qquad \exp\bigg(\frac{2\log\log(n-p_*-L)(p_*+L)}{n-2p_*-2L-2} - \frac{2\log\log(n-p_*)p_*}{n-2p_*-2}\bigg) - 1\bigg)\bigg\}.
\end{aligned}
$$

*3B.9. General Case*

Consider a model selection criterion, say $\text{MSC}_p$, of the form $\log(\text{SSE}) + \alpha(n,p)$. $\alpha(n,p)$ is the penalty function of $\text{MSC}_p$. MSC overfits if $\text{MSC}_{p_*+L} < \text{MSC}_{p_*}$. For finite $n$, $P\{\text{overfit}\}$ is

$$
\begin{aligned}
&P\left\{\text{MSC}_{p_*+L} < \text{MSC}_{p_*}\right\} \\
&= P\left\{\log\left(\text{SSE}_{p_*+L}\right) + \alpha(n, p_*+L) < \log\left(\text{SSE}_{p_*+L}\right) + \alpha(n, p_*+L)\right\} \\
&= P\left\{\log\left(\frac{\text{SSE}_{p_*+L}}{\text{SSE}_{p_*}}\right) < \alpha(n,p_*) - \alpha(n,p_*+L)\right\} \\
&= P\left\{\frac{\chi^2_{2L}}{\chi^2_{n-2p_*-2L}} > \exp\big(\alpha(n,p_*+L) - \alpha(n,p_*)\big) - 1\right\} \\
&= P\left\{F_{2L,n-2p_*-2L} > \frac{n-2p_*-2L}{2L}\Big(\exp\Big(\alpha(n,p_*+L) - \alpha(n,p_*)\Big) - 1\Big)\right\}.
\end{aligned}
$$

## Appendix 3C. Asymptotic Results

### *3C.1. Asymptotic Probabilities of Overfitting*

#### *3C.1.1. AICc*

For AICc, as $n \to \infty$,

$$
\begin{aligned}
\frac{n-2p_*-2L}{2L}&\left(\frac{n-p_*}{n-p_*-L}\exp\left(\frac{2Ln}{(n-2p_*-2L-2)(n-2p_*-2)}\right)-1\right)\\
=&\frac{n-2p_*-2L}{2L}\left(\frac{n-p_*}{n-p_*-L}\times\right.\\
&\left.\left(1+\frac{2Ln}{(n-2p_*-2L-2)(n-2p_*-2)}\right)-1\right)\\
\to&\frac{n}{2L}\left(\frac{L}{n}+\frac{2L}{n}\right)\\
\to&1.5.
\end{aligned}
$$

Thus we have $P\{\text{AICc overfits by } L\} = P\{\chi^2_{2L} > 3L\}$.

#### *3C.1.2. AICu*

For AICu, as $n \to \infty$,

$$
\begin{aligned}
\frac{n-2p_*-2L}{2L}&\left(\frac{n-2p_*}{n-2p_*-2L}\exp\left(\frac{2Ln}{(n-2p_*-2L-2)(n-2p_*-2)}\right)-1\right)\\
=&\frac{n-2p_*-2L}{2L}\left(\frac{n-2p_*}{n-2p_*-2L}\times\right.\\
&\left.\left(1+\frac{2Ln}{(n-2p_*-2L-2)(n-2p_*-2)}\right)-1\right)\\
\to&\frac{n}{2L}\left(\frac{2L}{n}+\frac{2L}{n}\right)\\
\to&2.
\end{aligned}
$$

Thus we have $P\{\text{AICu overfits by } L\} = P\{\chi^2_{2L} > 4L\}$.

#### *3C.1.3. FPEu*

In FPEu, as $n \to \infty$,

$$1+\frac{n-2p_*}{n-2p_*-2L} \to 2.$$

Thus we have $P\{\text{FPEu overfits by } L\} = P\{\chi^2_{2L} > 4L\}$.

### 3C.1.4. Cp

For Cp,

$$F_{2L,n-2P} \to \frac{\chi^2_{2L}}{2L}.$$

Thus we have $P\{\text{Cp overfits by } L\} = P\{\chi^2_{2L} > 3L\}$.

### 3C.1.5. HQ

For HQ, as $n \to \infty$,

$$\begin{aligned}
&\frac{n-2p_*-2L}{2L}\left(\frac{n-p_*}{n-p_*-L}\exp\left(\frac{2\log\log(n-p_*-L)(p_*+L)}{n-p_*-L}\right.\right. \\
&\qquad\left.\left. - \frac{2\log\log(n-p_*)p_*}{n-p_*}\right)-1\right) \\
&=\frac{n-2p_*-2L}{2L}\left(\frac{n-p_*}{n-p_*-L}\left(1+\frac{2\log\log(n-p_*-L)(p_*+L)}{n-p_*-L}\right.\right. \\
&\qquad\left.\left. - \frac{2\log\log(n-p_*)p_*}{n-p_*}\right)-1\right) \\
&\to\frac{n}{2L}\left(\frac{L}{n}+\frac{2\log\log(n)L}{n}\right) \\
&\to\infty.
\end{aligned}$$

Thus we have $P\{\text{HQ overfits by } L\} = 0$.

### 3C.1.6. HQc

In HQc, as $n \to \infty$,

$$\begin{aligned}
&\frac{n-2p_*-2L}{2L}\left(\frac{n-p_*}{n-p_*-L}\exp\left(\frac{2\log\log(n-p_*-L)(p_*+L)}{n-2p_*-2L-2}\right.\right. \\
&\qquad\left.\left. - \frac{2\log\log(n-p_*)p_*}{n-2p_*-2}\right)-1\right) \\
&=\frac{n-2p_*-2L}{2L}\left(\frac{n-p_*}{n-p_*-L}\left(1+\frac{2\log\log(n-p_*-L)(p_*+L)}{n-2p_*-2L-2}\right.\right. \\
&\qquad\left.\left. - \frac{2\log\log(n-p_*)p_*}{n-2p_*-2}\right)-1\right) \\
&\to\frac{n}{2L}\left(\frac{L}{n}+\frac{2\log\log(n)L}{n}\right) \\
&\to\infty.
\end{aligned}$$

Thus we have $P\{\text{HQc overfits by } L\} = 0$.

### 3C.2. Asymptotic Signal-To-noise Ratios

#### 3C.2.1. AICc

The asymptotic signal-to-noise ratio for AICc overfitting is

$$\lim_{n\to\infty} \frac{\sqrt{(n-2p_*-2L)(n-2p_*+2)}}{\sqrt{4L}} \left( \log\left(1-\frac{2L}{n-2p_*}\right) - \log\left(1-\frac{L}{n-p_*}\right) \right.$$
$$\left. - \frac{2L}{(n-2p_*-2L)(n-2p_*)} + \frac{2Ln}{(n-2p_*-2L-2)(n-2p_*-2)} \right)$$
$$= \lim_{n\to\infty} \frac{n}{2\sqrt{L}} \left( -\frac{2L}{n} + \frac{L}{n} - \frac{2L}{n^2} + \frac{2L}{n} \right)$$
$$= \frac{\sqrt{L}}{2}.$$

#### 3C.2.2. AICu

The asymptotic signal-to-noise ratio for AICu overfitting is

$$\lim_{n\to\infty} \frac{\sqrt{(n-2p_*-2L)(n-2p_*+2)}}{\sqrt{4L}} \left( -\frac{2L}{(n-2p_*-2L)(n-2p_*)} \right.$$
$$\left. + \frac{2Ln}{(n-2p_*-2L-2)(n-2p_*-2)} \right)$$
$$= \lim_{n\to\infty} \frac{n}{2\sqrt{L}} \left( -\frac{2L}{n^2} + \frac{2L}{n} \right)$$
$$= \frac{2L}{2\sqrt{L}}$$
$$= \sqrt{L}.$$

#### 3C.2.3. FPEu

The asymptotic signal-to-noise ratio for FPEu overfitting is

$$\lim_{n\to\infty} \frac{(n-2p_*)\sqrt{L(n-2p_*-2L)}}{\sqrt{(n-2p_*-2L)^3 + 8L(n-2p_*-L)^2}}$$
$$= \lim_{n\to\infty} \frac{L\sqrt{n^3}}{\sqrt{n^3}}$$
$$= \sqrt{L}.$$

*3C.2.4. Cp*

The asymptotic signal-to-noise ratio for Cp overfitting is

$$\lim_{n\to\infty} \frac{(n-2P)\left(\frac{-2L}{n-2P-2}\right)+3L}{(n-2P)\sqrt{\frac{4L+4L^2}{(n-2P-2)(n-2P-4)}-\frac{4L^2}{(n-2P-2)^2}}}$$

$$=\lim_{n\to\infty} \frac{L}{n\sqrt{\frac{4L}{n^2}}}$$

$$=\frac{\sqrt{L}}{2}.$$

*3C.2.5. HQ*

The asymptotic signal-to-noise ratio for HQ overfitting is

$$\lim_{n\to\infty} \frac{\sqrt{(n-2p_*-2L)(n-2p_*+2)}}{\sqrt{4L}}\left(\log\left(1-\frac{2L}{n-2p_*}\right)-\log\left(1-\frac{L}{n-p_*}\right)\right.$$
$$-\frac{2L}{(n-2p_*-2L)(n-2p_*)}+\frac{2\log\log(n-p_*-L)(p_*+L)}{n-p_*-L}$$
$$\left.-\frac{2\log\log(n-p_*)p_*}{n-p_*}\right)$$
$$=\lim_{n\to\infty} \frac{n}{2\sqrt{L}}\left(-\frac{2L}{n}+\frac{L}{n}-\frac{2L}{n^2}+\frac{2\log\log(n)L}{n}\right)$$
$$=\infty.$$

*3C.2.6. HQc*

The asymptotic signal-to-noise ratio for HQc overfitting is

$$\lim_{n\to\infty} \frac{\sqrt{(n-2p_*-2L)(n-2p_*+2)}}{\sqrt{4L}}\left(\log\left(1-\frac{2L}{n-2p_*}\right)-\log\left(1-\frac{L}{n-p_*}\right)\right.$$
$$-\frac{2L}{(n-2p_*-2L)(n-2p_*)}+\frac{2\log\log(n-p_*-L)(p_*+L)}{n-2p_*-2L-2}$$
$$\left.-\frac{2\log\log(n-p_*)p_*}{n-2p_*-2}\right)$$
$$=\lim_{n\to\infty} \frac{n}{2\sqrt{L}}\left(-\frac{2L}{n}+\frac{L}{n}-\frac{2L}{n^2}+\frac{2\log\log(n)L}{n}\right)$$
$$=\infty.$$

Chapter 4
# The Multivariate Regression Model

Multivariate modeling is a common method for describing the relationships between multiple variables. Rather than observing a single characteristic for $y_i$, as is done for univariate regression, it is often reasonable to observe several characteristics of $y_i$. If we observe $q$ characteristics, then $y_i$ becomes a vector of dimension $q$, where $q$ is often used to denote the dimension of the model. Hence multivariate regression is essentially $q$ univariate regressions using additional parameters, the covariances of the univariate errors for each characteristic. The dimension $q$ changes several key variables from scalars to matrices; for example, the variance of the errors, $\sigma^2$, becomes the covariance matrix, $\Sigma$. To use these matrices for model selection we must reduce them to scalars, and there are several functions that are commonly used to do this—the trace, the determinant, and the maximum root or eigenvalue. For our purposes we will concentrate on the determinant because of its relationship with the multivariate normal density. Our ability to test full versus reduced models in small samples is also affected by the matrix nature of multivariate regression. The standard F-test we used in Chapter 2 is no longer applicable. However, since most selection criteria are functions of the generalized variance, we can use the U-test, a generalization of the univariate F-test that is based on the generalized variance. When $q = 2$, U probabilities can be expressed in terms of the F-distribution.

The organization of this Chapter is similar to that of Chapter 2. We will discuss the Kullback–Leibler-based criteria AIC and AICc, the FPE and Cp criteria, which estimate the mean product prediction error matrix (similar to estimating $L_2$), and the consistent criteria SIC and HQ, in the context of multivariate model selection. Sparks, Coutsourides, and Troskie (1983) generalized Cp to multivariate regression. They showed that the determinant is not appropriate for multivariate Cp, and therefore we focus on the trace and the maximum root of Cp as scalars suitable for model selection. As we did for the univariate case, we will examine small-sample moments with respect to overfitting in order to suggest improved penalty functions for AIC and HQ, and use these to derive the signal-to-noise corrected variants AICu and HQc. We will cover overfitting and asymptotic properties, as well as small-sample properties, including underfitting using two special case models that vary in identifiability. Finally, we discuss underfitting using random $X$ regression, and examine criterion performance via a Monte Carlo simulation study using two special case models.

## 4.1. Model Description

### *4.1.1. Model Structure and Notation*

For univariate regression, at each index $i$ there is one observation. In other words, if $y_i$ is a vector of dimension $q$, $q = 1$ for the univariate case. For multivariate regression we observe several variables for each index. In order to study multivariate model selection, we first need to define the true model, the general model, and the fitted model. We define the *true model* to be

$$y_i = \mu_{*i} + \varepsilon_{*i} \text{ with } \varepsilon_{*i} \sim N(0, \Sigma_*),$$

where $y_i$ is a $q \times 1$ vector of responses and $\mu_{*i}$ is a $q \times 1$ vector of true unknown functions. We assume that the errors $\varepsilon_{*i}$ are independent and identically distributed, and have a multivariate normal distribution of dimension $q$ and variance $\Sigma_*$ for $i = 1, \ldots, n$.

Let $Y = (y_1, \ldots, y_n)'$ be an $n \times q$ matrix of observations. To compute its covariance, it is more convenient to express $Y$ as a vector. The $y_i$ column vectors that comprise $Y$ are stacked end-to-end into one large $nq \times 1$ vector referred to as $vec(Y)$. The covariance of $vec(Y)$ will be an $nq \times nq$ matrix which can be written as a Kronecker product $\Sigma_* \otimes I_n$. Thus, using the above notation, the *true regression model* can be expressed as

$$Y = X_* B_* + \varepsilon_* \tag{4.1}$$

and

$$vec(\varepsilon_*) \sim N(0, \Sigma_* \otimes I_n), \tag{4.2}$$

where $\varepsilon_* = (\varepsilon_{*1}, \ldots, \varepsilon_{*n})'$.

We next define the *general model,* which is

$$Y = XB + \varepsilon \tag{4.3}$$

and

$$vec(\varepsilon) \sim N(0, \Sigma \otimes I_n), \tag{4.4}$$

where $X$ is the known $n \times k$ design matrix of rank $k$, $B$ is a $k \times q$ matrix of unknown parameters, $\varepsilon$ is the $n \times q$ matrix of errors, and $\Sigma$ is a $q \times q$ matrix of the error covariance. If the constant (or intercept) is included in the model the first column of $X$ contains a column of 1's associated with the constant.

Finally we define the *fitted* or *candidate model* with respect to the general model. In order to classify candidate model types we will partition $X$ and $B$ such that $X = (X_0, X_1, X_2)$ and $B = (B_0', B_1', B_2')'$, where $X_0$, $X_1$, and $X_2$ are $n \times k_0$, $n \times k_1$ and $n \times k_2$ matrices, and $B_0, B_1$ and $B_2$ are $k_0 \times q$, $k_1 \times q$ and $k_2 \times q$ matrices, respectively.

If $\mu_*$ is a linear combination of unknown parameters such that $\mu_* = X_* B_*$, then underfitting will occur when $k = \text{rank}(X) < k_* = \text{rank}(X_*)$, and overfitting will occur when $k_* = \text{rank}(X_*) < k = \text{rank}(X)$, assuming both models are of full rank. Thus we can rewrite the model Eq. (4.3) in the following form:

$$\begin{aligned} Y &= X_0 B_0 + X_1 B_1 + X_2 B_2 + \varepsilon \\ &= X_* B_* + X_2 B_2 + \varepsilon, \end{aligned}$$

where $B_0 = (B_0', B_1')'$, $X_0$ is the design matrix for an underfitted model, $X_* = (X_0, X_1)$ is the design matrix for the true model, and $X = (X_0, X_1, X_2)$. Thus an underfitted model is written as

$$Y = X_0 B_0 + \varepsilon, \tag{4.5}$$

and an overfitted model is written as

$$Y = XB + \varepsilon X_0 B_0 + X_1 B_1 + X_2 B_2 + \varepsilon. \tag{4.6}$$

This overfitted model has the same structure as the general model. We will further assume that the method of least squares is used to fit models to the data, and that the candidate model (unless otherwise noted) will be of order $k$. The usual least squares parameter estimate of $B$ is

$$\hat{B} = (X'X)^{-1} X'Y.$$

This is also the maximum likelihood estimate (MLE) of $B$ since the errors $\varepsilon$ satisfy the assumption in Eq. (4.4). The unbiased and the maximum likelihood estimates of $\Sigma$ are given below:

$$S_k^2 = \frac{\text{SPE}_k}{n-k}, \tag{4.7}$$

and

$$\hat{\Sigma}_k = \frac{\text{SPE}_k}{n}, \tag{4.8}$$

where $\text{SPE}_k = (Y - \hat{Y})'(Y - \hat{Y})$, and $\hat{Y} = X\hat{B}$. Note that $\text{SPE}_k$ (or SPE), the *sum of product errors*, $S_k^2$, and $\hat{\Sigma}_k$ are all $q \times q$ matrices.

### *4.1.2. Distance Measures*

In order to evaluate how well a given candidate model approximates the true model given by Eq. (4.1) and Eq. (4.2), we will use the measures $L_2$ and the Kullback–Leibler discrepancy (K-L) to estimate the distance between the true and candidate models. These distances will then be used to compute observed efficiency.

Suppose there exists a true model $Y = X_*B_* + \varepsilon_*$, with $vec(\varepsilon_*) \sim N(0, \Sigma_* \otimes I_n)$. Then $L_2$, scaled by the sample size to express it as a rate or average distance per observation, is defined as

$$L_2 = \frac{1}{n}(\mu_* - \mu)'(\mu_* - \mu),$$

where $\mu_*$ denotes the expected value matrix of $Y$ under the true model and $\mu$ denotes the expected value matrix of $Y$ under the candidate model. Analogously, the $L_2$ distance measuring the difference between the estimated candidate model and the expectation of the true regression model is defined as

$$L_2 = \frac{1}{n}(X_*B_* - X\hat{B})'(X_*B_* - X\hat{B}). \tag{4.9}$$

$L_2$ is a $q \times q$ matrix, and as such there are many ways to form a distance function. The determinant and trace of $L_2$ are two such methods; however, there is little agreement as to which results in the better $L_2$ distance. The Kullback–Leibler information or discrepancy remains a scalar in multivariate regression, and in this respect is much more flexible than $L_2$.

In order to define the Kullback–Leibler discrepancy for multivariate regression we must first consider the density functions of the true and candidate models. Using the multivariate normality assumption, the log-likelihood of the true model, $f_*$, is

$$\log(f_*) = -\frac{nq}{2}\log(2\pi) - \frac{n}{2}\log|\Sigma_*| - \frac{1}{2}tr\{(Y - X_*B_*)\Sigma_*^{-1}(Y - X_*B_*)'\}.$$

The log-likelihood of the candidate model is

$$\log(f) = -\frac{nq}{2}\log(2\pi) - \frac{n}{2}\log|\Sigma| - \frac{1}{2}tr\{(Y - XB)\Sigma^{-1}(Y - XB)'\}.$$

As we did for $L_2$, we scale $\log(f_*)$ and $\log(f)$ by $2/n$ so that the resulting discrepancy represents the rate of change in distance with respect to the number of observations. Then the log-likelihood difference between the true model and the candidate model is

$$\log(f_*) - \log(f) = \log\left(\frac{|\Sigma|}{|\Sigma_*|}\right) + \frac{1}{n}tr\{(Y - XB)\Sigma^{-1}(Y - XB)'\}$$
$$- \frac{1}{n}tr\{(Y - X_*B_*)\Sigma_*^{-1}(Y - X_*B_*)'\}.$$

Next, taking the expectation with respect to the true model, we obtain the Kullback–Leibler discrepancy:

$$\text{K-L} = \log\left(\frac{|\Sigma|}{|\Sigma_*|}\right) + tr\{\Sigma^{-1}\Sigma_*\} + \frac{1}{n}tr\{(X_*B_* - XB)\Sigma^{-1}(X_*B_* - XB)'\} - q.$$

Substituting the maximum likelihood estimates into the Kullback–Leibler discrepancy, we define

$$\text{K-L} = \log\left(\frac{|\hat{\Sigma}_k|}{|\Sigma_*|}\right) + tr\{\hat{\Sigma}_k^{-1}\Sigma_*\} + tr\{\hat{\Sigma}_k^{-1}L_2\} - q, \tag{4.10}$$

where $\hat{\Sigma}_k$ and $L_2$ are defined in Eq. (4.8) and Eq. (4.9), respectively.

## 4.2. Selected Derivations of Model Selection Criteria

In this Section we will define our base (or foundation) and signal-to-noise corrected variant criteria within the multivariate regression framework. We will start with the $L_2$-based model selection criteria.

### *4.2.1. $L_2$-based Criteria FPE and Cp*

Although FPE was originally designed for autoregressive time series models, its derivation is straightforward for multivariate regression. Suppose we have $n$ observations from the overfitted regression model Eq. (4.6), and the resulting least squares estimate of $B$, $\hat{B}$. Now suppose we make $n$ new observations $Y_0 = (y_{10}, \ldots, y_{n0})' = XB + \varepsilon_0$, also obtained from Eq. (4.6). The predicted value of $Y_0$ is thus $\hat{Y}_0 = (\hat{y}_{10}, \ldots, \hat{y}_{n0})' = X\hat{B}$, and the mean product prediction error is

$$\frac{1}{n}E[(Y_0 - \hat{Y}_0)'(Y_0 - \hat{Y}_0)] = \frac{1}{n}E[(XB + \varepsilon_0 - X\hat{B})'(XB + \varepsilon_0 - X\hat{B})]$$
$$= \Sigma\left(1 + \frac{k}{n}\right).$$

The mean product prediction error, $\Sigma(1 + k/n)$, is also called the final prediction error, or MFPE, where the M is used to indicate a matrix. Akaike estimated $\Sigma$ with the unbiased estimate $S_k^2$, and substituting this into the above equation gives the unbiased estimate for MFPE, $\hat{\text{MFPE}} = S_k^2(1 + k/n)$. Rewriting in terms of the maximum likelihood estimate, $\hat{\Sigma}_k$, gives us the matrix form of $\hat{\text{MFPE}}$,

$$\text{MFPE}_k = \hat{\Sigma}_k \frac{n+k}{n-k}.$$

To identify the best model from among many candidates, selection criterion values are calculated for all and the model with the minimum value is chosen. To do this using MFPE we must transform it from a matrix into a scalar. The generalized variance is often used to do this by taking determinants and

$$\text{FPE}_k = |\hat{\Sigma}_k| \left(\frac{n+k}{n-k}\right)^q. \tag{4.11}$$

For the sake of simplicity, we sometimes use FPE to denote $\text{FPE}_k$.

When predicting new observations with the same $X$, the mean product prediction error includes the variance of the errors as well as the variance of the $X\hat{B}$. The influence of the generalized variance of the best linear predictor for $Y_0$ is balanced against the generalized variances of $X\hat{B}$ in $\text{FPE}_k$, and so minimizing $\text{FPE}_k$ maintains equilibrium between the two. Recall from above that the mean product prediction error is $\Sigma(1 + k/n)$, where $\Sigma$ is the variance of $\varepsilon_0$. The term $\Sigma k/n$ results from the variance of $X\hat{B}$, where $\Sigma$ is estimated by $S_k^2$. In underfitted models, $S_k^2$ is expected to be large, while $cov[X\hat{B}]$ may be small. By contrast, in overfitted models we expect $S_k^2$ to be small, and thus for $cov[X\hat{B}]$ to increase.

Mallows's Cp can also be generalized to the multivariate regression model, as Sparks, Coutsourides, and Troskie (1983) did using both the trace and maximum eigenvalue. Sparks *et al.* proposed multivariate Cp as

$$\text{Cp} = (n - K)\hat{\Sigma}_K^{-1}\hat{\Sigma}_k + (2k - n)I, \tag{4.12}$$

where $K$ denotes the order of the largest model, $k$ denotes the current model, and $I$ is the $q \times q$ identity matrix. Cp is a matrix and must be converted to a scalar so that it can be used for model selection. Sparks *et al.* point out that using the determinant is not appropriate, since $2k - n$ may be negative, causing a negative determinant. Often, such negative determinants lead to false minima for underfitted models. Hence, we focus on the trace and the maximum eigenvalue of Cp. The trace of Cp is defined as

$$\text{trCp} = (n - K)tr\{\hat{\Sigma}_K^{-1}\hat{\Sigma}_k\} + (2k - n)q. \tag{4.13}$$

and the maximum eigenvalue of Cp is defined as

$$\text{meCp} = \lambda_1,$$

where $\lambda_1$ is the largest eigenvalue for Eq. (4.12). Simulation results comparing the two forms of Cp can be found in Chapter 9.

### 4.2.2. Kullback–Leibler-based Criteria AIC and AICc

We now turn to AIC and AICc, which estimate the Kullback–Leibler discrepancy. Of all the criteria we derive, AIC is the most easily generalized:

$$\text{AIC} = -2\log(\textit{likelihood}) + 2 \times \textit{number of parameters}.$$

Using the maximum likelihood estimate, Eq. (4.8), we find that

$$-2\log(\textit{likelihood}) = nq\log(2\pi) + n\log|\hat{\Sigma}_k| + nq.$$

There are $kq$ parameters for $B$ and $0.5q(q+1)$ parameters for the error covariance matrix, $\Sigma_k$. Substituting,

$$\text{AIC} = nq\log(2\pi) + n\log|\hat{\Sigma}_k| + nq + 2kq + q(q+1).$$

The constants $nq\log(2\pi) + nq$ play no practical role in model selection and can be ignored, leaving

$$\text{AIC} = n\log|\hat{\Sigma}_k| + 2kq + q(q+1).$$

We scale AIC by $1/n$ to express it as an average per observation, and arrive at our definition of AIC for multivariate regression:

$$\text{AIC}_k = \log|\hat{\Sigma}_k| + \frac{2kq + q(q+1)}{n}. \tag{4.14}$$

As was the case for univariate regression, many authors have shown for the multivariate case that the small-sample properties of AIC lead to overfitting. Rather than estimating K-L itself and using asymptotic results, Bedrick and Tsai (1994) addressed this problem by estimating the expected Kullback–Leibler discrepancy, which can be computed for small samples. This resulted in AICc, a small-sample corrected version of AIC. To derive AICc for multivariate regression we again estimate the candidate model via maximum likelihood.

Bedrick and Tsai assumed that the true model is a member of the set of candidate models, and under this assumption

$$E_*[\text{K-L}] = E_*\left[\log\left(\frac{|\hat{\Sigma}_k|}{|\Sigma_*|}\right) + tr\{\hat{\Sigma}_k^{-1}\Sigma_*\} + tr\{\hat{\Sigma}_k^{-1}L_2\} - q\right],$$

where $E_*$ denotes expectations under the true model. These expectations can be simplified due to the fact that $E_*[tr\{\hat{\Sigma}_k^{-1}\Sigma_*\}] = nq/(n-k-q-1)$, $E_*[\hat{\Sigma}_k^{-1}] = n/(n-k-q-1)\Sigma_*^{-1}$, $X_*B_* = X\tilde{B}_*$, and that $\hat{\Sigma}_k$ and $L_2$ are independent:

$$\begin{aligned}
E_*[tr\{\hat{\Sigma}_k^{-1}L_2\}] &= E_*[tr\{\hat{\Sigma}_k^{-1}\frac{1}{n}(\hat{B}-\tilde{B}_*)'(X'X)(\hat{B}-\tilde{B}_*)\}] \\
&= \frac{1}{n}tr\{E_*[\hat{\Sigma}_k^{-1}]E_*[(\hat{B}-\tilde{B}_*)'(X'X)(\hat{B}-\tilde{B}_*)]\} \\
&= \frac{1}{n-k-q-1}tr\{\Sigma^{-1}E_*[(\hat{B}-\tilde{B}_*)'(X'X)(\hat{B}-\tilde{B}_*)]\} \\
&= \frac{1}{n-k-q-1}tr\{E_*[vec(\hat{B}-\tilde{B}_*)'(\Sigma_*^{-1}\otimes X'X)vec(\hat{B}-\tilde{B}_*)]\} \\
&= \frac{qk}{n-k-q-1},
\end{aligned}$$

where $\tilde{B}_* = (B_*', 0')'$ and 0 is a $k_2 \times q$ matrix of zeros. Substituting,

$$E_*[\text{K-L}] = E_*[\log|\hat{\Sigma}_k|] - \log|\Sigma_*| + \frac{nq}{n-k-q-1} + \frac{qk}{n-k-q-1} - q,$$

and simplifying,

$$E_*[\text{K-L}] = E_*[\log|\hat{\Sigma}_k|] - \log|\Sigma_*| + \frac{(n+k)q}{n-k-q-1} - q.$$

Noticing that $\log|\hat{\Sigma}_k|$ is unbiased for $E_*[\log|\hat{\Sigma}_k|]$, then

$$\log|\hat{\Sigma}_k| + \frac{(n+k)q}{n-k-q-1} - \log|\Sigma_*| - q$$

is unbiased for $E_*[\text{K-L}]$. The constants $-\log|\Sigma_*| - q$ do not contribute to model selection and can be ignored, yielding

$$\text{AICc}_k = \log|\hat{\Sigma}_k| + \frac{(n+k)q}{n-k-q-1}. \tag{4.15}$$

Bedrick and Tsai showed that AICc and AIC are asymptotically equivalent, although in small samples AICc outperforms AIC.

### *4.2.3. Consistent Criteria SIC and HQ*

Finally we generalize the consistent criteria SIC and HQ for multivariate regression. For univariate regression the $2k$ penalty in AIC was replaced by a $\log(n)k$ penalty in SIC, and the same substitution (factored by $q$) is made for multivariate regression, yielding

$$\text{SIC}_k = \log|\hat{\Sigma}_k| + \frac{\log(n)kq}{n}. \tag{4.16}$$

Lastly, Hannan and Quinn's HQ for multivariate regression is

$$\text{HQ}_k = \log|\hat{\Sigma}_k| + \frac{2\log\log(n)kq}{n}. \tag{4.17}$$

## 4.3. Moments of Model Selection Criteria

The quantity $|\text{SPE}|$ is difficult to work with in terms of model selection. While the distribution of $|\text{SPE}|$ is known, the distribution of $|\text{SPE}_{full}|-|\text{SPE}_{red}|$ is not, and as a result we cannot easily compute moments for FPE or Cp. An additional difficulty is presented by the need to reduce matrices to scalar values for the multivariate case, usually via either the trace or the determinant. Because the properties of traces are not well understood, we will not discuss moments involving traces. However, the distribution of the generalized multivariate ANOVA test $|\text{SPE}_{full}|/|\text{SPE}_{red}|$, or the U-statistic, is known, and we can use it to obtain signal-to-noise ratios for the $\log|\text{SPE}|$-based model selection criteria AIC, AICc, SIC, and HQ, since $\log|\text{SPE}_{full}| - \log|\text{SPE}_{red}| = \log(|\text{SPE}_{full}|/|\text{SPE}_{red}|)$.

For all model selection criteria (MSC) we choose model $k$ over model $k+L$ if $\text{MSC}_{k+L} > \text{MSC}_k$, and we define $\Delta\text{MSC} = \text{MSC}_{k+L} - \text{MSC}_k$. We will use the approximation for $E[\log|\text{SPE}_k|]$ given by Eq. (4A.4) and $sd[\Delta\log(|\text{SPE}_k|)] = sd[\log(|\text{SPE}_{k+L}|/|\text{SPE}_k|)]$ given by Eq. (4A.5) to estimate the signal and noise, respectively, for each criterion. Using the approximations we obtain

$$E[\log|\text{SPE}_k|] \doteq \log|\Sigma| + q\log(n-k-(q-1)/2) - \frac{q}{n-k-(q-1)/2}$$

and

$$sd\left[\log\left(\frac{|\text{SPE}_{k+L}|}{|\text{SPE}_k|}\right)\right] \doteq \frac{\sqrt{2Lq}}{\sqrt{(n-k-L-(q-1)/2)(n-k-(q-1)/2+2)}}.$$

See Appendix 4A.1 for details of the calculations.

### *4.3.1. AIC and AICc*

We will first look at the K-L-based criteria AIC and AICc. Applying Eq. (4A.4), the signal is

$$\begin{aligned}E[\Delta\text{AIC}] =& q\log\left(\frac{n-k-L-(q-1)/2}{n-k-(q-1)/2}\right)\\ &-\frac{Lq}{(n-k-L-(q-1)/2)(n-k-(q-1)/2)}+\frac{2Lq}{n},\end{aligned}$$

and from Eq. (4A.5) the noise is

$$\begin{aligned}sd[\Delta\text{AIC}] &= sd[\Delta\log|\text{SPE}|]\\ &= \frac{\sqrt{2Lq}}{\sqrt{(n-k-L-(q-1)/2)(n-k-(q-1)/2+2)}}.\end{aligned}$$

Therefore the signal-to-noise ratio is

$$\begin{aligned}\frac{E[\Delta\text{AIC}]}{sd[\Delta\text{AIC}]} =& \frac{\sqrt{(n-k-L-(q-1)/2)(n-k-(q-1)/2+2)}}{\sqrt{2Lq}}\times\\ &\left(q\log\left(\frac{n-k-L-(q-1)/2}{n-k-(q-1)/2}\right)\right.\\ &\left.-\frac{Lq}{(n-k-L-(q-1)/2)(n-k-(q-1)/2)}+\frac{2Lq}{n}\right).\end{aligned}$$

We will examine the behavior of the signal-to-noise ratio one term at a time. The first term,

$$\begin{aligned}&\frac{\sqrt{(n-k-L-(q-1)/2)(n-k-(q-1)/2+2)}}{\sqrt{2Lq}}\times\\ &\left(q\log\left(\frac{n-k-L-(q-1)/2}{n-k-(q-1)/2}\right)\right.\\ &\left.-\frac{Lq}{(n-k-L-(q-1)/2)(n-k-(q-1)/2)}\right),\end{aligned}$$

decreases to $-\infty$ as $L$ increases due to $\log|\hat{\Sigma}|$. The last term,

$$\frac{\sqrt{(n-k-L-(q-1)/2)(n-k-(q-1)/2+2)}}{\sqrt{2Lq}}\left(\frac{2Lq}{n}\right),$$

which results from the penalty function, increases for small $L$, then decreases to 0 as the number of extra variables $L$ increases.

The result of the behavior of these two terms is that typically the signal-to-noise ratio of AIC increases for small $L$, but as $L \to n-k-q$, the signal-to-noise ratio of AIC $\to -\infty$, resulting in AIC's well-known small-sample overfitting tendencies.

Next we look at the signal and noise for AIC's small-sample correction, AICc. Following the same procedure as above, the signal is

$$E[\Delta \text{AICc}] = q \log\left(\frac{n-k-L-(q-1)/2}{n-k-(q-1)/2}\right) - \frac{Lq}{(n-k-L-(q-1)/2)(n-k-(q-1)/2)} + \frac{Lq(2n-q-1)}{(n-k-L-q-1)(n-k-q-1)},$$

and the noise is

$$sd[\Delta \text{AICc}] = \frac{\sqrt{2Lq}}{\sqrt{(n-k-L-(q-1)/2)(n-k-(q-1)/2+2)}}.$$

Thus the signal-to-noise ratio is

$$\frac{E[\Delta \text{AICc}]}{sd[\Delta \text{AICc}]} = \frac{\sqrt{(n-k-L-(q-1)/2)(n-k-(q-1)/2+2)}}{\sqrt{2Lq}} \times \left(q \log\left(\frac{n-k-L-(q-1)/2}{n-k-(q-1)/2}\right) - \frac{Lq}{(n-k-L-(q-1)/2)(n-k-(q-1)/2)} + \frac{Lq(2n-q-1)}{(n-k-L-q-1)(n-k-q-1)}\right),$$

which increases as $L$ increases. This is because as $L$ increases, the last term increases much faster than the first term decreases. In fact, the signal-to-noise ratio for AICc is large in the overfitting case, indicating that AICc should perform well from an overfitting perspective.

AIC has a penalty function that is linear in $k$ whereas AICc has a penalty function that is superlinear in $k$. The following theorem shows that criteria with penalty functions similar to AICc, of the form $\alpha k/(n-k-q-1)$, have signal-to-noise ratios that increase as the amount of overfitting increases. Such criteria should overfit less than criteria with weaker (linear) penalty functions. The proof to Theorem 4.1 can be found in Appendix 4B.1.

**Theorem 4.1**

Given the $q$-dimensional multivariate regression model in Eqs. (4.1)–(4.2) and the criterion of the form $\log(\hat{\Sigma}_k)+\alpha kq/(n-k-q-1)$, for all $n \geq q+5$, $\alpha \geq 1$, $0 < L < n-k-q-1$, and for the overfitting case where $0 < k_* \leq k < n-q-2$, the signal-to-noise ratio of this criterion increases as the amount of overfitting $L$ increases.

### *4.3.2. SIC and HQ*

The same methods used for AIC and AICc to derive the signal and noise are used for SIC and HQ. SIC and HQ share with AIC the basic structure of an additive penalty function that is linear in $k$, and as we have previously noted, such linear penalty functions lead to overfitting in small samples. For SIC the signal is

$$E[\Delta\text{SIC}] = q\log\left(\frac{n-k-L-(q-1)/2}{n-k-(q-1)/2}\right) - \frac{Lq}{(n-k-L-(q-1)/2)(n-k-(q-1)/2)} + \frac{\log(n)Lq}{n},$$

and the noise is

$$sd[\Delta\text{SIC}] = \frac{\sqrt{2Lq}}{\sqrt{(n-k-L-(q-1)/2)(n-k-(q-1)/2+2)}}.$$

Thus the signal-to-noise ratio is

$$\frac{E[\Delta\text{SIC}]}{sd[\Delta\text{SIC}]} = \frac{\sqrt{(n-k-L-(q-1)/2)(n-k-(q-1)/2+2)}}{\sqrt{2Lq}} \times \left(q\log\left(\frac{n-k-L-(q-1)/2}{n-k-(q-1)/2}\right) - \frac{Lq}{(n-k-L-(q-1)/2)(n-k-(q-1)/2)} + \frac{\log(n)Lq}{n}\right).$$

A term-by-term analysis indicates that SIC suffers from the same problems in small samples that AIC does. As we noted above, SIC's penalty function structure is very similar to that of AIC—in small samples, the $\log(n)$ term is not strong enough to overcome the small-sample variability in differences between log|SPE|. Thus, as for AIC, the signal-to-noise ratio of SIC increases for small $L$ to medium $L$, but as $L \to n-k-q$, the signal-to-noise ratio of SIC $\to -\infty$, leading to small-sample overfitting. However, the degree of overfitting in small samples for SIC is less serious than that of AIC. In larger samples with $k << n$, SIC should not suffer from excessive overfitting.

Lastly, for HQ the signal is

$$E[\Delta \text{HQ}] = q \log \left( \frac{n-k-L-(q-1)/2}{n-k-(q-1)/2} \right) - \frac{Lq}{(n-k-L-(q-1)/2)(n-k-(q-1)/2)} + \frac{2 \log\log(n) Lq}{n},$$

and the noise is

$$sd[\Delta \text{HQ}] = \frac{\sqrt{2Lq}}{\sqrt{(n-k-L-(q-1)/2)(n-k-(q-1)/2+2)}}.$$

Thus the signal-to-noise ratio is

$$\frac{E[\Delta \text{HQ}]}{sd[\Delta \text{HQ}]} = \frac{\sqrt{(n-k-L-(q-1)/2)(n-k-(q-1)/2+2)}}{\sqrt{2Lq}} \times \left( q \log \left( \frac{n-k-L-(q-1)/2}{n-k-(q-1)/2} \right) - \frac{Lq}{(n-k-L-(q-1)/2)(n-k-(q-1)/2)} + \frac{2 \log\log(n) Lq}{n} \right).$$

As for AIC, the first term

$$\frac{\sqrt{(n-k-L-(q-1)/2)(n-k-(q-1)/2+2)}}{\sqrt{2Lq}} \times \left( q \log \left( \frac{n-k-L-(q-1)/2}{n-k-(q-1)/2} \right) - \frac{Lq}{(n-k-L-(q-1)/2)(n-k-(q-1)/2)} \right)$$

decreases to $-\infty$ as $L$ increases. Also like AIC, the signal-to-noise ratio of HQ increases for small $L$, and then decreases to $-\infty$ resulting in small-sample overfitting. Thus it is not surprising that in small samples HQ's performance is similar to that of AIC. As was the case for univariate regression, an examination of the moments of log(|SPE|) suggests that the penalty function should be superlinear (increasing much faster than $ck$ where $c$ is some constant) in $k$.

## 4.4. Signal-to-noise Corrected Variants

In the previous Section we were limited to discussing only log |SPE|-based model selection criteria, excluding the $L_2$-based criteria FPE or Cp. By the same token, in this Section we will not consider the signal-to-noise corrected variant FPEu for the multivariate case, only AICu and HQc.

As we did for the base criteria AIC and AICc, we will derive AICu and HQc under the regression model Eqs. (4.3)–(4.4), assuming general models of order $k$ and $k+L$, where $k \geq k_*$ and $L > 0$. We also assume $k$ and $k+L$ form nested models for all $L$.

### *4.4.1. AICu*

In Chapter 2, to obtain AICu we substituted the unbiased estimate $s_k^2$ for the $\hat{\sigma}_k^2$ term in AICc. Analogously, in multivariate regression $S_k^2$, Eq. (4.7), is the usual unbiased estimator for $\Sigma$. However, $|S_k^2|$ is not unbiased for $|\Sigma|$—the unbiased estimator for $|\Sigma|$ is the more complicated $|\tilde{S}_k^2|$, where $|\tilde{S}_k^2| = |\text{SPE}_k|/\prod(n-k-q+1)$. However, since $S_k^2$ is a well-known estimate for $\Sigma$, for simplicity we continue to substitute $S_k^2$ in the derivation of AICu for multivariate regression.

Although AIC is one of the most versatile model selection criteria, it was derived under asymptotic conditions. In small samples, it has a weak signal-to-noise ratio that results in overfitting. As we did for univariate regression and autoregressive models, therefore, we will examine its penalty function and attempt to modify it to improve its signal-to-noise ratio under multivariate regression conditions.

For $\text{AIC} = \log|\hat{\Sigma}_k| + 2kq/n$, let us consider substituting $\log|S_k^2|$ for $\log|\hat{\Sigma}_k|$ and substituting the penalty function $(2kq + q(q+1))/(n-k-q-1)$ for $(2kq + q(q+1))/n$ in order to impose a greater penalty for overfitting as the sample size increases. This yields

$$\text{AICu}_k = \log|S_k^2| + \frac{2kq + q(q+1)}{n-k-q-1},$$

or, equivalently, by adding $q$,

$$\text{AICu}_k = \log|S_k^2| + \frac{(n+k)q}{n-k-q-1}. \tag{4.18}$$

The difference between AICc and AICu is that the former one uses $\log|\hat{\Sigma}_k|$ and the latter one uses $\log|S_k^2|$. We use the same procedure described in Section 4.3 to find the signal and noise for AICu. The signal is

$$
\begin{aligned}
E[\Delta\text{AICu}] =& q\log\left(\frac{n-k-L-(q-1)/2}{n-k-(q-1)/2}\frac{n-k}{n-k-L}\right) \\
&- \frac{Lq}{(n-k-L-(q-1)/2)(n-k-(q-1)/2)} \\
&+ \frac{Lq(2n-q-1)}{(n-k-L-q-1)(n-k-q-1)},
\end{aligned}
$$

and the noise is

$$
sd[\Delta\text{AICu}] = \frac{\sqrt{2Lq}}{\sqrt{(n-k-L-(q-1)/2)(n-k-(q-1)/2+2)}},
$$

and thus the signal-to-noise ratio for AICu is

$$
\begin{aligned}
\frac{E[\Delta\text{AICu}]}{sd[\Delta\text{AICu}]} =& \frac{\sqrt{(n-k-L-(q-1)/2)(n-k-(q-1)/2+2)}}{\sqrt{2Lq}} \times \\
&\left(q\log\left(\frac{n-k-L-(q-1)/2}{n-k-(q-1)/2}\frac{n-k}{n-k-L}\right)\right. \\
&- \frac{Lq}{(n-k-L-(q-1)/2)(n-k-(q-1)/2)} \\
&\left.+ \frac{Lq(2n-q-1)}{(n-k-L-q-1)(n-k-q-1)}\right). \qquad (4.19)
\end{aligned}
$$

Is AICu an improvement over AIC? In the following theorem we state that the signal-to-noise ratio of AICu increases as $L$ increases, and is greater than the signal-to-noise ratio of AIC. The proof can be found in Appendix 4B.2.

**Theorem 4.2**

Given the $q$-dimensional multivariate regression model in Eqs. (4.1)–(4.2), for all $n \geq q+3$, $0 < L < n-k-q-1$, and for the overfitting case where $0 < k_* \leq k < n-q-2$, the signal-to-noise ratio of AICu increases as the amount of overfitting $L$ increases and is greater than the signal-to-noise ratio of AIC.

### *4.4.2. HQc*

As we saw earlier, the signal-to-noise ratio for HQ is weak and eventually decreases as the number of extra variables $L$ increases. Consider the relationship between the penalty functions of AIC, Eq. (4.14), and AICc, Eq. (4.15). The penalty function $-q$ in AICc is

$$\frac{q(n+k)}{n-k-q-1} - q = \frac{2kq + q(q+1)}{n-k-q-1}$$
$$= \frac{2kq + q(q+1)}{n} \times \frac{n}{n-k-q-1},$$

which is the penalty function for AIC scaled by a factor of $n/(n-k-q-1)$. As we did in Chapter 2, in order to derive HQc by analogy, for $\text{HQ} = \log|\hat{\Sigma}_k| + 2\log\log(n)kq/n$ we consider scaling the penalty function by the same factor to yield

$$\text{HQc}_k = \log|\hat{\Sigma}_k^2| + \frac{2\log\log(n)kq}{n-k-q-1}. \tag{4.20}$$

We can ask the same question for HQc that we did for AICu—is it really any improvement over its parent criterion, HQ? We examine the moments of HQc to answer this question. The signal is

$$\begin{aligned} E[\Delta\text{HQc}] =& q\log\left(\frac{n-k-L-(q-1)/2}{n-k-(q-1)/2}\right) \\ &- \frac{Lq}{(n-k-L-(q-1)/2)(n-k-(q-1)/2)} \\ &+ \frac{2\log\log(n)Lq(n-q-1)}{(n-k-L-q-1)(n-k-q-1)}, \end{aligned}$$

and the noise is

$$sd[\Delta\text{HQc}] = \frac{\sqrt{2Lq}}{\sqrt{(n-k-L-(q-1)/2)(n-k-(q-1)/2+2)}},$$

and thus the signal-to-noise ratio is

$$\begin{aligned} \frac{E[\Delta\text{HQc}]}{sd[\Delta\text{HQc}]} =& \frac{\sqrt{(n-k-L-(q-1)/2)(n-k-(q-1)/2+2)}}{\sqrt{2Lq}} \times \\ & \left(q\log\left(\frac{n-k-L-(q-1)/2}{n-k-(q-1)/2}\right)\right. \\ &- \frac{Lq}{(n-k-L-(q-1)/2)(n-k-(q-1)/2)} \\ &\left. + \frac{2\log\log(n)Lq(n-q-1)}{(n-k-L-q-1)(n-k-q-1)}\right). \end{aligned} \tag{4.21}$$

We see from the following theorem that the signal-to-noise ratio of HQc increases as $L$ increases, and is in fact greater than that of HQ. The proof can be found in Appendix 4B.3.

**Theorem 4.3**

Given the $q$-dimensional multivariate regression model in Eqs. (4.1)–(4.2), for all $n \geq q+5$, $0 < L < n-k-q-1$, and for the overfitting case where $0 < k_* \leq k < n-q-2$, the signal-to-noise ratio of HQc increases as the amount of overfitting $L$ increases and is greater than the signal-to-noise ration of HQ.

## 4.5. Overfitting Properties

In this Section we will look at the overfitting properties, both in small samples and asymptotically, of the five base model selection criteria (FPE, AIC, AICc, SIC, and HQ) and the two signal-to-noise corrected variants (AICu and HQc) considered in this Chapter. Asymptotic probabilities and signal-to-noise ratios of varying model dimensions $q$ have also been calculated to allow us to evaluate the relationship between these two properties. In $q$-dimensional models, if one variable is added to the model, the number of parameters increases by $q$. The higher the dimension, the more parameters involved. We expect that the probability of overfitting should decrease and the signal-to-noise ratio should increase for higher $q$.

We do not discuss asymptotic underfitting for multivariate regression, since the asymptotic noncentrality parameter is infinite for fixed $B$, assuming a true model of finite dimension exists. Also, extra assumptions are required to calculate noncentral moments involving logs of the determinant of a multivariate random variable.

### *4.5.1. Small-sample Probabilities of Overfitting*

To examine small-sample probabilities of overfitting, we make our usual supposition that there is a true order $k_*$, and we fit a candidate model of order $k_* + L$ where $L > 0$. Assume that only the two models $k_* + L$ and $k_*$ are compared and that they form nested models. To compute the probability of overfitting by $L$ extra variables, we will compute the probability of selecting a model with $\text{rank}(X) = k_* + L$ over the true model of $\text{rank}(X_*) = k_*$. We will present the detailed calculations for one criterion, AIC. Only the results of the analogous calculations will be given in this Section, but the details can be found in Appendix 4C. For any model selection criterion MSC and finite $n$, a criterion overfits if $\text{MSC}_{k_*+L} < \text{MSC}_{k_*}$.

As we noted at the beginning of the chapter, small-sample probabilities can also be expressed in terms of the U-statistic (Anderson, 1984, Lemma 8.4.2, p. 299), a generalized version of the F-statistic we used for univariate regression that is applicable in the multivariate case. $|\text{SPE}_{k_*+L}|$ results from the full model and $|\text{SPE}_{k_*}|$ results the reduced model. From Lemma 8.4.2 in Anderson,

$$\frac{|\text{SPE}_{k_*+L}|}{|\text{SPE}_{k_*}|} \sim U_{q,L,n-k_*-L}.$$

When $q = 2$, U probabilities can be derived in terms of the usual F distribution (Anderson, 1984, Theorem 8.4.5, p. 305) as follows:

$$P\{U_{2,L,n-k_*-L} < u\} = P\left\{F_{2L,2(n-k_*-L)} > \frac{1-\sqrt{u}}{\sqrt{u}}\frac{n-k_*-L-1}{L}\right\}.$$

For a model selection criterion of the form $\log|\text{SPE}_k| + p(n,k)$, probabilities of overfitting can be expressed as $P\{U_{q,L,n-k_*-L} < \exp\big(p(n,k_*) - p(n,k_*+L)\big)\}$. Hence, we give the overfitting probability expressions in terms of the U-statistic. Also included are probabilities expressed as sums of independent log-$\chi^2$. This form is useful for computing asymptotic probabilities.

**AIC**

We know that AIC overfits if $\text{AIC}_{k_*+L} < \text{AIC}_{k_*}$. For finite $n$, the probability that AIC prefers the overfitted model $k_* + L$ is

$$\begin{aligned}
&P\{\text{AIC}_{k_*+L} < \text{AIC}_{k_*}\}\\
&= P\left\{\log|\hat{\Sigma}_{k_*+L}| + \frac{2(k_*+L)q + q(q+1)}{n} < \log|\hat{\Sigma}_{k_*}| + \frac{2k_*q + q(q+1)}{n}\right\}\\
&= P\left\{\log|\text{SPE}_{k_*+L}| - q\log(n) + \frac{2(k_*+L)q + q(q+1)}{n}\right.\\
&\qquad\left. < \log|\text{SPE}_{k_*}| - q\log(n) + \frac{2k_*q + q(q+1)}{n}\right\}\\
&= P\left\{\log\left(\frac{|\text{SPE}_{k_*+L}|}{|\text{SPE}_{k_*}|}\right) < \frac{2k_*q + q(q+1)}{n} - \frac{2(k_*+L)q + q(q+1)}{n}\right\}\\
&= P\left\{\sum_{i=1}^{q}\text{log-Beta}\left(\frac{n-k_*-L-q+i}{2}, \frac{L}{2}\right) < -\frac{2Lq}{n}\right\}\\
&= P\left\{n\sum_{i=1}^{q}\log\left(1 + \frac{\chi^2_L}{\chi^2_{n-k_*-L+i}}\right) > 2Lq\right\}.
\end{aligned}$$

In terms of $U_{q,L,n-k_*-L}$,

$$P\{\text{overfit}\} = P\left\{U_{q,L,n-k_*-L} < \exp\Big(-\frac{2Lq}{n}\Big)\right\}.$$

**AICc**

$$P\{\text{AICc}_{k_*+L} < \text{AICc}_{k_*}\}$$
$$= P\left\{n\sum_{i=1}^{q}\log\left(1+\frac{\chi_L^2}{\chi_{n-k_*-L+i}^2}\right) > \frac{Lq(2n-q-1)n}{(n-k_*-L-q-1)(n-k_*-q-1)}\right\}.$$

In terms of $U_{q,L,n-k_*-L}$,

$$P\{\text{overfit}\} = P\left\{U_{q,L,n-k_*-L} < \exp\left(-\frac{Lq(2n-q-1)}{(n-k_*-L-q-1)(n-k_*-q-1)}\right)\right\}.$$

**AICu**

$$P\{\text{AICu}_{k_*+L} < \text{AICu}_{k_*}\}$$
$$= P\left\{n\sum_{i=1}^{q}\log\left(1+\frac{\chi_L^2}{\chi_{n-k_*-L+i}^2}\right) > -nq\log\left(\frac{n-k_*-L}{n-k_*}\right) + \frac{Lq(2n-q-1)n}{(n-k_*-L-q-1)(n-k_*-q-1)}\right\}.$$

In terms of $U_{q,L,n-k_*-L}$,

$$P\{\text{overfit}\} = P\left\{U_{q,L,n-k_*-L} < \left(\frac{n-k_*-L}{n-k_*}\right)^q \times \exp\left(-\frac{Lq(2n-q-1)}{(n-k_*-L-q-1)(n-k_*-q-1)}\right)\right\}.$$

**FPE**

Note that in this case we use $\log(\text{FPE}_k)$ for convenience in computing probabilities.

$$P\{\text{FPE}_{k_*+L} < \text{FPE}_{k_*}\}$$
$$= P\left\{n\sum_{i=1}^{q}\log\left(1+\frac{\chi_L^2}{\chi_{n-k_*-L}^2}\right) > nq\log\left(\frac{n-k_*}{n+k_*}\frac{n+k_*+L}{n-k_*-L}\right)\right\}.$$

In terms of $U_{q,L,n-k_*-L}$,

$$P\{\text{overfit}\} = P\left\{U_{q,L,n-k_*-L} < \left(\frac{(n+k_*)(n-k_*-L)}{(n-k_*)(n+k_*+L)}\right)^q\right\}.$$

**SIC**

$$P\{\text{SIC}_{k_*+L} < \text{SIC}_{k_*}\}$$
$$= P\left\{n\sum_{i=1}^{q}\log\left(1+\frac{\chi^2_L}{\chi^2_{n-k_*-L+i}}\right) > \log(n)Lq\right\}.$$

In terms of $U_{q,L,n-k_*-L}$,

$$P\{\text{overfit}\} = P\left\{U_{q,L,n-k_*-L} < \exp\Big(-\frac{\log(n)Lq}{n}\Big)\right\}.$$

**HQ**

$$P\{\text{HQ}_{k_*+L} < \text{HQ}_{k_*}\}$$
$$= P\left\{n\sum_{i=1}^{q}\log\left(1+\frac{\chi^2_L}{\chi^2_{n-k_*-L+i}}\right) > 2\log\log(n)Lq\right\}.$$

In terms of $U_{q,L,n-k_*-L}$,

$$P\{\text{overfit}\} = P\left\{U_{q,L,n-k_*-L} < \exp\Big(-\frac{2\log\log(n)Lq}{n}\Big)\right\}.$$

**HQc**

$$P\{\text{HQc}_{k_*+L} < \text{HQc}_{k_*}\}$$
$$= P\left\{n\sum_{i=1}^{q}\log\left(1+\frac{\chi^2_L}{\chi^2_{n-k_*-L+i}}\right) > \frac{2\log\log(n)Lq(n-q-1)n}{(n-k-L-q-1)(n-k-q-1)}\right\}.$$

In terms of $U_{q,L,n-k_*-L}$,

$$P\{\text{overfit}\} = P\left\{U_{q,L,n-k_*-L} < \exp\Big(-\frac{2\log\log(n)Lq(n-q-1)}{(n-k-L-q-1)(n-k-q-1)}\Big)\right\}.$$

### *4.5.2. Asymptotic Probabilities of Overfitting*

Using the above small-sample probabilities of overfitting as a basis, we will now look at overfitting in large samples by obtaining asymptotic probabilities of overfitting. Again, assume that only the two models $k_* + L$ and $k_*$ are compared and that they form nested models. We will need to use the following facts: as $n \to \infty$, with $k_*$, $L$, and $q$ fixed and $0 \le i \le q$, then $\chi^2_{n-k_*-L-q+i}/n \to 1$ a.s.; and $\log(1+z) \doteq z$ when $|z|$ is small. For independent $\chi^2$ we have

$$\begin{aligned} n\sum_{i=1}^{q}\log\left(1+\frac{\chi^2_L}{\chi^2_{n-k_*-L-q+i}}\right) &= n\sum_{i=1}^{q}\log\left(1+\frac{\frac{1}{n}\chi^2_L}{\frac{1}{n}\chi^2_{n-k_*-L-q+i}}\right) \\ &\to n\sum_{i=1}^{q}\frac{\frac{1}{n}\chi^2_L}{\frac{1}{n}\chi^2_{n-k_*-L-q+i}} \\ &\to \sum_{i=1}^{q}\chi^2_L \\ &= \chi^2_{qL}. \end{aligned}$$

We note that the $\chi^2$ distributions in the multivariate log-Beta distribution are in fact independent, and thus obtaining the asymptotic probabilities of overfitting are a matter of evaluating the limit of the critical value. We will show the calculations for AIC as an example, and only the results of the analogous calculations for the other criteria. However, the details can be found in Appendix 4D.

**AIC**

We have shown that, in small samples, AIC overfits with probability

$$P\left\{n\sum_{i=1}^{q}\log\left(1+\frac{\chi^2_L}{\chi^2_{n-k_*-L}}\right) > 2Lq\right\}.$$

Now, as $n \to \infty$,

$$n\sum_{i=1}^{q}\log\left(1+\frac{\chi^2_L}{\chi^2_{n-k_*-L}}\right) \to \chi^2_{qL},$$

thus the asymptotic probability of overfitting for AIC is

$$P\{\text{AIC overfits by } L\} = P\left\{\chi^2_{qL} > 2qL\right\}.$$

**AICc**

$$P\{\text{AICc overfits by } L\} = P\left\{\chi^2_{qL} > 2qL\right\}.$$

**AICu**

$$P\{\text{AICu overfits by } L\} = P\left\{\chi^2_{qL} > 3qL\right\}.$$

**FPE**

$$P\{\text{FPE overfits by } L\} = P\left\{\chi^2_{qL} > 2qL\right\}.$$

**SIC**

$$P\{\text{SIC overfits by } L\} = 0.$$

**HQ**

$$P\{\text{HQ overfits by } L\} = 0.$$

**HQc**

$$P\{\text{HQc overfits by } L\} = 0.$$

While SIC, HQc and HQ are asymptotically equivalent, it is important to note that even for very large sample sizes, say 200,000, the $\log(n)$ term for SIC is much larger than the $2\log\log(n)$ for HQ and HQc. Therefore the behavior of HQ and HQc is much different than that of SIC although they are all consistent criteria. In Tables 4.1 and 4.2 we give the calculated asymptotic probabilities of overfitting for model dimensions $q = 2$ and $q = 5$.

Table 4.1. Asymptotic probability of overfitting by $L$ variables for $q = 2$. Probabilities refer to selecting one particular overfit model over the true model.

| $L$ | AIC | AICc | AICu | SIC | HQ | HQc | FPE |
|---|---|---|---|---|---|---|---|
| 1 | 0.1353 | 0.1353 | 0.0498 | 0 | 0 | 0 | 0.1353 |
| 2 | 0.0916 | 0.0916 | 0.0174 | 0 | 0 | 0 | 0.0916 |
| 3 | 0.0620 | 0.0620 | 0.0062 | 0 | 0 | 0 | 0.0620 |
| 4 | 0.0424 | 0.0424 | 0.0023 | 0 | 0 | 0 | 0.0424 |
| 5 | 0.0293 | 0.0293 | 0.0009 | 0 | 0 | 0 | 0.0293 |
| 6 | 0.0203 | 0.0203 | 0.0003 | 0 | 0 | 0 | 0.0203 |
| 7 | 0.0142 | 0.0142 | 0.0001 | 0 | 0 | 0 | 0.0142 |
| 8 | 0.0100 | 0.0100 | 0.0000 | 0 | 0 | 0 | 0.0100 |
| 9 | 0.0071 | 0.0071 | 0.0000 | 0 | 0 | 0 | 0.0071 |
| 10 | 0.0050 | 0.0050 | 0.0000 | 0 | 0 | 0 | 0.0050 |

From Table 4.1 we see that the consistent criteria, as their definition demands, have zero probabilities of overfitting. By contrast, the efficient criteria have nonzero probabilities of overfitting. We note also that AICu has a much smaller asymptotic probability of overfitting than does AICc or AIC, as we would expect from the differences in their penalty functions.

Results in Table 4.2 for $q = 5$ are similar to those for $q = 2$ except that all the probabilities of overfitting are much smaller. In $q$-dimensional models, if one variable is added to the model, the number of parameters increases by $q$ and the chance of overfitting decreases.

Table 4.2. Asymptotic probability of overfitting by $L$ variables for $q = 5$. Probabilities refer to selecting one particular overfit model over the true model.

| $L$ | AIC | AICc | AICu | SIC | HQ | HQc | FPE |
|---|---|---|---|---|---|---|---|
| 1 | 0.0752 | 0.0752 | 0.0104 | 0 | 0 | 0 | 0.0752 |
| 2 | 0.0293 | 0.0293 | 0.0009 | 0 | 0 | 0 | 0.0293 |
| 3 | 0.0119 | 0.0119 | 0.0001 | 0 | 0 | 0 | 0.0119 |
| 4 | 0.0050 | 0.0050 | 0.0000 | 0 | 0 | 0 | 0.0050 |
| 5 | 0.0021 | 0.0021 | 0.0000 | 0 | 0 | 0 | 0.0021 |
| 6 | 0.0009 | 0.0009 | 0.0000 | 0 | 0 | 0 | 0.0009 |
| 7 | 0.0004 | 0.0004 | 0.0000 | 0 | 0 | 0 | 0.0004 |
| 8 | 0.0002 | 0.0002 | 0.0000 | 0 | 0 | 0 | 0.0002 |
| 9 | 0.0001 | 0.0001 | 0.0000 | 0 | 0 | 0 | 0.0001 |
| 10 | 0.0000 | 0.0000 | 0.0000 | 0 | 0 | 0 | 0.0000 |

### *4.5.3. Asymptotic Signal-to-noise Ratio*

Small-sample signal-to-noise ratios were computed in Section 4.3. We will use these small-sample ratios as a basis for obtaining the asymptotic signal-to-noise ratios, and we will present the detailed calculations for one criterion from the K-L (AIC) and consistent (SIC) categories. To do this we will need the following facts: assuming $k_*$ and $L$ are fixed, and that $n \to \infty$, we first note that

$$\frac{\sqrt{(n-k-L-(q-1)/2)(n-k-(q-1)/2+2)}}{\sqrt{2Lq}} \to \frac{n}{\sqrt{2Lq}}$$

and that $\log(1 - L/n) \doteq -L/n$ when $L << n$. This leads to

$$\log\left(\frac{n-k-L-(q-1)/2}{n-k-(q-1)/2}\right) = \log\left(1 - \frac{L}{n-k-(q-1)/2}\right) \to \frac{-L}{n}.$$

*4.5.3.1. K-L-based Criteria AIC, AICc, AICu, and FPE*

Starting with the finite-sample signal-to-noise ratio for AIC from Section 4.3, the asymptotic signal-to-noise ratio for AIC is

$$\begin{aligned}
\lim_{n\to\infty} & \frac{\sqrt{(n-k-L-(q-1)/2)(n-k-(q-1)/2+2)}}{\sqrt{2Lq}} \times \\
& \left(q\log\left(\frac{n-k-L-(q-1)/2}{n-k-(q-1)/2}\right)\right. \\
& \left. -\frac{Lq}{(n-k-L-(q-1)/2)(n-k-(q-1)/2)} + \frac{2Lq}{n}\right) \\
= & \lim_{n\to\infty} \frac{n}{\sqrt{2Lq}}\left(-\frac{Lq}{n} - \frac{Lq}{n^2} + \frac{2Lq}{n}\right) \\
= & \sqrt{\frac{Lq}{2}}.
\end{aligned}$$

Since they are asymptotically equivalent, the asymptotic signal-to-noise ratios for AICc and FPE are also $\sqrt{Lq/2}$. For AICu the result is $\sqrt{2Lq}$. We note that when $q=1$, these are equivalent to the univariate case.

*4.5.3.2. Consistent Criteria SIC, HQ, and HQc*

Note that

$$\begin{aligned}
\lim_{n\to\infty} & \frac{\sqrt{(n-k-L-(q-1)/2)(n-k-(q-1)/2+2)}}{\sqrt{2Lq}} \times \\
& \left(q\log\left(\frac{n-k-L-(q-1)/2}{n-k-(q-1)/2}\right)\right. \\
& \left. -\frac{Lq}{(n-k-L-(q-1)/2)(n-k-(q-1)/2)} + \frac{\log(n)Lq}{n}\right) \\
= & \lim_{n\to\infty} \frac{n}{\sqrt{2Lq}}\left(-\frac{Lq}{n} - \frac{Lq}{n^2} + \frac{\log(n)Lq}{n}\right) \\
= & \infty.
\end{aligned}$$

By definition, all consistent criteria will have infinite asymptotic signal-to-noise ratios, and so the results for HQ and HQc are the same as that for SIC.

In Tables 4.3 and 4.4 we present the asymptotic signal-to-noise ratios of model $k_*$ versus model $k_*+L$ for $q=2$ and $q=5$ for the criteria under consideration.

Table 4.3. Asymptotic signal-to-noise ratios for overfitting by $L$ variables for $q = 2$.

| $L$ | AIC | AICc | AICu | SIC | HQ | HQc | FPE |
|---|---|---|---|---|---|---|---|
| 1 | 1.000 | 1.000 | 2.000 | $\infty$ | $\infty$ | $\infty$ | 1.000 |
| 2 | 1.414 | 1.414 | 2.828 | $\infty$ | $\infty$ | $\infty$ | 1.414 |
| 3 | 1.732 | 1.732 | 3.464 | $\infty$ | $\infty$ | $\infty$ | 1.732 |
| 4 | 2.000 | 2.000 | 4.000 | $\infty$ | $\infty$ | $\infty$ | 2.000 |
| 5 | 2.236 | 2.236 | 4.472 | $\infty$ | $\infty$ | $\infty$ | 2.236 |
| 6 | 2.449 | 2.449 | 4.899 | $\infty$ | $\infty$ | $\infty$ | 2.449 |
| 7 | 2.646 | 2.646 | 5.292 | $\infty$ | $\infty$ | $\infty$ | 2.646 |
| 8 | 2.828 | 2.828 | 5.657 | $\infty$ | $\infty$ | $\infty$ | 2.828 |
| 9 | 3.000 | 3.000 | 6.000 | $\infty$ | $\infty$ | $\infty$ | 3.000 |
| 10 | 3.162 | 3.162 | 6.325 | $\infty$ | $\infty$ | $\infty$ | 3.162 |

We see from Table 4.3 that the signal-to-noise ratios of AICc and AIC are equivalent, as are those for SIC, HQ, and HQc. We also see that AICu has a much larger signal-to-noise ratio than either AICc or AIC. The results in Table 4.4 parallel those in Table 4.3 and show the positive relationship between the signal-to-noise ratio and $q$. The signal-to-noise ratios increase as the dimension of the model $q$ increases. This agrees with what we observed in Tables 4.1 and 4.2, where as $q$ increased, the probabilities of overfitting decreased. Higher signal-to-noise ratios indicate smaller probabilities of overfitting. In multivariate regression with dimension $q$, adding one variable to the model increases the parameter count by $q$. These higher parameter counts decrease the probability of overfitting as well as increasing the signal-to-noise ratio for overfitting.

Table 4.4. Asymptotic signal-to-noise ratios for overfitting by $L$ variables for $q = 5$.

| $L$ | AIC | AICc | AICu | SIC | HQ | HQc | FPE |
|---|---|---|---|---|---|---|---|
| 1 | 1.581 | 1.581 | 3.162 | $\infty$ | $\infty$ | $\infty$ | 1.581 |
| 2 | 2.236 | 2.236 | 4.472 | $\infty$ | $\infty$ | $\infty$ | 2.236 |
| 3 | 2.736 | 2.739 | 5.477 | $\infty$ | $\infty$ | $\infty$ | 2.739 |
| 4 | 3.162 | 3.162 | 6.325 | $\infty$ | $\infty$ | $\infty$ | 3.162 |
| 5 | 3.536 | 3.536 | 7.071 | $\infty$ | $\infty$ | $\infty$ | 3.536 |
| 6 | 3.873 | 3.873 | 7.746 | $\infty$ | $\infty$ | $\infty$ | 3.873 |
| 7 | 4.183 | 4.183 | 8.367 | $\infty$ | $\infty$ | $\infty$ | 4.183 |
| 8 | 4.472 | 4.472 | 8.944 | $\infty$ | $\infty$ | $\infty$ | 4.472 |
| 9 | 4.743 | 4.743 | 9.487 | $\infty$ | $\infty$ | $\infty$ | 4.743 |
| 10 | 5.000 | 5.000 | 10.000 | $\infty$ | $\infty$ | $\infty$ | 5.000 |

## 4.6. Underfitting

In previous Sections we have been concerned only with overfitting. Underfitting in the multivariate case involves noncentral Wishart distributions, and while the noncentrality parameter, $\Omega$, is easily computed, the moments of noncentral Wishart distributions are not. However, if we allow $X$ to be random, $\Omega$ is also random, which allows us to compute the moments for the underfitted model case. On the other hand, under the random $X$ structure $\mathrm{SPE}_k$ does not have a noncentral Wishart distribution in the underfitting case. Therefore we will need the distributions and moments for $\Omega$ and $\mathrm{SPE}_k$ to conduct our study of underfitting using two special case models.

### *4.6.1. Distributions for Underfitted Models*

The following two theorems give the distributions of $\Omega$ and $\mathrm{SPE}_k$. Theorem 4.4 states the distribution of $\Omega$, and Theorem 4.5 states the expected value of $\mathrm{SPE}_k$. The proofs for these theorems can be found in Appendices 4B.4 and 4B.5, respectively.

**Theorem 4.4**

Suppose $Y = X_* B_* + \varepsilon_*$ is the true model, where $X_* = (X_0, X_1) = (1, \tilde{X}_0, X_1)$, and $B_* = (B_0', B_1')'$. Ignoring the constant, the rows of $X_*$ are independent with the following multivariate normal distribution:

$$\tilde{X}_{*i} = \begin{pmatrix} \tilde{X}_{0i} \\ X_{1i} \end{pmatrix} \sim N\left( \begin{pmatrix} 0 \\ 0 \end{pmatrix}, \begin{pmatrix} \Sigma_{x00} & \Sigma_{x10} \\ \Sigma_{x01} & \Sigma_{x11} \end{pmatrix} \right).$$

If the model $Y = X_0 B_0$ is fit to the data, then the noncentrality parameter, $\Omega$, is random, and

$$\Omega \sim W_q(n - k_0, \Sigma_*^{-1} B_1' \Sigma_{x11\cdot 0} B_1),$$

where $k_0 = \mathrm{rank}(X_0)$, $\Sigma_{x11\cdot 0} = \Sigma_{x11} - \Sigma_{x01}' \Sigma_{00}^{-1} \Sigma_{x01}$ and $\Sigma_{x11} = var[\tilde{X}_{1i}]$, $\Sigma_{x00} = var[X_{0i}]$, and $\Sigma_{x01} = cov[\tilde{X}_{0i}, X_{1i}]$.

**Theorem 4.5**

Suppose $Y = X_* B_* + \varepsilon_*$ is the true model, where $X_* = (X_0, X_1) = (1, \tilde{X}_0, X_1)$, and $B_* = (B_0', B_1')'$. Ignoring the constant, the rows of $X_*$ are independent with the following multivariate normal distribution:

$$\tilde{X}_{*i} = \begin{pmatrix} \tilde{X}_{0i} \\ X_{1i} \end{pmatrix} \sim N\left( \begin{pmatrix} 0 \\ 0 \end{pmatrix}, \begin{pmatrix} \Sigma_{x00} & \Sigma_{x10} \\ \Sigma_{x01} & \Sigma_{x11} \end{pmatrix} \right).$$

If the model $Y = X_0 B_0$ is fit to the data, then

$$\mathrm{SPE}_{k_0} = Y'(I - H_0)Y \sim W_q(n - k_0, \Sigma_* + B_1' \Sigma_{x11\cdot 0} B_1),$$

where $k_0 = \mathrm{rank}(X_0)$, $\Sigma_{x11\cdot 0} = \Sigma_{x11} - \Sigma_{x01}' \Sigma_{00}^{-1} \Sigma_{x01}$ and $\Sigma_{x11} = var[\tilde{X}_{1i}]$, $\Sigma_{x00} = var[X_{0i}]$, and $\Sigma_{x01} = cov[\tilde{X}_{0i}, X_{1i}]$.

We can build on these results to compute (at least approximately) the expected values for several common model selection criteria. For the general model of order $k$, Eq. (4.3), with the noncentrality parameter $\Omega_k$ obtained from Theorem 4.5, we have that

$$\mathrm{SPE}_k \sim W_q(n-k, \Sigma(I_q+\Omega_k))$$

and

$$|\mathrm{SPE}_k| \sim |\Sigma(I_q+\Omega_k)| \prod_{i=1}^{q} \chi^2_{n-k-q+i},$$

with

$$E[|\mathrm{SPE}_k|] = |\Sigma(I_q+\Omega_k)| \prod_{i=1}^{q} (n-k-q+i). \tag{4.22}$$

Using the approximation for $\log|\mathrm{SPE}_k|$,

$$E[\log|\mathrm{SPE}_k|] = \log|\Sigma(I_q+\Omega_k)| + q\log(2) + \sum_{i=1}^{q} \psi\left(\frac{n-k-q+i}{2}\right). \tag{4.23}$$

The expected values for all of the model selection criteria, with the exception of FPE and Cp, can now be calculated using the approximation given in Eq. (4.23) for $\log|\mathrm{SPE}_k|$. FPE is the only nonlog-based criterion we examine here, and for it we apply Eq. (4.22) to obtain the moments. We have no convenient way to calculate Cp's signal-to-noise ratio and hence it is omitted from this Section.

**AIC**

$$\begin{aligned} E[\mathrm{AIC}_k] =& \log|\Sigma(I_q+\Omega_k)| + q\log(2) + \sum_{i=1}^{q} \psi\left(\frac{n-k-q+i}{2}\right) \\ &- q\log(n) + \frac{2kq+q(q+1)}{n}. \end{aligned}$$

**AICc**

$$\begin{aligned} E[\mathrm{AICc}_k] =& \log|\Sigma(I_q+\Omega_k)| + q\log(2) + \sum_{i=1}^{q} \psi\left(\frac{n-k-q+i}{2}\right) \\ &- q\log(n) + \frac{(n+k)q}{n-k-q-1}. \end{aligned}$$

**AICu**

$$\begin{aligned} E[\mathrm{AICu}_k] =& \log|\Sigma(I_q+\Omega_k)| + q\log(2) + \sum_{i=1}^{q} \psi\left(\frac{n-k-q+i}{2}\right) \\ &- q\log(n-k) + \frac{(n+k)q}{n-k-q-1}. \end{aligned}$$

**FPE**

$$E[\text{FPE}_k] = |\Sigma(I_q + \Omega_k)| \left(\frac{n+k}{n-k}\right)^q \prod_{i=1}^{q} \frac{n-k-q+i}{n}.$$

**SIC**

$$E[\text{SIC}_k] = \log|\Sigma(I_q + \Omega_k)| + q\log(2) + \sum_{i=1}^{q} \psi\left(\frac{n-k-q+i}{2}\right) - q\log(n) + \frac{\log(n)kq}{n}.$$

**HQ**

$$E[\text{HQ}_k] = \log|\Sigma(I_q + \Omega_k)| + q\log(2) + \sum_{i=1}^{q} \psi\left(\frac{n-k-q+i}{2}\right) - q\log(n) + \frac{2\log\log(n)kq}{n}.$$

**HQc**

$$E[\text{HQc}_k] = \log|\Sigma(I_q + \Omega_k)| + q\log(2) + \sum_{i=1}^{q} \psi\left(\frac{n-k-q+i}{2}\right) - q\log(n) + \frac{2\log\log(n)kq}{n-k-q-1}.$$

### *4.6.2. Expected Values for Two Special Case Models*

We now examine underfitting for two special case bivariate, $q = 2$, regression Models 7 and 8. In each case the true model has

$$n = 25, \; k_* = 5, \text{ independent columns of } X \text{ and } \Sigma_* = \begin{pmatrix} 1 & 0.7 \\ 0.7 & 1 \end{pmatrix}. \tag{4.24}$$

Model 7 has strongly identifiable parameters relative to the error, and Model 8 has much more weakly identifiable parameters. For Model 7,

$$y_i = \begin{pmatrix} 1 \\ 1 \end{pmatrix} + \begin{pmatrix} 1 \\ 1 \end{pmatrix} x_{i,1} + \begin{pmatrix} 1/4 \\ 1 \end{pmatrix} x_{i,2} + \begin{pmatrix} 0 \\ 1 \end{pmatrix} x_{i,3} + \begin{pmatrix} 0 \\ 1 \end{pmatrix} x_{i,4} + \varepsilon_{*i}. \tag{4.25}$$

For Model 8,

$$y_i = \begin{pmatrix} 1 \\ 1 \end{pmatrix} + \begin{pmatrix} 1 \\ 1 \end{pmatrix} x_{i,1} + \begin{pmatrix} 1/4 \\ 1/4 \end{pmatrix} x_{i,2} + \begin{pmatrix} 1/9 \\ 1/9 \end{pmatrix} x_{i,3} + \begin{pmatrix} 1/16 \\ 1/16 \end{pmatrix} x_{i,4} + \varepsilon_{*i}. \tag{4.26}$$

Models 7 and 8 differ only by the identifiability of the true model. The expected values for each criterion for each model are summarized in Tables 4.5–4.6. We also include the expected efficiencies for $\text{tr}\{L_2\}$, $\det(L_2)$ using the estimated expected trace of Eq. (4.9) and Eq. (1.3) and estimated expected determinant of Eq. (4.9) and Eq. (1.3), respectively, as well as estimated expected Kullback–Leibler (K-L) using Eq. (4.10) and Eq. (1.4). These estimated expected values are averaged based on 100,000 realizations.

For Model 7, all selection criteria attain their minimum expected values at

Table 4.5. Expected values and expected efficiency for Model 7.

| $k$ | AIC | AICc | AICu | SIC | HQ | HQc | FPE | $\mathrm{tr}\{L_2\}$ | $\det(L_2)$ | K-L |
|---|---|---|---|---|---|---|---|---|---|---|
| 1 | 1.962 | 4.038 | 4.120 | 1.819 | 1.749 | 1.784 | 6.099 | 0.081 | 0.006 | 0.363 |
| 2 | 1.374 | 3.514 | 3.681 | 1.330 | 1.189 | 1.282 | 3.406 | 0.134 | 0.054 | 0.457 |
| 3 | 1.150 | 3.377 | 3.633 | 1.202 | 0.991 | 1.168 | 2.737 | 0.200 | 0.075 | 0.508 |
| 4 | 0.738 | 3.081 | 3.429 | 0.888 | 0.607 | 0.898 | 1.828 | 0.344 | 0.113 | 0.631 |
| 5 | **-0.235** | **2.254** | **2.700** | **0.012** | **-0.340** | **0.100** | **0.698** | 1.000 | 1.000 | 1.000 |
| 6 | -0.186 | 2.489 | 3.037 | 0.159 | -0.264 | 0.367 | 0.743 | 0.833 | 0.667 | 0.798 |
| 7 | -0.144 | 2.763 | 3.420 | 0.299 | -0.195 | 0.678 | 0.789 | 0.714 | 0.476 | 0.646 |
| 8 | -0.109 | 3.085 | 3.857 | 0.431 | -0.133 | 1.043 | 0.836 | 0.624 | 0.356 | 0.527 |
| 9 | -0.082 | 3.468 | 4.361 | 0.555 | -0.079 | 1.475 | 0.884 | 0.555 | 0.277 | 0.433 |
| 10 | -0.065 | 3.928 | 4.950 | 0.670 | -0.035 | 1.992 | 0.933 | 0.499 | 0.221 | 0.356 |
| 11 | -0.059 | 4.486 | 5.646 | 0.774 | -0.002 | 2.617 | 0.982 | 0.454 | 0.181 | 0.294 |
| 12 | -0.066 | 5.174 | 6.482 | 0.864 | 0.019 | 3.386 | 1.031 | 0.416 | 0.151 | 0.241 |
| 13 | -0.088 | 6.037 | 7.505 | 0.940 | 0.024 | 4.347 | 1.080 | 0.384 | 0.128 | 0.197 |
| 14 | -0.128 | 7.142 | 8.784 | 0.998 | 0.011 | 5.576 | 1.128 | 0.356 | 0.109 | 0.160 |
| 15 | -0.190 | 8.599 | 10.431 | 1.033 | -0.024 | 7.190 | 1.175 | 0.333 | 0.095 | 0.128 |
| 16 | -0.280 | 10.587 | 12.630 | 1.040 | -0.087 | 9.390 | 1.219 | 0.312 | 0.083 | 0.100 |

Boldface type indicates the minimum expectation.

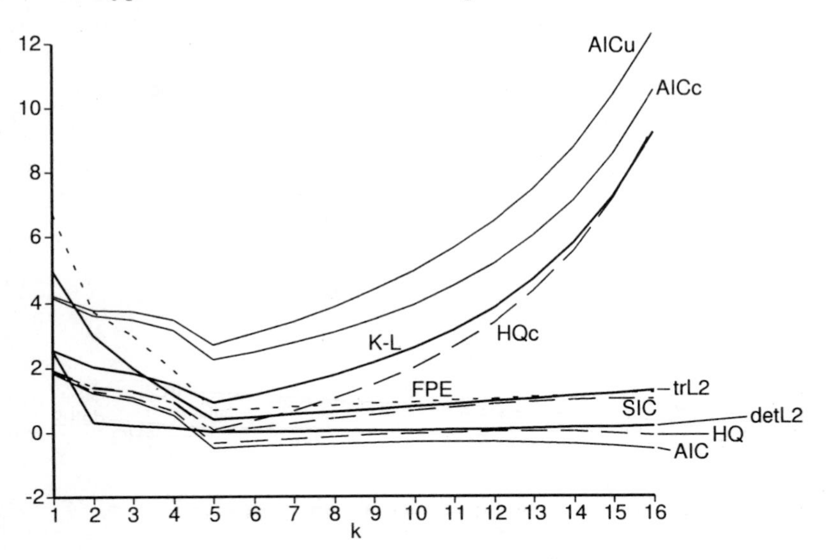


Figure 4.1. Expected values and expected distance for Model 7.

the correct order 5, but we see from their lower expected efficiencies that K-L and $\det(L_2)$ penalize overfitting more than $\text{tr}\{L_2\}$. With respect to underfitting, $\det(L_2)$ penalizes most harshly and K-L penalizes least. Thus, K-L should favor selection criteria that underfit slightly, $\text{tr}\{L_2\}$ should favor selection criteria that overfit slightly, and $\det(L_2)$ should penalize both underfitting

Table 4.6. Expected values and expected efficiency for Model 8.

| $k$ | AIC | AICc | AICu | SIC | HQ | HQc | FPE | $\text{tr}\{L_2\}$ | $\det(L_2)$ | K-L |
|---|---|---|---|---|---|---|---|---|---|---|
| 1 | 0.335 | 2.412 | 2.493 | 0.193 | 0.123 | 0.158 | 1.200 | 0.126 | 0.205 | 0.469 |
| 2 | -0.326 | **1.814** | **1.981** | **-0.371** | **-0.512** | **-0.419** | 0.622 | 0.885 | 1.000 | 1.000 |
| 3 | **-0.331** | 1.896 | 2.152 | -0.279 | -0.490 | -0.313 | **0.622** | 1.000 | 0.858 | 0.831 |
| 4 | -0.285 | 2.057 | 2.405 | -0.135 | -0.417 | -0.126 | 0.657 | 0.823 | 0.503 | 0.644 |
| 5 | -0.235 | 2.254 | 2.700 | 0.012 | -0.340 | 0.100 | 0.698 | 0.673 | 0.312 | 0.506 |
| 6 | -0.186 | 2.489 | 3.037 | 0.159 | -0.264 | 0.367 | 0.743 | 0.562 | 0.208 | 0.404 |
| 7 | -0.144 | 2.763 | 3.420 | 0.299 | -0.195 | 0.678 | 0.789 | 0.481 | 0.149 | 0.327 |
| 8 | -0.109 | 3.085 | 3.857 | 0.431 | -0.133 | 1.043 | 0.836 | 0.421 | 0.112 | 0.267 |
| 9 | -0.082 | 3.468 | 4.361 | 0.555 | -0.079 | 1.475 | 0.884 | 0.374 | 0.087 | 0.219 |
| 10 | -0.065 | 3.928 | 4.950 | 0.670 | -0.035 | 1.992 | 0.933 | 0.337 | 0.069 | 0.181 |
| 11 | -0.059 | 4.486 | 5.646 | 0.774 | -0.002 | 2.617 | 0.982 | 0.306 | 0.057 | 0.149 |
| 12 | -0.066 | 5.174 | 6.482 | 0.864 | 0.019 | 3.386 | 1.031 | 0.281 | 0.047 | 0.122 |
| 13 | -0.088 | 6.037 | 7.505 | 0.940 | 0.024 | 4.347 | 1.080 | 0.259 | 0.040 | 0.100 |
| 14 | -0.128 | 7.142 | 8.784 | 0.998 | 0.011 | 5.576 | 1.128 | 0.241 | 0.034 | 0.081 |
| 15 | -0.190 | 8.599 | 10.431 | 1.033 | -0.024 | 7.190 | 1.175 | 0.225 | 0.030 | 0.065 |
| 16 | -0.280 | 10.587 | 12.630 | 1.040 | -0.087 | 9.390 | 1.219 | 0.211 | 0.026 | 0.051 |

Boldface type indicates the minimum expectation.

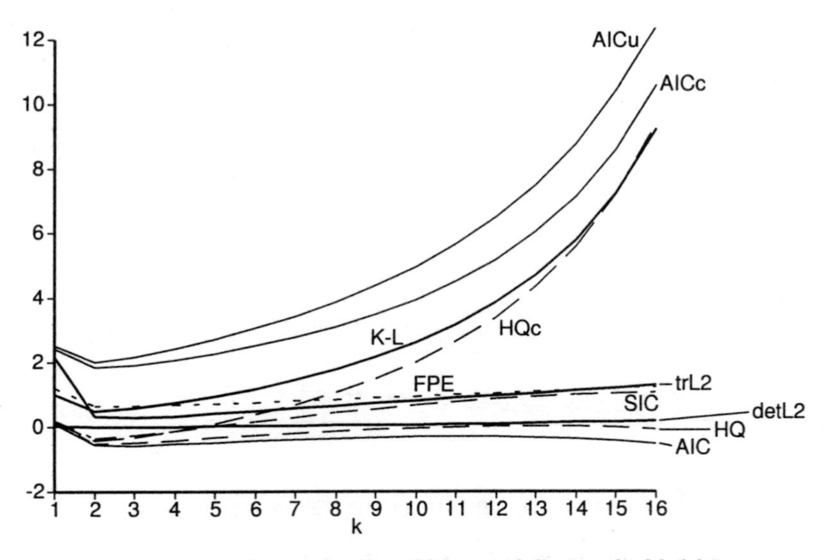


Figure 4.2. Expected values and expected distance for Model 8.

and overfitting tendencies. On the basis of these results we expect $\det(L_2)$ to yield the lowest efficiency in the simulation studies. Figure 4.1 displays the expected values for K-L, $\mathrm{tr}\{L_2\}$ and $\det(L_2)$ rather than the expected efficiencies presented in Table 4.5.

Model 8 has much more weakly identifiable parameters than Model 7, and thus the correct model will be difficult to detect in a sample size of only 25. Indeed, we see from Table 4.6 that none of the selection criteria have a well-defined minimum at the correct order 5; instead, all have minima at order 2 or 3. In other words, the closest candidate model according to the expected distance measures is no longer the true model. While underfitting is to be expected from the selection criteria in the case where the model parameters are only weakly identifiable, ideally this underfitting should not cause a loss in efficiency. The expected values for K-L, $\mathrm{tr}\{L_2\}$ and $\det(L_2)$ are presented in Figure 4.2 rather than the expected efficiencies in Table 4.6. Next we examine the approximate signal-to-noise ratios for these models.

### 4.6.3. Signal-to-noise Ratios for Two Special Case Models

The signal-to-noise ratio relates the expected difference between two nested models to its standard deviation. Candidate models are compared to the true model for the two special case models from the previous Section, given by Eqs. (4.24)–(4.26). Note that the signal-to-noise ratio is defined to be 0 when comparing the correct model to itself, and that signal-to-noise ratios for underfitted models are simulated on the basis of 100,000 realizations except for FPE. All signal-to-noise ratios for FPE are simulated, since the true noise is unknown. Also, for overfitting when $q = 2$, the noise for $\log|\mathrm{SPE}_k|$-based model selection criteria is

$$\sqrt{\sum_{i=1}^{2} \psi'\left(\frac{n-k_*-L-2+i}{2}\right) - \psi'\left(\frac{n-k_*-2+i}{2}\right)}.$$

Finally, the signal-to-noise ratios can be calculated from the expected values in Tables 4.5–4.6 and the above noise.

For Model 7, with its strongly identifiable parameters, we see that AICu has the highest signal-to-noise ratio with respect to overfitting and should overfit less than the other model selection criteria. AICc and HQc also have large signal-to-noise ratios. AIC, HQ, and FPE, which are known to overfit in practice, not surprisingly have weak signal-to-noise ratios. SIC has a moderate signal-to-noise ratio. The overfitting signal-to-noise ratios nicely illustrate the role of the penalty function. AICu, AICc, and HQc all have strong, superlinear penalty functions that increase rapidly as the number of variables $k$ increases. AIC, SIC, and HQ have weaker penalty functions, linear in $k$. As linear penalty functions go, SIC has a large penalty function, but in small samples, it is not nearly as strong as the penalty functions of AICu, AICc, and HQc.

Model 8 shows evidence of underfitting from all the selection criteria, as can be seen from the negative signal-to-noise ratios for $k < 5$. AICu has

Table 4.7. Approximate signal-to-noise ratios for Model 7.

| $k$ | AIC | AICc | AICu | SIC | HQ | HQc | FPE |
|---|---|---|---|---|---|---|---|
| 1 | 4.398 | 3.605 | 2.906 | 3.650 | 4.190 | 3.415 | 2.185 |
| 2 | 3.892 | 3.091 | 2.451 | 3.221 | 3.706 | 2.912 | 2.059 |
| 3 | 3.568 | 2.935 | 2.474 | 3.097 | 3.438 | 2.802 | 1.964 |
| 4 | 2.742 | 2.350 | 2.090 | 2.482 | 2.670 | 2.272 | 1.739 |
| 5 | 0 | 0 | 0 | 0 | 0 | 0 | 0 |
| 6 | 0.446 | 2.119 | 3.044 | 1.325 | 0.690 | 2.413 | 0.536 |
| 7 | 0.585 | 3.207 | 4.531 | 1.810 | 0.925 | 3.644 | 0.683 |
| 8 | 0.635 | 4.114 | 5.718 | 2.078 | 1.035 | 4.666 | 0.764 |
| 9 | 0.650 | 5.070 | 6.929 | 2.274 | 1.101 | 5.739 | 0.808 |
| 10 | 0.620 | 5.985 | 8.038 | 2.360 | 1.103 | 6.761 | 0.830 |
| 11 | 0.555 | 6.931 | 9.144 | 2.369 | 1.058 | 7.815 | 0.837 |
| 12 | 0.468 | 8.002 | 10.362 | 2.337 | 0.986 | 9.003 | 0.833 |
| 13 | 0.365 | 9.223 | 11.713 | 2.266 | 0.892 | 10.354 | 0.820 |
| 14 | 0.240 | 10.792 | 13.431 | 2.177 | 0.778 | 12.088 | 0.796 |
| 15 | 0.096 | 12.616 | 15.372 | 2.034 | 0.633 | 14.099 | 0.764 |
| 16 | -0.075 | 14.832 | 17.674 | 1.834 | 0.455 | 16.535 | 0.726 |

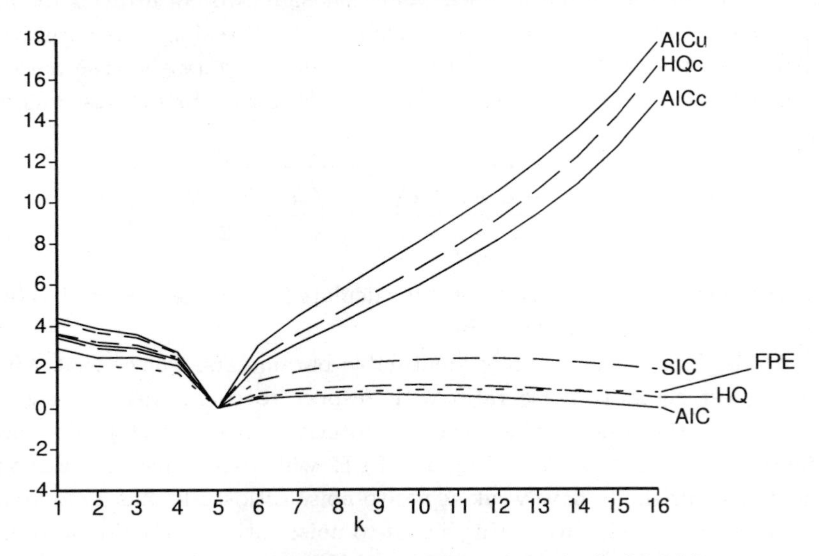


Figure 4.3. Approximate signal-to-noise ratios for Model 7.

the most negative signal-to-noise ratio, indicating that it will underfit more than the other selection criteria. While its strong penalty function prevents overfitting, it does so at the expense of some underfitting when the parameters are weak. AIC has a negative underfitting signal-to-noise ratio as well as a weak overfitting signal-to-noise ratio. Thus we expect AIC to underfit and also occasionally overfit excessively.

Table 4.8. Approximate signal-to-noise ratios for Model 8.

| $k$ | AIC | AICc | AICu | SIC | HQ | HQc | FPE |
|---|---|---|---|---|---|---|---|
| 1 | 1.567 | 0.434 | -0.565 | 0.498 | 1.270 | 0.161 | 1.312 |
| 2 | -0.426 | -2.069 | -3.384 | -1.802 | -0.807 | -2.438 | -0.503 |
| 3 | -0.614 | -2.291 | -3.511 | -1.862 | -0.960 | -2.643 | -0.663 |
| 4 | -0.462 | -1.814 | -2.711 | -1.358 | -0.711 | -2.081 | -0.528 |
| 5 | 0 | 0 | 0 | 0 | 0 | 0 | 0 |
| 6 | 0.446 | 2.119 | 3.044 | 1.325 | 0.690 | 2.413 | 0.536 |
| 7 | 0.585 | 3.207 | 4.531 | 1.810 | 0.925 | 3.644 | 0.683 |
| 8 | 0.635 | 4.114 | 5.718 | 2.078 | 1.035 | 4.666 | 0.764 |
| 9 | 0.650 | 5.070 | 6.929 | 2.274 | 1.101 | 5.739 | 0.808 |
| 10 | 0.620 | 5.985 | 8.038 | 2.360 | 1.103 | 6.761 | 0.830 |
| 11 | 0.555 | 6.931 | 9.144 | 2.369 | 1.058 | 7.815 | 0.837 |
| 12 | 0.468 | 8.002 | 10.362 | 2.337 | 0.986 | 9.003 | 0.833 |
| 13 | 0.365 | 9.223 | 11.713 | 2.266 | 0.892 | 10.354 | 0.820 |
| 14 | 0.240 | 10.792 | 13.431 | 2.177 | 0.778 | 12.088 | 0.796 |
| 15 | 0.096 | 12.616 | 15.372 | 2.034 | 0.633 | 14.099 | 0.764 |
| 16 | -0.075 | 14.832 | 17.674 | 1.834 | 0.455 | 16.535 | 0.726 |

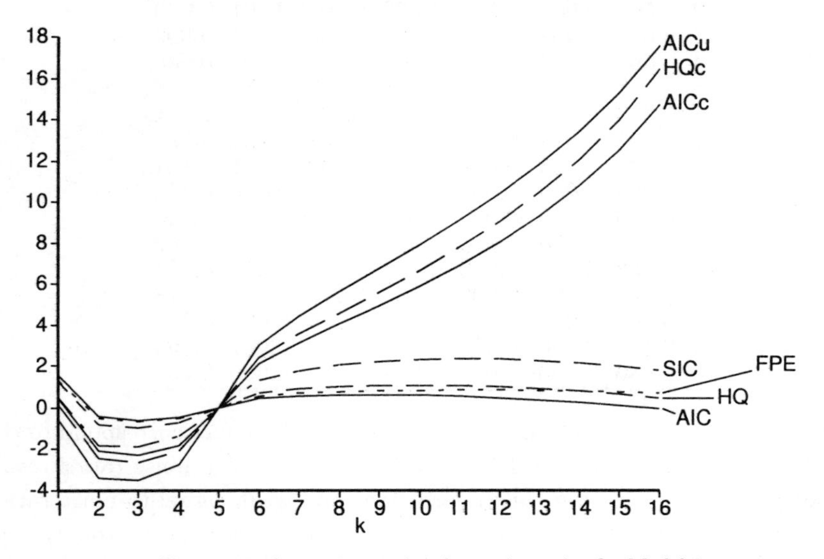


Figure 4.4. Approximate signal-to-noise ratios for Model 8.

For both models we expect that the signal-to-noise ratio results will correspond to probabilities of overfitting and underfitting. That is, high signal-to-noise ratios result in low probabilities of over- or underfitting, and vice versa. In the next Section we calculate these probabilities in order to see if our expectations are supported.

### 4.6.4. Probabilities for Two Special Case Models

Tables 4.9–4.10 summarize the approximate probabilities for selecting a particular candidate order $k$ over the correct order 5. Underfitting probabilities are approximated by simulation from the same 100,000 replications used to estimate underfitting signal-to-noise ratios. Probabilities for selecting the correct order 5 over itself are undefined and denoted by an asterisk. Overfitting probabilities are computed using the U-statistic and special case form for $q = 2$,

$$P\{U_{2,L,n-k_*-L} < u\} = P\left\{F_{2L,2(n-k_*-L)} > \frac{1-\sqrt{u}}{\sqrt{u}}\frac{n-k_*-L-1}{L}\right\}.$$

Table 4.9. Probability of selecting a particular candidate model of order $k$ over the true order 5 for Model 7.

| $k$ | AIC | AICc | AICu | SIC | HQ | HQc | FPE |
|---|---|---|---|---|---|---|---|
| 1 | 0.000 | 0.000 | 0.001 | 0.000 | 0.000 | 0.000 | 0.000 |
| 2 | 0.000 | 0.000 | 0.004 | 0.000 | 0.000 | 0.001 | 0.000 |
| 3 | 0.000 | 0.001 | 0.004 | 0.000 | 0.000 | 0.001 | 0.000 |
| 4 | 0.000 | 0.003 | 0.010 | 0.002 | 0.001 | 0.005 | 0.000 |
| 5 | * | * | * | * | * | * | * |
| 6 | 0.237 | 0.044 | 0.017 | 0.097 | 0.188 | 0.032 | 0.221 |
| 7 | 0.227 | 0.012 | 0.002 | 0.058 | 0.158 | 0.006 | 0.199 |
| 8 | 0.221 | 0.003 | 0.000 | 0.041 | 0.143 | 0.001 | 0.181 |
| 9 | 0.225 | 0.000 | 0.000 | 0.030 | 0.132 | 0.000 | 0.170 |
| 10 | 0.236 | 0.000 | 0.000 | 0.025 | 0.133 | 0.000 | 0.166 |
| 11 | 0.256 | 0.000 | 0.000 | 0.024 | 0.144 | 0.000 | 0.169 |
| 12 | 0.283 | 0.000 | 0.000 | 0.024 | 0.155 | 0.000 | 0.171 |
| 13 | 0.319 | 0.000 | 0.000 | 0.026 | 0.178 | 0.000 | 0.178 |
| 14 | 0.365 | 0.000 | 0.000 | 0.030 | 0.204 | 0.000 | 0.184 |
| 15 | 0.420 | 0.000 | 0.000 | 0.036 | 0.241 | 0.000 | 0.194 |
| 16 | 0.486 | 0.000 | 0.000 | 0.048 | 0.292 | 0.000 | 0.209 |

We see that the probabilities in Tables 4.9–4.10 do in fact correspond to the signal-to-noise ratios in Tables 4.7 and 4.8. Large signal-to-noise ratios result in small probabilities of over- or underfitting, and weak signal-to-noise ratios result in moderate probabilities of selecting the incorrect order. Strongly negative signal-to-noise ratios indicate probabilities of almost 1 for selecting the incorrect model order. For Model 7 all the selection criteria have small probabilities of underfitting, but there are significant differences in their overfitting probabilities. AICc, AICu, and HQc have the strongest penalty functions, the largest signal-to-noise ratios with respect to overfitting, and, consequently, the smallest probabilities of overfitting. Furthermore, as the amount of overfitting increases, their probabilities decrease. Compare this to AIC, HQ, and FPE, which have large probabilities of overfitting that do not decrease as the amount of overfitting increases. We therefore expect the performance of AIC, HQ, and FPE to be much worse than that of AICc, AICu, and HQc when extra or irrelevant variables are included in the study.

Table 4.10. Probability of selecting a particular candidate model of order $k$ over the true order 5 for Model 8.

| $k$ | AIC | AICc | AICu | SIC | HQ | HQc | FPE |
|---|---|---|---|---|---|---|---|
| 1 | 0.046 | 0.352 | 0.729 | 0.327 | 0.094 | 0.461 | 0.049 |
| 2 | 0.715 | 0.961 | 0.994 | 0.945 | 0.810 | 0.977 | 0.728 |
| 3 | 0.780 | 0.967 | 0.992 | 0.946 | 0.848 | 0.978 | 0.792 |
| 4 | 0.767 | 0.940 | 0.976 | 0.905 | 0.818 | 0.954 | 0.778 |
| 5 | * | * | * | * | * | * | * |
| 6 | 0.237 | 0.044 | 0.017 | 0.097 | 0.188 | 0.032 | 0.221 |
| 7 | 0.227 | 0.012 | 0.002 | 0.058 | 0.158 | 0.006 | 0.199 |
| 8 | 0.221 | 0.003 | 0.000 | 0.041 | 0.143 | 0.001 | 0.181 |
| 9 | 0.225 | 0.000 | 0.000 | 0.030 | 0.132 | 0.000 | 0.170 |
| 10 | 0.236 | 0.000 | 0.000 | 0.025 | 0.133 | 0.000 | 0.166 |
| 11 | 0.256 | 0.000 | 0.000 | 0.024 | 0.144 | 0.000 | 0.169 |
| 12 | 0.283 | 0.000 | 0.000 | 0.024 | 0.155 | 0.000 | 0.171 |
| 13 | 0.319 | 0.000 | 0.000 | 0.026 | 0.178 | 0.000 | 0.178 |
| 14 | 0.365 | 0.000 | 0.000 | 0.030 | 0.204 | 0.000 | 0.184 |
| 15 | 0.420 | 0.000 | 0.000 | 0.036 | 0.241 | 0.000 | 0.194 |
| 16 | 0.486 | 0.000 | 0.000 | 0.048 | 0.292 | 0.000 | 0.209 |

In the final Section in this Chapter we will also use these two special case regression models to evaluate selection criteria performance via simulation studies.

## 4.7. Monte Carlo Study

In practice, the researcher does not know the importance of the individual variables he or she is considering or the true order of the model. In the simulation study below, Model 7 is given by Eq. (4.24) with $B$ parameter structure defined by Eq. (4.25). Model 8 is given by Eq. (4.24) with the weaker $B$ parameter structure defined in Eq. (4.26). In both models $X$ is generated as $x_{ij}$ *i.i.d.* $N(0,1)$, $j = 1, \ldots, 9$, and $i = 1, \ldots, 25$ and $x_{i,0} = 1$ for $i = 1, \ldots, 25$. Generating a random $X$ for each replication reduces the effect of an unusual

$X$ matrix which may favor a particular model selection criterion. Random $X$ is also easy from a computational point of view. There are nine variables (eight $X_j$ plus the constant) in the largest candidate model. The method of least squares is used to calculate the regression parameters. The intercept or constant is always included so all 256 subsets are considered as potential candidate models. Since there are too many subsets to consider individually, subsets are summarized by the rank $k$ of their design matrix. For each of the 10,000 realizations, a new $X$ matrix and $\varepsilon$ matrix were generated. Although Tables 4.11 and 4.12 give counts for the selection of model order $k$, models of order $k = 5$ may include selections with the correct number of variables but that are not the true model. The "true" row lists the number of times the correct model was chosen out of the total count for order $k_* = 5$.

Observed efficiency is also computed for each realization. K-L observed efficiency is computed using Eq. (1.2) and Eq. (4.10). Since $L_2$ in Eq. (4.9) is a matrix, we continue to use the trace and determinant of $L_2$ as our two $L_2$ observed efficiency measures. $\mathrm{Tr}\{L_2\}$ observed efficiency is computed using Eq. (1.1) and the trace of Eq. (4.9). Similarly, $\det(L_2)$ observed efficiency is computed using the determinant of Eq. (4.9) and Eq. (1.1). The criteria select a model and the observed efficiency of this selection is recorded for each realization. Averages, medians and standard deviations are computed for the 10,000 K-L, $\mathrm{tr}\{L_2\}$, and $\det(L_2)$ observed efficiencies. Performance is based on the average observed efficiency with higher observed efficiency denoting better performance. The criterion with the highest observed efficiency is given rank 1 (best) while the criterion with the lowest observed efficiency is given rank 7 (worst of the 7 criteria considered here).

We know that Model 7 represents the case where a strong, easily identified true model belongs to the set of candidate models. The count results are one way to measure consistency, and we might therefore expect the consistent model selection criteria to have the highest counts. However, this is not always true in small samples. We see in Table 4.11 that for this case AICu has the highest count, identifying the correct model 81% of the time. Among the consistent criteria, which as a group have the next best performance, HQc is better than both SIC and HQ. AIC has the lowest count, correctly identifying the correct model only 32% of the time. The criteria with weak signal-to-noise ratios, *i.e.*, AIC, SIC, HQ, and FPE are seen to overfit. Of these, SIC has the largest penalty function and signal-to-noise ratio and overfits the least. Recall that AICc, AICu, and HQc all have superlinear penalty functions and strong signal-to-noise ratios, hence none of these criteria overfit much. Some underfitting is seen, particularly in AICu. The stronger penalty functions prevent overfitting at the expense of some underfitting. The distance measure values and observed efficiencies are also included to give an idea of which models are closest to the correct one.

The K-L and $\det(L_2)$ observed efficiencies in Table 4.11 parallel the counts; those selection criteria with higher counts for choosing the correct model also have higher observed efficiencies. On the other hand, the strong penalty function of AICu results in more underfitting under $\mathrm{tr}\{L_2\}$ than the other model selection criteria, reducing its observed efficiency in the $\mathrm{tr}\{L_2\}$ sense so that HQc and AICc now outperform AICu. AIC still has the lowest observed efficiency in small samples due to excessive overfitting. In fact, we see that the closest candidate model often has fewer variables than the true model. In terms of counts,

Table 4.11. Simulation results for Model 7. Counts and observed efficiency.

| | | | | | counts | | | | | |
|---|---|---|---|---|---|---|---|---|---|---|
| $k$ | AIC | AICc | AICu | SIC | HQ | HQc | FPE | $\mathrm{tr}\{L_2\}$ | $\det(L_2)$ | K-L |
| 1 | 0 | 0 | 1 | 0 | 0 | 0 | 0 | 0 | 0 | 0 |
| 2 | 0 | 0 | 26 | 0 | 0 | 2 | 0 | 0 | 2 | 0 |
| 3 | 0 | 24 | 132 | 14 | 1 | 41 | 1 | 1 | 73 | 41 |
| 4 | 21 | 401 | 1017 | 168 | 38 | 567 | 22 | 24 | 283 | 1111 |
| 5 | 3255 | 7995 | 8226 | 6358 | 4236 | 8192 | 3541 | 9962 | 9640 | 8441 |
| 6 | 3605 | 1453 | 568 | 2657 | 3558 | 1116 | 3769 | 12 | 2 | 345 |
| 7 | 2155 | 123 | 28 | 665 | 1601 | 79 | 1949 | 1 | 0 | 59 |
| 8 | 819 | 4 | 2 | 130 | 502 | 3 | 637 | 0 | 0 | 3 |
| 9 | 145 | 0 | 0 | 8 | 64 | 0 | 81 | 0 | 0 | 0 |
| true | 3209 | 7838 | 8100 | 6247 | 4177 | 8038 | 3491 | 9937 | 9553 | 7459 |
| | | | | | K-L observed efficiency | | | | | |
| | AIC | AICc | AICu | SIC | HQ | HQc | FPE | $\mathrm{tr}\{L_2\}$ | $\det(L_2)$ | K-L |
| ave | 0.606 | 0.827 | 0.850 | 0.746 | 0.651 | 0.839 | 0.622 | 0.940 | 0.932 | 1.000 |
| med | 0.561 | 1.000 | 1.000 | 0.910 | 0.607 | 1.000 | 0.578 | 1.000 | 1.000 | 1.000 |
| sd | 0.275 | 0.254 | 0.230 | 0.285 | 0.285 | 0.245 | 0.277 | 0.133 | 0.143 | 0.000 |
| rank | 7 | 3 | 1 | 4 | 5 | 2 | 6 | | | |
| | | | | | $\mathrm{tr}\{L_2\}$ observed efficiency | | | | | |
| | AIC | AICc | AICu | SIC | HQ | HQc | FPE | $\mathrm{tr}\{L_2\}$ | $\det(L_2)$ | K-L |
| ave | 0.723 | 0.897 | 0.892 | 0.840 | 0.761 | 0.901 | 0.737 | 1.000 | 0.974 | 0.839 |
| med | 0.737 | 1.000 | 1.000 | 1.000 | 0.818 | 1.000 | 0.763 | 1.000 | 1.000 | 1.000 |
| sd | 0.248 | 0.215 | 0.237 | 0.238 | 0.250 | 0.218 | 0.248 | 0.000 | 0.128 | 0.290 |
| rank | 7 | 2 | 3 | 4 | 5 | 1 | 6 | | | |
| | | | | | $\det(L_2)$ observed efficiency | | | | | |
| | AIC | AICc | AICu | SIC | HQ | HQc | FPE | $\mathrm{tr}\{L_2\}$ | $\det(L_2)$ | K-L |
| ave | 0.541 | 0.839 | 0.851 | 0.737 | 0.605 | 0.850 | 0.562 | 0.982 | 1.000 | 0.803 |
| med | 0.456 | 1.000 | 1.000 | 1.000 | 0.550 | 1.000 | 0.486 | 1.000 | 1.000 | 1.000 |
| sd | 0.347 | 0.299 | 0.298 | 0.345 | 0.357 | 0.293 | 0.349 | 0.096 | 0.000 | 0.340 |
| rank | 7 | 3 | 2 | 4 | 5 | 1 | 6 | | | |

K-L itself "underfits" significantly, again illustrating that K-L tends to favor underfitting more than overfitting. Of the three distance measures $\text{tr}\{L_2\}$ is the best, having a well-defined minimum at the true model, rarely "overfitting" or "underfitting."

We next look at the analogous Monte Carlo results for Model 8 in Table 4.12. The parameters in Model 8 are only weakly identifiable and difficult to determine, and hence underfitting is rampant. Table 4.12 shows that none of the selection criteria identify the correct model more than 1% of the time. In this case counts are clearly not a meaningful measure of performance, and here we must rely on observed efficiency. From the pattern of counts, the criteria with strong penalty functions choose smaller models than the criteria

Table 4.12. Simulation results for Model 8. Counts and observed efficiency.

| | | | | | counts | | | | | |
|---|---|---|---|---|---|---|---|---|---|---|
| $k$ | AIC | AICc | AICu | SIC | HQ | HQc | FPE | $\text{tr}\{L_2\}$ | $\det(L_2)$ | K-L |
| 1 | 4 | 45 | 213 | 93 | 11 | 74 | 4 | 0 | 1590 | 58 |
| 2 | 1362 | 4470 | 6969 | 4670 | 2214 | 5278 | 1418 | 1166 | 3151 | 4666 |
| 3 | 2625 | 3899 | 2410 | 3255 | 3125 | 3550 | 2766 | 5374 | 3742 | 4591 |
| 4 | 2588 | 1315 | 368 | 1343 | 2405 | 949 | 2650 | 3097 | 1355 | 642 |
| 5 | 1851 | 243 | 38 | 469 | 1370 | 136 | 1827 | 356 | 162 | 39 |
| 6 | 1021 | 27 | 2 | 143 | 598 | 12 | 926 | 6 | 0 | 4 |
| 7 | 405 | 1 | 0 | 23 | 218 | 1 | 322 | 0 | 0 | 0 |
| 8 | 124 | 0 | 0 | 4 | 50 | 0 | 77 | 1 | 0 | 0 |
| 9 | 20 | 0 | 0 | 0 | 9 | 0 | 10 | 0 | 0 | 0 |
| true | 91 | 14 | 1 | 25 | 75 | 6 | 87 | 205 | 115 | 12 |
| | | | | | K-L observed efficiency | | | | | |
| | AIC | AICc | AICu | SIC | HQ | HQc | FPE | $\text{tr}\{L_2\}$ | $\det(L_2)$ | K-L |
| ave | 0.435 | 0.644 | 0.770 | 0.641 | 0.495 | 0.685 | 0.443 | 0.898 | 0.855 | 1.000 |
| med | 0.362 | 0.655 | 0.920 | 0.672 | 0.423 | 0.788 | 0.371 | 0.976 | 0.991 | 1.000 |
| sd | 0.276 | 0.304 | 0.281 | 0.316 | 0.298 | 0.302 | 0.277 | 0.148 | 0.227 | 0.000 |
| rank | 7 | 3 | 1 | 4 | 5 | 2 | 6 | | | |
| | | | | | $\text{tr}\{L_2\}$ observed efficiency | | | | | |
| | AIC | AICc | AICu | SIC | HQ | HQc | FPE | $\text{tr}\{L_2\}$ | $\det(L_2)$ | K-L |
| ave | 0.490 | 0.608 | 0.661 | 0.601 | 0.525 | 0.627 | 0.496 | 1.000 | 0.788 | 0.870 |
| med | 0.450 | 0.614 | 0.684 | 0.605 | 0.492 | 0.638 | 0.457 | 1.000 | 0.979 | 0.974 |
| sd | 0.250 | 0.256 | 0.245 | 0.259 | 0.259 | 0.253 | 0.251 | 0.000 | 0.325 | 0.199 |
| rank | 7 | 3 | 1 | 4 | 5 | 2 | 6 | | | |
| | | | | | $\det(L_2)$ observed efficiency | | | | | |
| | AIC | AICc | AICu | SIC | HQ | HQc | FPE | $\text{tr}\{L_2\}$ | $\det(L_2)$ | K-L |
| ave | 0.226 | 0.414 | 0.529 | 0.414 | 0.279 | 0.450 | 0.233 | 0.713 | 1.000 | 0.747 |
| med | 0.102 | 0.303 | 0.505 | 0.303 | 0.139 | 0.362 | 0.108 | 0.926 | 1.000 | 0.960 |
| sd | 0.283 | 0.355 | 0.359 | 0.358 | 0.315 | 0.359 | 0.287 | 0.352 | 0.000 | 0.333 |
| rank | 7 | 3 | 1 | 4 | 5 | 2 | 6 | | | |

with weak penalty functions. Do these smaller models represent underfitting? Since counts may not be a reliable measure of performance, in order to tell we must examine observed efficiency.

For Kullback–Leibler (K-L) observed efficiency, AICu has the highest observed efficiency at 77%. Although AICu underfits in terms of counts of correct model choice, it does choose models that are closer to the true model in the K-L sense. HQc also performs well even though the true model is very weak and difficult to detect. We see that although HQc is a consistent selection criterion, it is competitive with efficient selection criteria in small samples. Finally, since the closest candidate models for Model 8 tend to have fewer variables than for Model 7, AIC and FPE do not overfit as excessively for Model 8 but AIC again has the lowest observed efficiency at 44%. The two $L_2$ observed efficiency results, parallel those for K-L.

From the Monte Carlo results for our two special case models we see that the parameter structure has a large impact on model selection in terms of selecting the correct model. However, its effect on observed efficiency is less dramatic since models with fewer than 5 variables tend to be closer to the true model with respect to both K-L and $L_2$ distances, and models closer to the true model have higher observed efficiency than those further from the true model. We will examine this issue in more detail in Chapter 9 with an expanded list of selection criteria and candidate models.

## 4.8. Summary

In developing multivariate model selection criteria, we build on the techniques discussed in Chapter 2 except that many of the basic components are no longer scalars, but matrices. For example, the sum of product errors ($\text{SPE}_k$) and the sum of product differences between the true mean vector and candidate mean vector ($L_2$) are both matrices under multivariate regression. One of the main questions for model selection under these circumstances is how to form suitable scalars from these matrices, and we have examined two of the most common methods, the determinant and the trace. The advantage for the determinant is that distributions involving $|\text{SPE}_k|$ are well-known, whereas distributions involving the trace are not as well-known. For this reason we have focused on the use of the determinant for the model selection criteria that are functions of $\log|\text{SPE}_k|$ or $|\text{SPE}_k|$.

The signal-to-noise ratio results continue to be consistent with expected values and probabilities of overfitting or underfitting. These are in turn consistent with simulation results—criteria with weak signal-to-noise ratios tend to overfit, and those with strong signal-to-noise ratios do not. The signal-to-noise ratio also can be used to suggest improved variants of common selection criteria for multivariate regression such as AICu and HQc. Both performed well in the simulations in this Chapter.

We find that underfitting is less of an issue than overfitting for multivariate regression when the model has strongly identifiable parameters, whereas underfitting becomes the main concern when model parameters are only weakly identifiable. The role of the penalty function in determining the signal-to-noise ratio and hence criterion behavior is reinforced.

More multivariate simulations can be found in Chapter 9. The two special case models are reexamined along with other multivariate selection criteria not discussed here. We also present a large-scale 504 model simulation study to investigate performance over a wide range of models. Lastly, two very large sample ($n = 25,000$) models are used to investigate the role of extra variables on overfitting.

Recently, Fujikoshi and Satoh (1997) proposed using modified AIC and Cp criteria when the family of candidate models includes both underspecified and overspecified models. While we do not include them in this Chapter, the interested reader can adopt the approaches discussed in this Chapter to examine the performance of these modified criteria.

## Chapter 4 Appendices

### Appendix 4A. Distributional Results in the Central Case

As for univariate regression, signal-to-noise ratios of multivariate model selection criteria depend on the distributions of $\mathrm{SPE}_k$ except that the $\chi^2$ distributions found in univariate regression are replaced by Wishart distributions. Here we derive some Wishart distribution moments that are useful in multivariate model selection.

Assume all distributions below are central. We take advantage of the well-known property in hierarchical (or nested) models that, for $k \geq k_*$ and $L > 0$,

$$\mathrm{SPE}_k - \mathrm{SPE}_{k+L} \sim W_q(L, \Sigma_*), \qquad (4A.1)$$

and

$$\mathrm{SPE}_k \sim W_q(n-k, \Sigma_*), \qquad (4A.2)$$

where $q$ is the dimension of the model.

The first moment of the Wishart distribution is a $q \times q$ matrix, the second moment is a four-dimensional $q \times q \times q \times q$ array, $D$, with elements $d_{i,j,r,s} =$

$cov[\text{SPE}_{ij}, \text{SPE}_{rs}]$. The first moment is straightforward:

$$E[\text{SPE}_k] = (n-k)\Sigma_*.$$

The second moment can be computed on an element-by-element basis as follows: let $\sigma_{i,j}$ represent the $i,j$th element of the covariance matrix $\Sigma_*$. Then,

$$cov[\text{SPE}_{i,j}, \text{SPE}_{r,s}] = \sigma_{ir}\sigma_{js} + \sigma_{is}\sigma_{jr}.$$

We will also need the moments of functions of $\text{SPE}_k$, particularly functions that can be used to reduce the $\text{SPE}_k$ matrix to a scalar. The trace is one such function. The actual distribution of the trace is not known in general; however,

$$E[tr\{\text{SPE}_k\}] = (n-k)tr\{\Sigma_*\}.$$

Another function that can be used to reduce the $\text{SPE}_k$ matrix to a scalar is the determinant. The distribution of the determinant $|\text{SPE}_k|$ of $\text{SPE}_k$ in Eq. (4A.2) is the distribution of a product of independent $\chi^2$ random variables:

$$\frac{|\text{SPE}_k|}{|\Sigma_*|} \sim \prod_{i=1}^{q} \chi^2_{n-k-q+i}. \tag{4A.3}$$

Note that Muirhead (1982, p. 100) gives a useful proof of this fact in his Theorem 3.2.15. Thus it follows that

$$E[|\text{SPE}_k|] = |\Sigma_*| \prod_{i=1}^{q} (n-k-q+i).$$

We will concentrate on the determinant, since most multivariate model selection criteria are functions of the determinant of $\text{SPE}_k$.

From Chapter 2 (Appendix 2A.1), if $X \sim \chi^2_{n-k}$ then

$$E[\log(X)] = \log(2) + \psi\left(\frac{n-k}{2}\right)$$

where $\psi$ is Euler's psi function. It follows from Eq. (4A.3) that

$$\log\left(\frac{|\text{SPE}_k|}{|\Sigma_*|}\right) \sim \sum_{i=1}^{q} \log \chi^2_{n-k-q+i}$$

and

$$E[\log|\text{SPE}_k|] = \log|\Sigma_*| + q\log(2) + \sum_{i=1}^{q}\psi\left(\frac{n-k-q+i}{2}\right).$$

Because no closed form exists we will approximate this expectation with a Taylor expansion as follows:

$$E[\log|\text{SPE}_k|] \doteq \log|\Sigma_*| + \sum_{i=1}^{q}\left[\log(n-k-q+i) - \frac{1}{n-k-q+i}\right].$$

However, when $q$ is large this expression is quite messy and offers little insight into the behavior of model selection criteria. If we assume that $n-k-q$ is much larger than $q$, then we can make a further simplification. For any $n$, $k$, $q$,

$$\sum_{i=1}^{q}(n-k-q+i) = q(n-k-(q-1)/2),$$

and applying this idea to the two terms in the summation we obtain

$$\sum_{i=1}^{q}\log(n-k-q+i) = q\log(n-k-(q-1)/2)$$

and

$$\sum_{i=1}^{q}\frac{1}{n-k-q+i} \doteq \frac{q}{n-k-(q-1)/2}.$$

Thus we have

$$E[\log|\text{SPE}_k|] \doteq \log|\Sigma_*| + q\log(n-k-(q-1)/2) - \frac{q}{n-k-(q-1)/2}. \quad (4A.4)$$

The distribution of differences between $\log|\text{SPE}_k|$ and $\log|\text{SPE}_{k+L}|$ is more involved. The ratio $|\text{SPE}_{k+L}|/|\text{SPE}_k|$ is the U-statistic with parameters $q$, $k$ and $k+L$. The distribution of a U-statistic is the product of independent Beta distributions. For nested models, $|\text{SPE}_{k+L}|/|\text{SPE}_k|$ has the following distribution:

$$\frac{|\text{SPE}_{k+L}|}{|\text{SPE}_k|} \sim \prod_{i=1}^{q}\text{Beta}\left(\frac{n-k-L-q+i}{2}, \frac{L}{2}\right).$$

Note that the Beta distributions are independent. Taking logs,

$$\log\left(\frac{|\text{SPE}_{k+L}|}{|\text{SPE}_k|}\right) \sim \sum_{i=1}^{q}\text{log-Beta}\left(\frac{n-k-L-q+i}{2}, \frac{L}{2}\right).$$

Using log-Beta facts discussed in Chapter 2 (Appendix 2A.1),

$$var\left[\log\left(\frac{|\mathrm{SPE}_{k+L}|}{|\mathrm{SPE}_k|}\right)\right] = \sum_{i=1}^{q} var\left[\text{log-Beta}\left(\frac{n-k-L-q+i}{2}, \frac{L}{2}\right)\right]$$
$$= \sum_{i=1}^{q}\left[\psi'\left(\frac{n-k-L-q+i}{2}\right) - \psi'\left(\frac{n-k-q+i}{2}\right)\right]$$

which has no closed form. Applying the first order Taylor expansion to each log-Beta term yields a convenient approximation:

$$var\left[\log\left(\frac{|\mathrm{SPE}_{k+L}|}{|\mathrm{SPE}_k|}\right)\right] \doteq \sum_{i=1}^{q} \frac{2L}{(n-k-L-q+i)(n-k-q+i+2)}.$$

However, once again this expression is not very useful for studying properties of multivariate model selection criteria, and thus we make a simplification similar to that used in Chapter 2, Appendix 2A.3:

$$var\left[\log\left(\frac{|\mathrm{SPE}_{k+L}|}{|\mathrm{SPE}_k|}\right)\right] \doteq \frac{2Lq}{(n-k-L-(q-1)/2)(n-k-(q-1)/2+2)},$$

and thus the standard deviation becomes

$$sd\left[\log\left(\frac{|\mathrm{SPE}_{k+L}|}{|\mathrm{SPE}_k|}\right)\right] \doteq \frac{\sqrt{2Lq}}{\sqrt{(n-k-L-(q-1)/2)(n-k-(q-1)/2+2)}}. \tag{4A.5}$$

Of course, Eq. (4A.4) and Eq. (4A.5) are only approximate, but the advantage of these approximations is that they are greatly simplified algebraic expressions for moments of the model selection criteria, and they are similar to the univariate Taylor expansions (when $q = 1$, the multivariate approximations are equivalent to the univariate approximations from Chapter 2 Appendix 2A). Note also that the approximations improve asymptotically.

## Appendix 4B. Proofs of Theorems 4.1 to 4.5

### *4B.1. Theorem 4.1*

Given the $q$-dimensional multivariate regression model in Eqs. (4.1)–(4.2) and the criterion of the form $\log(\hat{\Sigma}_k)+\alpha kq/(n-k-q-1)$, for all $n \geq q+5$, $\alpha \geq 1$, $0 < L < n-k-q-1$, and for the overfitting case where $0 < k_* \leq k < n-q-2$, the signal-to-noise ratio of this criterion increases as the amount of overfitting $L$ increases.

**Proof:**

Let $d = (q-1)/2$. The signal-to-noise ratio of this criterion can be written as a function of the amount of overfitting, $L$, as follows:

$$\mathrm{STN}(L) = \frac{\sqrt{(n-k-L-d)(n-k-d+2)}}{\sqrt{2Lq}} \left( q \log\left(\frac{n-k-L-d}{n-k-d}\right) - \frac{Lq}{(n-k-L-d)(n-k-d)} + \frac{\alpha Lq(n-q-1)}{(n-k-L-q-1)(n-k-q-1)} \right).$$

Since $\mathrm{STN}(L)$ is continuous in $L$, it is enough to show that its derivative is positive:

$$\begin{aligned}
\frac{d}{dL}\mathrm{STN}(L) = &-\frac{1}{2}\frac{\sqrt{n-k-d+2}}{\sqrt{2Lq(n-k-L-d)}} \left( q \log\left(\frac{n-k-L-d}{n-k-d}\right) \right. \\
&\left. - \frac{Lq}{(n-k-L-d)(n-k-d)} + \frac{\alpha Lq(n-q-1)}{(n-k-L-q-1)(n-k-q-1)} \right) \\
&- \frac{1}{2}\frac{\sqrt{(n-k-L-d)(n-k-d+2)}}{\sqrt{2L^3q}} \left( q \log\left(\frac{n-k-L-d}{n-k-d}\right) \right. \\
&\left. - \frac{Lq}{(n-k-L-d)(n-k-d)} + \frac{\alpha Lq(n-q-1)}{(n-k-L-q-1)(n-k-q-1)} \right) \\
&+ \frac{\sqrt{(n-k-L-d)(n-k-d+2)}}{\sqrt{2Lq}} \left( -\frac{q}{n-k-L-d} \right. \\
&- \frac{q}{(n-k-L-d)(n-k-d)} + \frac{\alpha q(n-q-1)}{(n-k-L-q-1)(n-k-q-1)} \\
&\left. - \frac{Lq}{(n-k-L-d)^2(n-k-d)} + \frac{\alpha Lq(n-q-1)}{(n-k-L-q-1)^2(n-k-q-1)} \right),
\end{aligned}$$

which can be written as $A(L)B(L)$, where

$$A(L) = -\frac{1}{2}\frac{\sqrt{(n-k-L-d)(n-k-d+2)}}{\sqrt{2Lq}} < 0$$

and

$$\begin{aligned}
B(L) = & \frac{q}{(n-k-L-d)} \log\left(\frac{n-k-L-d}{n-k-d}\right) - \frac{Lq}{(n-k-L-d)^2(n-k-d)} \\
& + \frac{\alpha L q(n-q-1)}{(n-k-L-d)(n-k-L-q-1)(n-k-q-1)} \\
& + \frac{q}{L} \log\left(\frac{n-k-L-d}{n-k-d}\right) - \frac{q}{(n-k-L-d)(n-k-d)} \\
& + \frac{\alpha q(n-q-1)}{(n-k-L-q-1)(n-k-q-1)} + \frac{2q}{n-k-L-d} \\
& + \frac{2q}{(n-k-L-d)(n-k-d)} - \frac{2\alpha q(n-q-1)}{(n-k-L-q-1)(n-k-q-1)} \\
& + \frac{2Lq}{(n-k-L-d)^2(n-k-d)} - \frac{2\alpha L q(n-q-1)}{(n-k-L-q-1)^2(n-k-q-1)}.
\end{aligned}$$

After rearranging terms,

$$\begin{aligned}
B(L) < & \frac{q(n-k-d)}{L(n-k-L-d)} \log\left(\frac{n-k-L-d}{n-k-d}\right) + \frac{2q}{n-k-L-d} \\
& + \frac{q}{(n-k-L-d)^2} - \frac{\alpha q(n-q-1)}{(n-k-L-q-1)(n-k-q-1)} \\
& - \frac{\alpha L q(n-q-1)}{(n-k-L-d)(n-k-L-q-1)(n-k-q-1)}.
\end{aligned}$$

Using the fact that

$$\begin{aligned}
& \frac{q(n-k-d)}{L(n-k-L-d)} \log\left(\frac{n-k-L-d}{n-k-d}\right) \\
& \qquad + \frac{2q}{n-k-L-d} - \frac{\alpha q(n-q-1)}{(n-k-L-q-1)(n-k-q-1)} \\
& \quad = - \frac{q(n-k-d)}{L(n-k-L-d)} \sum_{j=2}^{\infty} \frac{1}{j}\left(\frac{L}{n-k-d}\right)^j \\
& \qquad + \frac{q}{n-k-L-d} - \frac{\alpha q(n-q-1)}{(n-k-L-q-1)(n-k-q-1)} \\
& \quad < - \frac{q(n-k-d)}{L(n-k-L-d)} \sum_{j=2}^{\infty} \frac{1}{j}\left(\frac{L}{n-k-d}\right)^j \\
& \qquad + \frac{q}{n-k-L-d}\left(1 - \alpha q \frac{n-q-1}{n-k-q-1}\right) \\
& \quad < 0
\end{aligned}$$

and that

$$\begin{aligned}\frac{q}{(n-k-L-d)^2} &- \frac{\alpha Lq(n-q-1)}{(n-k-L-d)(n-k-L-q-1)(n-k-q-1)} \\ &< \frac{q}{(n-k-L-d)^2}\left(1-\alpha Lq\frac{n-q-1}{n-k-q-1}\right) \\ &< 0,\end{aligned}$$

we have $B(L) < 0$. Hence the signal-to-noise ratio increases as $L$ increases.

*4B.2. Theorem 4.2*

Given the $q$-dimensional multivariate regression model in Eqs. (4.1)–(4.2), for all $n \geq q+3$, $0 < L < n-k-q-1$, and for the overfitting case where $0 < k_* \leq k < n-q-2$, the signal-to-noise ratio of AICu increases as the amount of overfitting $L$ increases and is greater than the signal-to-noise ratio of AIC.

**Proof:**

First, as $L$ increases,

$$\frac{\sqrt{(n-k-L-(q-1)/2)(n-k-(q-1)/2+2)}}{\sqrt{2Lq}} \times \\ q\log\left(\frac{n-k-L-(q-1)/2}{n-k-(q-1)/2}\frac{n-k}{n-k-L}\right) \to 0$$

from below. Also,

$$\begin{aligned}&-\frac{Lq}{(n-k-L-(q-1)/2)(n-k-(q-1)/2)} \\ &\quad+\frac{Lq(2n-q-1)}{(n-k-L-q-1)(n-k-q-1)} \\ &\quad>\frac{Lq(2n-q-2)}{(n-k-L-q-1)(n-k-q-1)}\end{aligned}$$

and

$$\frac{\sqrt{(n-k-L-(q-1)/2)(n-k-(q-1)/2+2)}}{\sqrt{2Lq}} \times \\ \frac{Lq(2n-q-2)}{(n-k-L-q-1)(n-k-q-1)}$$

increases as $L$ increases. Hence, the signal-to-noise ratio of AICu increases as $L$ increases.

Since AIC and AICu have the same noise, comparing signal-to-noise ratios is equivalent to comparing signals. Note that for all $n$, $k < n$, $L < n-k-q-1$,

$$\log\left(\frac{n-k-L-(q-1)/2}{n-k-(q-1)/2}\right) < \log\left(\frac{n-k-L-(q-1)/2}{n-k-(q-1)/2}\frac{n-k}{n-k-L}\right)$$

and

$$\frac{2}{n} < \frac{2n-q-1}{n-k-q-1}.$$

Now,

$$\begin{aligned}E[\Delta\text{AIC}] =& q\log\left(\frac{n-k-L-(q-1)/2}{n-k-(q-1)/2}\right)\\ &-\frac{Lq}{(n-k-L-(q-1)/2)(n-k-(q-1)/2)}+\frac{2Lq}{n}\\ <& q\log\left(\frac{n-k-L-(q-1)/2}{n-k-(q-1)/2}\frac{n-k}{n-k-L}\right)\\ &-\frac{Lq}{(n-k-L-(q-1)/2)(n-k-(q-1)/2)}\\ &+\frac{Lq(2n-q-1)}{(n-k-L-q-1)(n-k-q-1)}\\ =& E[\Delta\text{AICu}],\end{aligned}$$

which completes the proof.

### *4B.3. Theorem 4.3*

Given the $q$-dimensional multivariate regression model in Eqs. (4.1)–(4.2), for all $n \geq q+5$, $0 < L < n-k-q-1$, and for the overfitting case where $0 < k_* \leq k < n-q-2$, the signal-to-noise ratio of HQc increases as the amount of overfitting $L$ increases and is greater than the signal-to-noise ration of HQ.

**Proof:**

Let $\alpha = 2\log\log(n)$ in Theorem 4.1. Then, by applying Theorem 4.1, we obtain that the signal-to-noise ratio of HQc increases as $L$ increases.

HQ and HQc have the same noise, comparing signal-to-noise ratios is equivalent to comparing signals. Note that for all $n$, $k < n$, $L < n-k-q-1$,

$$\frac{2\log\log(n)Lq}{n} < \frac{2\log\log(n)Lq(n-q-1)}{(n-k-L-q-1)(n-k-q-1)}.$$

We have,

$$
\begin{aligned}
E[\Delta\text{HQ}] =& q\log\left(\frac{n-k-L-(q-1)/2}{n-k-(q-1)/2}\right) \\
&- \frac{Lq}{(n-k-L-(q-1)/2)(n-k-(q-1)/2)} + \frac{2\log\log(n)Lq}{n} \\
<& q\log\left(\frac{n-k-L-(q-1)/2}{n-k-(q-1)/2}\right) \\
&- \frac{Lq}{(n-k-L-(q-1)/2)(n-k-(q-1)/2)} \\
&+ \frac{2\log\log(n)Lq(n-q-1)}{(n-k-L-q-1)(n-k-q-1)} \\
=& E[\Delta\text{HQc}],
\end{aligned}
$$

which completes the proof.

### *4B.4. Theorem 4.4*

Suppose $Y = X_*B_* + \varepsilon_*$ is the true model, where $X_* = (X_0, X_1) = (1, \tilde{X}_0, X_1)$, and $B_* = (B_0', B_1')'$. Ignoring the constant, the rows of $X_*$ are independent with the following multivariate normal distribution:

$$
\tilde{X}_{*i} = \begin{pmatrix} \tilde{X}_{0i} \\ X_{1i} \end{pmatrix} \sim N\left(\begin{pmatrix} 0 \\ 0 \end{pmatrix}, \begin{pmatrix} \Sigma_{x00} & \Sigma_{x10} \\ \Sigma_{x01} & \Sigma_{x11} \end{pmatrix}\right).
$$

If the model $Y = X_0B_0$ is fit to the data, then the noncentrality parameter, $\Omega$, is random, and

$$
\Omega \sim W_q(n-k_0, \Sigma_*^{-1}B_1'\Sigma_{x11\cdot 0}B_1),
$$

where $k_0 = \text{rank}(X_0)$, $\Sigma_{x11\cdot 0} = \Sigma_{x11} - \Sigma_{x01}'\Sigma_{00}^{-1}\Sigma_{x01}$ and $\Sigma_{x11} = var[\tilde{X}_{1i}]$, $\Sigma_{x00} = var[X_{0i}]$, and $\Sigma_{x01} = cov[\tilde{X}_{0i}, X_{1i}]$.

**Proof:**

An underfitted model results in the noncentrality parameter

$$
\Omega = \Sigma_*^{-1}E_*[Y'](I-H_0)E_*[Y],
$$

where

$$
H_0 = X_0(X_0'X_0)^{-1}X_0'.
$$

Conditioning on $X_*$, we have

$$
\Omega|X_* = \Sigma_*^{-1}B_1'X_1'(I-H_0)X_1B_1.
$$

Under the above partition of $X_{*i}$, $X_{1i}|\tilde{X}_{0i} \sim N(0, \Sigma_{x11\cdot0})$. $X_1B_1|X_0$ has the same distribution as $X_1B_1|\tilde{X}_0$. Since the rows of $\tilde{X}_*$ are independent, $X_1B_1|X_0 \sim N(0, B_1'\Sigma_{x11\cdot0}B_1 I)$. It follows that

$$\Omega|X_0 = \Sigma_*^{-1}B_1'X_1'(I - H_0)X_1B_1|X_0 \sim W_q(n - k_0, \Sigma_*^{-1}B_1'\Sigma_{x11\cdot0}B_1),$$

which does not depend on $X_0$. Unconditionally,

$$\Omega = \Sigma_*^{-1}B_1'X_1'(I - H_0)X_1B_1 \sim W_q(n - k_0, \Sigma_*^{-1}B_1'\Sigma_{x11\cdot0}B_1),$$

which completes the proof.

### *4B.5. Theorem 4.5*

Suppose $Y = X_*B_* + \varepsilon_*$ is the true model, where $X_* = (X_0, X_1) = (1, \tilde{X}_0, X_1)$, and $B_* = (B_0', B_1')'$. Ignoring the constant, the rows of $X_*$ are independent with the following multivariate normal distribution:

$$\tilde{X}_{*i} = \begin{pmatrix} \tilde{X}_{0i} \\ X_{1i} \end{pmatrix} \sim N\left(\begin{pmatrix} 0 \\ 0 \end{pmatrix}, \begin{pmatrix} \Sigma_{x00} & \Sigma_{x10} \\ \Sigma_{x01} & \Sigma_{x11} \end{pmatrix}\right).$$

If the model $Y = X_0B_0$ is fit to the data, then

$$\text{SPE}_{k_0} = Y'(I - H_0)Y \sim W_q(n - k_0, \Sigma_* + B_1'\Sigma_{x11\cdot0}B_1),$$

where $k_0 = \text{rank}(X_0)$, $\Sigma_{x11\cdot0} = \Sigma_{x11} - \Sigma_{x01}'\Sigma_{00}^{-1}\Sigma_{x01}$ and $\Sigma_{x11} = var[\tilde{X}_{1i}]$, $\Sigma_{x00} = var[X_{0i}]$, and $\Sigma_{x01} = cov[\tilde{X}_{0i}, X_{1i}]$.

**Proof:**

The true model can be written as

$$\begin{aligned} Y &= X_0B_0 + X_1B_1 + \varepsilon_* \\ &= X_0B_0 + \tilde{\varepsilon}_*. \end{aligned}$$

If we condition on $X_0$, then $\varepsilon_* \sim N(0, \Sigma_* \otimes I)$, $X_1B_1|X_0 \sim N(0, B_1'\Sigma_{x11\cdot0}B_1 \otimes I)$, and both are independent. Therefore, $\tilde{\varepsilon}_* \sim N(0, (\Sigma_* + B_1'\Sigma_{x11\cdot0}B_1) \otimes I)$. Hence,

$$\text{SPE}_{k_0} \sim W_q(n - k_0, \Sigma_* + B_1'\Sigma_{x11\cdot0}B_1),$$

which completes the proof.

## Appendix 4C. Small-sample Probabilities of Overfitting

### *4C.1. AICc*

AICc overfits if $\text{AICc}_{k_*+L} < \text{AICc}_{k_*}$. For finite $n$, the probability that AICc prefers the overfitted model $k_* + L$ is

$$
\begin{aligned}
&P\{\text{AICc}_{k_*+L} < \text{AICc}_{k_*}\} \\
&= P\left\{\log|\hat{\Sigma}_{k_*+L}| + \frac{(n+k_*+L)q}{n-k_*-L-q-1} < \log|\hat{\Sigma}_{k_*}| + \frac{(n+k_*)q}{n-k_*-q-1}\right\} \\
&= P\left\{\log\left(\frac{|\text{SPE}_{k_*+L}|}{|\text{SPE}_{k_*}|}\right) < \frac{(n+k_*)q}{n-k_*-q-1} - \frac{(n+k_*+L)q}{n-k_*-L-q-1}\right\} \\
&= P\left\{\sum_{i=1}^{q} \text{log-Beta}\left(\frac{n-k_*-L-q+i}{2}, \frac{L}{2}\right)\right. \\
&\qquad \left. < -\frac{Lq(2n-q-1)}{(n-k_*-L-q-1)(n-k_*-q-1)}\right\} \\
&= P\left\{n\sum_{i=1}^{q}\log\left(1+\frac{\chi^2_L}{\chi^2_{n-k_*-L+i}}\right) > \frac{Lq(2n-q-1)n}{(n-k_*-L-q-1)(n-k_*-q-1)}\right\}.
\end{aligned}
$$

### *4C.2. AICu*

AICu overfits if $\text{AICu}_{k_*+L} < \text{AICu}_{k_*}$. For finite $n$, the probability that AICu prefers the overfitted model $k_* + L$ is

$$
\begin{aligned}
&P\{\text{AICu}_{k_*+L} < \text{AICu}_{k_*}\} \\
&= P\left\{\log|S^2_{k_*+L}| + \frac{(n+k_*+L)q}{n-k_*-L-q-1} < \log|S^2_{k_*}| + \frac{(n+k_*)q}{n-k_*-q-1}\right\} \\
&= P\left\{\log\left(\frac{|\text{SPE}_{k_*+L}|}{|\text{SPE}_{k_*}|}\right) < -q\log\left(\frac{n-k_*}{n-k_*-L}\right)\right. \\
&\qquad \left. + \frac{(n+k_*)q}{n-k_*-q-1} - \frac{(n+k_*+L)q}{n-k_*-L-q-1}\right\} \\
&= P\left\{\sum_{i=1}^{q} \text{log-Beta}\left(\frac{n-k_*-L-q+i}{2}, \frac{L}{2}\right)\right. \\
&\qquad \left. < -q\log\left(\frac{n-k_*}{n-k_*-L}\right) - \frac{Lq(2n-q-1)}{(n-k_*-L-q-1)(n-k_*-q-1)}\right\} \\
&= P\left\{n\sum_{i=1}^{q}\log\left(1+\frac{\chi^2_L}{\chi^2_{n-k_*-L+i}}\right) > -nq\log\left(\frac{n-k_*-L}{n-k_*}\right)\right. \\
&\qquad \left. + \frac{Lq(2n-q-1)n}{(n-k_*-L-q-1)(n-k_*-q-1)}\right\}.
\end{aligned}
$$

### *4C.3. FPE*

FPE overfits if $\text{FPE}_{k_*+L} < \text{FPE}_{k_*}$. For finite $n$, the probability that FPE prefers the overfitted model $k_* + L$ is

$$
\begin{aligned}
&P\{\text{FPE}_{k_*+L} < \text{FPE}_{k_*}\} \\
&= P\{\log(\text{FPE}_{k_*+L}) < \log(\text{FPE}_{k_*})\} \\
&= P\left\{\log|\hat{\Sigma}_{k_*+L}| + q\log\left(\frac{n+k_*+L}{n-k_*-L}\right) < \log|\hat{\Sigma}_{k_*}| + q\log\left(\frac{n+k_*}{n-k_*}\right)\right\} \\
&= P\left\{\log\left(\frac{|\text{SPE}_{k_*+L}|}{|\text{SPE}_{k_*}|}\right) < q\log\left(\frac{n+k_*}{n-k_*}\frac{n-k_*-L}{n+k_*+L}\right)\right\} \\
&= P\left\{\sum_{i=1}^{q}\text{log-Beta}\left(\frac{n-k_*-L-q+i}{2}, \frac{L}{2}\right) < nq\log\left(\frac{n-k_*}{n+k_*}\frac{n+k_*+L}{n-k_*-L}\right)\right\} \\
&= P\left\{n\sum_{i=1}^{q}\log\left(1+\frac{\chi^2_L}{\chi^2_{n-k_*-L}}\right) > nq\log\left(\frac{n-k_*}{n+k_*}\frac{n+k_*+L}{n-k_*-L}\right)\right\}.
\end{aligned}
$$

### *4C.4. SIC*

SIC overfits if $\text{SIC}_{k_*+L} < \text{SIC}_{k_*}$. For finite $n$, the probability that SIC prefers the overfitted model $k_* + L$ is

$$
\begin{aligned}
&P\{\text{SIC}_{k_*+L} < \text{SIC}_{k_*}\} \\
&= P\left\{\log|\hat{\Sigma}_{k_*+L}| + \frac{\log(n)(k_*+L)q}{n} < \log|\hat{\Sigma}_{k_*}| + \frac{\log(n)k_*q}{n}\right\} \\
&= P\left\{\log\left(\frac{|\text{SPE}_{k_*+L}|}{|\text{SPE}_{k_*}|}\right) < \frac{\log(n)k_*q}{n} - \frac{\log(n)(k_*+L)q}{n}\right\} \\
&= P\left\{\sum_{i=1}^{q}\text{log-Beta}\left(\frac{n-k_*-L-q+i}{2}, \frac{L}{2}\right) < -\frac{\log(n)Lq}{n}\right\} \\
&= P\left\{n\sum_{i=1}^{q}\log\left(1+\frac{\chi^2_L}{\chi^2_{n-k_*-L+i}}\right) > \log(n)Lq\right\}.
\end{aligned}
$$

### *4C.5. HQ*

HQ overfits if $\text{HQ}_{k_*+L} < \text{HQ}_{k_*}$. For finite $n$, the probability that HQ

prefers the overfitted model $k_* + L$ is

$$
\begin{aligned}
&P\{\mathrm{HQ}_{k_*+L} < \mathrm{HQ}_{k_*}\} \\
&= P\left\{\log|\hat{\Sigma}_{k_*+L}| + \frac{2\log\log(n)(k_* + L)q}{n} < \log|\hat{\Sigma}_{k_*}| + \frac{2\log\log(n)k_*q}{n}\right\} \\
&= P\left\{\log\left(\frac{|\mathrm{SPE}_{k_*+L}|}{|\mathrm{SPE}_{k_*}|}\right) < \frac{2\log\log(n)k_*q}{n} - \frac{2\log\log(n)(k_* + L)q}{n}\right\} \\
&= P\left\{\sum_{i=1}^{q} \text{log-Beta}\left(\frac{n - k_* - L - q + i}{2}, \frac{L}{2}\right) < -\frac{2\log\log(n)Lq}{n}\right\} \\
&= P\left\{n\sum_{i=1}^{q}\log\left(1 + \frac{\chi^2_L}{\chi^2_{n-k_*-L+i}}\right) > 2\log\log(n)Lq\right\}.
\end{aligned}
$$

### *4C.6. HQc*

HQc overfits if $\mathrm{HQc}_{k_*+L} < \mathrm{HQc}_{k_*}$. For finite $n$, the probability that HQc prefers the overfitted model $k_* + L$ is

$$
\begin{aligned}
&P\{\mathrm{HQc}_{k_*+L} < \mathrm{HQc}_{k_*}\} \\
&= P\left\{\log|\hat{\Sigma}_{k_*+L}| + \frac{2\log\log(n)(k_* + L)q}{n - k_* - L - q - 1} < \log|\hat{\Sigma}_{k_*}| + \frac{2\log\log(n)k_*q}{n - k_* - q - 1}\right\} \\
&= P\left\{\log\left(\frac{|\mathrm{SPE}_{k_*+L}|}{|\mathrm{SPE}_{k_*}|}\right) < \frac{2\log\log(n)k_*q}{n - k_* - q - 1} - \frac{2\log\log(n)(k_* + L)q}{n - k_* - L - q - 1}\right\} \\
&= P\left\{\sum_{i=1}^{q} \text{log-Beta}\left(\frac{n - k_* - L - q + i}{2}, \frac{L}{2}\right)\right. \\
&\qquad \left. < -\frac{2\log\log(n)Lq(n - q - 1)}{(n - k - L - q - 1)(n - k - q - 1)}\right\} \\
&= P\left\{n\sum_{i=1}^{q}\log\left(1 + \frac{\chi^2_L}{\chi^2_{n-k_*-L+i}}\right) > \frac{2\log\log(n)Lq(n - q - 1)n}{(n - k - L - q - 1)(n - k - q - 1)}\right\}.
\end{aligned}
$$

### *4C.7. General Case*

Consider a model selection criterion, say $\mathrm{MSC}_k$, of the form $\log(\mathrm{SPE}) + \alpha(n, k, q)$, where $\alpha(n, k, q)$ is the penalty function of $\mathrm{MSC}_k$. MSC overfits if $\mathrm{MSC}_{k_*+L} < \mathrm{MSC}_{k_*}$. For finite $n$, the probability that MSC prefers the overfitted model $k_* + L$ is

$$\begin{aligned}
&P\{\text{MSC}_{k_*+L} < \text{MSC}_{k_*}\}\\
&= P\left\{\log\left(\frac{|\text{SPE}_{k_*+L}|}{|\text{SPE}_{k_*}|}\right) < \alpha(n, k_*, q) - \alpha(n, k_* + L, q)\right\}\\
&= P\left\{\sum_{i=1}^{q} \text{log-Beta}\left(\frac{n - k_* - L - q + i}{2}, \frac{L}{2}\right)\right.\\
&\qquad \left. < \alpha(n, k_*, q) - \alpha(n, k_* + L, q)\right\}\\
&= P\left\{n\sum_{i=1}^{q}\log\left(1 + \frac{\chi^2_L}{\chi^2_{n-k_*-L+i}}\right) > \alpha(n, k_* + L, q) - \alpha(n, k_*, q)\right\}.
\end{aligned}$$

## Appendix 4D. Asymptotic Probabilities of Overfitting

### *4D.1. AICc*

In small samples, AICc overfits with probability

$$P\left\{n\sum_{i=1}^{q}\log\left(1 + \frac{\chi^2_L}{\chi^2_{n-k_*-L+i}}\right) > \frac{Lq(2n - q - 1)n}{(n - k_* - L - q - 1)(n - k_* - q - 1)}\right\}$$

(see Appendix 4C.1). Now, as $n \to \infty$,

$$\frac{Lq(2n - q - 1)n}{(n - k_* - L - q - 1)(n - k_* - q - 1)} \to 2qL,$$

and thus the asymptotic probability of overfitting for AICc is

$$P\{\text{AICc overfits by } L\} = P\left\{\chi^2_{qL} > 2qL\right\}.$$

### *4D.2. AICu*

In small samples, AICu overfits with probability

$$\begin{aligned}
P\Bigg\{n\sum_{i=1}^{q}\log\left(1 + \frac{\chi^2_L}{\chi^2_{n-k_*-L+i}}\right) > &-nq\log\left(\frac{n - k_* - L}{n - k_*}\right)\\
&+ \frac{Lq(2n - q - 1)n}{(n - k_* - L - q - 1)(n - k_* - q - 1)}\Bigg\}
\end{aligned}$$

(see Appendix 4C.2). As $n \to \infty$,

$$\frac{Lq(2n-q-1)n}{(n-k_*-L-q-1)(n-k_*-q-1)} \to 2qL,$$

$$-nq\log\left(\frac{n-k_*-L}{n-k_*}\right) = -nq\log\left(1-\frac{L}{n-k_*}\right)$$
$$\to qL,$$

and thus the asymptotic probability of overfitting for AICu is

$$P\{\text{AICu overfits by } L\} = P\left\{\chi^2_{qL} > 3qL\right\}.$$

### 4D.3. FPE

In small samples, FPE overfits with probability

$$P\left\{n\sum_{i=1}^{q}\log\left(1+\frac{\chi^2_L}{\chi^2_{n-k_*-L}}\right) > nq\log\left(\frac{n-k_*}{n+k_*}\frac{n+k_*+L}{n-k_*-L}\right)\right\}.$$

(see Appendix 4C.3). Now, as $n \to \infty$,

$$nq\log\left(\frac{n-k_*}{n+k_*}\frac{n+k_*+L}{n-k_*-L}\right) = nq\log\left(1+\frac{2Ln}{(n+k_*)(n-k_*-L)}\right)$$
$$\to 2Lq,$$

and thus the asymptotic probability of overfitting for FPE is

$$P\{\text{FPE overfits by } L\} = P\left\{\chi^2_{qL} > 2qL\right\}.$$

### 4D.4. SIC

In small samples, SIC overfits with probability

$$P\left\{n\sum_{i=1}^{q}\log\left(1+\frac{\chi^2_L}{\chi^2_{n-k_*-L+i}}\right) > \log(n)Lq\right\}$$

(see Appendix 4C.4). Now, as $n \to \infty$,

$$\log(n)Lq \to \infty$$

and thus the asymptotic probability of overfitting for SIC is

$$P\{\text{SIC overfits by } L\} = 0.$$

### *4D.5. HQ*

In small samples, HQ overfits with probability

$$P\left\{n\sum_{i=1}^{q}\log\left(1+\frac{\chi^2_L}{\chi^2_{n-k_*-L+i}}\right) > 2\log\log(n)Lq\right\}$$

(see Appendix 4C.5). Now, as $n \to \infty$,

$$2\log\log(n)Lq \to \infty,$$

and thus the asymptotic probability of overfitting for HQ is

$$P\{\text{HQ overfits by } L\} = 0.$$

### *4D.6. HQc*

In small samples, HQc overfits with probability

$$P\left\{n\sum_{i=1}^{q}\log\left(1+\frac{\chi^2_L}{\chi^2_{n-k_*-L+i}}\right) > \frac{2\log\log(n)Lq(n-q-1)n}{(n-k-L-q-1)(n-k-q-1)}\right\}$$

(see Appendix 4C.6). Now, as $n \to \infty$,

$$\frac{2\log\log(n)Lq(n-q-1)n}{(n-k-L-q-1)(n-k-q-1)} \to \infty$$

and thus the asymptotic probability of overfitting for HQc is

$$P\{\text{HQc overfits by } L\} = 0.$$

## Appendix 4E. Asymptotic Signal-to-noise Ratios

### *4E.1. AICc*

Starting with the finite-sample signal-to-noise ratio for AICc from Section 4.3, the asymptotic signal-to-noise ratio for AICc is

$$\begin{aligned}
&\lim_{n\to\infty} \frac{\sqrt{(n-k-L-(q-1)/2)(n-k-(q-1)/2+2)}}{\sqrt{2Lq}} \times \\
&\quad \left( q\log\left(\frac{n-k-L-(q-1)/2}{n-k-(q-1)/2}\right) \right. \\
&\quad - \frac{Lq}{(n-k-L-(q-1)/2)(n-k-(q-1)/2)} \\
&\quad \left. + \frac{Lq(2n-q-1)}{(n-k-L-q-1)(n-k-q-1)} \right) \\
&= \lim_{n\to\infty} \frac{n}{\sqrt{2Lq}} \left( -\frac{Lq}{n} - \frac{Lq}{n^2} + \frac{2Lq}{n} \right) \\
&= \sqrt{\frac{Lq}{2}}.
\end{aligned}$$

### *4E.2. AICu*

Starting with the finite-sample signal-to-noise ratio for AICu from Section 4.3, the asymptotic signal-to-noise ratio for AICu is

$$\begin{aligned}
&\lim_{n\to\infty} \frac{\sqrt{(n-k-L-(q-1)/2)(n-k-(q-1)/2+2)}}{\sqrt{2Lq}} \times \\
&\quad \left( q\log\left(\frac{n-k-L-(q-1)/2}{n-k-(q-1)/2}\frac{n-k}{n-k-L}\right) \right. \\
&\quad - \frac{Lq}{(n-k-L-(q-1)/2)(n-k-(q-1)/2)} \\
&\quad \left. + \frac{Lq(2n-q-1)}{(n-k-L-q-1)(n-k-q-1)} \right) \\
&= \lim_{n\to\infty} \frac{n}{\sqrt{2Lq}} \left( -\frac{L(q-1)/2}{n^2} - \frac{Lq}{n^2} + \frac{2Lq}{n} \right) \\
&= \sqrt{2Lq}.
\end{aligned}$$

### *4E.3. HQ*

Starting with the finite-sample signal-to-noise ratio for HQ from Section 4.3, the asymptotic signal-to-noise ratio for HQ is

$$\begin{aligned}
&\lim_{n\to\infty} \frac{\sqrt{(n-k-L-(q-1)/2)(n-k-(q-1)/2+2)}}{\sqrt{2Lq}} \times \\
&\quad \left( q \log \left( \frac{n-k-L-(q-1)/2}{n-k-(q-1)/2} \right) \right. \\
&\quad \left. - \frac{Lq}{(n-k-L-(q-1)/2)(n-k-(q-1)/2)} + \frac{2\log\log(n)Lq}{n} \right) \\
&= \lim_{n\to\infty} \frac{n}{\sqrt{2Lq}} \left( -\frac{Lq}{n} - \frac{Lq}{n^2} + \frac{2\log\log(n)Lq}{n} \right) \\
&= \infty.
\end{aligned}$$

### *4E.4. HQc*

Starting with the finite-sample signal-to-noise ratio for HQc from Section 4.3, the asymptotic signal-to-noise ratio for HQc is

$$\begin{aligned}
&\lim_{n\to\infty} \frac{\sqrt{(n-k-L-(q-1)/2)(n-k-(q-1)/2+2)}}{\sqrt{2Lq}} \times \\
&\quad \left( q \log \left( \frac{n-k-L-(q-1)/2}{n-k-(q-1)/2} \right) \right. \\
&\quad - \frac{Lq}{(n-k-L-(q-1)/2)(n-k-(q-1)/2)} \\
&\quad \left. + \frac{2\log\log(n)Lq(n-q-1)}{(n-k-L-q-1)(n-k-q-1)} \right) \\
&= \lim_{n\to\infty} \frac{n}{\sqrt{2Lq}} \left( -\frac{Lq}{n} - \frac{Lq}{n^2} + \frac{2\log\log(n)Lq}{n} \right) \\
&= \infty.
\end{aligned}$$

Chapter 5
# The Vector Autoregressive Model

The vector autoregressive model or VAR, is probably one of the most common and straightforward methods for modeling multivariate time series data. The least squares VAR models in this Chapter are related to the multivariate regression models in Chapter 4, but the number of parameters involved in VAR models increases even more rapidly with model order. For VAR, increasing the order by one decreases the degrees of freedom by $q+1$ (where $q$ is the dimension of the model). This has two consequences: first, few candidate models can be considered if the sample size is small or the dimension of the model is large, and second, model selection criteria with strong penalty functions will be prone to excessive underfitting for these models. Of all the models we consider, VAR parameter counts increase most rapidly.

We will first describe the VAR model, its notation, parameter counts, and its relationship to multivariate regression. The FPE, AIC, and AICc criteria are readily adapted to this model, and we review their respective derivations. The other criteria discussed in Chapter 4 are rewritten in VAR notation. Small-sample moments and signal-to-noise ratios are derived next. As in previous chapters, two special case VAR models are discussed to illustrate the overfitting and underfitting behavior of the selection criteria. Finally, in order to see the relationship between each criterion's signal-to-noise ratios, probability of misselection, and its actual performance, we present simulation study results for the special case models.

## 5.1. Model Description

### *5.1.1. Vector Autoregressive Models*

We first define the vector autoregressive *general model* of order $p$, denoted VAR($p$), as

$$y_t = \Phi_1 y_{t-1} + \cdots + \Phi_p y_{t-p} + w_t, \quad t = p+1, \ldots, n \tag{5.1}$$

where

$$w_t \ \ i.i.d. \ \ N_q(0, \Sigma), \tag{5.2}$$

$y_t = (y_{1,t}, \ldots, y_{q,t})'$ is a $q \times 1$ observed vector at times $t = 1, \ldots, n$, and the $\Phi_j$ are $q \times q$ matrices of unknown parameters for $j = 1, \ldots, p$. As was the case for the univariate autoregressive model, we lose $p$ observations because we need $y_{t-p}$ to model $y_t$, and thus the effective series length for a candidate VAR($p$) model is $T = n - p$. For the sake of simplicity, where possible we will use the effective series length $T$. Note that no intercept is included in these VAR models. The assumption given by Eq. (5.2) is identical to that in Eq. (4.2) for the multivariate regression case. For now, assume that the true model is also a VAR($p$) model. If the true model is of finite order, $p_*$, then we further assume that the true VAR($p_*$) model belongs to the set of candidate models. In other words, this assumption will be true when VAR models of order 1 to $P$ are considered and where $P \geq p_*$.

First we will define the observation matrix $Y$ as $Y = (y_{p+1}, \ldots, y_n)'$ and consider fitting the candidate model VAR($p$). We can form a regression model from Eq. (5.1) by conditioning on the past and forming the design matrix, $X$, where ($X$) is of full rank $pq$. The rows of $X$ have the following structure:

$$X_t' = (y_{1,t-1}, \ldots, y_{q,t-1}, y_{1,t-2}, \ldots y_{q,t-2}, \ldots, y_{1,t-p}, \ldots, y_{q,t-p}).$$

By conditioning on the past, we assume $X$ is known and of dimension $(n-p) \times (pq)$. Since $X$ is formed by conditioning on past values of $Y$, least squares in this context are often referred to as conditional least squares.

Next we will obtain the least squares parameter estimates for the vector autoregressive case. Let $\Phi = (\Phi_1, \Phi_2, \ldots, \Phi_p)'$. The conditional least squares parameter estimate of $\Phi$ is

$$\hat{\Phi} = (X'X)^{-1}X'Y.$$

The unbiased and the maximum likelihood estimates of $\Sigma$ are given below:

$$S_p^2 = \frac{\text{SPE}_p}{T - pq} \tag{5.3}$$

and

$$\hat{\Sigma}_p = \frac{\text{SPE}_p}{T}, \tag{5.4}$$

where $\text{SPE}_p = \parallel Y - \hat{Y} \parallel^2$ and $\hat{Y} = X\hat{\Phi}$. For simplicity, we sometimes use SPE instead of $\text{SPE}_p$. Since all candidate models are of the VAR($p$) form given by Eq. (5.1), we will often refer to the candidate models by their order. Note that the VAR($p$) model has $n - p(q+1)$ degrees of freedom.

We next define the *true model*,

$$y_t = \mu_{*t} + w_{*t}, \quad t = 1, \ldots, n$$

with

$$w_{*t} \ i.i.d. \ N_q(0, \Sigma_*).$$

Usually, we will consider the true model of the form $\mu_{*t} = \Phi_{*1} y_{t-1} + \cdots + \Phi_{*p_*} y_{t-p_*}$ with true order $p_*$. If there is a true order $p_*$, then we can also define an overfitted and an underfitted model. Underfitting occurs when a VAR($p$) candidate model is fitted with $p < p_*$, and overfitting occurs when $p > p_*$. If the true model does not belong to the set of candidate models, then the definitions of underfitting and overfitting depend on the discrepancy used. For example, we define $\tilde{p}$ such that the VAR($\tilde{p}$) model is closest to the true model. Underfitting in the $L_2$ sense can now be stated as choosing the VAR($p$) model where $p < \tilde{p}$, and overfitting in the $L_2$ sense is stated as choosing the VAR($p$) model where $p > \tilde{p}$. These definitions can be obtained analogously for the Kullback–Leibler distance. In the next Section we consider how to use K-L and $L_2$ as distance measures with vector autoregressive models.

### 5.1.2. Distance Measures

We can define the Kullback–Leibler and $L_2$ distance measures for the vector autoregressive model described in Eq. (5.1) and Eq. (5.2) as follows. The $L_2$ distance between the true model and the candidate model VAR($p$) is defined as

$$L_2 = \frac{1}{T} \sum_{t=p+1}^{n} (\hat{y}_t - \mu_{*t})(\hat{y}_t - \mu_{*t})'. \tag{5.5}$$

Because $L_2$ is a $q \times q$ matrix, we use the determinant and trace of $L_2$ to reduce $L_2$ distance to a scalar. We denote the trace $L_2$ distance by

$$\mathrm{tr}\{L_2\} = tr\left\{\frac{1}{T} \sum_{t=p+1}^{n} (\hat{y}_t - \mu_{*t})(\hat{y}_t - \mu_{*t})'\right\} \tag{5.6}$$

and the determinant $L_2$ distance as

$$\det(L_2) = det\left(\frac{1}{T} \sum_{t=p+1}^{n} (\hat{y}_t - \mu_{*t})(\hat{y}_t - \mu_{*t})'\right). \tag{5.7}$$

Since the Kullback–Leibler information remains a scalar, it is preferred to $L_2$ in many cases.

To define the K-L, we use the multivariate normality assumption to obtain the log-likelihood of the true model, $f_*$:

$$\log(f_*) = -\frac{Tq}{2}\log(2\pi) - \frac{T}{2}\log|\Sigma_*| - \frac{1}{2}\sum_{t=p+1}^{n} tr\{(y_t - \mu_{*t})'\Sigma_*^{-1}(y_t - \mu_{*t})\}.$$

The log-likelihood of the candidate VAR($p$) model is

$$\log(f) = -\frac{Tq}{2}\log(2\pi) - \frac{T}{2}\log|\Sigma| \\ -\frac{1}{2}\sum_{t=p+1}^{n} tr\big\{(y_t - \Phi_1 y_{t-1}\cdots - \Phi_p y_{t-p})'\Sigma^{-1}(y_t - \Phi_1 y_{t-1}\cdots - \Phi_p y_{t-p})\big\}.$$

We know that the Kullback–Leibler information is defined as $E_*[\log(f_*) - \log(f)]$, here scaled by $2/T$ to express K-L as a rate or average distance and allowing us to compare models with different effective sample sizes. Substituting and scaling yields

$$\text{K-L} = \log\left(\frac{|\Sigma|}{|\Sigma_*|}\right) - \frac{1}{T}E_*\Bigg[\sum_{t=p+1}^{n} tr\{(y_t - \mu_{*t})'\Sigma_*^{-1}(y_t - \mu_{*t})\} \\ + \sum_{t=p+1}^{n} tr\big\{(y_t - \Phi_1 y_{t-1}\cdots - \Phi_p y_{t-p})'\Sigma^{-1}(y_t - \Phi_1 y_{t-1}\cdots - \Phi_p y_{t-p})\big\}\Bigg].$$

Taking expectations with respect to the true model yields

$$\text{K-L} = \log\left(\frac{|\Sigma|}{|\Sigma_*|}\right) + tr\{\Sigma^{-1}\Sigma_*\} - q \\ +\frac{1}{T}\sum_{t=p+1}^{n} tr\big\{(\mu_{*t} - \Phi_1 y_{t-1}\cdots - \Phi_p y_{t-p})'\Sigma^{-1}(\mu_{*t} - \Phi_1 y_{t-1}\cdots - \Phi_p y_{t-p})\big\}.$$

If we let the candidate model be the estimated VAR($p$) model with $\Phi$ of $\hat{\Phi}$ and variance $\hat{\Sigma}_p$, we have the Kullback–Leibler discrepancy for VAR:

$$\text{K-L} = \log\left(\frac{|\hat{\Sigma}_p|}{|\Sigma_*|}\right) + tr\{\hat{\Sigma}_p^{-1}\Sigma_*\} - q \\ +tr\left\{\hat{\Sigma}_p^{-1}\frac{1}{T}\sum_{t=p+1}^{n}(\mu_{*t} - \hat{\Phi}_1 y_{t-1}\cdots - \hat{\Phi}_p y_{t-p})(\mu_{*t} - \hat{\Phi}_1 y_{t-1}\cdots - \hat{\Phi}_p y_{t-p})'\right\}.$$

Simplifying in terms of $L_2$,

$$\text{K-L}^{\cdot} = \log\left(\frac{|\hat{\Sigma}_p|}{|\Sigma_*|}\right) + tr\{\hat{\Sigma}_p^{-1}\Sigma_*\} + tr\{\hat{\Sigma}_p^{-1}L_2\} - q, \tag{5.8}$$

with $\hat{\Sigma}_p$ from Eq. (5.4) and $L_2$ from Eq. (5.5)

## 5.2. Selected Derivations of Model Selection Criteria

Derivations for vector autoregressive models generally parallel those for multivariate regression with the exception that the effective sample size is a function of the candidate model order. A $q$-dimensional VAR($p$) model has $pq^2$ parameters, a potentially much bigger number than any of our other models. Here we will generalize our model selection criteria for the vector autoregressive case. Two model selection criteria, FPE and HQ, were originally derived in the autoregressive setting, and we will examine their VAR derivations in detail.

### *5.2.1. FPE*

The derivation of FPE for VAR begins by supposing that we observe the series $y_1, \ldots, y_n$ from the $q$-dimensional VAR($p$) model in Eqs. (5.1)–(5.2). Let $\{x_t\}$ be an observed series from another $q$-dimensional VAR($p$) model that is independent of $\{y_t\}$, but where $\{x_t\}$ and $\{y_t\}$ have the same statistical structure. Thus the model is

$$x_t = \Phi_1 x_{t-1} + \cdots + \Phi_p x_{t-p} + u_t, \quad t = p+1, \ldots, n$$

where

$$u_t \ \ i.i.d. \ \ N(0, \Sigma).$$

Note that $u_t$ and $w_t$ have the same distribution but are independent of each other. Akaike's (1969) approach to estimating the mean product prediction error for predicting the next $\{x_t\}$ observation, $x_{n+1}$, was to estimate the parameters from the $\{y_t\}$ data, and to then use these estimated parameters to make the prediction for $x_{n+1}$ using $\{x_t\}$. We generalize Akaike's approach to VAR models and obtain the prediction

$$\hat{x}_{n+1} = \hat{\Phi}_1 x_n + \cdots + \hat{\Phi}_p x_{n-p+1}.$$

In addition, we show that the mean product prediction error for large $T$ is

$$E[(\hat{x}_{n+1} - x_{n+1})(\hat{x}_{n+1} - x_{n+1})'] \doteq \Sigma\left(1 + \frac{qp}{T}\right),$$

and the expectation of $\hat{\Sigma}_p$ in Eq. (5.4) is

$$E[\hat{\Sigma}_p] = \left(1 - \frac{qp}{T}\right)\Sigma.$$

Hence, $\hat{\Sigma}_p/(1 - qp/T)$ is unbiased for $\Sigma$. Substituting this unbiased estimate for $\Sigma$ leads to an unbiased estimate of the mean product prediction error, and it is defined as MFPE:

$$\text{MFPE} = \hat{\Sigma}_p \left(\frac{T + qp}{T - qp}\right),$$

where the M denotes multivariate. We recall from Chapter 4 that in multivariate models the mean product prediction error is a matrix. Typically the determinant is used to reduce MFPE to a scalar that is a function of the generalized variance, as follows:

$$\text{FPE} = |\hat{\Sigma}_p| \left(\frac{T + qp}{T - qp}\right)^q. \tag{5.9}$$

Thus, the model minimizing FPE should have the minimum mean product prediction error among the candidates, and therefore should be the best model in terms of predicting future observations.

### *5.2.2. AIC*

AIC is easily generalized to VAR($p$) models as follows. We know that

$$\text{AIC} = -2\log(\textit{likelihood}) + 2 \times \textit{number of parameters}.$$

For the VAR case, making use of the MLE in Eq. (5.4),

$$-2\log(\textit{likelihood}) = Tq\log(2\pi) + T\log|\hat{\Sigma}_p| + Tq.$$

The parameter count is $pq^2$ for $\Phi$ and $0.5q(q+1)$ for the error covariance matrix, $\Sigma$. Substituting and then scaling by $1/T$,

$$\text{AIC}_p = q\log(2\pi) + \log|\hat{\Sigma}_p| + q + \frac{2pq^2 + q(q+1)}{T}.$$

The constants $q\log(2\pi) + q$ play no practical role in model selection and are ignored, yielding

$$\text{AIC}_p = \log|\hat{\Sigma}_p| + \frac{2pq^2 + q(q+1)}{T}. \tag{5.10}$$

As was the case for univariate regression, multivariate regression, and univariate autoregressive models, the small-sample properties of AIC lead to overfitting in the vector autoregressive case, and much model selection literature is devoted to corrections for this overfitting. We have noted that AICc and AICu are two such corrections, and SIC, due to its relationship to AIC, can be thought of as a Bayesian correction. We will address these three criteria for VAR models next.

### 5.2.3. AICc

Hurvich and Tsai (1993) derived AICc by estimating the expected Kullback–Leibler discrepancy in small-sample vector autoregressive models of the form Eqs. (5.1)–(5.2) using the assumption that the true model belongs to the set of candidate models. If we assume that the true model is VAR($p_*$), and that expectations are taken for the candidate model VAR($p$) where $p > p_*$, then

$$\begin{aligned} E_*[\text{K-L}] &= E_*\left[\log\left(\frac{|\hat{\Sigma}_p|}{|\Sigma_*|}\right) + tr[\hat{\Sigma}_p^{-1}\Sigma_*] + tr[\hat{\Sigma}_p^{-1}L_2] - q\right] \\ &= E_*[\log|\hat{\Sigma}_p|] - \log|\Sigma_*| + \frac{(T+qp)q}{T-qp-q-1} - q, \end{aligned}$$

where $E_*$ denotes expectations under the true model. This leads to

$$\log|\hat{\Sigma}_p| + \frac{(T+qp)q}{T-qp-q-1} - \log|\Sigma_*| - q$$

which is unbiased for $E_*[\text{K-L}]$. The constant $-\log|\Sigma_*| - q$ plays no role in model selection and can be ignored, yielding

$$\text{AICc}_p = \log|\hat{\Sigma}_p| + \frac{(T+qp)q}{T-qp-q-1}, \tag{5.11}$$

or equivalently,

$$\text{AICc}_p = \log|\hat{\Sigma}_p| + \frac{2pq^2 + q(q+1)}{T-qp-q-1},$$

by ignoring the constant $-\log|\Sigma_*|$ only.

### 5.2.4. AICu

Recall that AICu used for multivariate regression models is a signal-to-noise corrected variant of AIC developed in Chapter 4. In VAR notation, AICu can be written as

$$\text{AICu}_p = \log|S_p^2| + \frac{(T+qp)q}{T-qp-q-1}. \tag{5.12}$$

AICu is similar to AICc except for the use of $S_p^2$, Eq. (5.3), in place of the MLE $\hat{\Sigma}_p$, Eq. (5.4). Subtracting $q$ gives an equivalent form

$$\text{AICu}_p = \log|S_p^2| + \frac{2pq^2 + q(q+1)}{T - qp - q - 1}.$$

### *5.2.5. SIC*

Schwarz's SIC can be adapted to VAR models as follows. The $n$ in $\log(n)$ is the sample size used to compute the MLEs. For VAR, that $\log(n)$ becomes $\log(T)$ because only $T = n - p$ observations are used to compute the MLEs. Hence SIC is

$$\text{SIC}_p = \log|\hat{\Sigma}_p| + \frac{\log(T)pq^2}{T}. \tag{5.13}$$

### *5.2.6. HQ*

Hannan and Quinn's HQ for VAR is

$$\text{HQ}_p = \log|\hat{\Sigma}_p| + \frac{2\log\log(T)pq^2}{T}. \tag{5.14}$$

### *5.2.7. HQc*

HQc for VAR can be written as

$$\text{HQc}_p = \log|\hat{\Sigma}_p| + \frac{2\log\log(T)pq^2}{T - qp - q - 1}. \tag{5.15}$$

## 5.3. Small-sample Signal-to-noise Ratios

In order to calculate small-sample signal-to-noise ratios, we begin by computing some necessary expectations for signal and noise terms using approximations Eq. (5A.4) and Eq. (5A.5), respectively, from Appendix 5A. As in Chapter 4, because we use determinants to reduce the $\text{SPE}_p$ matrix to a scalar, discussion of FPE has been omitted (see Section 4.3). From Eq. (5A.4),

$$\begin{aligned} E[\log|\hat{\Sigma}_{p_*}|] &= \log|\Sigma_*| - q\log(n - p_*) + q\log\big(n - (q+1)p_* - (q-1)/2\big) \\ &\quad - \frac{q}{n - (q+1)p_* - (q-1)/2}, \\ E[\log|\hat{\Sigma}_{p_*+L}|] &= \log|\Sigma_*| - q\log(n - p_* - L) \\ &\quad + q\log\big(n - (q+1)(p_* + L) - (q-1)/2\big) \\ &\quad - \frac{q}{n - (q+1)(p_* + L) - (q-1)/2}, \end{aligned}$$

and

$$E[\log|\hat{\Sigma}_{p_*+L}| - \log|\hat{\Sigma}_{p_*}|] = q\log\left(\frac{n-p_*}{n-p_*-L}\right) + q\log\left(\frac{n-(q+1)(p_*+L)-(q-1)/2}{n-(q+1)p_*-(q-1)/2}\right) - \frac{Lq}{(n-p_*-L-(q-1)/2)(n-p_*-(q-1)/2)}. \quad (5.16)$$

Also,

$$E[\log|S^2_{p_*+L}| - \log|S^2_{p_*}|] = q\log\left(\frac{n-(q+1)p_*}{n-(q+1)(p_*+L)}\right) + q\log\left(\frac{n-(q+1)(p_*+L)-(q-1)/2}{n-(q+1)p_*-(q-1)/2}\right) - \frac{Lq}{(n-p_*-L-(q-1)/2)(n-p_*-(q-1)/2)}. \quad (5.17)$$

The standard deviations of $\log|\hat{\Sigma}_{p_*+L}| - \log|\hat{\Sigma}_{p_*}|$ and $\log|S^2_{p_*+L}| - \log|S^2_{p_*}|$ are identical, and are equal to the standard deviation of $\log|\text{SPE}_{p_*+L}| - \log|\text{SPE}_{p_*}|$. Using Eq. (5A.5) we find that the noise for AIC, AICc, AICu, SIC, HQ, and HQc is

$$\frac{\sqrt{2Lq(q+1)}}{\sqrt{(n-(q+1)(p+L)-(q-1)/2)(n-(q+1)p-(q-1)/2+2)}}.$$

We will begin by calculating the signal-to-noise ratios for the K-L-based criteria AIC, AICc, and AICu.

### 5.3.1. AIC

The signal for AIC is $E[\text{AIC}_{p_*+L} - \text{AIC}_{p_*}]$, and thus from Eq. (5.16) we have

$$q\log\left(\frac{n-p_*}{n-p_*-L}\right) + q\log\left(\frac{n-(q+1)(p_*+L)-(q-1)/2}{n-(q+1)p_*-(q-1)/2}\right) - \frac{Lq}{(n-p_*-L-(q-1)/2)(n-p_*-(q-1)/2)} + \frac{2Lq^2n+q(q+1)L}{(n-p_*-L)(n-p_*)}.$$

The noise is

$$\frac{\sqrt{2Lq(q+1)}}{\sqrt{(n-(q+1)(p+L)-(q-1)/2)(n-(q+1)p-(q-1)/2+2)}},$$

and thus the signal-to-noise ratio for AIC overfitting is

$$\frac{\sqrt{(n-(q+1)(p+L)-(q-1)/2)(n-(q+1)p-(q-1)/2+2)}}{\sqrt{2Lq(q+1)}}\times$$
$$\left(q\log\left(\frac{n-p_*}{n-p_*-L}\right)+q\log\left(\frac{n-(q+1)(p_*+L)-(q-1)/2}{n-(q+1)p_*-(q-1)/2}\right)\right.$$
$$-\frac{Lq}{(n-p_*-L-(q-1)/2)(n-p_*-(q-1)/2)}$$
$$\left.+\frac{2Lq^2n+q(q+1)L}{(n-p_*-L)(n-p_*)}\right).$$

The first three terms

$$\frac{\sqrt{(n-(q+1)(p+L)-(q-1)/2)(n-(q+1)p-(q-1)/2+2)}}{\sqrt{2Lq(q+1)}}\times$$
$$\left(q\log\left(\frac{n-p_*}{n-p_*-L}\right)+q\log\left(\frac{n-(q+1)(p_*+L)-(q-1)/2}{n-(q+1)p_*-(q-1)/2}\right)\right.$$
$$\left.-\frac{Lq}{(n-p_*-L-(q-1)/2)(n-p_*-(q-1)/2)}\right)$$

decrease as $L$ increases. These three terms result from using the MLE $\hat{\Sigma}$. As in Chapter 4, these terms $\to -\infty$ as $L$ increases towards the saturated model. The last term

$$\frac{\sqrt{(n-(q+1)(p+L)-(q-1)/2)(n-(q+1)p-(q-1)/2+2)}}{\sqrt{2Lq(q+1)}}\times$$
$$\frac{2Lq^2n+q(q+1)L}{(n-p_*-L)(n-p_*)}$$

increases as $L$ increases, but not nearly as fast as the first three terms decrease. Overall, the signal-to-noise ratio for AIC increases for small $L$ than decreases quickly, leading to the well-known overfitting tendencies of AIC.

### 5.3.2. AICc

From Eq. (5.16) the signal for AICc is

$$q\log\left(\frac{n-p_*}{n-p_*-L}\right)+q\log\left(\frac{n-(q+1)(p_*+L)-(q-1)/2}{n-(q+1)p_*-(q-1)/2}\right)$$
$$-\frac{Lq}{(n-p_*-L-(q-1)/2)(n-p_*-(q-1)/2)}$$
$$+\frac{2Lq(qn-(q-1)(q+1))}{(n-(q+1)(p_*+L)-q-1)(n-(q+1)p_*-q-1)},$$

and thus the signal-to-noise ratio for AICc overfitting is

$$\frac{\sqrt{(n-(q+1)(p+L)-(q-1)/2)(n-(q+1)p-(q-1)/2+2)}}{\sqrt{2Lq(q+1)}}\times$$
$$\left(q\log\left(\frac{n-p_*}{n-p_*-L}\right)+q\log\left(\frac{n-(q+1)(p_*+L)-(q-1)/2}{n-(q+1)p_*-(q-1)/2}\right)\right.$$
$$-\frac{Lq}{(n-p_*-L-(q-1)/2)(n-p_*-(q-1)/2)}$$
$$\left.+\frac{2Lq(qn-(q-1)(q+1))}{(n-(q+1)(p_*+L)-q-1)(n-(q+1)p_*-q-1)}\right).$$

The signal-to-noise ratio of AICc shares the first three terms with the signal-to-noise ratio of AIC. However, the last term

$$\frac{\sqrt{(n-(q+1)(p+L)-(q-1)/2)(n-(q+1)p-(q-1)/2+2)}}{\sqrt{2Lq(q+1)}}\times$$
$$\frac{2Lq(qn-(q-1)(q+1))}{(n-(q+1)(p_*+L)-q-1)(n-(q+1)p_*-q-1)}$$

increases much faster than the first three terms decrease. AICc has a strong signal-to-noise ratio that increases with increasing $L$.

### 5.3.3. AICu

For AICu, from Eq. (5.17) the signal is

$$q\log\left(\frac{n-(q+1)p_*}{n-(q+1)(p_*+L)}\right)+q\log\left(\frac{n-(q+1)(p_*+L)-(q-1)/2}{n-(q+1)p_*-(q-1)/2}\right)$$
$$-\frac{Lq}{(n-p_*-L-(q-1)/2)(n-p_*-(q-1)/2)}$$
$$+\frac{2Lq(qn-(q-1)(q+1))}{(n-(q+1)(p_*+L)-q-1)(n-(q+1)p_*-q-1)},$$

and thus the signal-to-noise ratio for AICu overfitting is

$$\frac{\sqrt{(n-(q+1)(p+L)-(q-1)/2)(n-(q+1)p-(q-1)/2+2)}}{\sqrt{2Lq(q+1)}}\times$$
$$\left(q\log\left(\frac{n-(q+1)p_*}{n-(q+1)(p_*+L)}\right)+q\log\left(\frac{n-(q+1)(p_*+L)-(q-1)/2}{n-(q+1)p_*-(q-1)/2}\right)\right.$$
$$-\frac{Lq}{(n-p_*-L-(q-1)/2)(n-p_*-(q-1)/2)}$$
$$\left.+\frac{2Lq(qn-(q-1)(q+1))}{(n-(q+1)(p_*+L)-q-1)(n-(q+1)p_*-q-1)}\right).$$

AICu has different terms in its signal-to-noise ratio. The first three terms

$$\frac{\sqrt{(n-(q+1)(p+L)-(q-1)/2)(n-(q+1)p-(q-1)/2+2)}}{\sqrt{2Lq(q+1)}}\times$$
$$\left(q\log\left(\frac{n-(q+1)p_*}{n-(q+1)(p_*+L)}\right)+q\log\left(\frac{n-(q+1)(p_*+L)-(q-1)/2}{n-(q+1)p_*-(q-1)/2}\right)\right.$$
$$\left.-\frac{Lq}{(n-p_*-L-(q-1)/2)(n-p_*-(q-1)/2)}\right)$$

are larger than the first three terms in the signal-to-noise ratio of AICc. AICu and AICc have the same penalty function and hence share the last term

$$\frac{\sqrt{(n-(q+1)(p+L)-(q-1)/2)(n-(q+1)p-(q-1)/2+2)}}{\sqrt{2Lq(q+1)}}\times$$
$$\frac{2Lq(qn-(q-1)(q+1))}{(n-(q+1)(p_*+L)-q-1)(n-(q+1)p_*-q-1)}.$$

This gives AICu a stronger signal-to-noise ratio than AICc.

We next look at the consistent criteria, beginning with SIC.

### *5.3.4. SIC*

From Eq. (5.16), the signal for SIC is

$$q\log\left(\frac{n-p_*}{n-p_*-L}\right)+q\log\left(\frac{n-(q+1)(p_*+L)-(q-1)/2}{n-(q+1)p_*-(q-1)/2}\right)$$
$$-\frac{Lq}{(n-p_*-L-(q-1)/2)(n-p_*-(q-1)/2)}$$
$$+\frac{\log(n-p_*-L)(p_*+L)q^2}{n-p_*-L}-\frac{\log(n-p_*)p_*q^2}{n-p_*},$$

and thus the signal-to-noise ratio for SIC overfitting is

$$\frac{\sqrt{(n-(q+1)(p+L)-(q-1)/2)(n-(q+1)p-(q-1)/2+2)}}{\sqrt{2Lq(q+1)}}\times$$
$$\left(q\log\left(\frac{n-p_*}{n-p_*-L}\right)+q\log\left(\frac{n-(q+1)(p_*+L)-(q-1)/2}{n-(q+1)p_*-(q-1)/2}\right)\right.$$
$$-\frac{Lq}{(n-p_*-L-(q-1)/2)(n-p_*-(q-1)/2)}$$
$$\left.+\frac{\log(n-p_*-L)(p_*+L)q^2}{n-p_*-L}-\frac{\log(n-p_*)p_*q^2}{n-p_*}\right).$$

Although SIC has a stronger penalty function than AIC, their structures are similar. Like AIC, the signal-to-noise ratio of SIC increases for small $L$ then decreases. However, unlike AIC, the signal-to-noise ratio of SIC increases rapidly as the effective sample size $T$ increases. Thus SIC suffers overfitting problems in small samples but not in large samples. In our small-sample special case models, we will see that the signal-to-noise ratio for SIC is similar to but larger than the signal-to-noise ratio for AIC.

### 5.3.5. HQ

For HQ, from Eq. (5.16) the signal is

$$q\log\left(\frac{n-p_*}{n-p_*-L}\right)+q\log\left(\frac{n-(q+1)(p_*+L)-(q-1)/2}{n-(q+1)p_*-(q-1)/2}\right)$$
$$-\frac{Lq}{(n-p_*-L-(q-1)/2)(n-p_*-(q-1)/2)}$$
$$+\frac{2\log\log(n-p_*-L)(p_*+L)q^2}{n-p_*-L}-\frac{2\log\log(n-p_*)p_*q^2}{n-p_*},$$

and thus the signal-to-noise ratio for HQ overfitting is

$$\frac{\sqrt{(n-(q+1)(p+L)-(q-1)/2)(n-(q+1)p-(q-1)/2+2)}}{\sqrt{2Lq(q+1)}}\times$$
$$\left(q\log\left(\frac{n-p_*}{n-p_*-L}\right)+q\log\left(\frac{n-(q+1)(p_*+L)-(q-1)/2}{n-(q+1)p_*-(q-1)/2}\right)\right.$$
$$-\frac{Lq}{(n-p_*-L-(q-1)/2)(n-p_*-(q-1)/2)}$$
$$\left.+\frac{2\log\log(n-p_*-L)(p_*+L)q^2}{n-p_*-L}-\frac{2\log\log(n-p_*)p_*q^2}{n-p_*}\right).$$

HQ's penalty function has a structure similar to those of both AIC and SIC. For effective sample sizes $T$ greater than 15, the magnitude of HQ's penalty function is between those of AIC and SIC, hence its signal-to-noise ratio will be between those of AIC and SIC.

### 5.3.6. HQc

Finally, from Eq. (5.16) we have the signal for HQc:

$$q \log\left(\frac{n-p_*}{n-p_*-L}\right) + q\log\left(\frac{n-(q+1)(p_*+L)-(q-1)/2}{n-(q+1)p_*-(q-1)/2}\right)$$
$$-\frac{Lq}{(n-p_*-L-(q-1)/2)(n-p_*-(q-1)/2)}$$
$$+\frac{2\log\log(n-p_*-L)(p_*+L)q^2}{n-(q+1)(p_*+L)-q-1} - \frac{2\log\log(n-p_*)p_*q^2}{n-(q+1)p_*-q-1},$$

and thus the signal-to-noise ratio for HQc overfitting is

$$\frac{\sqrt{(n-(q+1)(p+L)-(q-1)/2)(n-(q+1)p-(q-1)/2+2)}}{\sqrt{2Lq(q+1)}} \times$$
$$\left(q \log\left(\frac{n-p_*}{n-p_*-L}\right) + q\log\left(\frac{n-(q+1)(p_*+L)-(q-1)/2}{n-(q+1)p_*-(q-1)/2}\right)\right.$$
$$-\frac{Lq}{(n-p_*-L-(q-1)/2)(n-p_*-(q-1)/2)}$$
$$\left.+\frac{2\log\log(n-p_*-L)(p_*+L)q^2}{n-(q+1)(p_*+L)-q-1} - \frac{2\log\log(n-p_*)p_*q^2}{n-(q+1)p_*-q-1}\right).$$

HQc and AICc are similar. They have the first three terms of their signal-to-noise ratios in common, and they have penalty functions with similar structures. Consequently, both have similar signal-to-noise ratios. Except in very small samples, the penalty function and signal-to-noise ratio of HQc are greater than those of AICc. The improved penalty function in HQc gives it very good small-sample performance, the best of the consistent criteria.

What do we expect these signal-to-noise ratios to tell us about the behavior of these criteria? In general, the larger the signal-to-noise ratio, the smaller we expect the probability of overfitting to be, since the signal-to-noise ratio and probability of overfitting both depend on the penalty function. For the K-L criteria and VAR models, the signal-to-noise ratio of AIC is less than the signal-to-noise ratio of AICc, which in turn is less than that of AICu. For the consistent criteria, the penalty function of SIC is larger than the penalty function of HQ, and thus the signal-to-noise ratio of SIC is larger than the signal-to-noise ratio of HQ. In small samples, the signal-to-noise ratio of HQc is larger than the signal-to-noise ratio of SIC; however, in large samples the reverse is true. This makes it difficult to generalize the relationship between the signal-to-noise ratios of HQc and SIC.

## 5.4. Overfitting

In this Section we will look at the overfitting properties for the five base selection criteria and the two signal-to-noise corrected variants considered in this Chapter. We will be able to see if our expectations from the previous Section compare favorably with the overfitting probabilities in this Section.

### *5.4.1. Small-sample Probabilities of Overfitting*

As before, we assume that there is a true order $p_*$ and we fit a candidate model of order $p_* + L$ where $L > 0$. We will compute the probability of overfitting by $L$ extra variables by obtaining the probability of selecting the model of order $p_* + L$ over the model of order $p_*$. Remember that for VAR models, model VAR$(p_* + L)$ has $Lq^2$ more parameters than model VAR$(p_*)$, and its degrees of freedom are decreased by $L(q + 1)$. We can also express small-sample probabilities in terms of the U-statistic, where for VAR models

$$\frac{|\text{SPE}_{p_*+L}|}{|\text{SPE}_{p_*}|} \sim U_{q,Lq,n-(q+1)(p_*+L)}.$$

For model selection criteria of the form $\log|\text{SPE}_p| + \alpha(n,k)$, probabilities of overfitting can be expressed as $P\{U_{q,Lq,n-(q+1)(p_*+L)} < \exp(\alpha(n,p_*) - \alpha(n,p_* + L))\}$. A given model selection criterion MSC overfits if $\text{MSC}_{p*+L} < \text{MSC}_{p_*}$. We will present the calculations for only one criterion, AIC, as an example. Only the results for the other criteria will be given, but details can be found in Appendix 5B. Results are presented in the U-statistic form. When $q = 2$, U probabilities simplify to the usual F distribution, and we have

$$P\{U_{2,qL,n-(q+1)(p_*+L)} < u\}$$
$$= P\left\{F_{2qL,2(n-(q+1)(p_*+L))} > \frac{1-\sqrt{u}}{\sqrt{u}}\frac{n-(q+1)(p_*+L)-1}{qL}\right\}.$$

We present probabilities in terms of the U distribution as well as in terms of independent $\chi^2$,

$$n\sum_{i=1}^{q}\log\left(1 + \frac{\chi^2_{(q+1)L}}{\chi^2_{n-(q+1)(p_*+L)-q+i}}\right).$$

This second form will be more useful in deriving asymptotic probabilities of overfitting.

### AIC

AIC overfits if $\text{AIC}_{p_*+L} < \text{AIC}_{p_*}$. For finite $n$, the probability that AIC prefers the overfitted model $p_* + L$ in terms of the independent $\chi^2$ is

$$
\begin{aligned}
&P\{\text{AIC}_{p_*+L} < \text{AIC}_{p_*}\}\\
&= P\left\{\log|\hat{\Sigma}_{p_*+L}| + \frac{2(p_*+L)q^2 + q(q+1)}{n - p_* - L} < \log|\hat{\Sigma}_{p_*}| + \frac{2p_*q^2 + q(q+1)}{n - p_*}\right\}\\
&= P\left\{\log|\text{SPE}_{p_*+L}| - q\log(n - p_* - L) + \frac{2(p_*+L)q^2 + q(q+1)}{n - p_* - L}\right.\\
&\qquad \left. < \log|\text{SPE}_{p_*}| - q\log(n - p_*) + \frac{2p_*q^2 + q(q+1)}{n - p_*}\right\}\\
&= P\left\{\log\left(\frac{|\text{SPE}_{p_*+L}|}{|\text{SPE}_{p_*}|}\right) < -q\log\left(\frac{n - p_*}{n - p_* - L}\right)\right.\\
&\qquad \left. + \frac{2p_*q^2 + q(q+1)}{n - p*} - \frac{2(p_*+L)q^2 + q(q+1)}{n - p_* - L}\right\}\\
&= P\left\{\sum_{i=1}^{q} \text{log-Beta}\left(\frac{n - (q+1)(p_*+L) - q + i}{2}, \frac{(q+1)L}{2}\right)\right.\\
&\qquad \left. < -q\log\left(\frac{n - p_*}{n - p_* - L}\right) - \frac{2Lq^2n + q(q+1)L}{(n - p_* - L)(n - p_*)}\right\}\\
&= P\left\{n\sum_{i=1}^{q} \log\left(1 + \frac{\chi^2_{(q+1)L}}{\chi^2_{n-(q+1)(p_*+L)-q+i}}\right) > qn\log\left(\frac{n - p_*}{n - p_* - L}\right)\right.\\
&\qquad \left. + \frac{2Lq^2n^2 + q(q+1)Ln}{(n - p_* - L)(n - p_*)}\right\}.
\end{aligned}
$$

Expressed in terms of $U_{q,qL,n-(q+1)(p_*+L)}$, the $P\{\text{overfit}\}$ for AIC is

$$
P\left\{U_{q,qL,n-(q+1)(p_*+L)} < \left(\frac{n - p_* - L}{n - p_*}\right)^q \exp\left(-\frac{2Lq^2n + q(q+1)L}{(n - p_* - L)(n - p_*)}\right)\right\}.
$$

### AICc

The two forms for the probability that AICc overfits by $L$ are

$$
\begin{aligned}
&P\{\text{AICc}_{p_*+L} < \text{AICc}_{p_*}\}\\
&= P\left\{n\sum_{i=1}^{q} \log\left(1 + \frac{\chi^2_{(q+1)L}}{\chi^2_{n-(q+1)(p_*+L)-q+i}}\right) > qn\log\left(\frac{n - p_*}{n - p_* - L}\right)\right.\\
&\qquad \left. + \frac{2Lqn(qn - (q-1)(q+1))}{(n - (q+1)(p_*+L) - q - 1)(n - (q+1)p_* - q - 1)}\right\},
\end{aligned}
$$

and

$$P\left\{U_{q,qL,n-(q+1)(p_*+L)} < \left(\frac{n-p_*-L}{n-p_*}\right)^q \times \right.$$
$$\left. \exp\left(-\frac{2Lq(qn-(q-1)(q+1))}{(n-(q+1)(p_*+L)-q-1)(n-(q+1)p_*-q-1)}\right)\right\}.$$

**AICu**

The two forms for the probability that AICu overfits by $L$ are

$$P\{\text{AICu}_{p_*+L} < \text{AICu}_{p_*}\}$$
$$= P\left\{n\sum_{i=1}^{q}\log\left(1+\frac{\chi^2_{(q+1)L}}{\chi^2_{n-(q+1)(p_*+L)-q+i}}\right) > qn\log\left(\frac{n-(q+1)p_*}{n-(q+1)(p_*+L)}\right)\right.$$
$$\left. + \frac{2Lqn(qn-(q-1)(q+1))}{(n-(q+1)(p_*+L)-q-1)(n-(q+1)p_*-q-1)}\right\},$$

and

$$P\left\{U_{q,qL,n-(q+1)(p_*+L)} < \left(\frac{n-(q+1)(p_*+L)}{n-(q+1)p_*}\right)^q \times \right.$$
$$\left. \exp\left(-\frac{2Lq(qn-(q-1)(q+1))}{(n-(q+1)(p_*+L)-q-1)(n-(q+1)p_*-q-1)}\right)\right\}.$$

**FPE**

Note that we use log(FPE) for convenience in computing probabilities. The two forms for the probability that FPE overfits by $L$ are

$$P\{\text{FPE}_{p_*+L} < \text{FPE}_{p_*}\}$$
$$= P\left\{n\sum_{i=1}^{q}\log\left(1+\frac{\chi^2_{(q+1)L}}{\chi^2_{n-(q+1)(p_*+L)-q+i}}\right) > \right.$$
$$\left. qn\log\left(\frac{(n+(q-1)(p_*+L))(n-(q+1)p_*)(n-p_*)}{(n+(q-1)p_*)(n-(q+1)(p_*+L))(n-p_*-L)}\right)\right\},$$

and

$$P\left\{U_{q,qL,n-(q+1)(p_*+L)} < \left(\frac{(n+(q-1)(p_*+L))(n-(q+1)p_*)(n-p_*)}{(n+(q-1)p_*)(n-(q+1)(p_*+L))(n-p_*-L)}\right)^q\right\}.$$

**SIC**

The two forms for the probability that SIC overfits by $L$ are

$$P\{\mathrm{SIC}_{p_*+L} < \mathrm{SIC}_{p_*}\}$$
$$= P\left\{ n\sum_{i=1}^{q} \log\left(1 + \frac{\chi^2_{(q+1)L}}{\chi^2_{n-(q+1)(p_*+L)-q+i}}\right) > qn\log\left(\frac{n-p_*}{n-p_*-L}\right) - \frac{n\log(n-p_*)p_*q^2}{n-p_*} + \frac{n\log(n-p_*-L)(p_*+L)q^2}{n-p_*-L}\right\},$$

and

$$P\left\{ U_{q,qL,n-(q+1)(p_*+L)} < \left(\frac{n-p_*-L}{n-p_*}\right)^q \times \exp\left(\frac{\log(n-p_*)p_*q^2}{n-p_*} - \frac{\log(n-p_*-L)(p_*+L)q^2}{n-p_*-L}\right)\right\}.$$

**HQ**

The two forms for the probability that HQ overfits by $L$ are

$$P\{\mathrm{HQ}_{p_*+L} < \mathrm{HQ}_{p_*}\}$$
$$= P\left\{ n\sum_{i=1}^{q} \log\left(1 + \frac{\chi^2_{(q+1)L}}{\chi^2_{n-(q+1)(p_*+L)-q+i}}\right) > qn\log\left(\frac{n-p_*}{n-p_*-L}\right) - \frac{n2\log\log(n-p_*)p_*q^2}{n-p_*} + \frac{n2\log\log(n-p_*-L)(p_*+L)q^2}{n-p_*-L}\right\},$$

and

$$P\left\{ U_{q,qL,n-(q+1)(p_*+L)} < \left(\frac{n-p_*-L}{n-p_*}\right)^q \times \exp\left(\frac{2\log\log(n-p_*)p_*q^2}{n-p_*} - \frac{2\log\log(n-p_*-L)(p_*+L)q^2}{n-p_*-L}\right)\right\}.$$

**HQc**

Finally, the two forms for the probability that AICc overfits by $L$ are

$$P\{\mathrm{HQc}_{p_*+L} < \mathrm{HQc}_{p_*}\}$$
$$= P\left\{ n\sum_{i=1}^{q} \log\left(1 + \frac{\chi^2_{(q+1)L}}{\chi^2_{n-(q+1)(p_*+L)-q+i}}\right) > qn\log\left(\frac{n-p_*}{n-p_*-L}\right) - \frac{n2\log\log(n-p_*)p_*q^2}{n-(q+1)p_*-q-1} + \frac{n2\log\log(n-p_*-L)(p_*+L)q^2}{n-(q+1)(p_*+L)-q-1}\right\},$$

and

$$P\left\{U_{q,qL,n-(q+1)(p_*+L)} < \left(\frac{n-p_*-L}{n-p_*}\right)^q \times \exp\left(\frac{2\log\log(n-p_*)p_*q^2}{n-(q+1)p_*-q-1} - \frac{2\log\log(n-p_*-L)(p_*+L)q^2}{n-(q+1)(p_*+L)-q-1}\right)\right\}.$$

### 5.4.2. Asymptotic Probabilities of Overfitting

Using the above small-sample probabilities of overfitting, we can now derive asymptotic probabilities of overfitting. We will make use of the following facts: As $n \to \infty$, with $p_*$, $L$, and $q$ fixed and $0 \le i \le q$, $\chi^2_{n-(q+1)(p_*+L)-q+i}/n \to 1$ a.s.; and $\log(1+z) \doteq z$ when $|z|$ is small. For independent $\chi^2$ we have

$$\begin{aligned} n\sum_{i=1}^{q}\log\left(1+\frac{\chi^2_{(q+1)L}}{\chi^2_{n-(q+1)(p_*+L)-q+i}}\right) &= n\sum_{i=1}^{q}\log\left(1+\frac{\frac{1}{n}\chi^2_{(q+1)L}}{\frac{1}{n}\chi^2_{n-(q+1)(p_*+L)-q+i}}\right) \\ &\to n\sum_{i=1}^{q}\frac{\frac{1}{n}\chi^2_{(q+1)L}}{\frac{1}{n}\chi^2_{n-(q+1)(p_*+L)-q+i}} \\ &\to \sum_{i=1}^{q}\chi^2_{(q+1)L} \\ &= \chi^2_{q(q+1)L}. \end{aligned}$$

Since the $\chi^2$ distributions within the multivariate log-Beta distribution are independent, obtaining the asymptotic probabilities of overfitting is thus a matter of evaluating the limit of the critical value. To do this we also need to make use of two other limits: as $n \to \infty$, with $p_*$, $L$, and $q$ fixed,

$$\begin{aligned} nq\log\left(\frac{n-p_*}{n-p_*-L}\right) &= nq\log\left(1+\frac{L}{n-p_*-L}\right) \\ &\to qL, \end{aligned}$$

and

$$\begin{aligned} nq\log\left(\frac{n-(q+1)p_*}{n-(q+1)(p_*+L)}\right) &= nq\log\left(1+\frac{(q+1)L}{n-(q+1)(p_*+L)}\right) \\ &\to q(q+1)L. \end{aligned}$$

As before, we will show calculations for AIC and give details for the other criteria in Appendix 5C.

**AIC**

In small samples, AIC overfits with probability

$$P\left\{n\sum_{i=1}^{q}\log\left(1+\frac{\chi^2_{(q+1)L}}{\chi^2_{n-(q+1)(p_*+L)-q+i}}\right) > qn\log\left(\frac{n-p_*}{n-p_*-L}\right)\right.$$

$$\left.+\frac{2Lq^2n^2+q(q+1)Ln}{(n-p_*-L)(n-p_*)}\right\}.$$

As $n \to \infty$,

$$n\sum_{i=1}^{q}\log\left(1+\frac{\chi^2_{(q+1)L}}{\chi^2_{n-(q+1)(p_*+L)-q+i}}\right) \to \chi^2_{q(q+1)L}$$

and

$$qn\log\left(\frac{n-p_*}{n-p_*-L}\right)+\frac{2Lq^2n^2+q(q+1)Ln}{(n-p_*-L)(n-p_*)} \to Lq+2Lq^2,$$

thus the asymptotic probability of overfitting for AIC is

$$P\{\text{AIC overfits by } L\} = P\left\{\chi^2_{q(q+1)L} > (2q^2+q)L\right\}.$$

**AICc**

$$P\{\text{AICc overfits by } L\} = P\left\{\chi^2_{q(q+1)L} > (2q^2+q)L\right\}.$$

**AICu**

$$P\{\text{AICu overfits by } L\} = P\left\{\chi^2_{q(q+1)L} > (3q^2+q)L\right\}.$$

**FPE**

$$P\{\text{FPE overfits by } L\} = P\left\{\chi^2_{q(q+1)L} > (2q^2+q)L\right\}.$$

**SIC**

$$P\{\text{SIC overfits by } L\} = 0.$$

**HQ**

$$P\{\text{HQ overfits by } L\} = 0.$$

**HQc**

$$P\{\text{HQc overfits by } L\} = 0.$$

The above results show that SIC, HQ and HQc are asymptotically equivalent. However, HQ and HQc behave much differently than SIC even when the sample size is very large. This is mostly due to the size of the penalty function: for example, when $n$ = 100,000 and $p$ is small, the $2\log\log(n)$ term = 4.89 for HQ and HQc, whereas the $\log(n)$ term for SIC = 11.5—more than twice that of HQ. For $n$ = 10,000, $2\log\log(n) = 4.44$. For SIC $\log(n) = 9.2$, which is also much larger. For the $\alpha$ variants (see Bhansali and Downham, 1977), the recommended range of $\alpha$ is 2 to 5. For $n$ = 10,000, HQ falls in this range, but SIC does not. In practice, HQ and SIC behave differently. Tables 5.1 and 5.2 summarize the asymptotic probabilities of overfitting for model dimensions $q = 2$ and 5.

Table 5.1. Asymptotic probability of overfitting by $L$ variables for $q = 2$.

| $L$ | AIC | AICc | AICu | SIC | HQ | HQc | FPE |
|---|---|---|---|---|---|---|---|
| 1 | 0.1247 | 0.1247 | 0.0296 | 0 | 0 | 0 | 0.1247 |
| 2 | 0.0671 | 0.0671 | 0.0055 | 0 | 0 | 0 | 0.0671 |
| 3 | 0.0374 | 0.0374 | 0.0011 | 0 | 0 | 0 | 0.0374 |
| 4 | 0.0214 | 0.0214 | 0.0002 | 0 | 0 | 0 | 0.0214 |
| 5 | 0.0124 | 0.0124 | 0.0000 | 0 | 0 | 0 | 0.0124 |
| 6 | 0.0073 | 0.0073 | 0.0000 | 0 | 0 | 0 | 0.0073 |
| 7 | 0.0043 | 0.0043 | 0.0000 | 0 | 0 | 0 | 0.0043 |
| 8 | 0.0026 | 0.0026 | 0.0000 | 0 | 0 | 0 | 0.0026 |
| 9 | 0.0015 | 0.0015 | 0.0000 | 0 | 0 | 0 | 0.0015 |
| 10 | 0.0009 | 0.0009 | 0.0000 | 0 | 0 | 0 | 0.0009 |

Table 5.2. Asymptotic probability of overfitting by $L$ variables for $q = 5$.

| $L$ | AIC | AICc | AICu | SIC | HQ | HQc | FPE |
|---|---|---|---|---|---|---|---|
| 1 | 0.0035 | 0.0035 | 0.0000 | 0 | 0 | 0 | 0.0035 |
| 2 | 0.0001 | 0.0001 | 0.0000 | 0 | 0 | 0 | 0.0001 |
| 3 | 0.0000 | 0.0000 | 0.0000 | 0 | 0 | 0 | 0.0000 |
| 4 | 0.0000 | 0.0000 | 0.0000 | 0 | 0 | 0 | 0.0000 |
| $\geq$5 | 0.0000 | 0.0000 | 0.0000 | 0 | 0 | 0 | 0.0000 |

The patterns we have established in previous chapters are evident for vector autoregressive models as well. The consistent criteria have 0 probabilities of overfitting, and the signal-to-noise corrected variant AICu has probabilities of overfitting that lie between those for the efficient and consistent criteria for a given level of $q$. In addition, we see again that the probability of overfitting decreases as the dimension of the model $q$ increases.

### 5.4.3. Asymptotic Signal-to-noise Ratios

We derived the expressions for small-sample signal-to-noise ratios in Section 5.3, and we will use them here as the basis for obtaining the asymptotic signal-to-noise ratios. We will present calculations for one K-L-based criterion (AIC) and one consistent criterion (SIC). Only the results of the derivations will be presented for the other criteria, but details can be found in Appendix 5D.

Table 5.3. Asymptotic signal-to-noise ratios for overfitting by $L$ variables for $q = 2$.

| $L$ | AIC | AICc | AICu | SIC | HQ | HQc | FPE |
|---|---|---|---|---|---|---|---|
| 1 | 1.155 | 1.155 | 2.309 | $\infty$ | $\infty$ | $\infty$ | 1.155 |
| 2 | 1.633 | 1.633 | 3.266 | $\infty$ | $\infty$ | $\infty$ | 1.633 |
| 3 | 2.000 | 2.000 | 4.000 | $\infty$ | $\infty$ | $\infty$ | 2.000 |
| 4 | 2.309 | 2.309 | 4.619 | $\infty$ | $\infty$ | $\infty$ | 2.309 |
| 5 | 2.582 | 2.582 | 5.164 | $\infty$ | $\infty$ | $\infty$ | 2.582 |
| 6 | 2.828 | 2.828 | 5.657 | $\infty$ | $\infty$ | $\infty$ | 2.828 |
| 7 | 3.055 | 3.055 | 6.110 | $\infty$ | $\infty$ | $\infty$ | 3.055 |
| 8 | 3.266 | 3.266 | 6.532 | $\infty$ | $\infty$ | $\infty$ | 3.266 |
| 9 | 3.464 | 3.464 | 6.928 | $\infty$ | $\infty$ | $\infty$ | 3.464 |
| 10 | 3.651 | 3.651 | 7.303 | $\infty$ | $\infty$ | $\infty$ | 3.651 |

Table 5.4. Asymptotic signal-to-noise ratios for overfitting by $L$ variables for $q = 5$.

| $L$ | AIC | AICc | AICu | SIC | HQ | HQc | FPE |
|---|---|---|---|---|---|---|---|
| 1 | 3.227 | 3.227 | 6.455 | $\infty$ | $\infty$ | $\infty$ | 3.227 |
| 2 | 4.564 | 4.564 | 9.129 | $\infty$ | $\infty$ | $\infty$ | 4.564 |
| 3 | 5.590 | 5.590 | 11.180 | $\infty$ | $\infty$ | $\infty$ | 5.590 |
| 4 | 6.455 | 6.455 | 12.910 | $\infty$ | $\infty$ | $\infty$ | 6.455 |
| 5 | 7.217 | 7.217 | 14.434 | $\infty$ | $\infty$ | $\infty$ | 7.217 |
| 6 | 7.906 | 7.906 | 15.811 | $\infty$ | $\infty$ | $\infty$ | 7.906 |
| 7 | 8.539 | 8.539 | 17.078 | $\infty$ | $\infty$ | $\infty$ | 8.539 |
| 8 | 9.129 | 9.129 | 18.257 | $\infty$ | $\infty$ | $\infty$ | 9.129 |
| 9 | 9.682 | 9.682 | 19.365 | $\infty$ | $\infty$ | $\infty$ | 9.682 |
| 10 | 10.206 | 10.206 | 20.412 | $\infty$ | $\infty$ | $\infty$ | 10.206 |

We will also make use of the following facts: Assuming $p_*, L$ fixed and $n$ large,

$$\frac{\sqrt{(n-(q+1)(p+L)-(q-1)/2)(n-(q+1)p-(q-1)/2+2)}}{\sqrt{2Lq(q+1)}}$$

$$\to \frac{n}{\sqrt{2Lq(q+1)}},$$

$$q \log\left(\frac{n-(q+1)(p_*+L)-(q-1)/2}{n-(q+1)p_*-(q-1)/2}\right) \to \frac{-Lq(q+1)}{n},$$

and

$$\begin{aligned} q \log\left(\frac{n-p_*}{n-p_*-L}\right) &= q \log\left(1+\frac{L}{n-p_*-L}\right) \\ &\to \frac{Lq}{n}. \end{aligned}$$

### 5.4.3.1. K-L-based Criteria AIC, AICc, AICu, and FPE

Our detailed example for the K-L criteria will be AIC. Starting with the finite signal-to-noise ratio from Section 5.3, the corresponding asymptotic signal-to-noise ratio is

$$\begin{aligned} &\lim_{n\to\infty} \frac{\sqrt{(n-(q+1)(p+L)-(q-1)/2)(n-(q+1)p-(q-1)/2+2)}}{\sqrt{2Lq(q+1)}} \times \\ &\quad \left(q \log\left(\frac{n-p_*}{n-p_*-L}\right) + q \log\left(\frac{n-(q+1)(p_*+L)-(q-1)/2}{n-(q+1)p_*-(q-1)/2}\right)\right. \\ &\quad - \frac{Lq}{(n-p_*-L-(q-1)/2)(n-p_*-(q-1)/2)} \\ &\quad \left. + \frac{2Lq^2 n + q(q+1)L}{(n-p_*-L)(n-p_*)}\right) \\ &= \lim_{n\to\infty} \frac{n}{\sqrt{2Lq(q+1)}} \left(\frac{Lq}{n} - \frac{Lq(q+1)}{n} - \frac{Lq}{n^2} + \frac{2Lq^2}{n}\right) \\ &= \frac{q^2}{\sqrt{2q(q+1)}} \sqrt{L}. \end{aligned}$$

The asymptotic signal-to-noise ratios for AICc and FPE are also $q^2\sqrt{L}/\sqrt{2q(q+1)}$, but for AICu the value is twice as large, $2q^2\sqrt{L}/\sqrt{2q(q+1)}$.

### 5.4.3.2. Consistent Criteria SIC, HQ, and HQc

Our detailed example for the consistent criteria will be SIC. Once again, starting with the small-sample signal-to-noise ratio for SIC from Section 5.3, the corresponding asymptotic signal-to-noise ratio is

$$
\begin{aligned}
&\lim_{n\to\infty} \frac{\sqrt{(n-(q+1)(p+L)-(q-1)/2)(n-(q+1)p-(q-1)/2+2)}}{\sqrt{2Lq(q+1)}} \times \\
&\quad \left( q\log\left(\frac{n-p_*}{n-p_*-L}\right) + q\log\left(\frac{n-(q+1)(p_*+L)-(q-1)/2}{n-(q+1)p_*-(q-1)/2}\right) \right. \\
&\quad - \frac{Lq}{(n-p_*-L-(q-1)/2)(n-p_*-(q-1)/2)} \\
&\quad \left. + \frac{\log(n-p_*-L)(p_*+L)q^2}{n-p_*-L} - \frac{\log(n-p_*)p_*q^2}{n-p_*} \right) \\
&= \lim_{n\to\infty} \frac{n}{\sqrt{2Lq(q+1)}} \left( \frac{Lq}{n} - \frac{Lq(q+1)}{n} - \frac{Lq}{n^2} + \frac{\log(n)Lq^2}{n} \right) \\
&= \infty.
\end{aligned}
$$

Since consistent criteria have the same asymptotic signal-to-noise ratio, the ratios for HQ and HQc are the same as that for SIC.

Tables 5.3 and 5.4 given below present asymptotic signal-to-noise ratios of model $p_*$ versus model $p_* + L$ for $q = 2$ and 5, respectively. We see from Table 5.3 that, as we calculated above, the signal-to-noise ratios for AICc and AIC are equivalent. Also, the corrected variant AICu has a signal-to-noise ratio much larger than that of either AICc or AIC, but smaller than the infinite values for the consistent criteria. The results from Table 5.4 parallel those from Table 5.3 except that, as was the case in Chapter 4 for multivariate regression, the signal-to-noise ratios increase with $q$.

## 5.5. Underfitting in Two Special Case Models

In this Section, we will evaluate the performance of the selection criteria and distance measures we have discussed by using two special case VAR models. To make comparisons, we will compute expected values for the selected criteria, approximate and exact signal-to-noise ratios, and probabilities of overfitting for the true model versus candidate models. We begin by defining our special case models as follows: In both cases the true model has

$$
n = 35,\ p_* = 4, \text{ and } cov[\varepsilon_{*t}] = \Sigma_* = \begin{pmatrix} 1 & 0.7 \\ 0.7 & 1 \end{pmatrix}. \tag{5.18}
$$

The largest model order considered is $P = 8$. Model 9 is

$$
\begin{aligned}
y_t = &\begin{pmatrix} 0.090 & 0 \\ 0 & 0.090 \end{pmatrix} y_{t-1} + \begin{pmatrix} 0 & 0 \\ 0 & 0 \end{pmatrix} y_{t-2} + \begin{pmatrix} 0 & 0 \\ 0 & 0 \end{pmatrix} y_{t-3} \\
&+ \begin{pmatrix} 0 & 0.900 \\ 0.900 & 0 \end{pmatrix} y_{t-4} + w_{*t},
\end{aligned} \tag{5.19}
$$

and Model 10 is

$$y_t = \begin{pmatrix} 0.024 & 0.241 \\ 0.024 & 0.241 \end{pmatrix} y_{t-1} + \begin{pmatrix} 0 & 0.241 \\ 0 & 0.241 \end{pmatrix} y_{t-2} + \begin{pmatrix} 0 & 0.241 \\ 0 & 0.241 \end{pmatrix} y_{t-3}$$
$$+ \begin{pmatrix} 0 & 0.241 \\ 0 & 0.241 \end{pmatrix} y_{t-4} + w_{*t}. \tag{5.20}$$

Model 9 represents a strongly identifiable model. The $\Phi_4$ or VAR(4) parameters are large and should be easy to detect. Model 10 is similar but with much more weakly identifiable parameters.

### *5.5.1. Expected Values for Two Special Case Models*

Tables 5.5 and 5.6 summarize the expected values for the selected criteria as well as expected efficiencies for the distance measures, where maximum efficiency (1) corresponds to selecting the correct model order. Note that efficiency is defined to be 1 where the distance measures attain their minimum. Here, efficiencies and underfitting expectations are computed from 100,000 realizations of Models 9 and 10. In each realization, a new time series $Y$ is generated starting at $y_{-50}$ with $y_t = 0$ for all $t < -50$, but only observations $y_1, \ldots, y_{35}$ are kept. In Tables 5.5 and 5.6, $\text{tr}\{L_2\}$ is the trace of $L_2$ Eq. (5.6), $\det(L_2)$ is the determinant of $L_2$ Eq. (5.7), and K-L is the Kullback–Leibler distance measure given by Eq. (5.8). The $\text{tr}\{L_2\}$ and $\det(L_2)$ expected efficiencies are computed using the estimated expectation of Eq. (5.6) in Eq. (1.3) and estimated expectation of Eq. (5.7) in Eq. (1.3), respectively. K-L expected efficiency is computed from estimated expectations of Eq. (5.8) in Eq. (1.4). Figures 5.1 and 5.2 plot expected values for $L_2$ and K-L distance rather than expected efficiencies.

We first look at Table 5.5, the expected values and efficiencies for the model with strongly identifiable parameters. We see that all selection criteria attain minima at the correct order 4. As we first predicted in our analysis of the signal-to-noise ratios of AIC and HQ (Sections 5.3.1 and 5.3.5, respectively), the expectation increases for small amounts of overfitting, then decreases for excessive amounts of overfitting. We also recall that $\text{tr}\{L_2\}$ favors selection criteria that overfit slightly, and K-L favors selection criteria that underfit slightly. This can be seen in the overfitting rows $p > 4$ where $\text{tr}\{L_2\}$ efficiency is higher than either K-L or $\det(L_2)$ efficiency. Conversely, we see from its efficiency value that $\det(L_2)$ penalizes underfitting most (having the smallest efficiency values) and K-L the least (having the largest efficiency values). Det$(L_2)$ penalizes both underfitting and overfitting fairly harshly, and thus we expect $\det(L_2)$ to yield the lowest efficiency in the simulation studies in Section 5.6.

We next look at Table 5.6, the expected values and efficiencies for the model with weakly identifiable parameters. We expect the correct model to be difficult to detect, since the sample size is only 35 and the parameters are weak. In fact none of the selection criteria have a well-defined minimum at the correct order—AIC and FPE have a minimum at order 3, and most of the other selection criteria have minima at order 1. Based on the distance measures we expect to observe some underfitting, because both K-L and $\det(L_2)$ attain a minimum at some order less than 3. We noted that $\operatorname{tr}\{L_2\}$ penalizes underfitting most harshly, and in fact $\operatorname{tr}\{L_2\}$ underfits less severely than

Table 5.5. Expected values and expected efficiency for Model 9.

| $p$ | AIC | AICc | AICu | SIC | HQ | HQc | FPE | $\operatorname{tr}\{L_2\}$ | $\det(L_2)$ | K-L |
|---|---|---|---|---|---|---|---|---|---|---|
| 1 | 2.485 | 4.732 | 4.854 | 2.664 | 2.546 | 2.597 | 16.332 | 0.061 | 0.003 | 0.349 |
| 2 | 2.405 | 4.766 | 5.024 | 2.768 | 2.527 | 2.690 | 15.302 | 0.072 | 0.004 | 0.371 |
| 3 | 2.232 | 4.786 | 5.202 | 2.782 | 2.414 | 2.779 | 13.199 | 0.090 | 0.006 | 0.402 |
| 4 | **-0.348** | **2.519** | **3.116** | **0.392** | **-0.107** | **0.593** | **0.790** | 1.000 | 1.000 | 1.000 |
| 5 | -0.268 | 3.105 | 3.916 | 0.666 | 0.031 | 1.279 | 0.893 | 0.773 | 0.582 | 0.678 |
| 6 | -0.212 | 3.990 | 5.058 | 0.919 | 0.142 | 2.295 | 1.012 | 0.621 | 0.372 | 0.459 |
| 7 | -0.197 | 5.440 | 6.826 | 1.136 | 0.211 | 3.931 | 1.156 | 0.514 | 0.253 | 0.304 |
| 8 | -0.260 | 8.120 | 9.916 | 1.276 | 0.197 | 6.911 | 1.324 | 0.434 | 0.178 | 0.188 |

Boldface type indicates the minimum expectation.

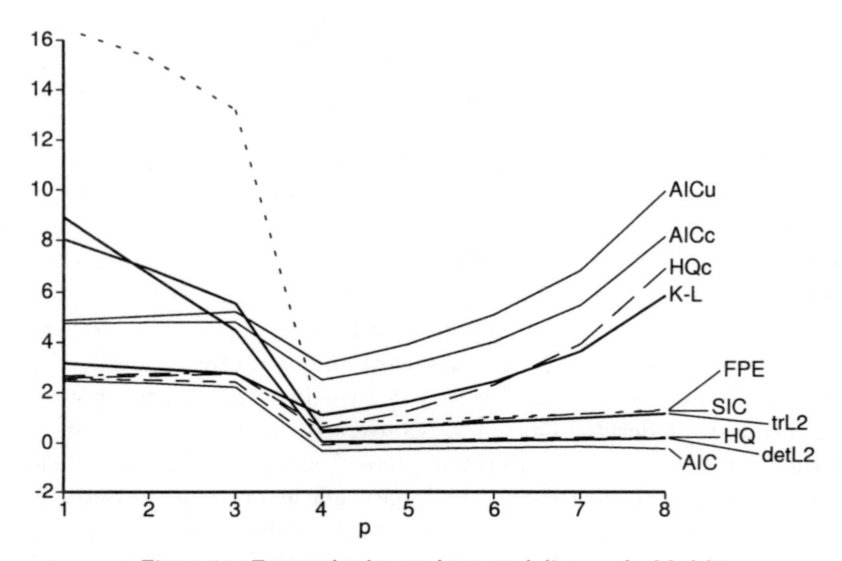


Figure 5.1. Expected values and expected distance for Model 9.

the other two measures for this model, attaining a minimum at order 3. While a criterion that performs well under K-L may also perform well in tr$\{L_2\}$ if the model is strongly identifiable and little underfitting is present, when the model is weakly identifiable a selection criterion that underfits may do well in the K-L sense but not as well in the tr$\{L_2\}$ sense. We see this for AICu, which performs well under all distance measures in Table 5.5 for the strongly identifiable model, but performs well only under K-L for the weakly identifiable model in Table 5.6.

Table 5.6. Expected values and expected efficiency for Model 10.

| $p$ | AIC | AICc | AICu | SIC | HQ | HQc | FPE | tr$\{L_2\}$ | det$(L_2)$ | K-L |
|---|---|---|---|---|---|---|---|---|---|---|
| 1 | -0.302 | **1.945** | **2.066** | **-0.123** | -0.241 | **-0.190** | 0.802 | 0.526 | 1.000 | 0.986 |
| 2 | -0.387 | 1.974 | 2.233 | -0.024 | **-0.265** | -0.101 | 0.737 | 0.886 | 0.895 | 1.000 |
| 3 | **-0.389** | 2.165 | 2.581 | 0.161 | -0.207 | 0.158 | **0.742** | 1.000 | 0.716 | 0.765 |
| 4 | -0.348 | 2.519 | 3.116 | 0.392 | -0.107 | 0.593 | 0.790 | 0.915 | 0.487 | 0.535 |
| 5 | -0.268 | 3.105 | 3.916 | 0.666 | 0.031 | 1.279 | 0.893 | 0.713 | 0.289 | 0.366 |
| 6 | -0.212 | 3.990 | 5.058 | 0.919 | 0.142 | 2.295 | 1.012 | 0.577 | 0.185 | 0.249 |
| 7 | -0.197 | 5.440 | 6.826 | 1.136 | 0.211 | 3.931 | 1.156 | 0.479 | 0.126 | 0.165 |
| 8 | -0.260 | 8.120 | 9.916 | 1.276 | 0.197 | 6.911 | 1.324 | 0.405 | 0.089 | 0.103 |

Boldface type indicates the minimum expectation.

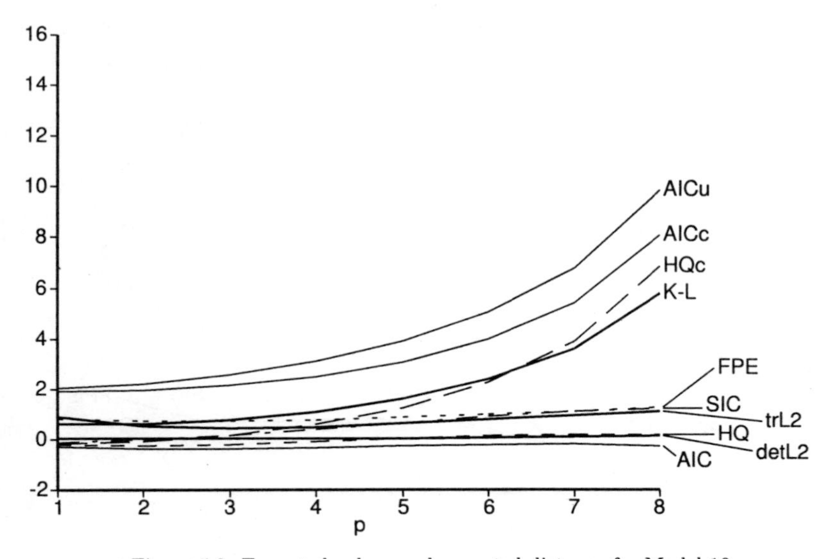


Figure 5.2. Expected values and expected distance for Model 10.

### *5.5.2. Signal-to-noise Ratios for Two Special Case Models*

We next look at the approximate signal-to-noise ratios for Models 9 and 10 for the above selection criteria. In all cases the correct order 4 is compared to candidate orders $p$, and the signal-to-noise ratio is defined to be 0 when comparing the correct model to itself. Note that underfitted signal-to-noise ratios were simulated on the basis of 100,000 realizations, except for FPE in which all signal-to-noise ratios were simulated. Note also that for overfitting when $q = 2$, the noise for log|SPE|-based model selection criteria is

$$\sqrt{\sum_{i=1}^{2} \psi'\left(\frac{n-(q+1)(p_*+L)-2+i}{2}\right) - \psi'\left(\frac{n-(q+1)p_*-2+i}{2}\right)}.$$

Table 5.7. Approximate signal-to-noise ratios for Model 9.

| $p$ | AIC | AICc | AICu | SIC | HQ | HQc | FPE |
|---|---|---|---|---|---|---|---|
| 1 | 3.536 | 2.761 | 2.166 | 2.835 | 3.311 | 2.500 | 1.089 |
| 2 | 3.482 | 2.841 | 2.412 | 3.004 | 3.331 | 2.651 | 1.045 |
| 3 | 3.291 | 2.890 | 2.658 | 3.047 | 3.215 | 2.787 | 0.995 |
| 4 | 0 | 0 | 0 | 0 | 0 | 0 | 0 |
| 5 | 0.485 | 3.526 | 4.815 | 1.654 | 0.832 | 4.131 | 0.668 |
| 6 | 0.532 | 5.753 | 7.597 | 2.064 | 0.975 | 6.660 | 0.809 |
| 7 | 0.438 | 8.431 | 10.709 | 2.148 | 0.918 | 9.635 | 0.854 |
| 8 | 0.194 | 12.279 | 14.908 | 1.939 | 0.666 | 13.852 | 0.830 |

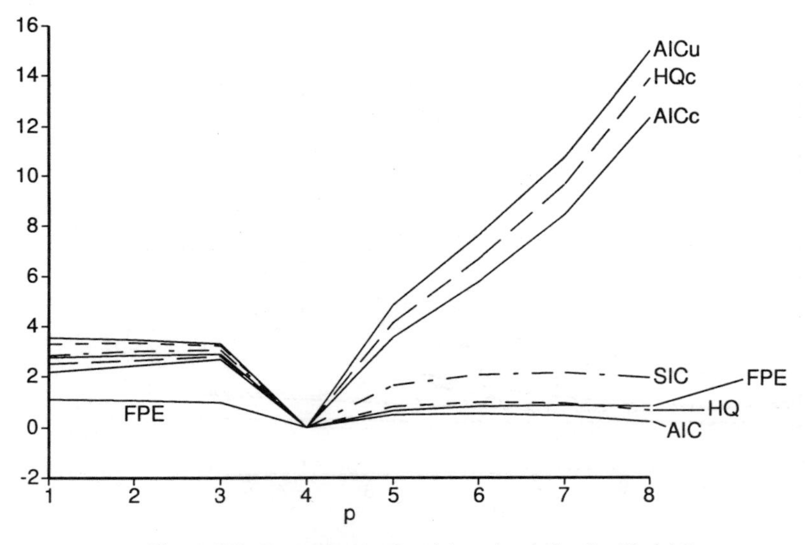


Figure 5.3. Approximate signal-to-noise ratios for Model 9.

In general we observe that the larger the expected values (see Tables 5.5 and 5.6), the larger the signal-to-noise ratios (see Tables 5.7 and 5.8). Table 5.7 shows that AICu has the highest signal-to-noise ratio with respect to overfitting, thus it should overfit less than the other criteria in this scenario. However, it also has one of the smallest signal-to-noise ratios with respect to underfitting due to its large penalty function. Furthermore, AICc and HQc have large overfitting signal-to-noise ratios, and AIC, HQ, and FPE, which are known to overfit, have weak overfitting signal-to-noise ratios. SIC has a moderate signal-to-noise ratio.

A clear pattern emerges from Figure 5.3. The criteria AICc, AICu and

Table 5.8. Approximate signal-to-noise ratios for Model 10.

| $p$ | AIC | AICc | AICu | SIC | HQ | HQc | FPE |
|---|---|---|---|---|---|---|---|
| 1 | 0.170 | -1.941 | -3.560 | -1.738 | -0.443 | -2.652 | 0.061 |
| 2 | -0.144 | -2.249 | -3.656 | -1.712 | -0.640 | -2.871 | -0.260 |
| 3 | -0.215 | -2.046 | -3.108 | -1.328 | -0.560 | -2.522 | -0.322 |
| 4 | 0 | 0 | 0 | 0 | 0 | 0 | 0 |
| 5 | 0.485 | 3.526 | 4.815 | 1.654 | 0.832 | 4.131 | 0.668 |
| 6 | 0.532 | 5.753 | 7.597 | 2.064 | 0.975 | 6.660 | 0.809 |
| 7 | 0.438 | 8.431 | 10.709 | 2.148 | 0.918 | 9.635 | 0.854 |
| 8 | 0.194 | 12.279 | 14.908 | 1.939 | 0.666 | 13.852 | 0.830 |

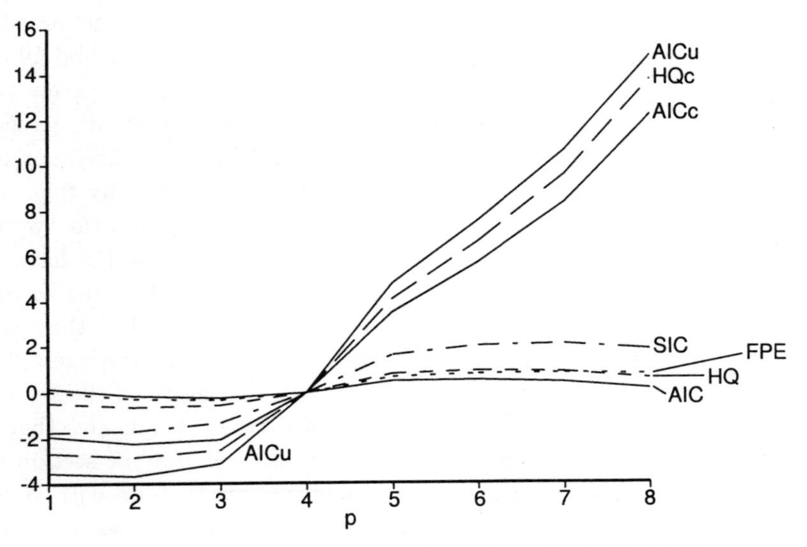


Figure 5.4. Approximate signal-to-noise ratios for Model 10.

HQc, with strong penalty functions that increase with $p$, turn out to have much larger signal-to-noise ratios than the other criteria. As discussed earlier, the signal-to-noise ratio of AICu is larger than the signal-to-noise ratio for AICc in the overfitting case. With a sample size of 35 and effective sample size being greater than 25, the penalty function of HQc is greater than the penalty function of AICc, and hence HQc has a slightly larger signal-to-noise ratio. By contrast AIC, FPE, HQ, and SIC all have similar penalty function structures that are conducive to overfitting in small samples, and consequently result in weak signal-to-noise ratios. However, of these four SIC is the best. Since the VAR(4) parameters are easily identifiable in Model 9, the underfitting signal-to-noise ratios are all large except for FPE.

In contrast to Model 9, Model 10 has weak parameters. Table 5.8 shows all the selection criteria have negative signal-to-noise ratios for underfitted candidate models. AICu has the strongest negative signal-to-noise ratio, thus we expect it to underfit more than the other selection criteria. As we have noted before, the strong penalty function in AICu discourages overfitting at the expense of some underfitting when the parameters are weak. AIC has a negative underfitting signal-to-noise ratio as well as a weak overfitting signal-to-noise ratio, thus AIC is expected to underfit and occasionally overfit excessively. Since Model 10 has the same true model order as Model 9, both have the same signal-to-noise ratios for overfitting.

Figure 5.4 contrasts nicely with Figure 5.3 in terms of underfitting. The two figures are identical with respect to overfitting since Models 9 and 10 differ only in the relative size of their parameters. Due to the weak parameters in Model 10, almost all of the signal-to-noise ratios for underfitting are negative and the minimum expected values for all criteria are at an order less than the true order 4. AICc, AICc, and HQc have strong penalty functions that prevent overfitting at the expense of some underfitting. In other words, these criteria have the strongest overfitting signal-to-noise ratios, but have the weakest signal-to-noise ratios in underfitting. In contrast, AIC, FPE, SIC, and HQ criteria have weak overfitting signal-to-noise ratios, but they have stronger (but still negative) underfitting signal-to-noise ratios. Of these, SIC has the strongest overfitting signal-to-noise ratio and it also has the weakest underfitting signal-to-noise ratio. The pattern in Figure 5.4 indicates that all the criteria should underfit somewhat, but overfitting still may be seen in the criteria with weak penalty functions. In conclusion, the criteria with strong penalty functions should underfit but not overfit, and those with weak penalty functions will sometimes underfit and sometimes overfit.

### 5.5.3. Probabilities for Two Special Case Models

Finally, we will look at the approximate probabilities for selecting the candidate order $p$ over the correct order 4 for our special case models to see if they are what we expect based on the above results. Underfitting probabilities are only approximate, since they are simulated from the same 100,000 realizations used to estimate the underfitting signal-to-noise ratios. Probabilities for selecting the correct order over itself are undefined, and are denoted by an asterisk. Overfitting probabilities are computed using the special case form of the U-statistic for $q = 2$:

$$P\{U_{2,2L,n-3p_*-3L} < u\} = P\left\{F_{4L,2(n-3p_*-3L)} > \frac{1-\sqrt{u}}{\sqrt{u}}\frac{n-3p_*-3L-1}{2L}\right\}.$$

We can see from Tables 5.9 and 5.10 that the probabilities do indeed mirror the signal-to-noise ratios. Large signal-to-noise ratios result in small probabilities of selecting the incorrect order, and weak signal-to-noise ratios result in moderate probabilities. Strongly negative signal-to-noise ratios give probabilities of almost 1 for selecting an incorrect model order.

Table 5.9. Probability of selecting order $p$ over the true order 4 for Model 9.

| $p$ | AIC | AICc | AICu | SIC | HQ | HQc | FPE |
|---|---|---|---|---|---|---|---|
| 1 | 0.000 | 0.001 | 0.011 | 0.001 | 0.000 | 0.004 | 0.000 |
| 2 | 0.000 | 0.001 | 0.004 | 0.000 | 0.000 | 0.001 | 0.000 |
| 3 | 0.000 | 0.000 | 0.001 | 0.000 | 0.000 | 0.000 | 0.000 |
| 4 | * | * | * | * | * | * | * |
| 5 | 0.262 | 0.006 | 0.001 | 0.068 | 0.180 | 0.003 | 0.217 |
| 6 | 0.264 | 0.000 | 0.000 | 0.037 | 0.158 | 0.000 | 0.183 |
| 7 | 0.297 | 0.000 | 0.000 | 0.030 | 0.171 | 0.000 | 0.167 |
| 8 | 0.384 | 0.000 | 0.000 | 0.041 | 0.234 | 0.000 | 0.172 |

For the model with strongly identifiable parameters, all the selection criteria have small probabilities of underfitting, and the most significant differences lie in their overfitting probabilities. AICc, AICu, and HQc have the strongest penalty functions, the largest signal-to-noise ratios with respect to overfitting, and consequently the smallest probabilities of overfitting. Furthermore, as the potential for selecting an overfitted model order increases, the probabilities of overfitting decrease. By comparison AIC, HQ, and FPE have large probabilities of overfitting that do not decrease as the amount of overfitting increases. Thus we expect AIC, HQ, and FPE to perform much worse than AICc, AICu, and HQc when extra, irrelevant variables are included in the study. In the next

Section we will evaluate the performance of our selection criteria using simulation results with these same special case VAR models, and compare them to the outcomes we expect based on the above theoretical conclusions.

Table 5.10. Probability of selecting order $p$ over the true order 4 for Model 10.

| $p$ | AIC | AICc | AICu | SIC | HQ | HQc | FPE |
|---|---|---|---|---|---|---|---|
| 1 | 0.462 | 0.963 | 0.998 | 0.948 | 0.693 | 0.990 | 0.493 |
| 2 | 0.601 | 0.975 | 0.998 | 0.941 | 0.761 | 0.991 | 0.634 |
| 3 | 0.651 | 0.960 | 0.991 | 0.897 | 0.753 | 0.979 | 0.679 |
| 4 | * | * | * | * | * | * | * |
| 5 | 0.262 | 0.006 | 0.001 | 0.068 | 0.180 | 0.003 | 0.217 |
| 6 | 0.264 | 0.000 | 0.000 | 0.037 | 0.158 | 0.000 | 0.183 |
| 7 | 0.297 | 0.000 | 0.000 | 0.030 | 0.171 | 0.000 | 0.167 |
| 8 | 0.384 | 0.000 | 0.000 | 0.041 | 0.234 | 0.000 | 0.172 |

## 5.6. Vector Autoregressive Monte Carlo Study

In this Section we will use the two special case VAR models from the previous Section, Models 9 and 10 ( Eqs. (5.18)–(5.20)), to examine the performance of model selection criteria in a simulation setting. Ten thousand realizations were generated, each with a new $w$ error matrix and hence a new $Y$ (starting at $y_{-50}$ with $y_t = 0$ for all $t < -50$, but only observations $y_1, \ldots, y_{35}$ are kept), and the selection criteria selected a model for each realization. The $\text{tr}\{L_2\}$, $\det(L_2)$, and K-L observed efficiencies were computed for each candidate model, and the observed efficiency of each selected model then computed under each of these three measures. The $\text{tr}\{L_2\}$ observed efficiency is computed from Eq. (5.7) and Eq. (1.1), $\det(L_2)$ observed efficiency from Eq. (5.8) and Eq. (1.1), and K-L observed efficiency is computed from Eq. (5.8) and Eq. (1.2). Averages, medians and standard deviations were next computed for the 10,000 observed efficiencies, and the criteria are ranked according to their average observed efficiency. The criteria with the highest observed efficiency is given rank 1 (best), the criteria with the lowest observed efficiency is given rank 7 (worst). Tied criteria are awarded the best rank from among the tied group. We also present counts of the number of times each candidate order was selected by each criterion out of 10,000 trials. As for multivariate regression in Chapter 4, the least squares method, conditioning on the past, was used to obtain parameter estimates. However, unlike multivariate regression, increasing the order by one for a VAR model results in increasing the number of parameters by $q^2$. Because parameter counts for VAR models increase very rapidly as the order of the candidate model increases, the probability of overfitting decreases expeditiously. However, criteria with strong penalty functions to prevent overfitting can tend to underfit as a result. We see this from the behavior of AICu, HQc, and AICc, all of which are designed to reduce small-sample overfitting. The true model is of order 4, and unlike the all-subsets regression used in Monte Carlo studies of Chapters 2 and 4, the candidate models are nested. Table 5.11 gives the results for Model 9, the strongly identifiable model. Table 5.12 summarizes the results for Model 10.

Because the VAR(4) component is strong in Model 9, the counts in Table 5.11 will be a meaningful measure of performance. By looking at the three scalar distance measures, we can estimate which model is closest to the correct one as identified by each distance measure. For both $L_2$ measures the closest model is never less than the true order, but for K-L, the closest model

Table 5.11. Simulation Results for Model 9. Counts and observed efficiency.

| | | | | counts | | | | | | |
|---|---|---|---|---|---|---|---|---|---|---|
| $p$ | AIC | AICc | AICu | SIC | HQ | HQc | FPE | tr$\{L_2\}$ | det$(L_2)$ | K-L |
| 1 | 0 | 10 | 104 | 7 | 1 | 38 | 0 | 0 | 0 | 55 |
| 2 | 0 | 11 | 15 | 6 | 1 | 13 | 1 | 0 | 0 | 78 |
| 3 | 0 | 5 | 5 | 1 | 0 | 4 | 0 | 0 | 0 | 166 |
| 4 | 4464 | 9906 | 9864 | 8743 | 6059 | 9918 | 6041 | 9326 | 9570 | 9559 |
| 5 | 1062 | 67 | 12 | 551 | 987 | 27 | 1322 | 528 | 370 | 137 |
| 6 | 796 | 1 | 0 | 206 | 622 | 0 | 845 | 119 | 54 | 5 |
| 7 | 950 | 0 | 0 | 166 | 635 | 0 | 715 | 14 | 4 | 0 |
| 8 | 2728 | 0 | 0 | 320 | 1695 | 0 | 1076 | 13 | 2 | 0 |
| | | | | K-L observed efficiency | | | | | | |
| | AIC | AICc | AICu | SIC | HQ | HQc | FPE | tr$\{L_2\}$ | det$(L_2)$ | K-L |
| ave | 0.576 | 0.989 | 0.988 | 0.901 | 0.696 | 0.990 | 0.712 | 0.982 | 0.989 | 1.000 |
| med | 0.510 | 1.000 | 1.000 | 1.000 | 1.000 | 1.000 | 1.000 | 1.000 | 1.000 | 1.000 |
| sd | 0.396 | 0.073 | 0.074 | 0.255 | 0.385 | 0.064 | 0.368 | 0.075 | 0.055 | 0.000 |
| rank | 7 | 2 | 3 | 4 | 6 | 1 | 5 | | | |
| | | | | tr$\{L_2\}$ observed efficiency | | | | | | |
| | AIC | AICc | AICu | SIC | HQ | HQc | FPE | tr$\{L_2\}$ | det$(L_2)$ | K-L |
| ave | 0.706 | 0.989 | 0.984 | 0.933 | 0.792 | 0.989 | 0.805 | 1.000 | 0.996 | 0.971 |
| med | 0.731 | 1.000 | 1.000 | 1.000 | 1.000 | 1.000 | 1.000 | 1.000 | 1.000 | 1.000 |
| sd | 0.291 | 0.060 | 0.094 | 0.177 | 0.276 | 0.068 | 0.264 | 0.000 | 0.022 | 0.142 |
| rank | 7 | 1 | 3 | 4 | 6 | 1 | 5 | | | |
| | | | | det$(L_2)$ observed efficiency | | | | | | |
| | AIC | AICc | AICu | SIC | HQ | HQc | FPE | tr$\{L_2\}$ | det$(L_2)$ | K-L |
| ave | 0.571 | 0.988 | 0.982 | 0.902 | 0.696 | 0.987 | 0.708 | 0.989 | 1.000 | 0.966 |
| med | 0.478 | 1.000 | 1.000 | 1.000 | 1.000 | 1.000 | 1.000 | 1.000 | 1.000 | 1.000 |
| sd | 0.400 | 0.081 | 0.114 | 0.253 | 0.386 | 0.086 | 0.373 | 0.058 | 0.000 | 0.168 |
| rank | 7 | 1 | 3 | 4 | 6 | 2 | 5 | | | |

occasionally is of lower order. This may explain AICu's relatively mediocre performance, with 104 selections of order 1. The consistent criterion HQc has the highest count, correctly identifying the correct order 9918 times out of 10,000 or 99% of the time, outperforming both SIC and HQ. AICc performs comparably, also identifying the correct order 99% of the time. AIC, with the lowest count, drastically overfits with 2728 selections of order 8. We find that the pattern in the counts parallels the pattern in the signal-to-noise ratios (see Figure 5.3.) Criteria with strong penalty functions overfit very little, whereas the four criteria with weaker penalty functions overfit more severely, and some excessively. Since the model is strongly identifiable, all the signal-to-noise ratios for underfitting are large for Model 9 and little underfitting is evident in the counts. AICc, AICu, and HQc, with their strong penalty functions,

Table 5.12. Simulation Results for Model 10. Counts and observed efficiency.

| | | | | | counts | | | | | |
|---|---|---|---|---|---|---|---|---|---|---|
| $p$ | AIC | AICc | AICu | SIC | HQ | HQc | FPE | tr$\{L_2\}$ | det$(L_2)$ | K-L |
| 1 | 1641 | 5782 | 8120 | 6915 | 3378 | 6994 | 1974 | 159 | 5196 | 3582 |
| 2 | 2054 | 3452 | 1777 | 2324 | 2718 | 2677 | 2586 | 2102 | 2571 | 4747 |
| 3 | 1565 | 686 | 99 | 487 | 1417 | 306 | 1965 | 4273 | 1481 | 1484 |
| 4 | 1044 | 79 | 4 | 161 | 761 | 23 | 1287 | 3200 | 714 | 183 |
| 5 | 539 | 1 | 0 | 30 | 315 | 0 | 610 | 225 | 32 | 4 |
| 6 | 499 | 0 | 0 | 15 | 223 | 0 | 436 | 28 | 4 | 0 |
| 7 | 640 | 0 | 0 | 20 | 289 | 0 | 431 | 9 | 2 | 0 |
| 8 | 2018 | 0 | 0 | 48 | 899 | 0 | 711 | 4 | 0 | 0 |
| | | | | | K-L observed efficiency | | | | | |
| | AIC | AICc | AICu | SIC | HQ | HQc | FPE | tr$\{L_2\}$ | det$(L_2)$ | K-L |
| ave | 0.485 | 0.812 | 0.834 | 0.800 | 0.634 | 0.825 | 0.580 | 0.772 | 0.897 | 1.000 |
| med | 0.478 | 0.855 | 0.864 | 0.845 | 0.720 | 0.859 | 0.630 | 0.804 | 0.997 | 1.000 |
| sd | 0.356 | 0.192 | 0.166 | 0.208 | 0.329 | 0.177 | 0.330 | 0.209 | 0.146 | 0.000 |
| rank | 7 | 3 | 1 | 4 | 5 | 2 | 6 | | | |
| | | | | | tr$\{L_2\}$ observed efficiency | | | | | |
| | AIC | AICc | AICu | SIC | HQ | HQc | FPE | tr$\{L_2\}$ | det$(L_2)$ | K-L |
| ave | 0.608 | 0.631 | 0.552 | 0.585 | 0.634 | 0.590 | 0.669 | 1.000 | 0.720 | 0.763 |
| med | 0.598 | 0.618 | 0.506 | 0.545 | 0.638 | 0.550 | 0.692 | 1.000 | 0.739 | 0.793 |
| sd | 0.256 | 0.237 | 0.223 | 0.233 | 0.248 | 0.233 | 0.243 | 0.000 | 0.249 | 0.213 |
| rank | 4 | 3 | 7 | 6 | 2 | 5 | 1 | | | |
| | | | | | det$(L_2)$ observed efficiency | | | | | |
| | AIC | AICc | AICu | SIC | HQ | HQc | FPE | tr$\{L_2\}$ | det$(L_2)$ | K-L |
| ave | 0.398 | 0.715 | 0.773 | 0.729 | 0.545 | 0.748 | 0.477 | 0.622 | 1.000 | 0.799 |
| med | 0.266 | 0.793 | 0.917 | 0.831 | 0.551 | 0.863 | 0.425 | 0.655 | 1.000 | 0.993 |
| sd | 0.364 | 0.303 | 0.271 | 0.300 | 0.369 | 0.286 | 0.361 | 0.336 | 0.000 | 0.276 |
| rank | 7 | 4 | 1 | 3 | 5 | 2 | 6 | | | |

underfit slightly, whereas AICu underfits more than the others. Signal-to-noise ratios seem to be a good indicator of performance in terms of counts.

Table 5.11 also presents the observed efficiency results for each criterion and distance measure. At the high end, the results for each distance measure are identical except that under the $L_2$ measures AICc ranks first and HQc ranks second, while under K-L the order is reversed. Another pattern can be seen as well. The observed efficiencies for the top three ranking criteria are all close to 1 and very close to each other. However, there is a large drop in observed efficiency between these top three and the criterion ranking fourth, SIC. An even larger drop is evident between SIC and 5th ranked criterion FPE. At the low end, AIC has the lowest observed efficiency under all three measures, but since overfitting is not as heavily penalized using $\mathrm{tr}\{L_2\}$, its observed efficiency is best under this distance measure. In general, the observed efficiencies parallel the counts. Those criteria with higher counts of selecting the true model tend to have higher observed efficiency. All these patterns agree well with the signal-to-noise ratios.

For Model 9, all criteria behave nearly identically across the three distance measures. However, this is not true for Model 10. The criteria behave differently under the $\mathrm{tr}\{L_2\}$ measure than under the other two measures, mainly due to Model 10's weakly identifiable parameters, which have resulted in extensive underfitting. The counts from Table 5.12 show that none of the selection criteria identify the correct model more than 15% of the time, so clearly counts are not the best measure of performance. Thus we turn to the observed efficiency results which show relative performances of the criteria. They are quite different for Model 10 than for Model 9. In terms of K-L and $\det(L_2)$, orders 1 and 2 tend to be the closest to the true model. However, for the $\mathrm{tr}\{L_2\}$ distance the closest orders tend to be 3 and 4. This is what we would expect since we know that underfitting is much more heavily penalized by $\mathrm{tr}\{L_2\}$ than by K-L or the determinant. This can be seen in the results for AICu, which underfits with respect to the trace but not with respect to K-L. Because the results differ substantially between distance measures, we will examine each individually.

Whereas AICc and HQc were the top performers under K-L for Model 9, here AICu has the highest observed efficiency at 83%, HQc ranks second, and AICc ranks third. HQc is a consistent selection criterion, but as we can see, it is competitive with efficient selection criteria in small samples. Since the closest candidate models tend to have fewer variables and thus the opportunities for overfitting are lessened, AIC and FPE do not overfit as excessively in Model 10 as in Model 9. However, this does not redeem AIC, as it again has the lowest observed efficiency at 48.5%. As was the case for Model 9, we see that the results for $\det(L_2)$ observed efficiency are similar to those for K-L with the exception that SIC now performs better than AICc.

For $\text{tr}\{L_2\}$ observed efficiency, we see drastic differences from K-L and $\det(L_2)$, as AICu drops to last place at 55% while FPE ranks first at 67%. Since $\text{tr}\{L_2\}$ penalizes underfitting much more than overfitting, selection criteria with large penalty functions, like AICu and HQc, perform poorly under $\text{tr}\{L_2\}$. The overfitting in FPE is not penalized heavily and thus FPE performs well under $\text{tr}\{L_2\}$.

## 5.7. Summary

In this Chapter we again must deal with the fact that many of the basic components of VAR models are matrices. However, unlike multivariate regression, parameter counts increase even more rapidly in VAR models. When a single variable is added to a multivariate regression, the parameters increase the dimension of $y_i$ by $q$. In a $q$-dimensional VAR model, the parameters increase by $q^2$ when the order increases by 1, potentially a very large number. A further problem for VAR models is that while the number of parameters to be modeled increases quickly, we also lose $p$ observations for conditioning on the past. In multivariate regression with $\text{rank}(X) = p$, we had $n - p$ degrees of freedom. In a VAR model of the same order we are reduced to $n - p - qp$ degrees of freedom. This means that probabilities of overfitting are much smaller for VAR models, and consequently the importance of underfitting tendencies is emphasized. This has a particular impact on the performance of selection criteria (such as AICu) whose strong penalty functions are designed to prevent overfitting but consequently can cause underfitting for VAR models.

The role of model identifiability in VAR model selection has been examined in the same way as in previous chapters. Two models that differ only in the ease with which their parameters may be identified were used to test the criteria, and we see that the behavior patterns for VAR are the same as those for univariate, multivariate regression, and AR. Although we have noted that overfitting is less common in VAR models, it is still a problem for criteria with weak penalty functions, such as AIC. Criteria with strong penalty functions neither overfit nor underfit and are seen to identify the correct order with very high probability when the true model is strongly identifiable.

One difference between VAR models and the previous model categories considered is that for VAR we saw that there can be significant differences in a criterion's relative performance under the K-L, $\text{tr}\{L_2\}$, and $\det(L_2)$ distance measures. In earlier chapters there was a fair amount of agreement between the three observed efficiencies. For VAR models the choice of the trace or determinant for reducing $L_2$ to a scalar can make a substantial difference to criterion performance. Thus for VAR models a good criterion needs to perform well with respect to all three observed efficiency measures. The expanded simulations in Chapter 9 will explore this issue further.

Finally, there are several other VAR model selection criteria that are not addressed in this Chapter. The interested reader can refer to Lütkepohl (1985 and 1991) and Brockwell and Davis (1991, p. 432).

## Chapter 5 Appendices

### Appendix 5A. Distributional Results in the Central Case

As in multivariate regression models, the derivations of some model selection criteria, in the vector autoregressive setting, as well as their signal-to-noise corrected variants, depend on distributions of $\text{SPE}_p$; in this case, Wishart distributions. Recall that in conditional VAR($p$) models the effective sample size is $T = n - p$ and that we lose $q + 1$ degrees of freedom whenever $p$ increases by one. Assume all distributions below are central. We know that for hierarchical (or nested) models for $p > p_*$ and $L > 0$,

$$\text{SPE}_p - \text{SPE}_{p+L} \sim W_q((q+1)L, \Sigma_*), \tag{5A.1}$$

and

$$\text{SPE}_p \sim W_q(n - (q+1)p, \Sigma_*) \tag{5A.2}$$

where $q$ is the dimension of the model.

The first moment of the Wishart distribution is a $q \times q$ matrix, and the second moment is four-dimensional; a $q \times q \times q \times q$ array, $D$, with elements $d_{i,j,r,s} = cov[\text{SPE}_{ij}, \text{SPE}_{rs}]$.

The first moment is straightforward:

$$E[\text{SPE}_p] = (n - (q+1)p)\Sigma_*.$$

The second moment can be computed on an element-by-element basis. Let $\sigma_{i,j}$ represent the $i, j$th element of the covariance matrix $\Sigma_*$. Then,

$$cov[\text{SPE}_{i,j}, \text{SPE}_{r,s}] = \sigma_{ir}\sigma_{js} + \sigma_{is}\sigma_{jr}.$$

We will also need the moments of certain functions of $\text{SPE}_p$, particularly those that reduce the $\text{SPE}_p$ matrix to a scalar. The trace is one such function, and while the actual distribution of the trace is not known, the mean is

$$E[tr\{\text{SPE}_p\}] = (n - (q+1)p)tr\{\Sigma_*\}.$$

Another function used to reduce $\text{SPE}_p$ to a scalar is the determinant $|\text{SPE}_p|$. We emphasize the determinant, since most multivariate model selection criteria are functions of the determinant of $\text{SPE}_p$. The distribution of the determinant is the distribution of a product of independent $\chi^2$ random variables, or

$$|\text{SPE}_p| \sim |\Sigma_*| \prod_{i=1}^{q} \chi^2_{n-(q+1)p-q+i}. \tag{5A.3}$$

It follows that

$$E[|\text{SPE}_p|] = |\Sigma_*| \prod_{i=1}^{q} (n - (q+1)p - q + i).$$

Taking logs of Eq. (5A.3), we have

$$\log |\text{SPE}_p| \sim \log |\Sigma_*| + \sum_{i=1}^{q} \log \chi^2_{n-(q+1)p-q+i}$$

and

$$E[\log |\text{SPE}_p|] = \log |\Sigma_*| + q\log(2) + \sum_{i=1}^{q} \psi\left(\frac{n-(q+1)p-q+i}{2}\right),$$

where $\psi$ is Euler's psi function. No closed form exists, and thus we approximate this expectation with a Taylor expansion:

$$E[\log |\text{SPE}_p|] \doteq \log |\Sigma_*| + \sum_{i=1}^{q} \left[\log\bigl(n-(q+1)p-q+i\bigr) - \frac{1}{n-(q+1)p-q+i}\right].$$

However, this expression is quite messy for large values of $q$, and it offers little insight into the behavior of model selection criteria. If one assumes that $n-(q+1)p$ is much larger than $q$, then following from the fact that (see Chapter 4, Appendix 4A with $k=(q+1)p$)

$$\sum_{i=1}^{q} \log(n-(q+1)p-q+i) \doteq q\log(n-(q+1)p-(q-1)/2)$$

and

$$\sum_{i=1}^{q} \frac{1}{n-(q+1)p-q+i} \doteq \frac{q}{n-(q+1)p-(q-1)/2},$$

we can make a further simplification:

$$E[\log |\text{SPE}_p|] \doteq \log |\Sigma_*| + q \log\bigl(n - (q+1)p - (q-1)/2\bigr) - \frac{q}{n - (q+1)p - (q-1)/2}. \tag{5A.4}$$

The distribution of differences between $\log |\text{SPE}_p|$ and $\log |\text{SPE}_{p+L}|$ is more complicated. Note that from Eqs. (5A.1)–(5A.2)

$$\frac{|\text{SPE}_{p+L}|}{|\text{SPE}_p|} \sim \prod_{i=1}^{q} \text{Beta}\left(\frac{n-(q+1)(p+L)-q+i}{2}, \frac{(q+1)L}{2}\right),$$

where the Beta distributions are independent. Taking logs,

$$\log\left(\frac{|\text{SPE}_{p+L}|}{|\text{SPE}_p|}\right) \sim \sum_{i=1}^{q} \text{log-Beta}\left(\frac{n-(q+1)(p+L)-q+i}{2}, \frac{(q+1)L}{2}\right).$$

Applying facts from Chapter 2 Appendix 2A,

$$var\left[\log\left(\frac{|\text{SPE}_{p+L}|}{|\text{SPE}_p|}\right)\right] = \sum_{i=1}^{q}\left[\psi'\left(\frac{n-(q+1)(p+L)-q+i}{2}\right) - \psi'\left(\frac{n-(q+1)p-q+i}{2}\right)\right],$$

which has no closed form. Applying the first order Taylor expansion to each log-Beta term yields a convenient approximation:

$$var\left[\log\left(\frac{|\text{SPE}_{p+L}|}{|\text{SPE}_p|}\right)\right] \doteq \sum_{i=1}^{q} \frac{2(q+1)L}{(n-(q+1)(p+L)-q+i)(n-(q+1)p-q+i)}.$$

Once again, this expression is not very useful for studying properties of multivariate model selection criteria. If we make a simplification similar to the one used in Eq. (5A.4) we find

$$var\left[\log\left(\frac{|\text{SPE}_{p+L}|}{|\text{SPE}_p|}\right)\right] \doteq \frac{2Lq(q+1)}{(n-(q+1)(p+L)-(q-1)/2)(n-(q+1)p-(q-1)/2+2)},$$

and the standard deviation becomes

$$sd\left[\log\left(\frac{|\text{SPE}_{p+L}|}{|\text{SPE}_p|}\right)\right] \doteq \frac{\sqrt{2Lq(q+1)}}{\sqrt{(n-(q+1)(p+L)-(q-1)/2)(n-(q+1)p-(q-1)/2+2)}}. \tag{5A.5}$$

Of course, Eq. (5A.4) and Eq. (5A.5) are approximate; however, they have the advantage of being simplified algebraic expressions for moments of the model selection criteria that are very similar to univariate Taylor expansions (when $q = 1$, the multivariate approximations are equivalent to the univariate approximations from Chapter 3). Also, these approximations improve asymptotically.

## Appendix 5B. Small-sample Probabilities of Overfitting

Small-sample probabilities of overfitting for VAR models can be expressed in terms of independent $\chi^2$ distributions or in terms of the U distribution. When converting from $n\sum_{i=1}^{q}\log(1+\chi^2_{(q+1)L}/\chi^2_{n-(q+1)(p_*+L)-q+i})$ probabilities to U probabilities, we use an exponential transformation to remove the log term and express everything in terms of $|\mathrm{SPE}_{p_*+L}|/|\mathrm{SPE}_{p_*}|$.

### *5B.1. AICc*

Rewriting AICc in terms of the original sample size $n$, we have $\mathrm{AICc}_p = \log|\hat{\Sigma}_p| + (n+(q-1)p)q/(n-(q+1)p-q-1)$. AICc overfits if $\mathrm{AICc}_{p*+L} < \mathrm{AICc}_{p_*}$. For finite $n$, the probability that AICc prefers the overfitted model $p_* + L$ is

$$
\begin{aligned}
&P\{\mathrm{AICc}_{p_*+L} < \mathrm{AICc}_{p_*}\}\\
&= P\left\{\log|\hat{\Sigma}_{p_*+L}| + \frac{nq - q(q-1)(p_*+L)}{n-(q+1)(p_*+L)-q-1}\right.\\
&\qquad \left. < \log|\hat{\Sigma}_{p_*}| + \frac{nq-q(q-1)p_*}{n-(q+1)p_*-q-1}\right\}\\
&= P\left\{\log\left(\frac{|\mathrm{SPE}_{p_*+L}|}{|\mathrm{SPE}_{p_*}|}\right) < -q\log\left(\frac{n-p_*}{n-p_*-L}\right)\right.\\
&\qquad \left. + \frac{nq-q(q-1)p_*}{n-(q+1)p_*-q-1} - \frac{nq-q(q-1)(p_*+L)}{n-(q+1)(p_*+L)-q-1}\right\}\\
&= P\left\{\sum_{i=1}^{q}\text{log-Beta}\left(\frac{n-(q+1)(p_*+L)-q+i}{2}, \frac{(q+1)L}{2}\right)\right.\\
&\qquad < -q\log\left(\frac{n-p_*}{n-p_*-L}\right)\\
&\qquad \left. - \frac{2Lq(qn-(q-1)(q+1))}{(n-(q+1)(p_*+L)-q-1)(n-(q+1)p_*-q-1)}\right\}.
\end{aligned}
$$

Expressed in terms of chi-square distributions, the $P\{\text{overfit}\}$ for AICc is

$$P\left\{n\sum_{i=1}^{q}\log\left(1+\frac{\chi^2_{(q+1)L}}{\chi^2_{n-(q+1)(p_*+L)-q+i}}\right) > qn\log\left(\frac{n-p_*}{n-p_*-L}\right) + \frac{2Lqn(qn-(q-1)(q+1))}{(n-(q+1)(p_*+L)-q-1)(n-(q+1)p_*-q-1)}\right\}.$$

Expressed in terms of $U_{q,qL,n-(q+1)(p_*+L)}$, the $P\{\text{overfit}\}$ for AICc is

$$P\left\{U_{q,qL,n-(q+1)(p_*+L)} < \left(\frac{n-p_*-L}{n-p_*}\right)^q \times \exp\left(-\frac{2Lq(qn-(q-1)(q+1))}{(n-(q+1)(p_*+L)-q-1)(n-(q+1)p_*-q-1)}\right)\right\}.$$

### *5B.2. AICu*

Rewriting AICu in terms of the original sample size $n$, we have $\text{AICu}_p = \log|S_p| + (n+(q-1)p)q/(n-(q+1)p-q-1)$. AICu overfits if $\text{AICu}_{p*+L} < \text{AICu}_{p_*}$. For finite $n$, the probability that AICu prefers the overfitted model $p_*+L$ is

$$\begin{aligned}
&P\{\text{AICu}_{p_*+L} < \text{AICu}_{p_*}\}\\
&= P\left\{\log|S^2_{p_*+L}| + \frac{nq-q(q-1)(p_*+L)}{n-(q+1)(p_*+L)-q-1} < \log|S^2_{p_*}| + \frac{nq-q(q-1)p_*}{n-(q+1)p_*-q-1}\right\}\\
&= P\left\{\log\left(\frac{|\text{SPE}_{p_*+L}|}{|\text{SPE}_{p_*}|}\right) < -q\log\left(\frac{n-(q+1)p_*}{n-(q+1)(p_*+L)}\right) + \frac{nq-q(q-1)p_*}{n-(q+1)p_*-q-1} - \frac{nq-q(q-1)(p_*+L)}{n-(q+1)(p_*+L)-q-1}\right\}\\
&= P\left\{\sum_{i=1}^{q}\text{log-Beta}\left(\frac{n-(q+1)(p_*+L)-q+i}{2}, \frac{(q+1)L}{2}\right) < -q\log\left(\frac{n-(q+1)p_*}{n-(q+1)(p_*+L)}\right) - \frac{2Lq(qn-(q-1)(q+1))}{(n-(q+1)(p_*+L)-q-1)(n-(q+1)p_*-q-1)}\right\}.
\end{aligned}$$

Expressed in terms of chi-square distributions, the $P\{\text{overfit}\}$ for AICu is

$$P\Big\{n\sum_{i=1}^{q}\log\Big(1+\frac{\chi^2_{(q+1)L}}{\chi^2_{n-(q+1)(p_*+L)-q+i}}\Big) > qn\log\left(\frac{n-(q+1)p_*}{n-(q+1)(p_*+L)}\right) + \frac{2Lqn(qn-(q-1)(q+1))}{(n-(q+1)(p_*+L)-q-1)(n-(q+1)p_*-q-1)}\Big\}.$$

Expressed in terms of $U_{q,qL,n-(q+1)(p_*+L)}$, the $P\{\text{overfit}\}$ for AICu is

$$P\Big\{U_{q,qL,n-(q+1)(p_*+L)} < \left(\frac{n-(q+1)(p_*+L)}{n-(q+1)p_*}\right)^q \times \exp\left(-\frac{2Lq(qn-(q-1)(q+1))}{(n-(q+1)(p_*+L)-q-1)(n-(q+1)p_*-q-1)}\right)\Big\}.$$

### *5B.3. FPE*

Rewriting FPE in terms of the original sample size $n$, we have $\text{FPE}_p = |\hat{\Sigma}_p|((n+(q-1)p)/(n-(q+1)p))^q$. FPE overfits if $\text{FPE}_{p*+L} < \text{FPE}_{p_*}$. For finite $n$, the probability that FPE prefers the overfitted model $p_*+L$ is

$$\begin{aligned}
&P\{\text{FPE}_{p_*+L} < \text{FPE}_{p_*}\}\\
&= P\Big\{\log|\hat{\Sigma}_{p_*+L}| + q\log\left(\frac{n+(q-1)(p_*+L)}{n-(q+1)(p_*+L)}\right)\\
&\qquad < \log|\hat{\Sigma}_{p_*}| + q\log\left(\frac{n+(q-1)p_*}{n-(q+1)p_*}\right)\Big\}\\
&= P\Big\{\log\left(\frac{|\text{SPE}_{p_*+L}|}{|\text{SPE}_{p_*}|}\right)\\
&\qquad < -q\log\left(\frac{(n+(q-1)(p_*+L))(n-(q+1)p_*)(n-p_*)}{(n+(q-1)p_*)(n-(q+1)(p_*+L))(n-p_*-L)}\right)\Big\}\\
&= P\Big\{\sum_{i=1}^{q}\text{log-Beta}\left(\frac{n-(q+1)(p_*+L)-q+i}{2},\frac{(q+1)L}{2}\right) <\\
&\qquad -q\log\left(\frac{(n+(q-1)(p_*+L))(n-(q+1)p_*)(n-p_*)}{(n+(q-1)p_*)(n-(q+1)(p_*+L))(n-p_*-L)}\right)\Big\}.
\end{aligned}$$

Expressed in terms of chi-square distributions, the $P\{\text{overfit}\}$ for FPE is

$$P\Big\{n\sum_{i=1}^{q}\log\Big(1+\frac{\chi^2_{(q+1)L}}{\chi^2_{n-(q+1)(p_*+L)-q+i}}\Big) > qn\log\left(\frac{(n+(q-1)(p_*+L))(n-(q+1)p_*)(n-p_*)}{(n+(q-1)p_*)(n-(q+1)(p_*+L))(n-p_*-L)}\right)\Big\}.$$

Expressed in terms of $U_{q,qL,n-(q+1)(p_*+L)}$, the

$$P\left\{U_{q,qL,n-(q+1)(p_*+L)} < \left(\frac{(n+(q-1)(p_*+L))(n-(q+1)p_*)(n-p_*)}{(n+(q-1)p_*)(n-(q+1)(p_*+L))(n-p_*-L)}\right)^q\right\}.$$

### *5B.4. SIC*

Rewriting SIC in terms of the original sample size $n$, we have $\text{SIC}_p = \log|\hat{\Sigma}_p| + \log(n-p)pq^2/(n-p)$. SIC overfits if $\text{SIC}_{p*+L} < \text{SIC}_{p_*}$. For finite $n$, the probability that SIC prefers the overfitted model $p_* + L$ is

$$\begin{aligned}
&P\{\text{SIC}_{p_*+L} < \text{SIC}_{p_*}\} \\
&= P\left\{\log|\hat{\Sigma}_{p_*+L}| + \frac{\log(n-p_*-L)(p_*+L)q^2}{n-p_*-L}\right. \\
&\qquad \left. < \log|\hat{\Sigma}_{p_*}| + \frac{\log(n-p_*)p_*q^2}{n-p_*}\right\} \\
&= P\left\{\log\left(\frac{|\text{SPE}_{p_*+L}|}{|\text{SPE}_{p_*}|}\right) < -q\log\left(\frac{n-p_*}{n-p_*-L}\right)\right. \\
&\qquad \left. + \frac{\log(n-p_*)p_*q^2}{n-p_*} - \frac{\log(n-p_*-L)(p_*+L)q^2}{n-p_*-L}\right\} \\
&= P\left\{\sum_{i=1}^{q}\text{log-Beta}\left(\frac{n-(q+1)(p_*+L)-q+i}{2}, \frac{(q+1)L}{2}\right)\right. \\
&\qquad < -q\log\left(\frac{n-p_*}{n-p_*-L}\right) \\
&\qquad \left. + \frac{\log(n-p_*)p_*q^2}{n-p_*} - \frac{\log(n-p_*-L)(p_*+L)q^2}{n-p_*-L}\right\}.
\end{aligned}$$

Expressed in terms of chi-square distributions, the $P\{\text{overfit}\}$ for SIC is

$$\begin{aligned}
P\Bigg\{n\sum_{i=1}^{q}\log\left(1+\frac{\chi^2_{(q+1)L}}{\chi^2_{n-(q+1)(p_*+L)-q+i}}\right) &> qn\log\left(\frac{n-p_*}{n-p_*-L}\right) \\
&- \frac{n\log(n-p_*)p_*q^2}{n-p_*} + \frac{n\log(n-p_*-L)(p_*+L)q^2}{n-p_*-L}\Bigg\}.
\end{aligned}$$

Expressed in terms of $U_{q,qL,n-(q+1)(p_*+L)}$, the $P\{\text{overfit}\}$ for SIC is

$$\begin{aligned}
P\Bigg\{U_{q,qL,n-(q+1)(p_*+L)} &< \left(\frac{n-p_*-L}{n-p_*}\right)^q \times \\
&\exp\left(\frac{\log(n-p_*)p_*q^2}{n-p_*} - \frac{\log(n-p_*-L)(p_*+L)q^2}{n-p_*-L}\right)\Bigg\}.
\end{aligned}$$

*5B.5. HQ*

Rewriting HQ in terms of the original sample size $n$, we have $\text{HQ}_p = \log|\hat{\Sigma}_p| + 2\log\log(n-p)pq^2/(n-p)$. HQ overfits if $\text{HQ}_{p*+L} < \text{HQ}_{p_*}$. For finite $n$, the probability that HQ prefers the overfitted model $p_* + L$ is

$$\begin{aligned}
&P\{\text{HQ}_{p_*+L} < \text{HQ}_{p_*}\} \\
&= P\bigg\{\log|\hat{\Sigma}_{p_*+L}| + \frac{2\log\log(n-p_*-L)(p_*+L)q^2}{n-p_*-L} \\
&\qquad < \log|\hat{\Sigma}_{p_*}| + \frac{2\log\log(n-p_*)p_*q^2}{n-p_*}\bigg\} \\
&= P\bigg\{\log\left(\frac{|\text{SPE}_{p_*+L}|}{|\text{SPE}_{p_*}|}\right) < -q\log\left(\frac{n-p_*}{n-p_*-L}\right) \\
&\qquad + \frac{2\log\log(n-p_*)p_*q^2}{n-p_*} - \frac{2\log\log(n-p_*-L)(p_*+L)q^2}{n-p_*-L}\bigg\} \\
&= P\bigg\{\sum_{i=1}^{q}\text{log-Beta}\left(\frac{n-(q+1)(p_*+L)-q+i}{2}, \frac{(q+1)L}{2}\right) \\
&\qquad < -q\log\left(\frac{n-p_*}{n-p_*-L}\right) \\
&\qquad + \frac{2\log\log(n-p_*)p_*q^2}{n-p_*} - \frac{2\log\log(n-p_*-L)(p_*+L)q^2}{n-p_*-L}\bigg\}.
\end{aligned}$$

Expressed in terms of chi-square distributions, the $P\{\text{overfit}\}$ for HQ is

$$\begin{aligned}
&P\bigg\{n\sum_{i=1}^{q}\log\left(1+\frac{\chi^2_{(q+1)L}}{\chi^2_{n-(q+1)(p_*+L)-q+i}}\right) > qn\log\left(\frac{n-p_*}{n-p_*-L}\right) \\
&\qquad - \frac{n2\log\log(n-p_*)p_*q^2}{n-p_*} + \frac{n2\log\log(n-p_*-L)(p_*+L)q^2}{n-p_*-L}\bigg\}.
\end{aligned}$$

Expressed in terms of $U_{q,qL,n-(q+1)(p_*+L)}$, the $P\{\text{overfit}\}$ for HQ is

$$\begin{aligned}
&P\bigg\{U_{q,qL,n-(q+1)(p_*+L)} < \left(\frac{n-p_*-L}{n-p_*}\right)^q \times \\
&\qquad \exp\left(\frac{2\log\log(n-p_*)p_*q^2}{n-p_*} - \frac{2\log\log(n-p_*-L)(p_*+L)q^2}{n-p_*-L}\right)\bigg\}.
\end{aligned}$$

## 5B.6. HQc

Rewriting HQc in terms of the original sample size $n$, we have $\mathrm{HQc}_p = \log|\hat{\Sigma}_p| + 2\log\log(n-p)pq^2/(n-(q+1)p-q-1)$. HQc overfits if $\mathrm{HQc}_{p*+L} < \mathrm{HQc}_{p_*}$. For finite $n$, the probability that HQc prefers the overfitted model $p_* + L$ is

$$
\begin{aligned}
&P\{\mathrm{HQc}_{p_*+L} < \mathrm{HQc}_{p_*}\}\\
&= P\bigg\{\log|\hat{\Sigma}_{p_*+L}| + \frac{2\log\log(n-p_*-L)(p_*+L)q^2}{n-(q+1)(p_*+L)-q-1}\\
&\qquad < \log|\hat{\Sigma}_{p_*}| + \frac{2\log\log(n-p_*)p_*q^2}{n-(q+1)p_*-q-1}\bigg\}\\
&= P\bigg\{\log\left(\frac{|\mathrm{SPE}_{p_*+L}|}{|\mathrm{SPE}_{p_*}|}\right) < -q\log\left(\frac{n-p_*}{n-p_*-L}\right)\\
&\qquad + \frac{2\log\log(n-p_*)p_*q^2}{n-(q+1)p_*-q-1} - \frac{2\log\log(n-p_*-L)(p_*+L)q^2}{n-(q+1)(p_*+L)-q-1}\bigg\}\\
&= P\bigg\{\sum_{i=1}^{q}\text{log-Beta}\left(\frac{n-(q+1)(p_*+L)-q+i}{2}, \frac{(q+1)L}{2}\right)\\
&\qquad < -q\log\left(\frac{n-p_*}{n-p_*-L}\right)\\
&\qquad + \frac{2\log\log(n-p_*)p_*q^2}{n-(q+1)p_*-q-1} - \frac{2\log\log(n-p_*-L)(p_*+L)q^2}{n-(q+1)(p_*+L)-q-1}\bigg\}.
\end{aligned}
$$

Expressed in terms of chi-square distributions, the $P\{\text{overfit}\}$ for HQc

$$
\begin{aligned}
P\bigg\{&n\sum_{i=1}^{q}\log\left(1+\frac{\chi^2_{(q+1)L}}{\chi^2_{n-(q+1)(p_*+L)-q+i}}\right) > qn\log\left(\frac{n-p_*}{n-p_*-L}\right)\\
&- \frac{n2\log\log(n-p_*)p_*q^2}{n-(q+1)p_*-q-1} + \frac{n2\log\log(n-p_*-L)(p_*+L)q^2}{n-(q+1)(p_*+L)-q-1}\bigg\}.
\end{aligned}
$$

Expressed in terms of $U_{q,qL,n-(q+1)(p_*+L)}$, the $P\{\text{overfit}\}$ for HQc is

$$
\begin{aligned}
P\bigg\{&U_{q,qL,n-(q+1)(p_*+L)} < \left(\frac{n-p_*-L}{n-p_*}\right)^q \times\\
&\exp\left(\frac{2\log\log(n-p_*)p_*q^2}{n-(q+1)p_*-q-1} - \frac{2\log\log(n-p_*-L)(p_*+L)q^2}{n-(q+1)(p_*+L)-q-1}\right)\bigg\}.
\end{aligned}
$$

### *5B.7. General Case*

Consider a model selection criterion, say $\text{MSC}_p$, of the form $\log(\text{SPE}) + \alpha(n,p,q)$, where $\alpha(n,p,q)$ is the penalty function of $\text{MSC}_p$. MSC overfits if $\text{MSC}_{p_*+L} < \text{MSC}_{p_*}$. For finite $n$, the probability that MSC prefers the overfitted model $p_* + L$ is

$$\begin{aligned}
&P\{\text{MSC}_{p_*+L} < \text{MSC}_{p_*}\} \\
&= P\left\{\log\left(\frac{|\text{SPE}_{p_*+L}|}{|\text{SPE}_{p_*}|}\right) < \alpha(n,p_*,q) - \alpha(n,p_*+L,q)\right\} \\
&= P\left\{\sum_{i=1}^{q} \text{log-Beta}\left(\frac{n-(q+1)(p_*+L)-q+i}{2}, \frac{(q+1)L}{2}\right)\right. \\
&\qquad \left. < \alpha(n,p_*,q) - \alpha(n,p_*+L,q)\right\}.
\end{aligned}$$

Expressed in terms of chi-square distributions, the $P\{\text{overfit}\}$ for MSC is

$$P\left\{n\sum_{i=1}^{q}\log\left(1+\frac{\chi^2_{(q+1)L}}{\chi^2_{n-(q+1)(p_*+L)-q+i}}\right) > \alpha(n,p_*+L,q) - \alpha(n,p_*,q)\right\}.$$

Expressed in terms of $U_{q,qL,n-(q+1)(p_*+L)}$, the $P\{\text{overfit}\}$ for MSC is

$$P\left\{U_{q,qL,n-(q+1)(p_*+L)} < \left(\frac{n-p_*-L}{n-p_*}\right)^q \exp\big(\alpha(n,p_*,q) - \alpha(n,p_*+L,q)\big)\right\}.$$

## Appendix 5C. Asymptotic Probabilities of Overfitting

### *5C.1. AICc*

In small samples, AICc overfits with probability

$$\begin{aligned}
P\Bigg\{&n\sum_{i=1}^{q}\log\left(1+\frac{\chi^2_{(q+1)L}}{\chi^2_{n-(q+1)(p_*+L)-q+i}}\right) > qn\log\left(\frac{n-p_*}{n-p_*-L}\right) \\
&+ \frac{2Lqn(qn-(q-1)(q+1))}{(n-(q+1)(p_*+L)-q-1)(n-(q+1)p_*-q-1)}\Bigg\}.
\end{aligned}$$

As $n \to \infty$,

$$\begin{aligned}
qn\log\left(\frac{n-p_*}{n-p_*-L}\right) + &\frac{2Lqn(qn-(q-1)(q+1))}{(n-(q+1)(p_*+L)-q-1)(n-(q+1)p_*-q-1)} \\
&\to Lq + 2Lq^2,
\end{aligned}$$

and thus the asymptotic probability of overfitting for AICc is

$$P\{\text{AICc overfits by } L\} = P\left\{\chi^2_{q(q+1)L} > (2q^2+q)L\right\}.$$

### *5C.2. AICu*

In small samples, AICu overfits with probability

$$P\Bigg\{n\sum_{i=1}^{q}\log\left(1+\frac{\chi^2_{(q+1)L}}{\chi^2_{n-(q+1)(p_*+L)-q+i}}\right) > qn\log\left(\frac{n-(q+1)p_*}{n-(q+1)(p_*+L)}\right) + \frac{2Lqn(qn-(q-1)(q+1))}{(n-(q+1)(p_*+L)-q-1)(n-(q+1)p_*-q-1)}\Bigg\}.$$

As $n \to \infty$,

$$qn\log\left(\frac{n-(q+1)p_*}{n-(q+1)(p_*+L)}\right) + \frac{2Lqn(qn-(q-1)(q+1))}{(n-(q+1)(p_*+L)-q-1)(n-(q+1)p_*-q-1)} \to Lq(q+1)+2Lq^2,$$

and thus the asymptotic probability of overfitting for AICu is

$$P\{\text{AICu overfits by } L\} = P\left\{\chi^2_{q(q+1)L} > (3q^2+q)L\right\}.$$

### *5C.3. FPE*

In small samples, FPE overfits with probability

$$P\Bigg\{n\sum_{i=1}^{q}\log\left(1+\frac{\chi^2_{(q+1)L}}{\chi^2_{n-(q+1)(p_*+L)-q+i}}\right) > qn\log\left(\frac{(n+(q-1)(p_*+L))(n-(q+1)p_*)(n-p_*)}{(n+(q-1)p_*)(n-(q+1)(p_*+L))(n-p_*-L)}\right)\Bigg\}.$$

As $n \to \infty$,

$$qn\log\left(\frac{(n+(q-1)(p_*+L))(n-(q+1)p_*)(n-p_*)}{(n+(q-1)p_*)(n-(q+1)(p_*+L))(n-p_*-L)}\right) \to Lq+2Lq^2$$

and thus the asymptotic probability of overfitting for FPE is

$$P\{\text{FPE overfits by } L\} = P\left\{\chi^2_{q(q+1)L} > (2q^2+q)L\right\}.$$

*5C.4. SIC*

In small samples, SIC overfits with probability

$$P\Bigg\{n\sum_{i=1}^{q}\log\left(1+\frac{\chi^2_{(q+1)L}}{\chi^2_{n-(q+1)(p_*+L)-q+i}}\right) > qn\log\left(\frac{n-p_*}{n-p_*-L}\right)$$
$$-\frac{n\log(n-p_*)p_*q^2}{n-p_*}+\frac{n\log(n-p_*-L)(p_*+L)q^2}{n-p_*-L}\Bigg\}.$$

As $n\to\infty$,

$$qn\log\left(\frac{n-p_*}{n-p_*-L}\right)-\frac{n\log(n-p_*)p_*q^2}{n-p_*}+\frac{n\log(n-p_*-L)(p_*+L)q^2}{n-p_*-L}$$
$$\to Lq+\frac{\log(n)Lq^2n^2+q(q+1)Ln}{(n-p_*-L)(n-p_*)}$$
$$\to Lq+\log(n)$$
$$\to\infty$$

and thus the asymptotic probability of overfitting for SIC is

$$P\{\text{SIC overfits by } L\}=0.$$

*5C.5. HQ*

In small samples, HQ overfits with probability

$$P\Bigg\{n\sum_{i=1}^{q}\log\left(1+\frac{\chi^2_{(q+1)L}}{\chi^2_{n-(q+1)(p_*+L)-q+i}}\right) > qn\log\left(\frac{n-p_*}{n-p_*-L}\right)$$
$$-\frac{n2\log\log(n-p_*)p_*q^2}{n-p_*}+\frac{n2\log\log(n-p_*-L)(p_*+L)q^2}{n-p_*-L}\Bigg\}.$$

As $n\to\infty$,

$$qn\log\left(\frac{n-p_*}{n-p_*-L}\right)-\frac{n2\log\log(n-p_*)p_*q^2}{n-p_*}$$
$$+\frac{n2\log\log(n-p_*-L)(p_*+L)q^2}{n-p_*-L}$$
$$\to Lq+\frac{\log(n)Lq^2n^2+q(q+1)Ln}{(n-p_*-L)(n-p_*)}$$
$$\to Lq+\log(n)$$
$$\to\infty$$

and thus the asymptotic probability of overfitting for HQ is

$$P\{\text{HQ overfits by } L\} = 0.$$

### *5C.6. HQc*

In small samples, HQc overfits with probability

$$P\Bigg\{n\sum_{i=1}^{q}\log\left(1+\frac{\chi^2_{(q+1)L}}{\chi^2_{n-(q+1)(p_*+L)-q+i}}\right) > qn\log\left(\frac{n-p_*}{n-p_*-L}\right) \\ -\frac{n2\log\log(n-p_*)p_*q^2}{n-(q+1)p_*-q-1}+\frac{n2\log\log(n-p_*-L)(p_*+L)q^2}{n-(q+1)(p_*+L)-q-1}\Bigg\}.$$

As $n \to \infty$,

$$\begin{aligned} & qn\log\left(\frac{n-p_*}{n-p_*-L}\right)-\frac{n2\log\log(n-p_*)p_*q^2}{n-(q+1)p_*-q-1} \\ & +\frac{n2\log\log(n-p_*-L)(p_*+L)q^2}{n-(q+1)(p_*+L)-q-1} \\ & \quad \to Lq+\frac{2\log\log(n)Lqn(qn-(q-1)(q+1))}{(n-(q+1)(p_*+L)-q-1)(n-(q+1)p_*-q-1)} \\ & \quad \to Lq+2\log\log(n) \\ & \quad \to \infty \end{aligned}$$

and thus the asymptotic probability of overfitting for HQc is

$$P\{\text{HQc overfits by } L\} = 0.$$

## Appendix 5D. Asymptotic Signal-to-noise Ratios

### *5D.1. AICc*

Starting with the small-sample signal-to-noise ratio for AICc from Section 5.3, the corresponding asymptotic signal-to-noise ratio is

$$\lim_{n\to\infty} \frac{\sqrt{(n-(q+1)(p+L)-(q-1)/2)(n-(q+1)p-(q-1)/2+2)}}{\sqrt{2Lq(q+1)}} \times$$

$$\left(q\log\left(\frac{n-p_*}{n-p_*-L}\right)+q\log\left(\frac{n-(q+1)(p_*+L)-(q-1)/2}{n-(q+1)p_*-(q-1)/2}\right)\right.$$

$$-\frac{Lq}{(n-p_*-L-(q-1)/2)(n-p_*-(q-1)/2)}$$

$$\left.+\frac{2Lq(qn-(q-1)(q+1))}{(n-(q+1)(p_*+L)-q-1)(n-(q+1)p_*-q-1)}\right)$$

$$=\lim_{n\to\infty}\frac{n}{\sqrt{2Lq(q+1)}}\left(\frac{Lq}{n}-\frac{Lq(q+1)}{n}-\frac{Lq}{n^2}+\frac{2Lq^2}{n}\right)$$

$$=\frac{q^2}{\sqrt{2q(q+1)}}\sqrt{L}.$$

### *5D.2. AICu*

Starting with the small-sample signal-to-noise ratio for AICu from Section 5.3, the corresponding asymptotic signal-to-noise ratio is

$$\lim_{n\to\infty} \frac{\sqrt{(n-(q+1)(p+L)-(q-1)/2)(n-(q+1)p-(q-1)/2+2)}}{\sqrt{2Lq(q+1)}} \times$$

$$\left(q\log\left(\frac{n-(q+1)p_*}{n-(q+1)(p_*+L)}\right)+q\log\left(\frac{n-(q+1)(p_*+L)-(q-1)/2}{n-(q+1)p_*-(q-1)/2}\right)\right.$$

$$-\frac{Lq}{(n-p_*-L-(q-1)/2)(n-p_*-(q-1)/2)}$$

$$\left.+\frac{2Lq(qn-(q-1)(q+1))}{(n-(q+1)(p_*+L)-q-1)(n-(q+1)p_*-q-1)}\right).$$

$$=\lim_{n\to\infty}\frac{n}{\sqrt{2Lq(q+1)}}\left(\frac{Lq(q+1)}{n}-\frac{Lq(q+1)}{n}-\frac{Lq}{n^2}+\frac{2Lq^2}{n}\right)$$

$$=\frac{2q^2}{\sqrt{2q(q+1)}}\sqrt{L}.$$

### *5D.3. HQ*

Starting with the small-sample signal-to-noise ratio for HQ from Section 5.3, the corresponding asymptotic signal-to-noise ratio is

$$\lim_{n\to\infty} \frac{\sqrt{(n-(q+1)(p+L)-(q-1)/2)(n-(q+1)p-(q-1)/2+2)}}{\sqrt{2Lq(q+1)}} \times$$
$$\left( q\log\left(\frac{n-p_*}{n-p_*-L}\right) + q\log\left(\frac{n-(q+1)(p_*+L)-(q-1)/2}{n-(q+1)p_*-(q-1)/2}\right)\right.$$
$$- \frac{Lq}{(n-p_*-L-(q-1)/2)(n-p_*-(q-1)/2)}$$
$$\left. + \frac{2\log\log(n-p_*-L)(p_*+L)q^2}{n-p_*-L} - \frac{2\log\log(n-p_*)p_*q^2}{n-p_*}\right).$$
$$= \lim_{n\to\infty} \frac{n}{\sqrt{2Lq(q+1)}}\left(\frac{Lq}{n} - \frac{Lq(q+1)}{n} - \frac{Lq}{n^2} + \frac{2\log\log(n)Lq^2}{n}\right)$$
$$=\infty.$$

### *5D.4. HQc*

Starting with the small-sample signal-to-noise ratio for HQc from Section 5.3, the corresponding asymptotic signal-to-noise ratio is

$$\lim_{n\to\infty} \frac{\sqrt{(n-(q+1)(p+L)-(q-1)/2)(n-(q+1)p-(q-1)/2+2)}}{\sqrt{2Lq(q+1)}} \times$$
$$\left( q\log\left(\frac{n-p_*}{n-p_*-L}\right) + q\log\left(\frac{n-(q+1)(p_*+L)-(q-1)/2}{n-(q+1)p_*-(q-1)/2}\right)\right.$$
$$- \frac{Lq}{(n-p_*-L-(q-1)/2)(n-p_*-(q-1)/2)}$$
$$\left. + \frac{2\log\log(n-p_*-L)(p_*+L)q^2}{n-(q+1)(p_*+L)-q-1} - \frac{2\log\log(n-p_*)p_*q^2}{n-(q+1)p_*-q-1}\right).$$
$$= \lim_{n\to\infty} \frac{n}{\sqrt{2Lq(q+1)}}\left(\frac{Lq}{n} - \frac{Lq(q+1)}{n} - \frac{Lq}{n^2} + \frac{2\log\log(n)Lq^2}{n}\right)$$
$$=\infty.$$

## Chapter 6
## Cross-validation and the Bootstrap

In previous chapters we used functions of the residuals from all the data to obtain model selection criteria by minimizing the discrepancy between the candidate and true models. Normality of the residuals played a key role in deriving these criteria, but in practice, the normality assumption may not be valid. In this Chapter, we discuss two nonparametric model selection techniques based on data resampling—cross-validation and bootstrap. Cross-validation involves dividing the data into two subsamples, using one (the training set) to choose a statistic and estimate a model, and using the second subsample (the validation set) to assess the model's predictive performance. Bootstrapping involves estimating the distribution of a statistic by constructing a distribution from subsamples (bootstrap pseudo-samples). We not only apply cross-validation and bootstrap procedures to regression model selection but also adapt them to AR and VAR time series models. In addition, we study how these procedures compare to the standard model selection criteria we have previously discussed when the normal error assumption holds.

### 6.1. Univariate Regression Cross-validation

#### *6.1.1. Withhold-1 Cross-validation*

One measure of the performance of a model is its ability to make predictions. In this context, we can define the model that minimizes the *mean squared error of prediction* (MSEP) as the best model. Consider the model

$$y_i = x_i'\beta + \varepsilon_i, \quad i = 1, \ldots, n,$$

where

$$\varepsilon_i \text{ are } i.i.d., \; E[\varepsilon_i] = 0, \; var[\varepsilon_i] = \sigma^2,$$

$x_i$ is a $k \times 1$ vector of known values, and $\beta$ is a $k \times 1$ vector of parameters. For now, no distributional assumptions other than independence are made for $\varepsilon_i$. The method of least squares discussed in Chapter 2 (Section 2.1.1) is used to estimate the $\beta$ parameters. Now suppose we have a new observation, $y_{n+1}$, independent of the original $n$ observations. The predicted value for the new observation is $\hat{y}_{n+1}$, and the MSEP is defined as

$$\text{MSEP} = E\big[(y_{n+1} - \hat{y}_{n+1})^2\big].$$

We recall that Akaike derived the FPE criterion, which estimates the MSEP, under the normal distribution assumption of $\varepsilon_i$. What if the distribution is unknown? FPE may no longer be an appropriate model selection tool since its derivation assumes normal errors. Allen (1974) suggested a cross-validation-type approach to estimating the MSEP by resampling from the original data. In particular, he suggested leaving 1 observation out.

Suppose one observation, say $(x_i', y_i)$, is withheld from the data set. This leaves a new data set with $n-1$ observations. Under the independent errors assumption, we know that $y_i$ is independent of this new data set. Using the remaining $n-1$ observations, least squares estimation yields $\hat{\beta}_{(i)}$, where the subscript $(i)$ denotes the data set with the $i$th observation withheld. Thus the prediction for $y_i$ is $\hat{y}_{(i)} = x_i'\hat{\beta}_{(i)}$. If $e_{(i)} = y_i - \hat{y}_{(i)}$ is the prediction error for $y_i$ when $(x_i', y_i)$ is withheld, then $e_{(i)}^2$ is unbiased for MSEP. This procedure can be repeated for $i = 1, \ldots, n$, yielding $e_{(1)}, \ldots, e_{(n)}$. In this setting, Allen defined PRESS as

$$\text{PRESS} = \sum_{i=1}^{n} e_{(i)}^2.$$

We will refer to PRESS as the equivalent cross-validation criterion CV(1), defined as

$$\text{CV}(1) = \frac{1}{n}\sum_{i=1}^{n} e_{(i)}^2,$$

where the (1) denotes that 1 observation is withheld. PRESS/$n$ and CV(1) are unbiased for MSEP, and the model with the minimum value for CV(1) is considered to be the best. At the first glance it appears that $n$ additional regression models of size $n-1$ must be computed to obtain CV(1). However, this task can be greatly simplified, as we will explain next.

An advantage to using least squares to compute parameter estimates is that it greatly reduces the computation necessary to obtain CV(1). Let $X = (x_1, \ldots, x_n)'$, $Y = (y_1, \ldots, y_n)'$ and $\hat{\beta} = (X'X)^{-1}X'Y$ be the parameter estimate. If we withhold $(x_i', y_i)$, then

$$\hat{\beta}_{(i)} = (X'X - x_i x_i')^{-1}(X'Y - x_i y_i).$$

It is known that

$$(A - bb')^{-1} = A^{-1} + A^{-1}bb'A^{-1}/d, \tag{6.1}$$

where $A$ is a matrix, $b$ is a vector and $d = (1 - b'A^{-1}b)$ is a scalar. Substituting $X'X$ for $A$ and $x_i$ for $b$ in Eq. (6.1), we have

$$\hat{\beta}_{(i)} = \Big((X'X)^{-1} + (X'X)^{-1}x_i x_i'(X'X)^{-1}/(1-h_i)\Big)\Big(X'Y - x_i y_i\Big),$$

where

$$h_i = x_i'(X'X)^{-1}x_i. \tag{6.2}$$

Thus the cross-validated prediction error for $y_i$ is

$$\begin{aligned}
e_{(i)} &= y_i - \hat{y}_{(i)} \\
&= y_i - x_i'\Big((X'X)^{-1} + (X'X)^{-1}x_i x_i'(X'X)^{-1}/(1-h_i)\Big)\Big(X'Y - x_i y_i\Big) \\
&= y_i - \big(1 + x_i'(X'X)^{-1}x_i/(1-h_i)\big)x_i'(X'X)^{-1}(X'Y - x_i y_i) \\
&= y_i - x_i'(X'X)^{-1}(X'Y - x_i y_i)/(1-h_i) \\
&= y_i\big(1 + x_i'(X'X)^{-1}x_i/(1-h_i)\big) - x_i'\hat{\beta}/(1-h_i) \\
&= (y_i - x_i'\hat{\beta})/(1-h_i) \\
&= e_i/(1-h_i),
\end{aligned}$$

where $e_i = y_i - x_i'\hat{\beta}$. Hence, CV(1) can be defined as

$$\mathrm{CV}(1) = \frac{1}{n}\sum_{i=1}^{n}\frac{e_i^2}{(1-h_i)^2}. \tag{6.3}$$

No additional regressions are required other than the usual least squares regression on all the data, but we must compute the diagonal elements of the projection or hat matrix, $H = X(X'X)^{-1}X'$. Unfortunately, small-sample properties of CV(1) are quite difficult to compute. However, asymptotic properties can be computed using a convenient approximation, as we will show.

It is known that $tr\{H\} = k$. Hence, $\sum_{i=1}^{n} h_i = k$ and the average value of the $h_i$ is $k/n$. For $n >> k$, we can use this average value, and substituting we obtain

$$\begin{aligned}
\mathrm{CV}(1) &\doteq \frac{1}{n}\sum_{i=1}^{n} e_i^2/(1-k/n)^2 \\
&= \frac{1}{n}\mathrm{SSE}_k/(1-k/n)^2 \\
&= \mathrm{SSE}_k\frac{n}{(n-k)^2}.
\end{aligned} \tag{6.4}$$

Shao (1995) showed that FPE in Eq. (2.11) and CV(1) are asymptotically equivalent. This can be seen from Eq. (6.4) as follows. Suppose that a true order $k_*$ exists and we compare the model of true order with another candidate model that overfits by $L$ variables. CV(1) will overfit if $\text{CV}(1)_{k_*+L} < \text{CV}(1)_{k_*}$. For finite $n$ and $n >> k_*$,

$$\begin{aligned}
&P\{\text{CV}(1)_{k_*+L} < \text{CV}(1)_{k_*}\} \\
&\doteq P\left\{\text{SSE}_{k_*+L}\frac{n}{(n-k_*-L)^2} < \text{SSE}_{k_*}\frac{n}{(n-k_*)^2}\right\} \\
&= P\left\{\frac{\text{SSE}_{k_*}}{\text{SSE}_{k_*+L}} > \left(\frac{n-k_*}{n-k_*-L}\right)^2\right\} \\
&= P\left\{\frac{\text{SSE}_{k_*}-\text{SSE}_{k_*+L}}{\text{SSE}_{k_*+L}} > \left(\frac{n-k_*}{n-k_*-L}\right)^2 - 1\right\} \\
&= P\left\{\frac{\chi^2_L}{\chi^2_{n-k_*-L}} > \frac{2nL-2k_*L-L^2}{(n-k_*-L)^2}\right\} \\
&= P\left\{F_{L,n-k_*-L} > \frac{2n-2k_*-L}{n-k_*-L}\right\}.
\end{aligned}$$

We recall from Chapter 2 that $F_{L,n-k_*-L} \to \chi^2_L/L$, and we know that

$$\lim_{n\to\infty}\frac{2n-2k_*-L}{n-k_*-L} = 2.$$

Given the above, the asymptotic probability of CV(1) overfitting by $L$ variables is $P\{\chi^2_L > 2L\}$, which is asymptotically equivalent to that for FPE under univariate regression. Therefore, CV(1) is an efficient model selection criterion and behaves similarly to FPE in large samples when the errors are normally distributed.

### *6.1.2. Delete-d Cross-validation*

It is also possible to obtain a consistent cross-validated selection criterion. The usual CV(1) is asymptotically efficient, as we have just seen. A delete-$d$ cross-validation selection criterion improves the consistency of CV(1). The delete-$d$ cross-validation, CV($d$), was first suggested by Geisser (1977) and later studied by Burman (1989) and Shao (1993, 1995). Shao (1993) showed that CV($d$) is consistent if $d/n \to 1$ and $n-d \to \infty$ as $n \to \infty$. For this scheme the training set should be much smaller than the validation set—just the opposite of CV(1), where the difference is only one observation. We must also address the problem of choosing $d$. Actually, any $d$ of the form $d = n - n^a$ where $a < 1$

will satisfy the conditions $d/n \to 1$, and $n - d \to \infty$ as $n \to \infty$. However, Shao recommends a different value for $d$.

In addition to showing that FPE and CV(1) are asymptotically equivalent, Shao (1995) also showed that SIC, Eq. (2.15), and CV($d$) are asymptotically equivalent when

$$d = n\left(1 - \frac{1}{\log(n) - 1}\right),$$

where the size of the training set is $n - d = n/(\log(n) - 1)$. For a sample size of $n$ this implies that there are $C^n_{n-d}$ training sets, a potentially huge number. For example, when $n = 20$, then $d = 10$ and there are 184,756 subsets. In practice, some smaller number of training sets $M$ is required. Unfortunately, the best choice of $M$ is unknown. The larger the chosen value of $M$, the more computations are required. Generally, the accuracy of CV($d$) increases as $M$ increases; however, we will see from simulations later in this Chapter that $M$ as small as 25 can lead to satisfactory performance. Thus if we let $\mathcal{M}_m$ be the $m$th training set of size $n - d$, we can define CV($d$) as

$$\text{CV}(d) = \frac{1}{Md}\sum_{m=1}^{M}\sum_{i \notin \mathcal{M}_m}(y_i - \hat{y}_i^d)^2, \tag{6.5}$$

where $\hat{y}_i^d$ is the predicted value for $y_i$ from training set $\mathcal{M}_m$. Detailed discussions of various alternative CV($d$) methods can be found in Shao (1993).

## 6.2. Univariate Autoregressive Cross-validation

### *6.2.1. Withhold-1 Cross-validation*

One of the key assumptions in computing the MSEP is that the new observation, $y_{n+1}$, is independent of the original $n$ observations. Likewise, in order to estimate the MSEP via cross-validation, a key assumption is that the observation withheld is independent of the remaining $n - 1$ observations. This assumption will fail for the following autoregressive models:

$$y_t = \phi_1 y_{t-1} + \cdots + \phi_p y_{t-p} + w_t, \quad t = p+1, \ldots, n$$

where

$$w_t \text{ are } i.i.d. \ E[w_t] = 0, \text{ and } var[w_t] = \sigma^2.$$

Clearly, $y_i$ and $y_j$ are not independent. In finite samples, we may assume that there exists a constant $l$ such that $y_i$ and $y_j$ are approximately independent for $|i - j| > l$. Therefore, to circumvent the nonindependence in autoregressive models, when withholding $y_t$ we should withhold the block $y_{t-l}, \ldots, y_t, \ldots, y_{t+l}$. Such cross-validation, withholding $\pm l$ additional observations around 1 observation, can be defined as

$$\mathrm{CV}_l(1) = \frac{1}{n-p} \sum_{t=p+1}^{n} (y_t - \hat{y}^l_{(t)})^2,$$

where

$$\hat{y}^l_{(t)} = \hat{\phi}^l_1 y_{t-1} + \cdots + \hat{\phi}^l_p y_{t-p}$$

and the superscript denotes that $\pm l$ neighboring observations are also withheld. Obviously cross-validating in this manner may not be possible for small sample sizes. Our simulation study in Section 6.9.2 shows that the usual withhold-1 cross-validation, CV(1), works well and can be used in small samples. It is also more flexible than the withhold-l procedure, particularly in small samples.

Let $Y = (y_{p+1}, \ldots, y_n)'$, $\phi = (\phi_1, \ldots, \phi_p)'$, $(X)_{t,j} = x_{t,j}$ for $j = 1, \ldots, p$ and $t = p+1, \ldots, n$, and $x_t$ is the $t$th row of $X$. In addition, let $\hat{y}_{(t)} = x'_t \hat{\phi}_{(t)}$ and $\hat{\phi}_{(t)}$ be the least squares estimator of $\phi$ obtained when the $t$th row of $Y$ and $X$ are withheld. Then CV(1) for AR models can be defined as

$$\mathrm{CV}(1) = \frac{1}{n-p} \sum_{t=p+1}^{n} (y_t - \hat{y}_{(t)})^2. \tag{6.6}$$

It is asymptotically equivalent to FPE in Eq. (3.12).

### *6.2.2. Delete-d Cross-validation*

The consistent cross-validated selection criteria can be generalized to the AR model. For an autoregressive model of order $p$ we use a training set of size $n - d = (n-p)/(\log(n-p) - 1)$, randomly selected from $t = p+1, \ldots, n$. Let $\mathcal{M}_m$ be the $m$th training set. The validation set is then all $t = p+1, \ldots, n$ such that $t \notin \mathcal{M}_m$. Thus for AR($p$) models, we define CV($d$) as

$$\mathrm{CV}(d) = \frac{1}{Md} \sum_{m=1}^{M} \sum_{t>p, t\notin\mathcal{M}_m} (y_t - \hat{y}^d_t)^2, \tag{6.7}$$

where $\hat{y}^d_t$ is the $t$th prediction from training set $\mathcal{M}_m$.

## 6.3. Multivariate Regression Cross-validation

### 6.3.1. Withhold-1 Cross-validation

Cross-validation can also be generalized to the multivariate regression model, where the MSEP is now a matrix called MPEP, the *mean product error of prediction.* In order to define the MPEP, consider the regression model

$$y_i' = x_i'B + \varepsilon_i', \quad i = 1, \dots, n$$

where

$$\varepsilon_i \text{ are } i.i.d., \ E[\varepsilon_i] = 0, \ Cov[\varepsilon_i] = \Sigma,$$

$y_i$ is a $q \times 1$ vector of responses, $x_i$ is a $k \times 1$ vector of known values, and $B$ is a $k \times q$ matrix of parameters. Again, we will make no distributional assumptions about $\varepsilon_i$. As in the univariate case, suppose that $y_{n+1}$ is a new observation and it is independent of the original $n$ observations. Then the MPEP is defined as the $q \times q$ matrix

$$\text{MPEP} = E\big[(y_{n+1} - \hat{y}_{n+1})(y_{n+1} - \hat{y}_{n+1})'\big].$$

We now can generalize Allen's (1974) cross-validation-type approach to estimate MPEP.

Assume that observations $(x_1', y_1'), \dots, (x_n', y_n')$ are independent and the observation $(x_i', y_i')$ is withheld. Then multivariate least squares can be applied to the remaining $n-1$ observations to yield the parameter estimate $\hat{B}_{(i)}$ for $B$. The prediction for $y_i$, based on the sample with $y_i$ withheld, is $x_i'\hat{B}_{(i)}$. If we let $e_{(i)}' = y_i' - x_i'\hat{B}_{(i)}$ be the prediction error vector for $y_i'$ when $y_i$ is withheld, the multivariate withhold-1 cross-validated estimate of the MPEP is

$$\text{CV}(1) = \frac{1}{n}\sum_{i=1}^{n} e_{(i)}e_{(i)}'.$$

The CV(1) for the multivariate case can be computed in much the same manner as for the univariate case, yielding

$$\text{CV}(1) = \frac{1}{n}\sum_{i=1}^{n} \frac{e_i e_i'}{(1-h_i)^2},$$

where $e_i' = y_i' - x_i'\hat{B}$ and $h_i$ is defined as for Eq. (6.2). However, CV(1) in this form is not suitable for model selection use, and must first be transformed into a scalar. In Chapter 4 we used two common methods for transforming matrices to scalars, the determinant and trace. We define the determinant of CV(1) to be

$$\text{deCV} = |\text{CV}(1)| = \left| \frac{1}{n} \sum_{i=1}^{n} \frac{e_i e_i'}{(1-h_i)^2} \right|, \tag{6.8}$$

where $|\cdot|$ denotes the determinant. In addition, we define the trace of CV(1) to be

$$\text{trCV} = tr\{\text{CV}(1)\} = tr\left\{ \frac{1}{n} \sum_{i=1}^{n} \frac{e_i e_i'}{(1-h_i)^2} \right\}. \tag{6.9}$$

We will show that deCV is asymptotically equivalent to the multivariate FPE in Eq. (4.11). Let SPE be the usual sum of product errors, $\text{SPE}_k = \sum_{i=1}^{n} e_i e_i'$. For $n >> k$, we can use the average value of $h_i$, $k/n$, to obtain

$$\begin{aligned} \text{deCV} &\doteq \left| \frac{1}{n} \sum_{i=1}^{n} e_i e_i' / (1 - k/n)^2 \right| \\ &= \left| \frac{1}{n} \text{SPE}_k / (1 - k/n)^2 \right| \\ &= |\text{SPE}_k| \frac{n^q}{(n-k)^{2q}}. \end{aligned} \tag{6.10}$$

From Eq. (6.10) we can show that deCV is asymptotically equivalent to the usual multivariate FPE, assuming that the errors are normally distributed. As we have done before, we assume that a true model order $k_*$ exists and compare it to one candidate model of order $k_* + L$. The deCV criterion overfits if $\text{deCV}_{k_*+L} < \text{deCV}_{k_*}$; however, it is algebraically more convenient to use log(deCV) and the log of the approximation in Eq. (6.10). Doing so, we find that for finite $n$ and $n >> k_*$,

$$\begin{aligned} &P\{\text{deCV}_{k_*+L} < \text{deCV}_{k_*}\} \\ &= P\Big\{ \log(\text{deCV}_{k_*+L}) < \log(\text{deCV}_{k_*}) \Big\} \\ &= P\Big\{ \log|\text{SPE}_{k_*+L}| + q\log(n) - 2q\log(n - k_* - L) \\ &\qquad < \log|\text{SPE}_{k_*}| + q\log(n) - 2q\log(n - k_*) \Big\} \\ &= P\left\{ \log\left( \frac{|\text{SPE}_{k_*+L}|}{|\text{SPE}_{k_*}|} \right) < 2q\log\left( \frac{n - k_* - L}{n - k_*} \right) \right\} \\ &= P\left\{ n \sum_{i=1}^{q} \log\left( 1 + \frac{\chi^2_L}{\chi^2_{n-k_*-L}} \right) > n2q\log\left( \frac{n-k_*}{n-k_*-L} \right) \right\}. \end{aligned}$$

We know from Chapter 4 that as $n \to \infty$,

$$n \sum_{i=1}^{q} \log \left(1 + \frac{\chi_L^2}{\chi_{n-k_*-L}^2}\right) \to \chi_{qL}^2.$$

Here,

$$\begin{aligned}\lim_{n\to\infty} n2q \log\left(\frac{n-k_*}{n-k_*-L}\right) &= \lim_{n\to\infty} n2q\log\left(1+\frac{L}{n-k_*-L}\right) \\ &= \lim_{n\to\infty} \frac{2nqL}{n-k_*-L} \\ &= 2qL.\end{aligned}$$

Thus the asymptotic probability of deCV overfitting by $L$ variables is $P\{\chi_{qL}^2 > 2qL\}$, which is asymptotically equivalent to that for FPE (see Section 4.5.2), and hence deCV is an efficient model selection criterion.

### 6.3.2. Delete-d Cross-validation

A multivariate version of the delete-$d$ cross-validated approximation to MPEP is also possible. Each training set has $n - d = n/(\log(n) - 1)$ observations. Let $\mathcal{M}_m$ be the $m$th training set, and let

$$e_{(i)} = y_i - \hat{y}_i^d, \quad i \notin \mathcal{M}_m,$$

be the vector of $d$ cross-validated errors. An unbiased estimator for MPEP is then

$$\frac{1}{d} \sum_{i \notin \mathcal{M}_m} e_{(i)} e'_{(i)}.$$

For $M$ training sets, $\mathrm{CV}(d)$ can be defined as

$$\mathrm{CV}(d) = \frac{1}{Md} \sum_{m=1}^{M} \sum_{i \notin \mathcal{M}_m} e_{(i)} e'_{(i)},$$

where $e_{(i)}$ is the $i$th prediction error from training set $\mathcal{M}_m$. Using the determinant and the trace of $\mathrm{CV}(d)$, two scalars suitable for model selection can be obtained as follows:

$$\mathrm{deCV}(d) = |\mathrm{CV}(d)| = \left| \frac{1}{Md} \sum_{m=1}^{M} \sum_{i \notin \mathcal{M}_m} e_{(i)} e'_{(i)} \right| \tag{6.11}$$

and

$$\text{trCV}(d) = tr\{\text{CV}(d)\} = tr\left\{\frac{1}{Md}\sum_{m=1}^{M}\sum_{i\notin\mathcal{M}_m} e_{(i)}e'_{(i)}\right\}. \tag{6.12}$$

## 6.4. Vector Autoregressive Cross-validation

### *6.4.1. Withhold-1 Cross-validation*

We saw in Section 6.2 that one of the key assumptions in computing the MSEP matrix is that the new observation $y_{n+1}$ is independent of the original $n$ observations. Likewise, a key assumption when estimating the MPEP via cross-validation is that the withheld observation is independent of the remaining $n-1$ observations. This assumption failed in the univariate autoregressive model, and of course also fails in the vector autoregressive model. We must again find a way around this problem.

Consider the vector autoregressive VAR($p$) model

$$y_t = \Phi_1 y_{t-1} + \cdots + \Phi_p y_{t-p} + w_t, \quad t = p+1, \ldots, n$$

where

$$w_t \text{ are } i.i.d.\ E[w_t] = 0,\ Cov[w_t] = \Sigma,$$

the $\Phi_j$ are $q \times q$ matrices of unknown parameters, and $y_t$ is a $q \times 1$ vector observed at times $t = 1, \ldots, n$. Clearly, $y_i$ and $y_j$ are not independent. As we did in Section 6.2, we find a constant $l$ such that $y_i$ and $y_j$ are approximately independent for $|i-j| > l$, and withhold the block $y_{t-l}, \ldots, y_t, \ldots, y_{t+l}$. Cross-validation can then be defined as

$$\text{CV}_l(1) = \frac{1}{n-p}\sum_{t=p+1}^{n}(y_t - \hat{y}^l_{(t)})(y_t - \hat{y}^l_{(t)})',$$

where

$$y^l_{(t)} = \hat{\Phi}^l_1 y_{t-1} + \cdots + \hat{\Phi}^l_p y_{t-p}.$$

The $l$ superscript denotes that $\pm l$ neighboring observations are also withheld. Once again, in small samples $n$ may be too small to cross-validate in this manner, and CV(1) may be more practical. In this context, CV(1) is

$$\text{CV}(1) = \frac{1}{n-p}\sum_{t=p+1}^{n}(y_t - \hat{y}_{(t)})(y_t - \hat{y}_{(t)})'.$$

Then, using the determinant and trace of CV(1), we form the model selection criteria

$$\text{deCV} = |\text{CV}(1)| = \left| \frac{1}{n-p} \sum_{t=p+1}^{n} (y_t - \hat{y}_{(t)})(y_t - \hat{y}_{(t)})' \right| \tag{6.13}$$

and

$$\text{trCV} = tr\{\text{CV}(1)\} = tr\left\{ \frac{1}{n-p} \sum_{t=p+1}^{n} (y_t - \hat{y}_{(t)})(y_t - \hat{y}_{(t)})' \right\}. \tag{6.14}$$

### *6.4.2. Delete-d Cross-validation*

The delete-$d$ cross-validation selection criterion CV($d$) can also be applied to the VAR model. For VAR($p$), the model is cast into the multivariate regression setting by forming the $(n-p) \times q$ matrix $Y$ and the $(n-p) \times pq$ matrix $X$ as described in Chapter 5, Section 5.1.1. We next split the $n-p$ observations randomly into training sets of size $n - d = (n-p)/(\log(n-p) - 1)$ and the withheld validation set of size $d$. If we let $\mathcal{M}_m$ be the $m$th training set, then the validation set is all $t = p+1, \ldots, n$ such that $t \notin \mathcal{M}_m$. Therefore we can define CV($d$) as

$$\text{CV}(d) = \frac{1}{Md} \sum_{m=1}^{M} \sum_{t>p, t \notin \mathcal{M}_m} (y_t - \hat{y}_t^d)(y_t - \hat{y}_t^d)',$$

where $\hat{y}_t^d$ is the $t$th prediction from training set $\mathcal{M}_m$. Finally, for use in model selection we define the determinant of CV($d$) as

$$\text{deCV}(d) = |\text{CV}(d)|, \tag{6.15}$$

and the trace of CV($d$) as

$$\text{trCV}(d) = tr\{\text{CV}(d)\}. \tag{6.16}$$

## 6.5. Univariate Regression Bootstrap

### *6.5.1. Overview of the Bootstrap*

The bootstrap is another data resampling procedure that can be very useful when the underlying parameter distributions are unknown. It was first proposed by Efron (1979), and then was adapted to model selection by Linhart and Zucchini (1986). We will begin with a brief overview of the use of the bootstrap for estimating a simple statistic for *i.i.d.* observations. Then we will examine the bootstrap with respect to model selection and discuss some of the problems in applying bootstrapping to model selection using least squares regression.

Suppose the following *i.i.d.* observations, $x_1, \ldots, x_n$, are observed from a population with cdf $F$. The sample average is unbiased for the mean of $F$ and sample variance is unbiased for the variance of $F$. What if the sample statistic in mind is the median and the underlying distribution $F$ is unknown? Now variances are impossible to compute. This kind of situation is where the bootstrap technique is most useful. Since the data are *i.i.d.* from $F$, the empirical cdf $\hat{F}$ represents $F$. The idea behind bootstrapping is to sample from $\hat{F}$ and determine the empirical properties of $\hat{F}$, and in turn use these empirically determined properties to estimate the true properties of $F$.

Now we use a simple example to illustrate the bootstrap procedure, estimating the variance of a sample median. First we randomly sample $n$ observations, with replacement, from $x_1, \ldots, x_n$. In effect, the empirical distribution $\hat{F}$ puts a probability of $1/n$ on each of the observed values $x_i$, $i = 1, \ldots, n$. Call this new bootstrap sample $\mathbf{x}^* = (x_1^*, \ldots, x_n^*)'$. Next we generate a large number of independent bootstrap samples $\mathbf{x}^{*1}, \ldots, \mathbf{x}^{*R}$, each of size n, and compute the median of each bootstrap sample. Lastly we compute the sample variance from the $R$ bootstrap medians to obtain an estimate for the true variance of the sample median.

While bootstrapping is obviously simple to apply to a variety of statistical problems, it is computationally intensive. Also it requires us to assume that the original sample represents the true population, which is not always true in least squares regression. The regression model $Y = X\beta + \varepsilon$ is often an approximation to the true model, which can be further affected by underfitting and overfitting. Furthermore, regression residuals are linear combinations of the errors $\varepsilon_i$, and thus may only approximate the true error distribution.

These difficulties leave us with two issues around the bootstrap that must be resolved for it to be useful in univariate regression: bias and sample selection. Recall that in the univariate regression model in Eqs. (2.3)–(2.4), the errors $\varepsilon$ are unobserved. Estimates of the errors may be obtained from the least squares equations, but these estimates may be biased. The estimates are often much smaller than the true errors. This bias requires that the bootstrap be modified to work in this context. Also, there are two possibilities for what will constitute the sample: first, both $x$ and $y$ may be random and the sample represented by $(x', y)$. Or, $x$ may be fixed and the errors are sampled from $\varepsilon$.

In the first case the focus is on the data itself and not on the errors. In the second case, all the emphasis is placed on the errors. In regression, sampling pairs (as in the first case) is much more computationally intensive than sampling errors (as in the second case). However, sampling errors requires inflating the errors by some suitable function, as we will discuss later.

In order to compare the sampling pairs and sampling errors approaches to bootstrapping, we first define the target function that we will use to select the best model. Assume that $(x'_1, y_1), \ldots, (x'_n, y_n)$ are *i.i.d.* $F$, where $F$ is some joint distribution of $x$ and $y$. Let $z = (x', y)$ represent the data, $\hat{F}$ be the empirical distribution of the data, and $\eta_z(x_0) = \eta(x_0, \hat{F})$ be the prediction function of $y$ at $x = x_0$ for the given function $\eta$. Under these conditions, Efron and Tibshirani (1993) define the prediction error for the new independent observation $(x'_0, y_0)$ as

$$\mathrm{err}(z, F) \equiv E_{0F}\left[Q(y_0, \eta_z(x_0))\right],$$

where $Q(y_0, \eta_z(x_0))$ is a measure of error between $y_0$ and the prediction $\eta_z(x_0)$.

In multiple regression, this becomes

$$\mathrm{err}(z, F) = E_{0F}\left[(y_0 - x'_0\hat{\beta})^2\right], \tag{6.17}$$

which is our target function. The notation $E_{0F}$ indicates expectation over a new observation $(x'_0, y_0)$ from the distribution $F$. The model that has minimum prediction error will be considered the best model. This expectation requires knowledge of $F$ and $\beta$, which in practice are typically unknown. Some common estimates of Eq. (6.17) are

1) the parametric estimate, FPE,

$$\mathrm{FPE} = \hat{\sigma}_k^2 \frac{n+k}{n-k};$$

2) the cross-validated estimate,

$$\mathrm{CV}(1) = \frac{1}{n}\sum_{i=1}^{n}\left[(y_i - x'_i\hat{\beta}_{(i)})^2\right]; \text{ and}$$

3) the bootstrap, discussed below.

Before discussing the bootstrap, we need to introduce the following notation. Let

$$\mathrm{err}(z^*, \hat{F}) = \frac{1}{n}\sum_{i=1}^{n}\left[(y_i - x'_i\hat{\beta}^*)^2\right] \tag{6.18}$$

represent the prediction error from one bootstrapped sample using the original data and $\hat{\beta}^*$, the bootstrapped estimate of $\beta$. Let

$$\text{err}(z^*, \hat{F}^*) = \frac{1}{n}\sum_{i=1}^{n}\left[(y_i^* - x_i^{*\prime}\hat{\beta}^*)^2\right]$$

represent the prediction error from one bootstrapped sample using the bootstrapped data as well as the bootstrapped estimate of $\beta$. Finally, let

$$\text{err}(z, \hat{F}) = \frac{1}{n}\sum_{i=1}^{n}\left[(y_i - x_i'\hat{\beta})^2\right] = \hat{\sigma}^2,$$

where $\hat{\sigma}^2$ is the usual MLE for the residual variance. All three of the functions above are estimates of Eq. (6.17).

The first step of the bootstrap procedure is to randomly select, with replacement, a sample of size $n$ from $(x_1', y_1), \ldots, (x_n', y_n)$. Then compute $\hat{\beta}^*$ from this new sample and calculate

$$\text{err}(z^*, \hat{F}) = \frac{1}{n}\sum_{i=1}^{n}\left[(y_i - x_i'\hat{\beta}^*)^2\right].$$

This gives us an estimate from one pseudo-sample. A bootstrap estimate of Eq. (6.17) can be produced by repeatedly sampling the data $R$ times, computing the estimate of Eq. (6.18) for each pseudo-sample, and averaging all $R$ estimates to yield

$$\begin{aligned}\widetilde{\text{err}}(z^*, \hat{F}) &= \frac{1}{R}\sum_{r=1}^{R}\sum_{i=1}^{n}\text{err}(z_r^*, \hat{F}) \\ &= \frac{1}{R}\sum_{r=1}^{R}\sum_{i=1}^{n}\left[(y_i - x_i'\hat{\beta}_r^*)^2\right]/n.\end{aligned}$$

While this gives us an estimate for the prediction error, such a bootstrapped function may be biased. To address this bias, Efron and Tibshirani suggest using what they call "the more refined bootstrap approach," which includes an estimate of the bias. They first refined bootstrap estimate the bias in $\text{err}(z, \hat{F})$ for estimating $\text{err}(z, F)$, and then correct $\text{err}(z, \hat{F})$ by subtracting its estimated bias. Since $\text{err}(z, \hat{F})$ underestimates the true prediction error $\text{err}(z, F)$, Efron and Tibshirani define the average optimism as

$$ao(F) = E_F\left[\text{err}(z, F) - \text{err}(z, \hat{F})\right].$$

The bootstrapped estimate for $ao(F)$ is

$$\widehat{ao}(\hat{F}) = \frac{1}{R}\sum_{r=1}^{n}\left[\widetilde{\text{err}}(z^*, \hat{F}) - \widetilde{\text{err}}(z^*, \hat{F}^*)\right]$$
$$= \frac{1}{Rn}\left[\sum_{r=1}^{R}\sum_{i=1}^{n}[(y_i - x_i'\hat{\beta}_r^*)^2] - \sum_{r=1}^{R}\sum_{i=1}^{n}[(y_i^* - x_i^{*\prime}\hat{\beta}_r^*)^2]\right].$$

Therefore the refined bootstrapped estimate of Eq. (6.17) is

$$\widehat{\text{err}}(z, \hat{F}) = \text{err}(z, \hat{F}) + \widehat{ao}(\hat{F})$$
$$= \hat{\sigma}^2 + \frac{1}{Rn}\left[\sum_{r=1}^{R}\sum_{i=1}^{n}[(y_i - x_i'\hat{\beta}_r^*)^2] - \sum_{r=1}^{R}\sum_{i=1}^{n}[(y_i^* - x_i^{*\prime}\hat{\beta}_r^*)^2]\right]. \quad (6.19)$$

The model that minimizes Eq. (6.19) is considered the best model.

Now we can compare the two approaches to bootstrapping, randomly selecting pairs $(x_i', y_i)$ and randomly selecting the residuals $y_i - x_i'\hat{\beta}$ computed from the full data model. We will start with selecting pairs.

First we randomly select $n$ pairs of samples with replacement from the original observations $\{(x_1', y_1), \ldots, (x_n', y_n)\}$. The resulting bootstrap sample is $\{(x_1^{*\prime}, y_1^*), \ldots, (x_n^{*\prime}, y_n^*)\}$. Next, applying Eq. (6.19), we obtain the refined bootstrap estimate of the prediction error. The advantages of this approach are that no assumptions are required about the errors, and that this procedure is more flexible in nonlinear models. However, we have found in multiple regression that bootstrapping pairs overfits excessively, is computationally much slower than bootstrapping residuals, and is only applicable when both $x$ and $y$ are random. Thus, it may not be effective for model selection. Our other option is to bootstrap the residuals. This approach allows us to obtain great computational savings in the linear model setting, although flexibility for nonlinear models is sacrificed. Simulation studies later in this Chapter indicate that bootstrapping residuals seems to perform as well as bootstrapping pairs.

We have already noted that when bootstrapping residuals, additional care is needed since the residuals are themselves biased. The $i$th residual is

$$e_i = y_i - x_i'\hat{\beta}.$$

Furthermore,

$$\text{err}(z, \hat{F}) = \frac{1}{n}\sum_{i=1}^{n} e_i^2$$

is biased for the true residual variance. The $e_i$ tend to be smaller than the true errors and must be inflated. Shao (1996) has recommended inflating the residuals by $\sqrt{1/(1-k/n)}$. His reasoning follows from the fact that $s^2 = \sum_{i=1}^{n} e_i^2/(n-k)$ is unbiased for the true residual variance. On average, the $e_i$ are $1-k/n$-fold smaller than the true residuals. Let

$$\tilde{e}_i = e_i/\sqrt{1-k/n}.$$

Bootstrap samples $\tilde{e}_1^*, \ldots, \tilde{e}_n^*$ are generated by randomly selecting with replacement from $\tilde{e}_1, \ldots, \tilde{e}_n$ and forming $y_i^* = x_i'\hat{\beta} + \tilde{e}_i^*$. Using $Y^* = (y_1^*, \ldots, y_n^*)'$ as a dependent vector, the resulting parameter estimator is $\hat{\beta}^* = (X'X)^{-1}X'Y^*$. A bootstrapped estimate can be formed by substituting the above $y_*$ and $\hat{\beta}^*$ into Eq. (6.19). Here is where we obtain our computational savings, since $(X'X)^{-1}X'$ has already been computed. However, one disadvantage of this approach is that the inflation factor depends on a parameter count, and parameter counts may not be available (as in the case of nonlinear regression). Another disadvantage is that the residuals from regression do not have constant variance, violating a classic regression assumption required for computing $\hat{\beta}^*$. To accommodate these situations, Wu (1986) has suggested a different inflation factor based on the cross-validation weights for multiple regression, $1-h_i$ in Eq. (6.2). If we let

$$\hat{e}_i = e_i/\sqrt{1-h_i},$$

then the $\hat{e}_i$ have constant variance. Recall that $\sum h_i = k$ and $h_i = n/k$ on average. The $\sqrt{1-h_i}$ inflation should be similar to Shao's $\sqrt{1-k/n}$ residual inflation factor with the added property of having constant variance, and therefore we use Wu's inflated residuals for our bootstrapped model selection criteria. For simplicity, we call $\hat{e}_i$ the *adjusted residual.*

### *6.5.2. Doubly Cross-validated Bootstrap Selection Criterion*

In order to obtain a bootstrap criterion for regression we undertake the following steps:

Step 1: Run the usual regression model on the full data and compute $e_i$.

Step 2: Form $\hat{e}_i$ by inflating the $e_i$ by $\sqrt{1-h_i}$ to ensure that all residuals $\hat{e}_i$ have constant variance.

Step 3: Form a bootstrap pseudo-sample of size $n$, $\hat{e}_1^*, \ldots, \hat{e}_n^*$ by selecting with replacement from $\hat{e}_1, \ldots, \hat{e}_n$.

Step 4: Form $y_i^* = x_i'\hat{\beta} + \hat{e}_i^*$ and compute $\hat{\beta}^* = (X'X)^{-1}X'Y^*$.

Step 5: Compute new residuals $v_i = y_i - x_i'\hat{\beta}^*$.

Repeat steps 3–5 $R$ times to yield the naive bootstrap estimate

$$\widetilde{\text{err}}^{cv}(z^*, \hat{F}) = \frac{1}{Rn}\sum_{r=1}^{R}\sum_{i=1}^{n} v_{ir}^2, \tag{6.20}$$

where $v_{ir}$ is the $i$th residual obtained in the $r$th bootstrap sample.

However, simulations indicate that this procedure overfits. Note that even with inflated residuals, $\widetilde{\text{err}}^{cv}(z^*, \hat{F})$ underestimates the true residual variance. We have observed that $\widetilde{\text{err}}^{cv}(z^*, \hat{F}) \approx s^2$. This suggests that a penalty function of some sort is required to prevent overfitting. Hence we add a parametric penalty function to obtain the penalty weighted naive bootstrap estimate

$$\text{BFPE} = \widetilde{\text{err}}^{cv}(z^*, \hat{F})\frac{n+k}{n-k}, \tag{6.21}$$

which does in fact perform better in simulations than the nonpenalty weighted bootstrap. We can also derive a bootstrap with a penalty function that does not depend on parameter counts—the doubly cross-validated bootstrap.

To obtain the doubly cross-validated bootstrap, inflate the bootstrapped residuals by the cross-validation weights, $1 - h_i$ in Eq. (6.2). Apply cross-validation to form the residuals, $\hat{e}_i = e_i/\sqrt{1-h_i}$. Then, repeat steps 3–5 to form new residuals $v_{ir}$. Ordinarily $\widetilde{\text{err}}^{cv}(z^*, \hat{F})$ is computed using the $v_{ir}$. However, we will instead define a selection criterion based on minimizing $\frac{1}{n}\sum_{i=1}^{n}(v_{ir}/(1-h_i))^2$ for some bootstrap sample $\mathcal{R}_r$:

$$\text{DCVB} = \frac{1}{Rn}\sum_{r=1}^{R}\sum_{i=1}^{n}\frac{v_{ir}^2}{(1-h_i)^2}, \tag{6.22}$$

where the $v_{ir}$ are computed following steps 1–5 above. Simulation studies in Section 6.9 show that DCVB outperforms the other bootstraps, and is competitive with AICc. The additional weighting in Eq. (6.22) by $1/(1-h_i)^2$ results in a stronger penalty function than the $(n+k)/(n-k)$ term in Eq. (6.21). Since $1 - h_i \doteq 1 - k/n$, $1/(1-h_i)^2 \doteq n^2/(n-k)^2 > (n+k)/(n-k)$. BFPE, DCVB, and AICu are all asymptotically equivalent.

Most bootstrap model selection criteria try to obtain a bootstrap estimate for the squared prediction error. Other approaches include bootstrapped estimates for the likelihood and bootstrapped estimates for the Kullback–Leibler discrepancy. For the state-space time series model, Cavanaugh and Shumway

(1997) bootstrapped the likelihood function and obtained a refined bootstrap estimate for AIC, called AICb. One advantage of AICb is that the underlying distribution of the errors need not be known. Shibata (1997) proposed a bootstrapped estimate for the Kullback–Leibler discrepancy. We know from Chapter 2 that AIC and AICc estimate the Kullback–Leibler discrepancy, and that these two criteria and FPE (which estimates the mean squared prediction error) are all asymptotically equivalent. Since the approaches are all asymptotically equivalent, we will focus on bootstrapping the mean squared prediction error.

## 6.6. Univariate Autoregressive Bootstrap

Bootstrapping the univariate autoregressive model is similar to bootstrapping the univariate regression model. The key difference between the two is that the errors in underfitted AR models are correlated, and this can lead to biased estimates for the residual variance. However, underfitted models also tend to have omitted parameters and thus higher residual variance than the correct or even overfitted models. If we treat bootstrapping in AR as we did in regression, we must make the assumption that any bias due to bootstrapping correlated errors for underfitted models is much smaller than the increased size of the residuals due to the noncentrality parameter.

As described in Chapter 3, Section 3.1.1, we obtain our AR regression model by conditioning on the past and forming the design matrix $X$ with elements

$$x_{t,j} = y_{t-j} \text{ for } j = 1, \ldots, p \text{ and } t = p+1, \ldots, n.$$

We recall that the first $p$ observations will be lost due to conditioning on the past observations, and hence the dimensions of the design matrix $X$ are $(n-p) \times p$. Assuming no intercept is included, the AR($p$) model can now be written as

$$Y = X\phi + w,$$

where $Y$ is an $(n-p)\times 1$ vector, $X$ is an $(n-p)\times p$ matrix, $w = (w_{p+1}, \ldots, w_n)'$ is an $(n-p) \times 1$ vector, and the $w_t$ are *i.i.d.* $N(0, \sigma^2)$. Now, $x_t$ represents the past values $y_{t-1}, \ldots, t_{t-p}$ associated with $y_t$. This allows us to treat AR models as special regression models.

The refined bootstrap estimate from Eq. (6.19) can be adapted to the autoregressive model as follows:

$$\widehat{\text{err}}(z,\hat{F}) = \hat{\sigma}^2 + \frac{1}{R(n-p)}\left[\sum_{r=1}^{R}\sum_{t=p+1}^{n}[(y_t - x_t'\hat{\phi}_r^*)^2] - \sum_{r=1}^{R}\sum_{t=p+1}^{n}[(y_t^* - x_t^{*\prime}\hat{\phi}_r^*)^2]\right], \tag{6.23}$$

where

$$\begin{aligned}\hat{\sigma}^2 &= \frac{1}{n-p}\sum_{t=p+1}^{n}(y_t - \hat{\phi}_1 y_{t-1} - \cdots - \hat{\phi}_p y_{t-p})^2 \\ &= \frac{1}{n-p}\sum_{t=p+1}^{n}(y_t - x_t'\hat{\phi})^2.\end{aligned}$$

The model minimizing Eq. (6.23) is selected as the best model. The two methods for forming bootstrap samples we have discussed previously, sampling pairs and sampling residuals, can also be applied to AR models.

By conditioning on the past, we can form the pairs $(x_t', y_t)$ for $t = p+1, \ldots, n$. We draw samples of size $n-p$ by randomly selecting, with replacement, from the original $n-p$ observations. This bootstrap random sample is $\{(x_t^{*\prime}, y_t^*) : t = p+1, \ldots, n\}$, where $x_t^* = (y_{t-1}^*, \ldots, y_{t-p}^*)'$. A straightforward application of Eq. (6.23) follows. As was the case in multiple regression, simulations show that this procedure overfits excessively.

The other approach is to bootstrap the residuals. Since we can write the AR model in regression form, the computational savings also apply to the AR model. The $t$th residual is

$$e_t = y_t - x_t'\hat{\phi},$$

and

$$\text{err}(z,\hat{F}) = \frac{1}{n-p}\sum_{t=p+1}^{n} e_t^2$$

is biased for the true residual variance. The $e_t$ tend to be smaller than the true errors; following Shao's (1996) recommendation we could inflate the residuals by $\sqrt{1/(1-p/(n-p))}$. This follows from the fact that $s^2 = \sum_{t=p+1}^{n} e_t^2/(n-2p)$ is unbiased for the true residual variance. On average, the $e_t$ are $1-p/(n-p)$ fold smaller than the true residuals. Let

$$\tilde{e}_t = e_t/\sqrt{1-p/(n-p)}.$$

Bootstrap samples $\tilde{e}_{p+1}^*, \ldots, \tilde{e}_n^*$ are generated by randomly selecting with replacement from $\tilde{e}_{p+1}, \ldots, \tilde{e}_n$ and forming $y_t^* = x_t'\hat{\phi} + \tilde{e}_t^*$.

Analogously, the errors can also be inflated by an inflation factor based on the cross-validation weights from the conditioned design matrix $X$, $1 - h_t$ in Eq. (6.2). Let

$$\hat{e}_t = e_t / \sqrt{1 - h_t}.$$

Then, bootstrap samples $\hat{e}^*_{p+1}, \ldots, \hat{e}^*_n$ are generated by randomly selecting with replacement from $\hat{e}_{p+1}, \ldots, \hat{e}_n$ and forming $y^*_t = x_t\hat{\phi} + \hat{e}^*_t$. Now, $\hat{\phi}^* = (X'X)^{-1}X'Y^*$ and $Y^* = (y^*_{p+1}, \ldots, y^*_n)'$. The bootstrap residual $v_t$ is

$$\begin{aligned} v_t &= y^*_t - x_t\hat{\phi}^* \\ &= y^*_t - \phi^*_1 y_{t-1} - \cdots - \phi^*_p y_{t-p}. \end{aligned}$$

The five computational steps we used in Section 6.5.2 can be modified to the autoregressive setting. We can compute the naive bootstrap

$$\widetilde{\text{err}}^{cv}(z^*, \hat{F}) = \frac{1}{R(n-p)} \sum_{r=1}^{R} \sum_{t=p+1}^{n} v^2_{tr}. \tag{6.24}$$

However, this procedure also overfits, again suggesting that a penalty function is required to prevent overfitting. Adding a penalty function based on parameter counts, we obtain

$$\text{BFPE} = \widetilde{\text{err}}^{cv}(z^*, \hat{F}) \frac{n}{n - 2p}. \tag{6.25}$$

Simulation results in Section 6.9 show that Eq. (6.25) performs much better than the nonpenalty weighted bootstraps. We can also use the doubly cross-validated bootstrap, introduced in Section 6.5, which does not depend on parameter counts. DCVB for autoregressive models is formed by inflating the $v_{tr}$ in Eq. (6.24) through $1 - h_t$, resulting in

$$\text{DCVB} = \frac{1}{R(n-p)} \sum_{r=1}^{R} \sum_{t=p+1}^{n} \frac{v^2_{tr}}{(1 - h_t)^2}. \tag{6.26}$$

## 6.7. Multivariate Regression Bootstrap

When we model multivariate regression we must always address the fact that the target functions are matrices. In Section 6.3 we noted that MPEP is commonly used as a basis for multivariate model selection and for multivariate cross-validation model selection. Hence, we define the prediction error for the new independent observation $(x_0', y_0')$ as

$$\text{err}(z, F) = E_{0F}\left[(y_0' - x_0'B)'(y_0' - x_0'B)\right], \tag{6.27}$$

where $y_0$ and $x_0$ are $q \times 1$ and $k \times 1$ vectors, respectively. By analogy to Section 6.5, some common estimates of Eq. (6.27) are
1) the parametric estimate, FPE,

$$\text{FPE} = \hat{\Sigma}_k \frac{n+k}{n-k};$$

2) the cross-validated estimate,

$$\text{CV}(1) = \frac{1}{n}\sum_{i=1}^{n}\left[(y_i' - x_i'\hat{B}_{(i)})'(y_i' - x_i'\hat{B}_{(i)})\right]; \text{ and}$$

3) the bootstrap.

Before discussing the bootstrap in the context of multivariate regression, we need to introduce the following notation. Let

$$\text{err}(z^*, \hat{F}) = \frac{1}{n}\sum_{i=1}^{n}\left[(y_i' - x_i'\hat{B}^*)'(y_i' - x_i'\hat{B}^*)\right] \tag{6.28}$$

represent the prediction error from one bootstrapped sample using the original data and the bootstrapped estimate of $B$, $\hat{B}^*$. Let

$$\text{err}(z^*, \hat{F}^*) = \frac{1}{n}\sum_{i=1}^{n}\left[(y_i^{*\prime} - x_i^{*\prime}\hat{B}^*)'(y_i^{*\prime} - x_i^{*\prime}\hat{B}^*)\right]$$

represent the prediction error from one bootstrapped sample using the bootstrapped data as well as the bootstrapped estimate of $B$. Let

$$\text{err}(z, \hat{F}) = \frac{1}{n}\sum_{i=1}^{n}\left[(y_i' - x_i'\hat{B})'(y_i' - x_i'\hat{B})\right]$$

represent $\hat{\sigma}^2$, the usual MLE for the residual variance. All of these functions are estimates of Eq. (6.27).

First we randomly select, with replacement, a sample of size $n$ from $(x_1', y_1'), \ldots, (x_n', y_n')$, compute $\hat{B}^*$ from this new sample, and then calculate the following estimate from the first bootstrap sample:

$$\text{err}(z^*, \hat{F}) = \frac{1}{n}\sum_{i=1}^{n}\left[(y_i' - x_i'\hat{B}^*)'(y_i' - x_i'\hat{B}^*)\right].$$

As noted in Section 6.5, a bootstrap estimate of Eq. (6.27) can be produced by repeatedly sampling the data $R$ times. Specifically, compute the $err(z^*, \hat{F})$ in Eq. (6.28) for each pseudo-sample, and then average the $R$ estimates to obtain the following estimate for the prediction error:

$$\begin{aligned}\widetilde{\text{err}}(z, \hat{F}) &= \frac{1}{R}\sum_{r=1}^{R}\sum_{i=1}^{n}\text{err}(z_r^*, \hat{F}) \\ &= \frac{1}{Rn}\sum_{r=1}^{R}\sum_{i=1}^{n}\left[(y_i' - x_i'\hat{B}_r^*)'(y_i' - x_i'\hat{B}_r^*)\right].\end{aligned}$$

However, because such a bootstrapped function may be biased, we once again use Efron and Tibshirani's (1993) refined bootstrap, which estimates the bias in $\text{err}(z, \hat{F})$ for estimating $\text{err}(z, F)$. We know that $\text{err}(z, \hat{F})$ underestimates the true prediction error $\text{err}(z, F)$, and we can define the average optimism as

$$ao(F) = E_F\left[\text{err}(z, F) - \text{err}(z, \hat{F})\right].$$

The idea now is to obtain a bootstrapped estimate for $ao(F)$:

$$\begin{aligned}\widehat{ao}(\hat{F}) =& \frac{1}{R}\sum_{r=1}^{n}\left[\text{err}(z^*, \hat{F}) - \text{err}(z^*, \hat{F}^*)\right] \\ =& \frac{1}{Rn}\Bigg[\sum_{r=1}^{R}\sum_{i=1}^{n}\left[(y_i' - x_i'\hat{B}_r^*)'(y_i' - x_i'\hat{B}_r^*)\right] \\ & - \sum_{r=1}^{R}\sum_{i=1}^{n}\left[(y_i^{*\prime} - x_i^{*\prime}\hat{B}_r^*)'(y_i^{*\prime} - x_i^{*\prime}\hat{B}_r^*)\right]\Bigg].\end{aligned}$$

Thus the refined bootstrapped estimate of Eq. (6.27) is

$$\begin{aligned}\widehat{\text{err}}(z, \hat{F}) =& \hat{\Sigma} + \frac{1}{Rn}\Bigg[\sum_{r=1}^{R}\sum_{i=1}^{n}\left[(y_i' - x_i'\hat{B}_r^*)'(y_i' - x_i'\hat{B}_r^*)\right] \\ & - \sum_{r=1}^{R}\sum_{i=1}^{n}\left[(y_i^{*\prime} - x_i^{*\prime}\hat{B}_r^*)'(y_i^{*\prime} - x_i^{*\prime}\hat{B}_r^*)\right]\Bigg].\end{aligned} \tag{6.29}$$

The model minimizing the determinant or trace of Eq. (6.29) is selected as the best model.

We must next consider the issue of sampling pairs versus sampling errors to obtain bootstrap samples for the multivariate regression case. If we randomly select pairs $(x_i', y_i')$ as our bootstrapped samples, we again have the advantages that no assumptions are required about the errors and that this procedure is more flexible in nonlinear models. Bootstrapping pairs remains appropriate when both $x$ and $y$ are random. Here again bootstrapping pairs is computationally much slower than bootstrapping residuals, and the added flexibility is not worth the increased computational intensity. When selecting pairs, samples of size $n$ are drawn by randomly selecting, with replacement, from the original $n$ observations. This bootstrap sample is $\{(x_1^{*\prime}, y_1^{*\prime}), \ldots, (x_n^{*\prime}, y_n^{*\prime})\}$. A straightforward application of Eq. (6.29) gives the refined bootstrap estimate of Eq. (6.27). We will see in Section 6.9 that this procedure overfits excessively for multivariate as well as univariate regression.

By bootstrapping residuals, we again gain the advantage of computational savings in the linear model setting, while sacrificing flexibility. In this setting bootstrapping residuals again seems to perform as well as bootstrapping pairs, although additional care is needed since the residuals are themselves biased. The $i$th residual is

$$e_i' = y_i' - x_i'\hat{B}.$$

Furthermore,

$$\mathrm{err}(z, \hat{F}) = \frac{1}{n}\sum_{i=1}^{n} e_i e_i'$$

is biased for the true residual variance; the $e_i$ tend to be smaller than the true errors and it must be inflated. Shao's (1996) recommendation of inflating the residuals by $\sqrt{1/(1-k/n)}$ can again be applied as follows:

$$\tilde{e}_i = e_i/\sqrt{1-k/n}.$$

As stated in Section 6.5.1, there are two disadvantages in using $\tilde{e}_i$. Hence, we adopt Wu's (1986) inflation factor $1 - h_i$ to obtain the adjusted residual

$$\hat{e}_i = e_i/\sqrt{1-h_i}.$$

Then we generate bootstrap samples, $\hat{e}_1^*, \ldots, \hat{e}_n^*$, by randomly selecting, with replacement, from $\hat{e}_1, \ldots, \hat{e}_n$ and form $y_i^{*\prime} = x_i'\hat{B} + \hat{e}_i^{*\prime}$. Next, we obtain $\hat{B}^* = (X'X)^{-1}X'Y^*$ and $v_i{}' = y_i' - x_i'\hat{B}^*$, where $Y^* = (y_1^*, \ldots, y_n^*)'$. Finally, we have a naive bootstrap

$$\widetilde{\mathrm{err}}^{cv}(z^*, \hat{F}) = \frac{1}{Rn}\sum_{r=1}^{R}\sum_{i=1}^{n} v_{ir}v_{ir}'. \tag{6.30}$$

Even with inflated residuals, $\widetilde{\mathrm{err}}^{cv}(z^*, \hat{F})$ underestimates the true residual variance, suggesting that a penalty function is required to prevent overfitting.

One possible approach is to modify Eq. (6.30) by adding a parametric penalty function based on parameter counts, which leads to

$$\text{BFPE} = \widetilde{\text{err}}^{cv}(z^*, \hat{F})\frac{n+k}{n-k}.$$

Finally, two model selection criteria can be derived from BFPE using the determinant and trace, as follows. We let

$$\text{deBFPE} = \left|\widetilde{\text{err}}^{cv}(z^*, \hat{F})\frac{n+k}{n-k}\right| \tag{6.31}$$

and

$$\text{trBFPE} = tr\left\{\widetilde{\text{err}}^{cv}(z^*, \hat{F})\frac{n+k}{n-k}\right\}. \tag{6.32}$$

Simulations in Chapter 9 show that deBFPE performs much better than the nonpenalty weighted bootstraps. While this penalty function depends on a parameter count, we can also obtain a penalty function bootstrap that does not depend on parameter counts—the doubly cross-validated bootstrap, which inflates the bootstrapped residuals by the cross-validation weights, $1-h_i$, Eq. (6.2). In multivariate regression DCVB is

$$\text{DCVB} = \frac{1}{Rn}\sum_{r=1}^{R}\sum_{i=1}^{n}\frac{v_{ir}v_{ir}'}{(1-h_i)^2}.$$

Forming scalars, we can define

$$\text{deDCVB} = \left|\frac{1}{Rn}\sum_{r=1}^{R}\sum_{i=1}^{n}\frac{v_{ir}v_{ir}'}{(1-h_i)^2}\right| \tag{6.33}$$

and

$$\text{trDCVB} = tr\left\{\frac{1}{Rn}\sum_{r=1}^{R}\sum_{i=1}^{n}\frac{v_{ir}v_{ir}'}{(1-h_i)^2}\right\}. \tag{6.34}$$

Simulation studies in Chapter 9 show that DCVB outperforms the other bootstraps in this Section, and is competitive with AICc.

## 6.8. Vector Autoregressive Bootstrap

We generate vector autoregressive VAR models for bootstrapping in the same way as in Chapter 5, using the VAR model given by Eq. (5.1) and

Eq. (5.2), where we condition on the past and form the design matrix $X$. The rows of $X$ have the structure

$$x'_t = (y_{1,t-1}, \ldots, y_{q,t-1}, y_{1,t-2}, \ldots, y_{q,t-2}, \ldots, y_{1,t-p}, \ldots, y_{q,t-p}).$$

Because we have no observations before $y_1$ and we condition on the past of $Y$, the first $p$ observations at the beginning of the series are lost, and thus the design matrix $X$ has dimensions $(n-p) \times (pq)$. The usual VAR estimates are computed from the full data model to yield estimates $\hat{\Phi}$ and residuals $e$. By conditioning on the past, the VAR model can be written in terms of the pairs $(x'_t, y'_t)$ with residuals $e'_t = y'_t - x'_t\hat{\Phi}$. Applying the bootstrap techniques from Section 6.7, we can either resample the pairs $(x'_t, y'_t)$ or resample the adjusted residuals $\hat{e}_t = e_t/\sqrt{1-h_t}$ for $t = p+1, \ldots, n$. The refined bootstrap refined bootstrapped estimate of the mean product error of prediction matrix MPEP is

$$\begin{aligned}\widehat{\text{err}}(z, \hat{F}) =& \hat{\Sigma} + \frac{1}{R(n-p)}\left[\sum_{r=1}^{R}\sum_{t=p+1}^{n}[(y'_t - x'_t\hat{\Phi}^*_r)'(y'_t - x'_t\hat{\Phi}^*_r)]\right.\\ &\left. - \sum_{r=1}^{R}\sum_{t=p+1}^{n}[(y^{*\prime}_t - x^{*\prime}_t\hat{\Phi}^*_r)'(y^{*\prime}_t - x^{*\prime}_t\hat{\Phi}^*_r)]\right]. \qquad (6.35)\end{aligned}$$

The model minimizing either determinant or trace of Eq. (6.35) is selected as the best model.

In VAR models, the bootstrap residual is

$$\begin{aligned}v_t &= y'_t - x'_t\hat{\Phi}^* \\ &= y'_t - \hat{\Phi}^*_1 y_{t-1} - \cdots - \hat{\Phi}^*_p y_{t-p}.\end{aligned}$$

Hence, the naive bootstrap estimate is

$$\widetilde{\text{err}}^{cv}(z^*, \hat{F}) = \frac{1}{R(n-p)}\sum_{r=1}^{R}\sum_{t=p+1}^{n} v_{tr}v'_{tr}.$$

Simulation studies in Chapter 9 show that the naive estimate with inflated residuals performs about the same as the refined estimates; that is to say, both overfit.

Bootstrap selection criteria can be derived for VAR models with either a parameter count based penalty function or the doubly cross-validated penalty function. We first derive the bootstrap model selection by adding a parametric penalty function based on parameter counts:

$$\text{BFPE} = \widetilde{\text{err}}^{cv}(z^*, \hat{F})\frac{n + (q-1)p}{n - (q+1)p}.$$

Two model selection criteria can be derived from BFPE using the determinant and trace. We define

$$\text{deBFPE} = \left|\widetilde{\text{err}}^{cv}(z^*, \hat{F})\frac{n + (q-1)p}{n - (q+1)p}\right| \tag{6.36}$$

and

$$\text{trBFPE} = tr\left\{\widetilde{\text{err}}^{cv}(z^*, \hat{F})\frac{n + (q-1)p}{n - (q+1)p}\right\}. \tag{6.37}$$

Simulations in Chapter 9 show that, as expected, deBFPE performs much better than the nonpenalty weighted bootstrap version.

The doubly cross-validated bootstrap (DCVB) for vector autoregressive models is

$$\text{DCVB} = \frac{1}{R(n-p)}\sum_{r=1}^{R}\sum_{t=p+1}^{n}\frac{v_{tr}v_{tr}'}{(1-h_t)^2}.$$

Forming model selection criteria, we define

$$\text{deDCVB} = \left|\frac{1}{R(n-p)}\sum_{r=1}^{R}\sum_{t=p+1}^{n}\frac{v_{tr}v_{tr}'}{(1-h_t)^2}\right| \tag{6.38}$$

and

$$\text{trDCVB} = tr\left\{\frac{1}{R(n-p)}\sum_{r=1}^{R}\sum_{t=p+1}^{n}\frac{v_{tr}v_{tr}'}{(1-h_t)^2}\right\}. \tag{6.39}$$

Simulation studies in Chapter 9 show that DCVB outperforms the other bootstraps for VAR, including those with parametric penalty functions, and is competitive with AICc.

## 6.9. Monte Carlo Study

In this Section we will focus only on the univariate case. However, simulation studies for the multivariate case are given in Chapter 9.

One issue associated with both CV($d$) and bootstrapping is how to choose the number of replications or bootstrap samples. We will consider both regression and autoregressive models, first by evaluating the effect of the number of bootstrap samples, then by looking at special case models to assess the effect of the sample size. Finally, we test the performance of our usual selection criteria as well as the resampling criteria introduced in this Chapter via extensive simulation studies that vary many different model parameters.

### 6.9.1. Univariate Regression

#### 6.9.1.1. Model Formation

We first consider regression models

$$y_i = \beta_0 + \beta_1 x_{i,1} + \cdots + \beta_{k_*-1} x_{i,k_*-1} + \varepsilon_{*i}, \quad i = 1, \ldots, n,$$

where the $\varepsilon_{*i}$ are *i.i.d.* $N(0, \sigma_*^2)$ and $x_{i,j} \sim N(0,1)$ for $j \geq 1$ and $x_{i,0} = 1$. Further, for each observation $i$, the $x_{i,j}$ can be correlated according to the structure $corr(x_{i,j-1}, x_{i,j}) = \rho_x$, $j = 2, \ldots, k_* - 1$. The regression models we will consider can differ by the sample size $n$, true model order $k_*$, amount of overfitting $o$, correlation $\rho_x$, and parameter structure $\beta_1, \ldots, \beta_{k_*-1}$. The available choices for $k_*$ are $k_* = 3, 6$, where each of these has two possible levels of overfitting: $o = 2$ or 5 extra variables. Two parameter structures are used, and these are described in Table 6.2. Two levels of $X$-correlation are used, $\rho_x = 0, 0.9$, where all $x_{i,j}$ $(j \geq 1)$ are *i.i.d.* when $\rho_x = 0$. A value of $\rho_x = 0.9$ represents a high degree of multicollinearity in the design matrix. Performance will be examined in three sample sizes, $n$ = 15, 35, 100, each with error variance $\sigma_*^2 = 1$. Overall, this yields 3 sample sizes × 2 parameter structures × 2 values for the true order × 2 levels of overfitting × 2 levels of $X$-correlation, or 48 separate possible true regression models, summarized in Table 6.1.

Table 6.1. Summary of the regression models in simulation study.

| sample size $n$ | error variance $\sigma_*^2$ | parameter structure $\beta_j$ | true order $k_*$ | overfitting $o$ | $\rho_x$ |
|---|---|---|---|---|---|
| 15 | 1 | $1/j$, 1 | 3, 6 | 2, 5 | 0, 0.9 |
| 35 | 1 | $1/j$, 1 | 3, 6 | 2, 5 | 0, 0.9 |
| 100 | 1 | $1/j$, 1 | 3, 6 | 2, 5 | 0, 0.9 |

Table 6.2 Relationship between parameter structure and true order.

Parameter structure 1: $\beta_j = 1/j$
- $k_* = 3$ $\quad \beta_0 = 1, \beta_1 = 1, \beta_2 = 1/2$
- $k_* = 6$ $\quad \beta_0 = 1, \beta_1 = 1, \beta_2 = 1/2, \beta_3 = 1/3, \beta_4 = 1/4, \beta_5 = 1/5$

Parameter structure 2: $\beta_j = 1$
- $k_* = 3$ $\quad \beta_0 = 1, \beta_1 = 1, \beta_2 = 1$
- $k_* = 6$ $\quad \beta_0 = 1, \beta_1 = 1, \beta_2 = 1, \beta_3 = 1, \beta_4 = 1, \beta_5 = 1$

One hundred realizations are produced for each of the 48 models. For each realization Kullback–Leibler (K-L) and $L_2$ observed efficiency are computed for each candidate model, where K-L observed efficiency is defined using Eq. (1.2) and Eq. (2.10), and $L_2$ observed efficiency is defined using Eq. (1.1) and Eq. (2.9). The selection criteria each select a model, and the observed efficiencies for each are ranked, where lower rank indicates better performance. Details are presented in Tables 6.6 and 6.7.

#### 6.9.1.2. *Effect of Bootstrap Sample Size R*

We will first look at the effect of the number of bootstrap samples on each resampling criterion's performance. Each possible model is replicated 100 times. All candidate models include the intercept so the number of candidate models will depend on the choice of $k_*$ and $o$: for $k_* = 3, o = 2$ there are 16 candidate models, for $k_* = 3, o = 5$ or $k_* = 6, o = 2$ there are 128 candidate models, and for $k_* = 6, o = 5$ there are 1024 candidate models.

Table 6.3. Bootstrap relative K-L performance.
Smaller rank represents better performance.

| resampling criterion | | Bootstrap replications $R$ 5 | 10 | 25 | 50 |
|---|---|---|---|---|---|
| CV($d$) | Eq. (6.5) | 4 | 3 | 2 | 1 |
| bR | Eq. (6.19) residuals | 1 | 4 | 2 | 3 |
| bP | Eq. (6.19) pairs | 1 | 2 | 3 | 4 |
| nB | Eq. (6.20) | 4 | 3 | 1 | 2 |
| BFPE | Eq. (6.21) | 4 | 3 | 2 | 1 |
| DCVB | Eq. (6.22) | 4 | 3 | 2 | 1 |

Note that these rankings are for row comparisons only.

For model selection, only the mean is of interest rather than a representation of the entire distribution, or histogram. Histograms require much more detail than estimating a mean and thus more bootstrap replications. However, bootstrapping in model selection is similar to bootstrapping a mean in that we do not need details for the distribution of the criterion. Hence, fewer bootstrap replications may be needed; for example, in some cases as few as five bootstrap samples can give satisfactory results. We will examine four values (5, 10, 25 and 50) for the number of bootstrap samples, $R$. Note that computation time will increase substantially with $R$ particularly when sampling pairs and for CV($d$). Although we might expect performance to improve as $R$ increases, the simulation results in Table 6.3 show that this is not necessarily the case. Since the $L_2$ results are very similar to the K-L results, they are not presented here. The rankings in Table 6.3 represent the relative performance for each criterion among its different bootstrap sample sizes, $R$. Table 6.3 does not compare performance between the different criteria. Two procedures, bR and bP, actually performed worse with increased $R$. CV($d$), BFPE and DCVB perform best at $R = 50$, and the performance of nB is poor for small $R$. For the sake of simplicity, we use $R = 5$ for bR and bP and use $R = 50$ for CV($d$), BFPE, DCVB, and nB in the resampling procedures discussed in the next Subsection.

#### 6.9.1.3. Special Case Regression Models

Next we will look at simulation results for the two special case models, for which we vary only the sample size. Both models have $k_* = 6$, $o = 5$, and parameter structure $\beta_j = 1$ with independent columns of $X$. The errors in both models are standard normals. Model 11 has a small sample size of $n = 15$ and Model 12 has a moderate sample size of $n = 100$. Based on what we have already seen in this Chapter, we expect the criteria to perform worse with Model 11, since criteria with strong penalty functions will underfit badly and criteria with weak penalty functions will overfit badly. On the other hand, because of its larger sample size, criteria with strong penalty functions should perform much better than the criteria with weak penalty functions for Model 12. Tables 6.4 and 6.5 summarize the results for Models 11 and 12, respectively.

The selection criteria used in Tables 6.4–6.7 are:

| | |
|---|---|
| AICc | Eq. (2.14). |
| AICu | Eq. (2.18). |
| SIC | Eq. (2.15). |
| CV | Withhold-1 cross-validation, CV(1) Eq. (6.3). |
| CVd | delete-$d$ cross-validation, CV($d$) Eq. (6.5) using 50 subsamples. |
| bR | Refined bootstrap estimate Eq. (6.19) by resampling residuals using 5 pseudo-samples. |
| bP | Refined bootstrap estimate Eq. (6.19) by resampling pairs $(Y, X)$ using 5 pseudo-samples. |
| nB | Naive bootstrap estimate Eq. (6.20) using 50 pseudo-samples. |
| BFPE | Penalty weighted naive bootstrap estimate Eq. (6.21) using 50 pseudo-samples. |
| DCVB | Doubly cross-validated bootstrap Eq. (6.22) using 50 pseudo-samples. |

Table 6.4 shows a clear pattern in the counts. The criteria with strong

penalty functions, AICc and AICu, underfit badly, but hardly overfit at all. The two bootstraps, BFPE and DCVB, also have large penalty functions, but they underfit less severely even though the sample size is small. SIC has a weak penalty function in small samples and thus overfits. The three bootstrap selection criteria, bR, bP, and nB, all overfit excessively. Indeed, these three are refinements to $R^2_{adj}$ which was seen in Chapter 2 to have a very weak signal-to-noise ratio, therefore they behave as though they have very weak penalty functions (much like $R^2_{adj}$). The withhold-1 CV, which is asymptotically efficient, not surprisingly overfits a bit more than the delete-$d$ criterion CVd, which was designed to be consistent. $L_2$ has a well-defined minimum distance at the true model as can be seen from the counts. K-L does not have such a well-defined minimum (no particular model has high counts).

In terms of observed efficiency, the two distance measures perform differently for Model 11. As observed in earlier chapters, K-L seems to favor underfitting more than overfitting. This can be seen in the better performance

Table 6.4. Simulation results for Model 11. Counts and observed efficiency.

| | | | | | counts | | | | | | | |
|---|---|---|---|---|---|---|---|---|---|---|---|---|
| $k$ | AICc | AICu | SIC | CV | CVd | bR | bP | nB | BFPE | DCVB | $L_2$ | K-L |
| 1 | 2 | 14 | 0 | 0 | 1 | 0 | 0 | 0 | 0 | 2 | 0 | 0 |
| 2 | 24 | 97 | 1 | 1 | 4 | 0 | 0 | 0 | 3 | 20 | 0 | 1 |
| 3 | 119 | 212 | 7 | 6 | 30 | 1 | 0 | 0 | 19 | 90 | 0 | 2 |
| 4 | 253 | 293 | 20 | 36 | 103 | 1 | 0 | 1 | 50 | 151 | 3 | 9 |
| 5 | 297 | 224 | 67 | 122 | 227 | 16 | 0 | 15 | 135 | 251 | 14 | 111 |
| 6 | 254 | 143 | 192 | 292 | 316 | 82 | 34 | 116 | 319 | 288 | 901 | 474 |
| 7 | 47 | 16 | 233 | 259 | 236 | 194 | 177 | 236 | 243 | 136 | 51 | 246 |
| 8 | 4 | 1 | 224 | 186 | 73 | 297 | 424 | 303 | 141 | 42 | 24 | 118 |
| 9 | 0 | 0 | 151 | 76 | 10 | 255 | 293 | 219 | 64 | 18 | 7 | 35 |
| 10 | 0 | 0 | 79 | 20 | 0 | 119 | 70 | 87 | 22 | 2 | 0 | 4 |
| 11 | 0 | 0 | 26 | 2 | 0 | 35 | 2 | 23 | 4 | 0 | 0 | 0 |
| true | 156 | 89 | 100 | 133 | 101 | 22 | 1 | 54 | 167 | 154 | 857 | 234 |

| | | | | | K-L observed efficiency | | | | | | | |
|---|---|---|---|---|---|---|---|---|---|---|---|---|
| | AICc | AICu | SIC | CV | CVd | bR | bP | nB | BFPE | DCVB | $L_2$ | K-L |
| ave | 0.36 | 0.37 | 0.20 | 0.25 | 0.33 | 0.19 | 0.41 | 0.18 | 0.25 | 0.32 | 0.68 | 1.00 |
| med | 0.33 | 0.36 | 0.13 | 0.17 | 0.28 | 0.13 | 0.40 | 0.12 | 0.17 | 0.29 | 0.72 | 1.00 |
| sd | 0.22 | 0.19 | 0.21 | 0.23 | 0.22 | 0.19 | 0.21 | 0.18 | 0.24 | 0.22 | 0.30 | 0.00 |
| rank | 3 | 2 | 8 | 6 | 4 | 9 | 1 | 10 | 6 | 5 | | |

| | | | | | $L_2$ observed efficiency | | | | | | | |
|---|---|---|---|---|---|---|---|---|---|---|---|---|
| | AICc | AICu | SIC | CV | CVd | bR | bP | nB | BFPE | DCVB | $L_2$ | K-L |
| ave | 0.50 | 0.42 | 0.56 | 0.57 | 0.53 | 0.55 | 0.38 | 0.56 | 0.58 | 0.53 | 1.00 | 0.70 |
| med | 0.45 | 0.36 | 0.55 | 0.55 | 0.49 | 0.54 | 0.34 | 0.55 | 0.55 | 0.48 | 1.00 | 0.71 |
| sd | 0.28 | 0.26 | 0.23 | 0.25 | 0.26 | 0.22 | 0.22 | 0.22 | 0.26 | 0.28 | 0.00 | 0.26 |
| rank | 8 | 9 | 3 | 2 | 6 | 5 | 10 | 3 | 1 | 6 | | |

of AICc and AICu with respect to K-L as opposed to $L_2$. On the other hand, $L_2$ favors overfitting more than underfitting. Both penalize excessive underfitting or overfitting. In general, the criteria that overfit perform better with respect to $L_2$ while those that underfit perform better with respect to K-L. One exception is bP, which performs very well in Model 11 with respect to K-L observed efficiency. BFPE is best in the $L_2$ sense. The results vary greatly between count data and K-L and $L_2$ observed efficiencies, making it virtually impossible to identify an overall rank. None of the criteria do consistently well over both observed efficiency measures.

Next we look at the results for special case Model 12 with the sample size of 100. Counts and observed efficiencies are summarized in Table 6.5. We see that with the larger sample size the correct model becomes easier to detect, and there is also very good agreement on relative performance among the two observed efficiency measures. One hundred is a large enough sample size for the consistent properties of SIC to manifest themselves, and in fact SIC

Table 6.5. Simulation results for Model 12. Counts and observed efficiency.

| | | | | | counts | | | | | | | |
|---|---|---|---|---|---|---|---|---|---|---|---|---|
| $k$ | AICc | AICu | SIC | CV | CVd | bR | bP | nB | BFPE | DCVB | $L_2$ | K-L |
| 1 | 0 | 0 | 0 | 0 | 0 | 0 | 0 | 0 | 0 | 0 | 0 | 0 |
| 2 | 0 | 0 | 0 | 0 | 0 | 0 | 0 | 0 | 0 | 0 | 0 | 0 |
| 3 | 0 | 0 | 0 | 0 | 0 | 0 | 0 | 0 | 0 | 0 | 0 | 0 |
| 4 | 0 | 0 | 0 | 0 | 0 | 0 | 0 | 0 | 0 | 0 | 0 | 0 |
| 5 | 0 | 0 | 0 | 0 | 0 | 0 | 0 | 0 | 0 | 0 | 0 | 0 |
| 6 | 456 | 662 | 815 | 395 | 164 | 26 | 17 | 92 | 571 | 601 | 1000 | 949 |
| 7 | 381 | 284 | 170 | 391 | 441 | 155 | 121 | 281 | 329 | 304 | 0 | 43 |
| 8 | 142 | 48 | 14 | 173 | 306 | 328 | 306 | 324 | 88 | 81 | 0 | 8 |
| 9 | 18 | 6 | 1 | 38 | 83 | 291 | 323 | 224 | 10 | 14 | 0 | 0 |
| 10 | 3 | 0 | 0 | 3 | 6 | 163 | 188 | 63 | 2 | 0 | 0 | 0 |
| 11 | 0 | 0 | 0 | 0 | 0 | 37 | 45 | 16 | 0 | 0 | 0 | 0 |
| true | 456 | 662 | 815 | 395 | 163 | 26 | 17 | 92 | 571 | 601 | 1000 | 997 |
| | | | | | K-L observed efficiency | | | | | | | |
| | AICc | AICu | SIC | CV | CVd | bR | bP | nB | BFPE | DCVB | $L_2$ | K-L |
| ave | 0.74 | 0.83 | 0.90 | 0.72 | 0.79 | 0.67 | 0.68 | 0.61 | 0.79 | 0.81 | 1.00 | 1.00 |
| med | 0.76 | 1.00 | 1.00 | 0.73 | 0.85 | 0.69 | 0.71 | 0.60 | 1.00 | 1.00 | 1.00 | 1.00 |
| sd | 0.27 | 0.26 | 0.22 | 0.26 | 0.22 | 0.23 | 0.22 | 0.23 | 0.26 | 0.26 | 0.00 | 0.00 |
| rank | 6 | 2 | 1 | 7 | 4 | 9 | 8 | 10 | 4 | 3 | | |
| | | | | | $L_2$ observed efficiency | | | | | | | |
| | AICc | AICu | SIC | CV | CVd | bR | bP | nB | BFPE | DCVB | $L_2$ | K-L |
| ave | 0.74 | 0.83 | 0.90 | 0.72 | 0.79 | 0.66 | 0.68 | 0.61 | 0.79 | 0.81 | 1.00 | 1.00 |
| med | 0.77 | 1.00 | 1.00 | 0.74 | 0.85 | 0.68 | 0.70 | 0.61 | 1.00 | 1.00 | 1.00 | 1.00 |
| sd | 0.27 | 0.26 | 0.22 | 0.27 | 0.22 | 0.23 | 0.22 | 0.23 | 0.27 | 0.26 | 0.00 | 0.00 |
| rank | 6 | 2 | 1 | 7 | 4 | 9 | 8 | 10 | 4 | 3 | | |

has the best performance over all measures. Although CVd is asymptotically equivalent to SIC, this sample size still does not allow CVd to perform as well as SIC. Since the true model belongs to the set of candidate models, CVd outperforms CV. The overfitting tendencies of bR, bP and nB have become evident. Of the bootstrapping procedures, BFPE and DCVB perform the best since their penalty functions reduce overfitting tendencies.

In this moderate sample size example, a pattern can be seen in the counts. We relate the performance of all the criteria to overfitting probabilities discussed in Chapter 2. Recall that $R^2_{adj}$ is an $\alpha 1$ criterion and overfits with the highest probability. Here, bP, bR and nB all behave similarly to $R^2_{adj}$—all overfit excessively. Efficient criteria are $\alpha 2$. AICc and CV are efficient, and while they overfit moderately in Model 12, they overfit less severely than the refinements to $R^2_{adj}$, bP, bR, and nB. Note that AICu is $\alpha 3$ and overfits less than the efficient criteria. Although we do not give a proof here, BFPE and DCVB are asymptotically equivalent to AICu (this will be discussed further in Chapter 9), and these three criteria overfit less than the others previously mentioned. Finally, SIC and CVd overfit the least since they are both consistent criteria, although some overfitting is still seen. Overall the observed efficiency results parallel the count patterns.

Both $L_2$ and K-L have well-defined minima at the true model. Criteria with high counts of selecting the true model obviously will have higher observed efficiencies. Relative performance among the criteria is the same for both measures—the less overfitting, the higher the observed efficiency. Because the model is strongly identifiable no underfitting is evident, and all performance depends on overfitting properties.

### *6.9.1.4. Large-scale Simulation*

Results from the limited special case models do not give us enough information to make generalizations about the performance of the criteria, and so we will also look at the performance of our ten criteria over 100 realizations and all possible 48 models. The results are generated as follows: for each of the 4800 realizations, each selection criterion selects a candidate model. The observed efficiencies of all criteria are computed and ranked, with rank 1 going to the criterion with highest observed efficiency and rank 10 going to the criterion with lowest observed efficiency. Ties get the average rank. Then an average rank over all 4800 realizations is computed for each criterion. Performance is evaluated based on this average of the individual realization ranks. Tables 6.6 and 6.7 summarize the overall results.

The relative performances under K-L and $L_2$ are virtually identical. AICu has the best overall performance, and we recall that AICu was also a strong performer in Chapter 2. The bootstrap procedures without a penalty function all perform poorly—there is little difference between bP, bR, or nB, which finish in the bottom three positions. The two weighted bootstraps, BFPE and DCVB do perform well overall, and CV(1) and CV($d$) perform near the middle, as does SIC. In the next Section we will see how these criteria perform in the AR setting.

Table 6.6. Simulation results over 48 regression models—K-L observed efficiency ranks.

| rank | AICc | AICu | SIC | CV | CVd | bR | bP | nB | BFPE | DCVB |
|---|---|---|---|---|---|---|---|---|---|---|
| 1 | 30 | 140 | 92 | 55 | 708 | 311 | 522 | 149 | 11 | 60 |
| 2 | 279 | 444 | 221 | 147 | 418 | 295 | 340 | 157 | 108 | 238 |
| 3 | 590 | 767 | 521 | 268 | 366 | 237 | 211 | 178 | 614 | 722 |
| 4 | 1295 | 1256 | 1148 | 1045 | 523 | 223 | 220 | 351 | 1290 | 1310 |
| 5 | 1000 | 935 | 889 | 902 | 581 | 310 | 313 | 506 | 1030 | 999 |
| 6 | 638 | 529 | 590 | 670 | 365 | 306 | 252 | 345 | 707 | 607 |
| 7 | 498 | 347 | 460 | 668 | 372 | 419 | 340 | 380 | 544 | 436 |
| 8 | 307 | 211 | 420 | 544 | 382 | 665 | 514 | 631 | 361 | 269 |
| 9 | 121 | 111 | 288 | 346 | 407 | 908 | 811 | 971 | 119 | 104 |
| 10 | 42 | 60 | 171 | 155 | 678 | 1126 | 1277 | 1132 | 16 | 55 |
| ave rank | 4.71 | 4.31 | 5.08 | 5.49 | 5.29 | 6.91 | 6.68 | 7.09 | 4.86 | 4.60 |
| ranking | 3 | 1 | 5 | 7 | 6 | 9 | 8 | 10 | 4 | 2 |

Table 6.7. Simulation results over 48 regression models—$L_2$ observed efficiency ranks.

| rank | AICc | AICu | SIC | CV | CVd | bR | bP | nB | BFPE | DCVB |
|---|---|---|---|---|---|---|---|---|---|---|
| 1 | 30 | 105 | 94 | 68 | 564 | 370 | 473 | 208 | 15 | 50 |
| 2 | 267 | 367 | 253 | 194 | 372 | 363 | 353 | 257 | 145 | 225 |
| 3 | 555 | 682 | 565 | 323 | 328 | 294 | 259 | 269 | 643 | 690 |
| 4 | 1238 | 1191 | 1186 | 1112 | 499 | 260 | 236 | 448 | 1279 | 1248 |
| 5 | 915 | 857 | 849 | 888 | 549 | 362 | 299 | 569 | 967 | 908 |
| 6 | 580 | 481 | 567 | 621 | 325 | 320 | 287 | 379 | 651 | 568 |
| 7 | 525 | 399 | 495 | 622 | 386 | 421 | 343 | 440 | 538 | 470 |
| 8 | 410 | 328 | 391 | 516 | 409 | 616 | 496 | 616 | 387 | 354 |
| 9 | 201 | 215 | 255 | 308 | 453 | 806 | 844 | 812 | 141 | 180 |
| 10 | 79 | 175 | 145 | 148 | 915 | 988 | 1210 | 802 | 34 | 107 |
| ave rank | 4.91 | 4.73 | 4.96 | 5.33 | 5.76 | 6.53 | 6.64 | 6.46 | 4.86 | 4.82 |
| ranking | 3 | 1 | 5 | 6 | 7 | 8 | 10 | 9 | 4 | 2 |

### *6.9.2. Univariate Autoregressive Models*

We will again consider the effect of R, the effect of sample size in special case models, and extensive simulations varying many model parameters, to evaluate the performance of selection criteria for autoregressive models.

#### 6.9.2.1. *Model Formation*

The data is generated as we described in the simulation section of Chapter 3 by using autoregressive models of the form

$$y_t = \phi_1 y_{t-1} + \cdots + \phi_{p_*} y_{t-p_*} + w_{*t}, \quad t = p_* + 1, \ldots, n,$$

where the $w_{*t}$ are *i.i.d.* $N(0, \sigma_*^2)$. The AR models considered here assume that $\sigma_*^2 = 1$, but they can differ by sample size $n$ (15, 35, 100), true model order $p_*$ (2, 5), level of overfitting $o$ (2, 5), and parameter structure $\phi_1, \ldots, \phi_{p_*}$. Three parameter structures are used. Overall, this yields 3 sample sizes $\times$ 3 parameter structures $\times$ 2 values for the true order $\times$ 2 levels of overfitting, or 36 possible different AR models, summarized in Table 6.8. Table 6.9 presents the relationship between the candidate model parameter structures and the true model.

Table 6.8 Summary of the autoregressive models.
All models have $\sigma_* = 1$.

| Sample Size $n$ | Parameter Structure $\phi_j$ | True Model Order $p_*$ | Overfitting $o$ |
|---|---|---|---|
| 15 | 3 structures | 2, 5 | 2, 5 |
| 35 | 3 structures | 2, 5 | 2, 5 |
| 100 | 3 structures | 2, 5 | 2, 5 |

Table 6.9 Relationship between parameter structure and true model order.

| | |
|---|---|
| Parameter structure 1: $\phi_j \propto 1/j^2$ | |
| $p_* = 2$ | $\phi_1 = 0.792, \phi_2 = 0.198$ |
| $p_* = 5$ | $\phi_1 = 0.676, \phi_2 = 0.169, \phi_3 = 0.075, \phi_4 = 0.042, \phi_5 = 0.027$ |
| Parameter structure 2: $\phi_j \propto 1/\sqrt{j}$ | |
| $p_* = 2$ | $\phi_1 = 0.580, \phi_2 = 0.410$ |
| $p_* = 5$ | $\phi_1 = 0.306, \phi_2 = 0.217, \phi_3 = 0.177, \phi_4 = 0.153, \phi_5 = 0.137$ |
| Parameter structure 3: seasonal random walk | |
| $p_* = 2$ | $\phi_2 = 1$ |
| $p_* = 5$ | $\phi_5 = 1$ |

#### 6.9.2.2. *Effect of Bootstrap Sample Size R in AR Models*

Table 6.10 summarizes resampling criterion performance with respect to the number of bootstrap replications, $R$, over our 48 AR models. One hundred realizations are generated for each model. The number of candidate models depends on $p_* + o$ as well as the sample size. Here, the maximum order considered is $p_* + o$, which is also the number of candidate models. Because we lose the first $p$ observations, the maximum order must be less than $n/2 - 1$. When $n = 15$ the maximum order is 6, regardless of the choice of $p_*$ and $o$.

As in the univariate regression study, observed efficiency is used to compare the performance of the selection criteria. For each realization and candidate model, K-L and $L_2$ observed efficiencies are computed. K-L observed efficiency is computed using Eq. (1.2) and Eq. (3.8) while $L_2$ observed efficiency is computed using Eq. (1.1) and Eq. (3.7). The selection criteria select their model and the observed efficiencies are recorded. The observed efficiencies for the individual criteria for each realization are then ranked, and the performance of each criterion is determined from these rankings. As in earlier chapters, lower ranks denote higher observed efficiency and better performance.

Table 6.10. Bootstrap relative K-L performance.
Smaller rank represents better performance.

| resampling criterion | | Bootstrap replications $R$ 5 | 10 | 25 | 50 |
|---|---|---|---|---|---|
| CV($d$) | Eq. (6.7) | 1 | 3 | 4 | 1 |
| bR | Eq. (6.23) residuals | 1 | 4 | 2 | 3 |
| bP | Eq. (6.23) pairs | 1 | 2 | 3 | 4 |
| nB | Eq. (6.24) | 4 | 3 | 1 | 2 |
| BFPE | Eq. (6.25) | 3 | 4 | 2 | 1 |
| DCVB | Eq. (6.26) | 4 | 3 | 1 | 2 |

Note that these rankings are for row comparisons only.

We see that, as in the regression simulations, performance in the AR setting does not always improve as the number of bootstrap samples increases. Table 6.10 presents performance row-by-row. BR and bP perform best when $R = 5$. The delete-$d$ cross-validation performs best when $R = 5$ followed closely by $R = 50$. However, the DCVB and BFPE perform best when $R = 25$ and $R = 50$, respectively. Comparing the observed efficiencies from each realization shows that there is little difference form DCVB for $R = 25$ and $R = 50$. The difference between $R = 25$ and $R = 50$ is much larger for BFPE. In addition, nB does not perform well when $R$ is small. For the sake of simplicity, we use $R = 5$ for CV($d$), bR, and bP and use $R = 50$ for BFPE, DCVB, and nB in the resampling procedures given in the next Subsection.

### 6.9.2.3. *Special Case AR Models*

We now compare the performances of our ten criteria with respect to sample size via two autoregressive special case models. Both models represent a seasonal random walk with an easily identifiable correct order, and their parameter structures are $\phi_5 = 1$, $p_* = 5$, and $o = 5$. The errors are standard normal. They differ only in sample size, which is 25 for Model 13 and 100 for Model 14. Tables 6.11 and 6.12 summarizes the count and observed efficiency results for Models 13 and 14, respectively.

The selection criteria used in Tables 6.11–6.14 are

| | |
|---|---|
| AICc | Eq. (3.10). |
| AICu | Eq. (3.11). |
| SIC | Eq. (3.15). |
| CV | Withhold-1 cross-validation, CV(1) Eq. (6.6). |
| CVd | delete-$d$ cross-validation, CV($d$) Eq. (6.7) using 5 subsamples. |
| bR | Refined bootstrap estimate Eq. (6.23) by resampling residuals using 5 pseudo-samples. |
| bP | Refined bootstrap estimate Eq. (6.23) by resampling pairs $(Y, X)$ using 5 pseudo-samples. |
| nB | Naive bootstrap estimate Eq. (6.24) using 50 pseudo-samples. |
| BFPE | Penalty weighted naive bootstrap estimate Eq. (6.25) using 50 pseudo-samples. |
| DCVB | Doubly cross-validated bootstrap Eq. (6.26) using 50 pseudo-samples. |

Although the sample size is small, Model 13 is easily identified and little underfitting is observed in Table 6.11. Each candidate model is cast into a regression form and all resampling is done as if the model were indeed a regression model. This leads to an incomplete withholding for the cross-validation procedures as discussed earlier. Since the true model is a random walk, the size of the block that would have to be withheld to guarantee independence might exceed the sample size. Underfitting resulting from such incomplete withholding can be seen in CVd. CV performs about the same in Model 13 as it does in the regression models (at least relative to the other criteria). Since the sample size is small for SIC, it overfits. AICc and AICu both perform well. The three bootstrap criteria bP, bR, and nB all perform poorly in that they overfit excessively, just as they did in our regression models. The penalty function bootstrap criteria BFPE and DCVB perform the best of the resampling procedures. It seems that casting autoregressive models into a regression framework affects the performance of the bootstrap criteria. As for the regression model, a clear pattern emerges in Table 6.11 from both the count and observed efficiency results—criteria with strong penalty functions perform better than the criteria with weak penalty functions.

The best criteria, AICu, AICc, and DCVB, all have observed efficiencies near 90% in both the K-L and $L_2$ sense. There is good agreement between the two observed efficiency measures for Model 13. In general, $L_2$ observed efficiencies are slightly higher than the corresponding K-L observed efficiencies. This is again due to $L_2$ penalizing overfitting less than K-L. BFPE and SIC have observed efficiencies in the lower 80s followed by CV. The three worst performers (due to excessive overfitting), bR, bP, and nB, are only about 60% efficient in the $L_2$ sense and have K-L observed efficiency in the lower 50s. CVd is the only criterion that underfits, which accounts for its low observed efficiency. However, note that its K-L observed efficiency is higher than its $L_2$ observed efficiency, in contrast to bR, bP, and nB, which lose observed efficiency due to overfitting, but CVd does not underfit as excessively as they overfit. On balance, the underfitting in CVd is penalized by $L_2$ by about the same amount as the overfitting is penalized in nB. Next we look at the results for Model 14, summarized in Table 6.12.

Table 6.11. Simulation results for Model 13. Counts and observed efficiency.

| | counts | | | | | | | | | | | |
|---|---|---|---|---|---|---|---|---|---|---|---|---|
| $p$ | AICc | AICu | SIC | CV | CVd | bR | bP | nB | BFPE | DCVB | $L_2$ | K-L |
| 1 | 0 | 0 | 0 | 0 | 197 | 0 | 1 | 0 | 0 | 0 | 0 | 0 |
| 2 | 0 | 0 | 0 | 0 | 84 | 0 | 1 | 0 | 0 | 0 | 0 | 0 |
| 3 | 1 | 1 | 1 | 0 | 53 | 0 | 0 | 0 | 1 | 2 | 0 | 0 |
| 4 | 0 | 0 | 0 | 0 | 51 | 0 | 0 | 0 | 0 | 0 | 0 | 1 |
| 5 | 827 | 911 | 750 | 625 | 610 | 194 | 81 | 283 | 769 | 826 | 784 | 779 |
| 6 | 118 | 71 | 124 | 157 | 5 | 159 | 132 | 140 | 134 | 119 | 128 | 147 |
| 7 | 40 | 12 | 48 | 73 | 0 | 144 | 133 | 101 | 49 | 32 | 52 | 52 |
| 8 | 8 | 3 | 21 | 46 | 0 | 147 | 178 | 113 | 19 | 9 | 27 | 13 |
| 9 | 6 | 2 | 30 | 59 | 0 | 164 | 216 | 137 | 17 | 10 | 5 | 5 |
| 10 | 0 | 0 | 26 | 40 | 0 | 192 | 258 | 226 | 11 | 2 | 4 | 3 |
| | K-L observed efficiency | | | | | | | | | | | |
| | AICc | AICu | SIC | CV | CVd | bR | bP | nB | BFPE | DCVB | $L_2$ | K-L |
| ave | 0.88 | 0.92 | 0.82 | 0.75 | 0.66 | 0.54 | 0.53 | 0.52 | 0.84 | 0.87 | 0.99 | 1.00 |
| med | 1.00 | 1.00 | 1.00 | 0.97 | 0.98 | 0.49 | 0.48 | 0.42 | 1.00 | 1.00 | 1.00 | 1.00 |
| sd | 0.24 | 0.19 | 0.30 | 0.33 | 0.41 | 0.32 | 0.28 | 0.34 | 0.28 | 0.24 | 0.05 | 0.00 |
| rank | 2 | 1 | 5 | 6 | 7 | 8 | 9 | 10 | 4 | 3 | | |
| | $L_2$ observed efficiency | | | | | | | | | | | |
| | AICc | AICu | SIC | CV | CVd | bR | bP | nB | BFPE | DCVB | $L_2$ | K-L |
| ave | 0.89 | 0.93 | 0.85 | 0.79 | 0.61 | 0.62 | 0.59 | 0.61 | 0.86 | 0.89 | 1.00 | 0.99 |
| med | 1.00 | 1.00 | 1.00 | 0.98 | 0.97 | 0.61 | 0.57 | 0.58 | 1.00 | 1.00 | 1.00 | 1.00 |
| sd | 0.20 | 0.16 | 0.24 | 0.27 | 0.46 | 0.28 | 0.25 | 0.29 | 0.23 | 0.20 | 0.00 | 0.05 |
| rank | 2 | 1 | 5 | 6 | 8 | 7 | 10 | 8 | 4 | 2 | | |

In Table 6.12 there is good agreement between the count patterns and the two observed efficiency measures. The larger sample combined with a correct model that is easily identified favors SIC, and thus would be expected to favor CVd. Indeed, SIC and CVd now rank first and second, respectively. The consistent properties of SIC and CVd can been seen in their high counts for selecting the true model. The counts drop for both AICc and AICu. The penalty functions of these criteria actually decrease as the sample size increases, causing more overfitting in large samples than in small samples. As a result, AICc and AICu actually overfit more in Model 14 than in Model 13. The weighted bootstrap DCVB still performs well, and excessive overfitting is still seen for bR, bP, and nB.

The patterns in Table 6.12 are similar to those we saw in Table 6.5, where the amount of overfitting decreases with increasing $\alpha$. Criteria with similar overfitting properties have similar count patterns. Excessive overfitting is seen from the $\alpha 1$ criteria bP, bR, and nB. The $\alpha 2$ criteria AICc and CV are better, but overfit moderately. Next are the $\alpha 3$ criteria AICu, BFPE, and DCVB, and best are SIC and CVd ($\alpha\infty$ criteria), which overfit the least in large samples.

Table 6.12. Simulation results for Model 14. Counts and observed efficiency.

| | | | | | counts | | | | | | | |
|---|---|---|---|---|---|---|---|---|---|---|---|---|
| $p$ | AICc | AICu | SIC | CV | CVd | bR | bP | nB | BFPE | DCVB | $L_2$ | K-L |
| 1 | 0 | 0 | 0 | 0 | 0 | 0 | 0 | 0 | 0 | 0 | 0 | 0 |
| 2 | 0 | 0 | 0 | 0 | 0 | 0 | 0 | 0 | 0 | 0 | 0 | 0 |
| 3 | 0 | 0 | 0 | 0 | 0 | 0 | 0 | 0 | 0 | 0 | 0 | 0 |
| 4 | 0 | 0 | 0 | 0 | 0 | 0 | 0 | 0 | 0 | 0 | 0 | 0 |
| 5 | 690 | 838 | 915 | 625 | 744 | 175 | 126 | 290 | 786 | 808 | 839 | 818 |
| 6 | 148 | 107 | 62 | 145 | 208 | 148 | 137 | 139 | 122 | 116 | 105 | 129 |
| 7 | 65 | 35 | 16 | 76 | 44 | 149 | 136 | 114 | 45 | 36 | 32 | 33 |
| 8 | 46 | 13 | 7 | 67 | 4 | 164 | 162 | 123 | 27 | 24 | 15 | 10 |
| 9 | 34 | 6 | 0 | 48 | 0 | 189 | 224 | 158 | 14 | 11 | 6 | 6 |
| 10 | 17 | 1 | 0 | 39 | 0 | 175 | 215 | 176 | 6 | 5 | 3 | 4 |
| | | | | | K-L observed efficiency | | | | | | | |
| | AICc | AICu | SIC | CV | CVd | bR | bP | nB | BFPE | DCVB | $L_2$ | K-L |
| ave | 0.84 | 0.92 | 0.95 | 0.81 | 0.94 | 0.67 | 0.65 | 0.64 | 0.89 | 0.90 | 1.00 | 1.00 |
| med | 1.00 | 1.00 | 1.00 | 1.00 | 1.00 | 0.68 | 0.66 | 0.63 | 1.00 | 1.00 | 1.00 | 1.00 |
| sd | 0.26 | 0.19 | 0.15 | 0.27 | 0.14 | 0.27 | 0.25 | 0.29 | 0.22 | 0.21 | 0.02 | 0.00 |
| rank | 6 | 3 | 1 | 7 | 2 | 8 | 9 | 10 | 5 | 4 | | |
| | | | | | $L_2$ observed efficiency | | | | | | | |
| | AICc | AICu | SIC | CV | CVd | bR | bP | nB | BFPE | DCVB | $L_2$ | K-L |
| ave | 0.85 | 0.92 | 0.95 | 0.81 | 0.93 | 0.66 | 0.65 | 0.66 | 0.90 | 0.91 | 1.00 | 1.00 |
| med | 1.00 | 1.00 | 1.00 | 1.00 | 1.00 | 0.69 | 0.65 | 0.65 | 1.00 | 1.00 | 1.00 | 1.00 |
| sd | 0.25 | 0.18 | 0.15 | 0.27 | 0.14 | 0.27 | 0.25 | 0.29 | 0.21 | 0.20 | 0.00 | 0.03 |
| rank | 6 | 3 | 1 | 7 | 2 | 8 | 10 | 8 | 5 | 4 | | |

### 6.9.2.4. Large-scale Simulations

To gain an idea of the performance of these criteria over a wide range of different AR models, observed efficiency summaries for each criterion over all possible 36 AR models are presented in Tables 6.13 and 6.14 for 3600 realizations. Efficiencies are calculated in the usual manner.

Table 6.13. Simulation results over 36 autoregressive models—K-L observed efficiency ranks.

| rank | AICc | AICu | SIC | CV | CVd | bR | bP | nB | BFPE | DCVB |
|---|---|---|---|---|---|---|---|---|---|---|
| 1 | 13 | 12 | 26 | 36 | 198 | 216 | 265 | 62 | 2 | 9 |
| 2 | 122 | 205 | 111 | 115 | 323 | 203 | 239 | 160 | 47 | 85 |
| 3 | 338 | 414 | 200 | 146 | 329 | 120 | 176 | 107 | 302 | 373 |
| 4 | 889 | 892 | 753 | 723 | 604 | 181 | 197 | 191 | 929 | 947 |
| 5 | 1042 | 1024 | 968 | 972 | 832 | 455 | 321 | 626 | 1048 | 1061 |
| 6 | 549 | 507 | 498 | 551 | 394 | 266 | 253 | 367 | 575 | 547 |
| 7 | 396 | 316 | 390 | 446 | 267 | 259 | 241 | 297 | 435 | 395 |
| 8 | 146 | 111 | 271 | 337 | 165 | 400 | 305 | 389 | 234 | 144 |
| 9 | 99 | 98 | 298 | 221 | 199 | 755 | 468 | 719 | 27 | 37 |
| 10 | 6 | 21 | 85 | 53 | 289 | 745 | 1135 | 682 | 1 | 2 |
| ave rank | 4.82 | 4.64 | 5.34 | 5.34 | 4.96 | 6.70 | 6.74 | 6.79 | 4.92 | 4.74 |
| ranking | 3 | 1 | 6 | 6 | 5 | 8 | 9 | 10 | 4 | 2 |

Table 6.14. Simulation results over 36 autoregressive models—$L_2$ observed efficiency ranks.

| rank | AICc | AICu | SIC | CV | CVd | bR | bP | nB | BFPE | DCVB |
|---|---|---|---|---|---|---|---|---|---|---|
| 1 | 10 | 8 | 21 | 41 | 140 | 234 | 234 | 92 | 2 | 10 |
| 2 | 99 | 135 | 114 | 150 | 228 | 250 | 215 | 214 | 54 | 92 |
| 3 | 322 | 371 | 228 | 193 | 277 | 173 | 189 | 164 | 335 | 377 |
| 4 | 876 | 858 | 796 | 777 | 555 | 235 | 186 | 273 | 952 | 945 |
| 5 | 994 | 971 | 975 | 973 | 781 | 494 | 308 | 688 | 1025 | 1012 |
| 6 | 516 | 491 | 513 | 520 | 397 | 273 | 257 | 382 | 546 | 519 |
| 7 | 411 | 357 | 394 | 427 | 302 | 236 | 254 | 272 | 434 | 406 |
| 8 | 184 | 160 | 254 | 286 | 195 | 364 | 297 | 364 | 215 | 178 |
| 9 | 174 | 194 | 238 | 187 | 306 | 679 | 443 | 625 | 34 | 55 |
| 10 | 14 | 55 | 67 | 46 | 419 | 662 | 1217 | 526 | 3 | 6 |
| ave rank | 4.98 | 4.95 | 5.22 | 5.16 | 5.49 | 6.36 | 6.88 | 6.30 | 4.87 | 4.78 |
| ranking | 4 | 3 | 6 | 5 | 7 | 9 | 10 | 8 | 2 | 1 |

In terms of K-L observed efficiency, AICu is the best all-around performer. The two weighted bootstrap criteria also perform well, whereas bR and bP perform poorly, and nB performs the worst. The results for $L_2$ are similar except that the weighted bootstrap criteria, DCVB and BFPE, perform the best.

A pattern can be seen from these tables. The better performing criteria tend to have higher numbers of low to middle ranks (due to multiple ties). The worst performers tend to have large numbers of high ranks indicating poor performance. Note that the rankings for bR, bP, and nB all are distributed fairly evenly—sometimes they perform very well, yet other times they perform poorly. We can conclude from this pattern that their behavior is erratic over all our models considered here.

## 6.10. Summary

Of the two data resampling techniques we applied to our models from Chapters 2–5, cross-validation was easier to apply and computationally much less demanding than bootstrapping. The cross-validation criterion CV(1) is asymptotically equivalent to FPE, and the two procedures seem to perform similarly in small samples as well when the errors are normally distributed. However, simulations throughout the chapters so far have shown that both selection criteria overfit. Although consistent variants of cross-validation performed better than the usual CV, they also increased the computations required, and we feel that the degree of performance improvement does not offset the increased effort.

Several issues complicate bootstrapping that do not affect cross-validation, one of which is whether to bootstrap the residuals or bootstrap the pairs ($y_i$, $x_i$). We found little difference between these two techniques when it comes to model selection performance, and because a new design matrix is created and inverted for each bootstrap pseudo-sample for pairs, bootstrapping residuals is much easier. We therefore recommend bootstrapping residuals.

Another issue in bootstrapping is the size of the bootstrapped pseudo-sample. While it is commonly thought that a large number of replicates is needed for good selection performance, we found that this is not necessarily so. In some cases good performance can be obtained with as few as ten replications. However, in the very large samples we will discuss in Chapter 9, we will see that 100 replications performed best, suggesting that the number of bootstrap replications needs to increase with the sample size.

Most importantly, the common bootstrap approaches attempt to improve upon $s^2$ as a basis for selecting a model by fine tuning $\mathrm{R}^2_{adj}$. Since $\mathrm{R}^2_{adj}$ performs poorly to begin with, it may be wiser to attempt to bootstrap some criterion that performs better. Resampling by itself does not guarantee good model selection performance, and we saw that bootstrap criteria performed better when weighted with a penalty function. One approach to penalty functions is to "refine" FPEu for the bootstrap by adding a penalty function dependent on parameter count. However, we obtained the best results with the doubly cross-validated bootstrap, which has a strong penalty function that does not require a parameter count. Of all the techniques in this Chapter, DCVB is the most competitive with the other criteria.

Finally, the reader may refer to two articles which extend the work of Shao (1993). One is Robust Linear Model Selection by Cross-validation (Ronchetti, Field and Blanchard, 1997), and the other one is Variable Selection in Regression Via Repeated Data Splitting (Thall, Russell, and Simon, 1997).

# Chapter 7
# Robust Regression and Quasi-likelihood

In Chapter 6, we introduced one exception to the standard error assumptions of the model structure—nonnormal distribution of errors. In this Chapter we examine other exceptions to the assumption of normal errors, primarily robust regression with additive errors (Hampel, Ronchetti, Rousseeuw, and Stahel, 1986) and quasi-likelihood for non-additive errors (McCullagh and Nelder, 1989). We derive robust versions of the AIC family of criteria, Cp, FPE, and the Wald test, and examine the performance of these selection criteria with respect to each other and to some of the nonrobust criteria. We begin with the standard model structure and a well-behaved design matrix, by which we mean that only the error distribution departs from the usual regression model structure and assumptions. We will then discuss a further complication where both the errors and the design matrix have outliers.

Since robust methods are used, we will need a robust distance measure. The $L_2$ distance has been used in the earlier chapters due in part to its relationship to least squares regression. We introduce the $L_1$ distance, or absolute error norm, in this Chapter. Like $L_2$, $L_1$ does not require the error distribution to be known, but $L_1$ is associated with maximum likelihood estimation of models with double exponential errors *i.e.* robust $L_1$ regression. This makes $L_1$ more flexible than the Kullback–Leibler distance, and more robust than $L_2$.

## 7.1. Nonnormal Error Regression Models

During the past twenty years, much research has addressed the problem of finding robust regression statistics that do not depend on assumptions of normality for exactness. Such statistics are particularly useful when there are consequential outliers present in the data being evaluated. Because many nonnormal distributions are heavy-tailed, outliers can arise when the underlying distribution for the data or errors is nonnormal. Large error variances can also produce consequential outliers.

Of the many robust methods that have been adapted to regression, we will consider least absolute error (LAD or $L_1$) regression, trimmed regression and

M-estimates. We will assume that all models have an additive error structure (except the quasi-likelihood function discussed in Section 7.7). In other words, the generating or true model has the form

$$y_i = \mu_{*i} + \varepsilon_{*i}, \quad i = 1, \ldots, n, \tag{7.1}$$

and the candidate model has the form

$$y_i = x_i'\beta + \varepsilon_i, \quad i = 1, \ldots, n, \tag{7.2}$$

where $x_i$ and $\beta$ are $k \times 1$ vectors. If the true model is a regression model, then it has the form

$$y_i = x_{*i}'\beta_* + \varepsilon_{*i}, \quad i = 1, \ldots, n, \tag{7.3}$$

where $x_{*i}$ and $\beta_*$ are $k_* \times 1$ vectors. A key assumption is that the errors are additive but nonnormal.

### *7.1.1. $L_1$ Distance and Efficiency*

In Chapter 1 we defined the K-L and $L_2$ distance measures and their corresponding observed efficiencies. For regression models, the $L_2$ distance measure between the true and fitted candidate models can be defined as

$$L_2 = \frac{1}{n}\sum_{i=1}^{n}(\mu_{*i} - x_i'\hat{\beta})^2. \tag{7.4}$$

Analogously, we also can define $L_1$ distance to measure the difference between the true and fitted candidate models as

$$L_1(k) = \frac{1}{n}\sum_{i=1}^{n}|\mu_{*i} - x_i'\hat{\beta}|. \tag{7.5}$$

Given this, the $L_1$ observed efficiency can be defined as the ratio

$$L_1 \text{ observed efficiency} = \frac{\min_k L_1(k)}{L_1(\text{selected model})}. \tag{7.6}$$

We discuss the K-L distance measure and its observed efficiency in the next Section. Note that K-L requires the true distribution be known.

## 7.2. Least Absolute Deviations Regression

$L_1$ regression involves minimizing the $L_1$-norm in order to obtain a linear estimate of a vector of parameters. Although this concept probably predates that of least squares regression, which has optimum properties only under limited circumstances, least squares has been by far the more popular. This is because the computations for $L_1$ regression have no closed form and hence are more difficult to compute than least squares estimates. Recently, however, due to the relative insensitivity of $L_1$ estimators to outliers and the development of fast computational algorithms (see Bloomfield and Steiger, 1983), $L_1$ regression is becoming popular. Of all the robust nonleast squares techniques discussed in this book, $L_1$ regression is computationally the fastest. Although $L_1$ regression is iterative, it only requires one inversion of $X'X$ in the $L_1$ regression algorithm, where $X = (x_1, \ldots, x_n)'$. Because other methods require an inverse to be taken at each iteration, $L_1$ is computationally much faster. As is true for any type of regression, there is a need for data driven model selection in $L_1$ regression, and we next discuss the use of the Kullback–Leibler-based criteria AIC and AICc for this purpose.

### *7.2.1. L1AICc*

AIC is often used to select a model under K-L, but although it is an asymptotically unbiased estimate of the expected Kullback–Leibler information for each candidate model, in small samples it can be quite biased. AICc (Hurvich and Tsai, 1989), was developed in order to overcome this bias. However, its derivation relies heavily on the assumption of normally distributed errors and the method of least squares. If we are to apply it under circumstances that violate those conditions, it is important to obtain AICc for nonnormal distributions. Since the $L_1$ regression parameter estimators are the same as the maximum likelihood estimators of regression coefficients in the double exponential distribution (*e.g.*, Bloomfield and Steiger, 1983), and the resulting distributions of parameter estimators have nice pivotal properties, Hurvich and Tsai (1990) obtained AICc for this specific nonnormal distribution as shown below.

Suppose the data are generated by the true model given in Eq. (7.1), where the $\varepsilon_{*i}$ are independent and identically distributed with the double exponential distribution and $y_i$ has density

$$f_*(y_i; \mu_{*i}, \sigma_*) = \frac{1}{2\sigma_*} \exp\left( -\frac{|y_i - \mu_{*i}|}{\sigma_*} \right), \quad -\infty < y_i < \infty.$$

The $y_i$ have mean $\mu_{*i}$ and variance $2\sigma_*^2$, and in this case $\sigma_*$ is a scale parameter rather than the variance. Also suppose that the candidate models are of the form given by Eq. (7.2), where the $\varepsilon_i$ are *i.i.d.* with the double exponential distribution. The $y_i$ have density functions

$$f(y_i;\beta,\sigma) = \frac{1}{2\sigma}\exp\left(-\frac{|y_i - x_i'\beta|}{\sigma}\right), \quad -\infty < y_i < \infty.$$

Then the Kullback–Leibler discrepancy can be defined as

$$\text{K-L} = \frac{2}{n}E_*\left[\log\left(f_*(Y;\mu_*,\sigma_*)\right) - \log\left(f(Y;\beta,\sigma)\right)\right],$$

where $Y = (y_1,\ldots,y_n)'$, $\mu_* = (\mu_{1*},\ldots,\mu_{n*})'$ and $E_*$ denotes expectation under the true model. Using the assumption of double exponential errors, the Kullback–Leibler discrepancy is

$$\text{K-L} = \log\left(\frac{\sigma^2}{\sigma_*^2}\right) - 2 + \frac{2\sigma_*}{n\sigma}\sum_{i=1}^{n}\left(\frac{|\mu_{*i} - x_i'\beta|}{\sigma_*} + \exp\left(\frac{|\mu_{*i} - x_i'\beta|}{\sigma_*}\right)\right),$$

which is equivalent to

$$\text{K-L} = \log\left(\frac{\sigma^2}{\sigma_*^2}\right) - 2 + \frac{2\sigma_*}{n\sigma}\sum_{i=1}^{n}\left(\frac{|\mu_{*i} - x_i'\beta|}{\sigma_*} + \exp\left(\frac{|\mu_{*i} - x_i'\beta|}{\sigma_*}\right)\right).$$

For comparing the fitted model to the true model,

$$\text{K-L} = \log\left(\frac{\hat{\sigma}^2}{\sigma_*^2}\right) - 2 + \frac{2\sigma_*}{n\hat{\sigma}}\sum_{i=1}^{n}\left(\frac{|\mu_{*i} - x_i'\hat{\beta}|}{\sigma_*} + \exp\left(\frac{|\mu_{*i} - x_i'\hat{\beta}|}{\sigma_*}\right)\right), \quad (7.7)$$

where $\hat{\beta}$ and $\hat{\sigma}$ are MLEs of $\beta$ and $\sigma$, respectively. As we noted above, the MLEs are also the $L_1$ regression estimates, $\hat{\beta}$ minimizes $\sum_{i=1}^{n}|y_i - x_i'\beta|$ and

$$\hat{\sigma} = \sum_{i=1}^{n}|y_i - x_i'\hat{\beta}|/n. \quad (7.8)$$

Since the term $-\log(\sigma_*^2) - 2$ in Eq. (7.7) plays no role in model selection, Hurvich and Tsai ignored it to obtain

$$\Delta(\hat{\beta},\hat{\sigma}) = \log(\hat{\sigma}^2) + \frac{2\sigma_*}{n\hat{\sigma}}\sum_{i=1}^{n}\left(\frac{|\mu_{*i} - x_i'\hat{\beta}|}{\sigma_*} + \exp\left(-\frac{|\mu_{*i} - x_i'\hat{\beta}|}{\sigma_*}\right)\right). \quad (7.9)$$

Given the family of candidate models, then, the one that minimizes $E_*[\Delta(\hat{\beta}, \hat{\sigma})]$ is preferred.

The development of L1AICc requires two further assumptions: first that the candidate model is of the form given by Eq. (7.2) where the $\varepsilon_i$ are *i.i.d.* double exponential, second that the true model belongs to the set of candidate models. In other words, a true model exists where the $\varepsilon_{*i}$ are *i.i.d.* double exponential. Under these assumptions, the quantities $\sigma_*/\hat{\sigma}$ and $(x'_{*i}\beta_* - x'_i\hat{\beta})/\sigma_*$ have distributions independent of $\beta_*$ and $\sigma_*$ (this can be derived from Antle and Bain, 1969). Hence, the second term in Eq. (7.9) can be obtained by Monte Carlo methods as follows. Generate pseudo-data sets from the model in Eq. (7.1) with $\mu_{*i} = 0$ and $\sigma_* = 1$, then obtain $L_1$ estimates $\tilde{\beta}_*$, $\tilde{\sigma}_*$, and finally evaluate the average value of the second term of Eq. (7.9) over many replicated pseudo-data sets (perhaps 100). We denote this average $\tilde{h}(k, X)$, where $X$ is the $n \times k$ design matrix. Hurvich and Tsai defined the criterion obtained in this way to be L1AICc:

$$\text{L1AICc} = \log(\hat{\sigma}^2) + \tilde{h}(k, X), \tag{7.10}$$

where L1AICc is an exact unbiased estimator of $E_*[\Delta(\hat{\beta}, \hat{\sigma})]$.

### 7.2.2. Special Case Models

How do the model selection criteria perform under $L_1$ regression with double exponential errors? Four special case models are simulated in this Section to illustrate model selection performance with respect to model identifiability and sample size. The criteria to be compared are AIC, AICc, L1AICc, SIC, HQ, HQc, and the three distance measures. AIC, AICc, SIC, HQ, and HQc have forms similar to those discussed in Chapter 2: AIC $= \log(\hat{\sigma}^2) + (2k+1)/n$, the small-sample correction AICc $= \log(\hat{\sigma}^2) + (n+k)/(n-k-2)$, the consistent criterion SIC $= \log(\hat{\sigma}^2) + \log(n)k/n$, another consistent criterion HQ $= \log(\hat{\sigma}^2) + 2\log\log(n)k/n$, and finally its small-sample correction HQc $= \log(\hat{\sigma}^2) + 2\log\log(n)k/(n-k-2)$. The only difference between these selection criteria and their counterparts found in Chapter 2 is the use of $\hat{\sigma}$ in Eq. (7.8). For comparison, we also include the robust Wald test criterion, RTp, which we will derive in Section 7.4, Eq. (7.12).

The simulation study models to which these criteria are applied all have the true model given by Eq. (7.3) with order $k_* = 6$ and $\sigma_* = 0.5$. Models 15 and 17 have a strongly identifiable true parameter structure $\beta_j = 1$, where $\beta_0 = 1, \beta_1 = 1, \ldots, \beta_5 = 1$. Models 16 and 18 have a weakly identifiable true parameter structure, $\beta_j = 1/j$, where $\beta_0 = 1$, $\beta_1 = 1$, $\beta_2 = 1/2$, $\beta_3 = 1/3$,

$\beta_4 = 1/4$, $\beta_5 = 1/5$. Note that these are the same two parameter structures used in Chapter 2. Models 15 and 16 have 25 observations and Models 17 and 18 have sample size 50. One thousand realizations are computed for each of the four special case models, and candidate models of the form Eq. (7.2) are then fit to the data.

For each candidate model, we compute the distance between the true model and the fitted candidate model and the Kullback–Leibler (K-L) observed efficiency (Eq. (7.7) and Eq. (1.2)), the $L_2$ observed efficiency (Eq. (7.4) and Eq. (1.1)), and $L_1$ observed efficiency (Eq. (7.5) and Eq. (7.6)). We will use all three measures to compare selection criteria performance on the basis of average observed efficiency over all the realizations, and we will also use counts of model order selected. These count patterns are a useful reference for discussing overfitting and underfitting. Results are summarized in Tables 7.1–7.4.

Table 7.1 summarizes the results for Model 15. Because the parameters are all strongly identifiable, little underfitting is evident even in the small sample size of 25. With little underfitting, relative performance will depend primarily on the overfitting properties of the criteria. The robust Wald test, RTp, overfits the least and performs the best in Model 15. HQc is next, followed by AICc. HQc has a structure similar to that of AICc, but its penalty function is slightly larger, reducing overfitting and improving its performance relative to AICc. Although AICc has a parametric penalty function and L1AICc has an estimated stochastic penalty function, their performances are essentially equivalent. The consistent SIC and HQ overfit badly due to their weaker penalty functions in small samples. Finally, AIC has the weakest penalty function in this group of selection criteria, and not surprisingly performs worst.

We first observed in Chapter 2 that when the true model is strongly identifiable, criteria with strong penalty functions perform much better than criteria with weak penalty functions. This is because little underfitting is evident for strong models, and so criterion performance depends on overfitting properties, which in turn depends on the strength of the penalty function. This is important in this Chapter as well, since even though we are using robust $L_1$ regression, the idea of a weak versus strong penalty function is still useful in predicting performance, as we see from the results for Model 15, which favors the criteria with stronger penalty functions. In such cases, observed efficiency is related to the probability of selecting the correct model, and higher counts are associated with higher observed efficiencies. As noted in earlier chapters, K-L penalizes overfitting more than $L_2$, resulting in lower K-L observed efficiencies compared to those under $L_2$ for a given criterion. In turn, $L_2$ penalizes overfitting more than $L_1$. Although no moments or graphs are presented for

$L_1$ distance, we see that it has a less well-defined minimum than either K-L or $L_2$. Also, $L_1$ distance does not increase as rapidly for overfitting, resulting in higher observed efficiencies. Even AIC has a high $L_1$ observed efficiency (81.5%). We next look at the results for a weakly identifiable model and small sample size.

The combination of a true model that is difficult to detect and a small sample size results in both underfitting as well as overfitting, as can be seen in the count data in Table 7.2. It is difficult to see any overall patterns for Model 16. For Model 15, RTp, HQc, and AICc/L1AICc were the top ranking

Table 7.1. Simulation results for Model 15. Counts and observed efficiency. $n = 25$, $\beta_j = 1$, $k_* = 6$.

| | | | | counts | | | | | | |
|---|---|---|---|---|---|---|---|---|---|---|
| $k$ | AIC | AICc | L1AICc | SIC | HQ | HQc | RTp | $L_1$ | $L_2$ | K-L |
| 1 | 0 | 0 | 0 | 0 | 0 | 0 | 0 | 0 | 0 | 0 |
| 2 | 0 | 0 | 0 | 0 | 0 | 0 | 0 | 0 | 0 | 0 |
| 3 | 0 | 0 | 0 | 0 | 0 | 0 | 0 | 0 | 0 | 0 |
| 4 | 0 | 0 | 0 | 0 | 0 | 1 | 6 | 0 | 0 | 0 |
| 5 | 0 | 5 | 4 | 2 | 0 | 8 | 17 | 0 | 0 | 0 |
| 6 | 248 | 631 | 633 | 467 | 306 | 681 | 775 | 403 | 1000 | 556 |
| 7 | 322 | 290 | 283 | 305 | 345 | 251 | 169 | 383 | 0 | 333 |
| 8 | 264 | 67 | 71 | 161 | 227 | 53 | 27 | 166 | 0 | 91 |
| 9 | 118 | 7 | 9 | 51 | 90 | 6 | 6 | 43 | 0 | 17 |
| 10 | 38 | 0 | 0 | 13 | 26 | 0 | 0 | 5 | 0 | 3 |
| 11 | 10 | 0 | 0 | 1 | 6 | 0 | 0 | 0 | 0 | 0 |
| true | 246 | 626 | 627 | 463 | 304 | 677 | 771 | 403 | 1000 | 556 |

| | | | | K-L observed efficiency | | | | | | |
|---|---|---|---|---|---|---|---|---|---|---|
| | AIC | AICc | L1AICc | SIC | HQ | HQc | RTp | $L_1$ | $L_2$ | K-L |
| ave | 0.645 | 0.808 | 0.809 | 0.729 | 0.669 | 0.828 | 0.867 | 0.97 | 0.99 | 1.00 |
| med | 0.627 | 0.988 | 0.989 | 0.768 | 0.660 | 0.994 | 0.998 | 1.00 | 1.00 | 1.00 |
| sd | 0.260 | 0.259 | 0.259 | 0.278 | 0.267 | 0.255 | 0.240 | 0.05 | 0.03 | 0.00 |
| rank | 7 | 4 | 3 | 5 | 6 | 2 | 1 | | | |

| | | | | $L_2$ observed efficiency | | | | | | |
|---|---|---|---|---|---|---|---|---|---|---|
| | AIC | AICc | L1AICc | SIC | HQ | HQc | RTp | $L_1$ | $L_2$ | K-L |
| ave | 0.687 | 0.835 | 0.836 | 0.764 | 0.709 | 0.854 | 0.886 | 0.98 | 1.00 | 0.98 |
| med | 0.698 | 1.000 | 1.000 | 0.845 | 0.723 | 1.000 | 1.000 | 1.00 | 1.00 | 1.00 |
| sd | 0.254 | 0.244 | 0.245 | 0.265 | 0.258 | 0.240 | 0.229 | 0.05 | 0.00 | 0.10 |
| rank | 7 | 4 | 3 | 5 | 6 | 2 | 1 | | | |

| | | | | $L_1$ observed efficiency | | | | | | |
|---|---|---|---|---|---|---|---|---|---|---|
| | AIC | AICc | L1AICc | SIC | HQ | HQc | RTp | $L_1$ | $L_2$ | K-L |
| ave | 0.815 | 0.897 | 0.897 | 0.855 | 0.827 | 0.907 | 0.950 | 1.00 | 0.99 | 0.98 |
| med | 0.845 | 0.983 | 0.982 | 0.928 | 0.859 | 0.986 | 0.990 | 1.00 | 1.00 | 1.00 |
| sd | 0.165 | 0.152 | 0.152 | 0.168 | 0.166 | 0.150 | 0.130 | 0.00 | 0.02 | 0.07 |
| rank | 7 | 3 | 3 | 5 | 6 | 2 | 1 | | | |

criteria. By contrast, for Model 16 RTp underfits the most, causing reduced observed efficiency. HQc also underfits, but not as much as RTp. K-L is the most lenient measure with respect to underfitting, and in fact HQc has the highest K-L observed efficiency. However, both $L_2$ and $L_1$ penalize HQc for underfitting. AICc and L1AICc perform near the middle in all three observed efficiency measures. AICc and L1AICc seem to balance overfitting and underfitting well in that they perform about the same under all three observed efficiency measures. AIC, SIC, and HQ still overfit the most, and hence they are penalized more by K-L than $L_2$ or $L_1$. In fact, HQ has the

Table 7.2. Simulation results for Model 16. Counts and observed efficiency.
$n = 25$, $\beta_j = 1/j$, $k_* = 6$.

| | | | | counts | | | | | | |
|---|---|---|---|---|---|---|---|---|---|---|
| $k$ | AIC | AICc | L1AICc | SIC | HQ | HQc | RTp | $L_1$ | $L_2$ | K-L |
| 1 | 0 | 0 | 0 | 0 | 0 | 0 | 4 | 0 | 0 | 0 |
| 2 | 1 | 6 | 6 | 6 | 3 | 9 | 67 | 0 | 0 | 0 |
| 3 | 19 | 74 | 73 | 65 | 26 | 97 | 267 | 1 | 1 | 0 |
| 4 | 84 | 251 | 254 | 192 | 118 | 282 | 330 | 20 | 19 | 27 |
| 5 | 188 | 340 | 333 | 291 | 220 | 338 | 212 | 109 | 116 | 204 |
| 6 | 280 | 239 | 236 | 245 | 285 | 204 | 90 | 412 | 800 | 514 |
| 7 | 209 | 70 | 77 | 130 | 197 | 57 | 27 | 301 | 53 | 204 |
| 8 | 137 | 18 | 19 | 50 | 93 | 12 | 5 | 132 | 11 | 41 |
| 9 | 62 | 2 | 2 | 17 | 44 | 1 | 1 | 22 | 0 | 9 |
| 10 | 19 | 0 | 0 | 4 | 13 | 0 | 0 | 3 | 0 | 1 |
| 11 | 1 | 0 | 0 | 0 | 1 | 0 | 0 | 0 | 0 | 0 |
| true | 75 | 69 | 63 | 71 | 75 | 64 | 31 | 283 | 673 | 306 |
| | | | | K-L observed efficiency | | | | | | |
| | AIC | AICc | L1AICc | SIC | HQ | HQc | RTp | $L_1$ | $L_2$ | K-L |
| ave | 0.509 | 0.554 | 0.552 | 0.536 | 0.519 | 0.558 | 0.547 | 0.92 | 0.94 | 1.00 |
| med | 0.480 | 0.537 | 0.537 | 0.513 | 0.490 | 0.549 | 0.548 | 0.98 | 1.00 | 1.00 |
| sd | 0.218 | 0.211 | 0.209 | 0.217 | 0.219 | 0.208 | 0.199 | 0.12 | 0.12 | 0.00 |
| rank | 7 | 2 | 3 | 5 | 6 | 1 | 4 | | | |
| | | | | $L_2$ observed efficiency | | | | | | |
| | AIC | AICc | L1AICc | SIC | HQ | HQc | RTp | $L_1$ | $L_2$ | K-L |
| ave | 0.563 | 0.560 | 0.557 | 0.559 | 0.564 | 0.555 | 0.499 | 0.97 | 1.00 | 0.92 |
| med | 0.535 | 0.530 | 0.528 | 0.531 | 0.535 | 0.527 | 0.472 | 0.99 | 1.00 | 1.00 |
| sd | 0.231 | 0.236 | 0.234 | 0.236 | 0.233 | 0.236 | 0.229 | 0.06 | 0.00 | 0.16 |
| rank | 2 | 3 | 5 | 4 | 1 | 7 | 6 | | | |
| | | | | $L_1$ observed efficiency | | | | | | |
| | AIC | AICc | L1AICc | SIC | HQ | HQc | RTp | $L_1$ | $L_2$ | K-L |
| ave | 0.733 | 0.728 | 0.726 | 0.728 | 0.733 | 0.724 | 0.710 | 1.00 | 0.99 | 0.95 |
| med | 0.736 | 0.733 | 0.728 | 0.738 | 0.740 | 0.727 | 0.726 | 1.00 | 1.00 | 0.99 |
| sd | 0.157 | 0.162 | 0.161 | 0.162 | 0.159 | 0.163 | 0.164 | 0.00 | 0.02 | 0.10 |
| rank | 1 | 3 | 5 | 3 | 1 | 6 | 7 | | | |

highest observed efficiency under both $L_2$ and $L_1$.

Although Models 15 and 16 differ greatly in terms of their identifiability, $L_1$ again has a shallow minimum and overall high observed efficiencies. Because misselection in $L_1$ is not heavily penalized and there is little difference in $L_1$ observed efficiencies among the criteria, we begin to suspect that it may not be a useful measure of performance. However, we will continue to include $L_1$ observed efficiency results for the purposes of comparison. We next see how these two models perform when the sample size is doubled. Results for Model 17 (Model 15 with $n = 50$) are given in Table 7.3.

Table 7.3. Simulation results for Model 17. Counts and observed efficiency. $n = 50$, $\beta_j = 1$, $k_* = 6$.

| | counts | | | | | | | | | |
|---|---|---|---|---|---|---|---|---|---|---|
| $k$ | AIC | AICc | L1AICc | SIC | HQ | HQc | RTp | $L_1$ | $L_2$ | K-L |
| 1 | 0 | 0 | 0 | 0 | 0 | 0 | 0 | 0 | 0 | 0 |
| 2 | 0 | 0 | 0 | 0 | 0 | 0 | 0 | 0 | 0 | 0 |
| 3 | 0 | 0 | 0 | 0 | 0 | 0 | 0 | 0 | 0 | 0 |
| 4 | 0 | 0 | 0 | 0 | 0 | 0 | 0 | 0 | 0 | 0 |
| 5 | 0 | 0 | 0 | 0 | 0 | 0 | 0 | 0 | 0 | 0 |
| 6 | 351 | 522 | 518 | 681 | 496 | 674 | 781 | 498 | 999 | 695 |
| 7 | 374 | 348 | 338 | 244 | 343 | 257 | 185 | 369 | 1 | 249 |
| 8 | 197 | 106 | 113 | 61 | 130 | 61 | 31 | 112 | 0 | 51 |
| 9 | 72 | 24 | 31 | 14 | 29 | 8 | 3 | 18 | 0 | 5 |
| 10 | 6 | 0 | 0 | 0 | 2 | 0 | 0 | 3 | 0 | 0 |
| 11 | 0 | 0 | 0 | 0 | 0 | 0 | 0 | 0 | 0 | 0 |
| true | 351 | 522 | 518 | 681 | 496 | 674 | 781 | 498 | 999 | 695 |

| | K-L observed efficiency | | | | | | | | | |
|---|---|---|---|---|---|---|---|---|---|---|
| | AIC | AICc | L1AICc | SIC | HQ | HQc | RTp | $L_1$ | $L_2$ | K-L |
| ave | 0.753 | 0.811 | 0.810 | 0.862 | 0.801 | 0.861 | 0.900 | 0.99 | 1.00 | 1.00 |
| med | 0.785 | 0.985 | 0.983 | 1.000 | 0.932 | 1.000 | 1.000 | 1.00 | 1.00 | 1.00 |
| sd | 0.234 | 0.232 | 0.234 | 0.221 | 0.235 | 0.220 | 0.200 | 0.02 | 0.01 | 0.00 |
| rank | 7 | 4 | 5 | 2 | 6 | 3 | 1 | | | |

| | $L_2$ observed efficiency | | | | | | | | | |
|---|---|---|---|---|---|---|---|---|---|---|
| | AIC | AICc | L1AICc | SIC | HQ | HQc | RTp | $L_1$ | $L_2$ | K-L |
| ave | 0.750 | 0.810 | 0.808 | 0.863 | 0.800 | 0.862 | 0.902 | 0.99 | 1.00 | 1.00 |
| med | 0.790 | 1.000 | 1.000 | 1.000 | 0.951 | 1.000 | 1.000 | 1.00 | 1.00 | 1.00 |
| sd | 0.246 | 0.242 | 0.243 | 0.228 | 0.244 | 0.226 | 0.204 | 0.02 | 0.00 | 0.02 |
| rank | 7 | 4 | 5 | 2 | 6 | 3 | 1 | | | |

| | $L_1$ observed efficiency | | | | | | | | | |
|---|---|---|---|---|---|---|---|---|---|---|
| | AIC | AICc | L1AICc | SIC | HQ | HQc | RTp | $L_1$ | $L_2$ | K-L |
| ave | 0.851 | 0.886 | 0.885 | 0.916 | 0.880 | 0.916 | 0.957 | 1.00 | 1.00 | 1.00 |
| med | 0.890 | 0.982 | 0.981 | 0.995 | 0.973 | 0.994 | 1.000 | 1.00 | 1.00 | 1.00 |
| sd | 0.157 | 0.151 | 0.152 | 0.141 | 0.153 | 0.140 | 0.134 | 0.00 | 0.01 | 0.01 |
| rank | 7 | 4 | 5 | 2 | 6 | 2 | 1 | | | |

With the same strongly identifiable model and the benefit of an increased sample size, none of the criteria underfit in terms of counts. As in Model 15, rank results are identical across the three observed efficiency measures, and RTp is the best overall performer. However, the performance of the consistent criterion SIC has improved to second from fifth for Model 15. Its penalty function increases rapidly with sample size, and with a sample size of 50 and a strongly identifiable true model, the consistent nature of SIC can be seen in its increased performance. While the ranks for HQ and HQc are about

Table 7.4. Simulation results for Model 18. Counts and observed efficiency. $n = 50$, $\beta_j = 1/j$, $k_* = 6$.

| | | | | | counts | | | | | |
|---|---|---|---|---|---|---|---|---|---|---|
| $k$ | AIC | AICc | L1AICc | SIC | HQ | HQc | RTp | $L_1$ | $L_2$ | K-L |
| 1 | 0 | 0 | 0 | 0 | 0 | 0 | 0 | 0 | 0 | 0 |
| 2 | 0 | 0 | 0 | 1 | 1 | 1 | 1 | 0 | 0 | 0 |
| 3 | 1 | 2 | 2 | 9 | 2 | 4 | 23 | 0 | 0 | 0 |
| 4 | 26 | 42 | 40 | 96 | 46 | 81 | 162 | 0 | 0 | 0 |
| 5 | 130 | 200 | 202 | 347 | 195 | 324 | 386 | 41 | 38 | 58 |
| 6 | 367 | 441 | 443 | 392 | 425 | 425 | 346 | 499 | 942 | 663 |
| 7 | 284 | 231 | 226 | 132 | 225 | 141 | 73 | 335 | 14 | 232 |
| 8 | 137 | 69 | 72 | 20 | 82 | 24 | 9 | 102 | 6 | 40 |
| 9 | 47 | 15 | 15 | 3 | 22 | 0 | 0 | 21 | 0 | 6 |
| 10 | 8 | 0 | 0 | 0 | 2 | 0 | 0 | 2 | 0 | 1 |
| 11 | 0 | 0 | 0 | 0 | 0 | 0 | 0 | 0 | 0 | 0 |
| true | 234 | 295 | 288 | 282 | 287 | 300 | 251 | 466 | 909 | 619 |
| | | | | | K-L observed efficiency | | | | | |
| | AIC | AICc | L1AICc | SIC | HQ | HQc | RTp | $L_1$ | $L_2$ | K-L |
| ave | 0.679 | 0.700 | 0.697 | 0.694 | 0.696 | 0.700 | 0.675 | 0.98 | 0.99 | 1.00 |
| med | 0.673 | 0.695 | 0.696 | 0.683 | 0.687 | 0.691 | 0.656 | 1.00 | 1.00 | 1.00 |
| sd | 0.242 | 0.246 | 0.246 | 0.245 | 0.247 | 0.247 | 0.245 | 0.05 | 0.04 | 0.00 |
| rank | 6 | 2 | 3 | 5 | 4 | 1 | 7 | | | |
| | | | | | $L_2$ observed efficiency | | | | | |
| | AIC | AICc | L1AICc | SIC | HQ | HQc | RTp | $L_1$ | $L_2$ | K-L |
| ave | 0.668 | 0.686 | 0.682 | 0.666 | 0.682 | 0.676 | 0.637 | 0.99 | 1.00 | 0.98 |
| med | 0.663 | 0.679 | 0.677 | 0.647 | 0.674 | 0.663 | 0.606 | 1.00 | 1.00 | 1.00 |
| sd | 0.259 | 0.266 | 0.266 | 0.273 | 0.266 | 0.273 | 0.276 | 0.03 | 0.00 | 0.07 |
| rank | 5 | 1 | 3 | 6 | 2 | 4 | 7 | | | |
| | | | | | $L_1$ observed efficiency | | | | | |
| | AIC | AICc | L1AICc | SIC | HQ | HQc | RTp | $L_1$ | $L_2$ | K-L |
| ave | 0.802 | 0.812 | 0.809 | 0.799 | 0.809 | 0.804 | 0.775 | 1.00 | 1.00 | 0.99 |
| med | 0.815 | 0.830 | 0.829 | 0.812 | 0.825 | 0.822 | 0.780 | 1.00 | 1.00 | 1.00 |
| sd | 0.170 | 0.173 | 0.174 | 0.179 | 0.174 | 0.178 | 0.181 | 0.00 | 0.01 | 0.04 |
| rank | 5 | 1 | 2 | 6 | 2 | 4 | 7 | | | |

the same between the two models, their observed efficiency values are larger. HQ and HQc are asymptotically equivalent, and in large samples we expect them to perform about the same. However, a sample size of 50 is not sufficiently large, and HQc still outperforms HQ. AICc and AIC also are asymptotically equivalent, as well as efficient. In large samples, we expect both to perform about the same and to overfit. However, here also $n = 50$ is not sufficiently large to change their relative performances, and their ranks are unchanged (7 and 4, respectively). The penalty function for AICc decreases slightly as the sample size increases, and so it overfits more in Model 17 than in Model 15.

Finally, we consider the effect of increased sample size on a weakly identifiable model, Model 18 (Model 16 with $n = 50$), which is summarized in Table 7.4. In Table 7.4 we see larger count values and observed efficiencies, and much less underfitting than for Model 16. However, otherwise the results of this simulation are quite surprising. From ranking first for both strongly identifiable models and 4th for Model 16, RTp has dropped to nearly last due to severe underfitting. SIC, which ranked second for Model 16, here drops to fifth place under K-L and sixth under both $L_2$ and $L_1$ for the same reason. Much less overfitting is seen from AIC and its performance is much closer to AICc, which here performs better than L1AICc. The gap between HQ and HQc is lessened, but HQc still outperforms HQ. Another difference seen in the counts is that both underfitting and overfitting in Model 18 is reduced from that in Model 16 across the board. Criteria with strong penalty functions underfit less in Model 18 due to the increased sample size. This large sample size also reduces overfitting in criteria with weak penalty functions. The increased sample size causes to count patterns to appear less extreme, which in turn results in improved observed efficiencies.

These four models give some insight into model selection with $L_1$ regression. The robust Wald test RTp is easily adopted to $L_1$ regression and performs well when the true model is strongly identifiable. However, RTp performs poorly when the true model is difficult to detect. Although it is an exact unbiased estimate of the Kullback–Leibler distance (minus constants), L1AICc does not perform much better than the approximate AICc. Considering the increased computational burden for computing L1AICc, AICc seems to be the better practical choice. Although four special case models are not enough to determine an overall selection criterion choice, since each criterion showed weaknesses in some of the models, AIC consistently performed worst and can be ruled out as a practical choice.

## 7.3. Robust Version of Cp

In Hurvich and Tsai's (1990) derivation of L1AICc, they considered $L_1$ regression when the error distribution was nonnormal and assumed that the errors were distributed as double exponential. The design matrix itself, $X$, had no outliers; all outliers resulted from the error distribution. What if we suppose both $X$ and the errors have outliers? Ronchetti and Staudte (1994) developed a robust version of Mallows's Cp, RCp, which can be used with a large variety of robust estimators, including M-estimators (a generalized form of MLEs), GM estimators (*e.g.*, bounded influence estimators) and one-step M-estimators with a high breakdown starting point. In their (1994) derivation of RCp, Ronchetti and Staudte give a useful example for the need for a robust method that identifies outliers and/or influential points so that they can be removed and a model fit to the majority of the data.

Consider a simple linear relationship between $x$ and $y$. Now suppose that one of the middle values of $x$ has an incorrect corresponding $y$, a gross error. Such errors in $y$ can cause the data to appear nonlinear, as in Ronchetti and Staudte's example where the model appears to require a quadratic term, favoring a quadratic model over the true linear one. Thus they make the point that one outlier can potentially distort the entire model. Such gross errors have less to do with the error distribution than with transcription errors or other sources of outliers.

### 7.3.1. Derivation of RCp

Consider the usual regression model in Eq. (7.2), where the rows of $X$ are *i.i.d.* from the marginal distribution $F$, and $F$ is the joint distribution of $\varepsilon_i$ with its corresponding row $x_i$. An M-estimator $\tilde{\beta}_M$ is the solution to

$$\sum_{i=1}^{n} \eta(x_i, y_i - x_i'\beta)x_i = 0,$$

where $\eta(x_i, \varepsilon_i)$ is some known function. Define the residual $e_i = y_i - x_i'\tilde{\beta}_M$ and the weight $\tilde{w}_i = \eta(x_i, e_i)/e_i$. The rescaled mean squared weighted prediction error, $\Gamma_k$, for candidate model $k$ is defined as

$$\Gamma_k = \frac{1}{\sigma^2} E_F \left[ \sum_{i=1}^{n} \tilde{w}_i^2 (\tilde{y}_i - E(y_i))^2 \right],$$

where $\tilde{y}_i$ is the fitted value for candidate model $k$, $E_F$ is the expected value evaluated under the joint distribution, and $E(y_i)$ is the expected value evaluated under the full, and assumed to be correct, model. The weights $\tilde{w}$ reduce the effects of outliers on the predictions. $\Gamma_k$ is a reasonable indication of model adequacy in the presence of outliers, and therefore a robust model selection criterion can be derived by finding a good estimate of $\Gamma_k$. To do this, first define the following quantities

$$R = E_F[w^2 xx'],$$

$$M = E_F\left[\frac{\partial}{\partial \varepsilon}\eta(x,\varepsilon)xx'\right],$$

$$N = E_F\left[\eta^2 \frac{\partial}{\partial \varepsilon}\eta(x,\varepsilon)xx'\right],$$

$$Q = E_F[\eta^2(x,\varepsilon)xx'],$$

and

$$L = E_F\left[\left(\left(\frac{\partial}{\partial \varepsilon}\eta(x,\varepsilon)\right)^2 + 2\frac{\partial}{\partial \varepsilon}\eta(x,\varepsilon)w - 3w^2\right)xx'\right],$$

where $w = w(x, \varepsilon)$ is some known weighting function. Now define

$$W_k = \sum \tilde{w}_i^2 e_i^2,$$

$$U_k = E_F\Big[\sum_i \eta^2(x_i,\varepsilon_i)\Big] - 2tr\{NM^{-1}\} + tr\{LM^{-1}QM^{-1}\} + tr\{RM^{-1}QM^{-1}\},$$

and

$$V_k = tr\{RM^{-1}QM^{-1}\}.$$

$W_k$ is the weighted SSE. As in Chapter 2, $k$ represents the candidate model of order $k$ and $K$ represents the full model including all variables.

Given the above, Ronchetti and Staudte (1994) define a robust version of Cp as

$$\text{RCp} = \frac{W_k}{\tilde{\sigma}^2} - (U_k - V_k), \tag{7.11}$$

where $\tilde{\sigma}^2 = W_K/U_K$ is a consistent and robust estimator of $\sigma^2$. If the model $k$ is as good as the full model, then it can be shown that $\tilde{\sigma}^2 \approx W_k/U_k$ and RCp $\approx V_k$. Plots of RCp versus $V_k$ are a new, robust procedure for model selection (see Ronchetti and Staudte, 1994). Cp versus $k$ plots have long been used in least squares regression as a diagnostic for determining better candidate models. A good subset should have Cp$\leq k$ (see Mallows, 1973). Ronchetti and Staudte's RCp extends this usefulness to the robust setting.

## 7.4. Wald Test Version of Cp

Another robust version of Cp can be derived from the Wald test as proposed by Sommer and Huggins (1996). They wanted to derive a selection criterion that is easy to compute and generalize to other nonadditive error model structures. Details of other Wald test criteria can also be found in Sommer and Huggins (1996).

Suppose one observes $y_1, \ldots, y_n$ for which the distribution depends on a $K$-dimensional vector of parameters $\theta = (\theta_0, \ldots, \theta_{K-1})'$. Let $\hat{\theta}$ be an estimator (often the MLE) of $\theta$. As $n - K \to \infty$,

$$\sqrt{n}(\hat{\theta} - \theta) \xrightarrow{d} N_K(0, \Sigma(\theta)),$$
$$\Sigma(\hat{\theta}) \xrightarrow{p} \Sigma(\theta).$$

To test a $k$-dimensional subset of $\theta$, $\theta_1$, partition $\theta$ as $\theta = (\theta_1', \theta_2')'$ and $\Sigma(\theta)$ as

$$\begin{pmatrix} \Sigma_{11}(\theta) & \Sigma_{12}(\theta) \\ \Sigma_{21}(\theta) & \Sigma_{22}(\theta) \end{pmatrix}.$$

The Wald test for $H_0$: $\theta_2 = 0$ is based on the statistic

$$A_k = n\hat{\theta}_2' \Sigma_{22}^{-1}(\hat{\theta})\hat{\theta}_2.$$

Under $H_0$, $A_k$ has a $\chi^2_{K-k}$ distribution. Now define the selection criteria Tp as

$$\mathrm{Tp} = A_k - K + 2k,$$

which is analogous to Cp. If $H_0$ is true, then in large samples we expect Tp $= k$. In addition, Sommer and Huggins have shown that for regression with normal errors, Tp is equivalent to Cp for any sample size $n$ and $\theta = \beta$.

One advantage of the Wald-based Tp criterion is that it is easily adapted to other classes of models, such as those for logistic regression, and thus is more flexible than Cp. It can be used when only its asymptotic distribution is known, as is often the case. Another advantage of Tp is that robust versions of the Wald test exist, and these can be applied to Tp to create a robust version of Tp such as RTp.

To derive RTp, first consider the full regression model $Y = X\beta + \varepsilon$, where $\beta$ is a $K \times 1$ vector and $\varepsilon = (\varepsilon_1, \ldots, \varepsilon_n)'$. Also consider the $k$-dimensional candidate model. Partition $\beta$ into $\beta = (\beta_1', \beta_2')'$ and let $\beta = (\beta_1', \beta_2')'$ represent the full model and $\beta = \beta_1$ represent the reduced candidate model. The full model is of order $K$ and the reduced candidate model is of order $k$. All candidate models have some $\beta_2 = 0$ except the one candidate model of order $K$ that is the largest model (in which case RTp= $K$). Thus, by analogy to Tp, RTp has the form

$$\text{RTp} = \tilde{\beta}_{2M}' \Sigma_{22}^{-1} \tilde{\beta}_{2M} - K + 2k, \tag{7.12}$$

where $\tilde{\beta}_{2M}$ is an M-estimate for $\beta_2$ and $\Sigma_{22}$ is an M-estimate for $cov(\tilde{\beta}_{2M})$. As was the case for Tp, if the candidate model is a good model then RTp should be close to $k$. This feature allows us to use RTp in exactly the same manner as we would use Tp and Cp.

In the past two sections we have reviewed robust versions of Mallows's Cp-type criteria. In the next Section we expand discussion to a robust version of the Akaike's prediction error criterion FPE.

## 7.5. FPE for Robust Regression

Burman and Nolan (1995) derived a robust version of Akaike's FPE criterion as follows. Let the generating model be of the form in Eq. (7.1) with distribution $F_*$ and variance $\sigma_*^2$, and the candidate models be of the form in Eq. (7.2). Let $y_{0i} = \mu_{*i} + \varepsilon_{0i}$, where $\varepsilon_{0i}$ $(i = 1, \ldots, n)$ are errors independent from $F_*$ and also independent of the $\varepsilon_{*i}$. The final prediction error, a measure of a model's predictive ability, is defined as

$$\sum_{i=1}^{n} E_*[(y_{0i} - x_i'\hat{\beta})^2] = \sum_{i=1}^{n} E_*[(y_i - x_i'\bar{\beta})^2] + k\sigma_*^2,$$

where $\hat{\beta}$ is the least squares estimator of $\beta$ and $\bar{\beta}$ is the minimizer of $\sum_{i=1}^{n} E_*[(y_i - x_i'\beta)^2]$. If we want to measure predictive performance using an arbitrary loss function $\rho(\cdot)$, then predictive ability becomes

$$\sum_{i=1}^{n} E_*[\rho(y_{0i} - x_i'\hat{\beta})]. \tag{7.13}$$

The best model is the model that minimizes Eq. (7.13).

This idea of estimating the prediction error from the model also can be applied to $L_1$ regression, where predictive ability is measured by the expected absolute prediction error or $\rho(\cdot) = |\cdot|$ in Eq. (7.13). Burman and Nolan showed that the expected absolute prediction error can be approximated by

$$\sum_{i=1}^{n} |y_i - x_i'\hat{\beta}| + \frac{k \sum_{i=1}^{n} 4F_*(x_i'\bar{\beta} - \mu_{*i})[1 - F_*(x_i'\bar{\beta} - \mu_{*i})]}{\sum_{i=1}^{n} 2f_*(\mu_{*i} - x_i'\bar{\beta})},$$

where $f_*$ is the density of $\varepsilon_{*i}$ and $\hat{\beta}$ is the usual $L_1$ estimator for $\beta$. Furthermore, if one assumes the model error $\mu_{*i} - x_i'\bar{\beta}$ is small, the above expression simplifies to

$$\sum_{i=1}^{n} |y_i - x_i'\hat{\beta}| + \frac{k}{2f_*(0)}.$$

This assumption can be checked by examining the estimated errors from the candidate model, and it greatly simplifies the derivation. However, it may be difficult to justify if the estimated errors are large.

In order to develop a general procedure for estimating the prediction error, Burman and Nolan (1995) first considered a $\rho$ function which satisfies the following conditions:

(1) $\rho(t)$ is convex with unique minimum at $t = 0$,

(2) $R(t) = E_*[\rho(\varepsilon_* + t)]$ is twice differentiable almost everywhere with second derivative $R_2$, and

(3) $E_*[\psi(\varepsilon_*)] = 0$ where $\psi(\varepsilon_*) = d\rho(\varepsilon_*)/d\varepsilon_*$. Given (1)–(3), they then defined the expected prediction error of $\rho$ for new observations $y_{0i}$ $(i = 1, \ldots, n)$ as

$$L(k) = \sum_{i=1}^{n} E_*[\rho(y_{0i} - x_i'\hat{\beta})],$$

where $k$ is the dimension of $x_i$, $\hat{\beta}$ minimizes $\sum_{i=1}^{n} \rho(y_i - x_i'\beta)$, and $\bar{\beta}_\rho$ minimizes $\sum_{i=1}^{n} E_*[\rho(y_i - x_i'\beta)]$. $L(k)$ can be approximated by

$$\bar{L}(k) = \sum_{i=1}^{n} \rho(e_i) + kC_\rho,$$

where

$$C_\rho = \frac{\sum_{i=1}^{n} var[\psi(y_i - x_i'\bar{\beta}_\rho)]}{\sum_{i=1}^{n} R_2(\mu_{*i} - x_i'\bar{\beta}_\rho)},$$

and $e_i = y_i - x_i'\hat{\beta}$.

$C_\rho$ needs to be estimated or approximated if the distribution $F_*$ is unknown. Some common approximations given in Burman and Nolan for the numerator and denominator are

$$\hat{\sigma}_1^2 = \frac{1}{n-k} \sum_{i=1}^{n} \psi(e_i)^2$$

and

$$\hat{r}_2 = \frac{1}{n-k} \sum_{i=1}^{n} \hat{R}_2(e_i),$$

where $\hat{R}_2$ is an approximation for $R_2$. Using these approximations, $\hat{C}_\rho = \hat{\sigma}_1^2/\hat{r}_2$, and

$$\hat{L}(k) = \sum_{i=1}^{n} \rho(e_i) + k\hat{C}_\rho. \quad (7.14)$$

Table 7.5 (taken from Table 4 in Burman and Nolan, 1995, used with permission from the Biometrika trustees) compares $\hat{L}(k)$ efficiencies under three estimation methods and two distributions. One thousand realizations were generated from the true model

$$y_i = 2\sin(2\pi x_i) + \varepsilon_{*i},$$

where $x_i = i/n$ and $n = 41$ and 81. For each realization, the $k$th degree polynomial regression in $x$ is fitted to the data, $k = 0, 1, \ldots, 7$. Least squares ($\rho(z) = z^2$) and $L_1$ ($\rho(z) = |z|$) estimation methods are presented, as well as the Huber case. The distributions are the $t_3$ distribution (Student's distribution with 3 degrees of freedom) and the following $G$ distribution from Burman and Nolan:

$$G(t) = \begin{cases} \frac{1}{2}(1-t)^{-2} & t \leq 0, \\ 1 - \frac{1}{2}(1+t)^{-2} & t > 0. \end{cases}$$

Table 7.5. Model selection performance of $\hat{L}(k)$.

| | Least Squares | | $L_1$ | | Huber | |
|---|---|---|---|---|---|---|
| | mean | sd | mean | sd | mean | sd |
| $n = 41$, $\varepsilon \sim t_3$ | 0.78 | 0.28 | 0.75 | 0.28 | 0.78 | 0.28 |
| $n = 81$, $\varepsilon \sim t_3$ | 0.81 | 0.25 | 0.77 | 0.24 | 0.79 | 0.24 |
| $n = 41$, $\varepsilon \sim G$ | — | — | 0.71 | 0.27 | 0.73 | 0.29 |
| $n = 81$, $\varepsilon \sim G$ | — | — | 0.69 | 0.26 | 0.74 | 0.24 |

We see from Table 7.5 that $\hat{L}(k)$ performs reasonably well in choosing the dimension of a model in terms of efficiency. It is interesting to note that the least squares version has good efficiency when the errors have a $t_3$ distribution. There is little difference in the efficiency of $\hat{L}(k)$ for the three forms of $\rho(\cdot)$. In general, $\hat{L}(k)$ has higher efficiency in the larger sample size.

## 7.6. Unification of AIC Criteria

So far, we have followed the development of the AIC criteria from AIC itself, widely used over the past thirty years, to Hurvich and Tsai's (1989) AICc (Eq. (2.14)), intended to overcome AIC's drastic bias in small samples. The AIC family is derived under normal errors, and when the normality assumption fails, one solution is to modify the Kullback–Leibler information so that it can measure the discrepancy of a robust function evaluated under both the true model and fitted models. To this end Shi and Tsai (1998) proposed a generalized Kullback–Leibler information, from which they obtained two generalized Akaike information criteria, AICR* and AICcR. These criteria allow us to choose the best model based on two considerations: good model fit for the majority of the data (discounting outliers and influential points), and accommodation of nonnormally distributed errors. Fitting most of the data is an idea favored by Ronchetti (1985), who proposed the robust AICR, a generalized Akaike information criterion that takes into account the presence of outliers. AICR was later expanded upon by Hurvich and Tsai (1990) and Hurvich, Shumway and Tsai (1990), who used Monte Carlo methods to obtain the model selection criteria L1AICc (discussed earlier in Section 7.1, Eq. (7.10)) for use with double exponential regression models, and AICi (Hurvich, Shumway and Tsai 1990) for autoregressive models. The relationships between the two unified criteria, AICR* and AICcR, and each of the AIC, AICc, AICR, and AICi criteria, are illustrated next with three analytical examples using location–scale regression models. In addition, Monte Carlo studies are presented to illustrate their relative performance.

### *7.6.1. The Unification of the AIC Family*

We will adapt Shi and Tsai's (1998) results to present a set of equations that unify most existing AIC criteria and which also allow greater freedom from distributional assumptions than the model selection techniques we have discussed in previous chapters. Shi and Tsai used the M-estimators for the parameters $\beta$ and $\sigma$, as well as an approximation of a generalized Kullback–Leibler discrepancy, to obtain a generalized description of the AIC family.

Suppose that data $y_1, \ldots, y_n$ are independently generated from a true model with probability density function $f_*(y_i; x'_{*i}\beta_*, \sigma_*)$, where $x_{*i}$ is a $k_* \times 1$ vector of explanatory variables, $\beta_*$ is a $k_* \times 1$ vector of parameters, and $\sigma_*$ is a scale parameter. Consider fitting the data with a family of candidate models with probability density functions $f(y_i; x'_i\beta, \sigma)$, where $x_i$ is a $k \times 1$ vector, $\beta$ is a $k \times 1$ vector of unknown parameters, and $\sigma$ is a scale parameter. Hence, the true model and the candidate models will differ both in their location (regression) and scale parameters.

After fitting the candidate models to the data, we next assume that the unknown regression parameter $\beta$ and the scale parameter $\sigma$ can be estimated by minimizing the function

$$\sum_{i=1}^{n} \rho\left(\frac{y_i - x_i'\beta}{\sigma}\right),$$

where $\rho$ is a suitable function chosen with respect to the data and the candidate model structure. This function can be scaled by the sample size and expressed as a rate:

$$D(\beta, \sigma) = \frac{1}{n} \sum_{i=1}^{n} \rho\left(\frac{y_i - x_i'\beta}{\sigma}\right). \tag{7.15}$$

The resulting generalized MLE parameter estimators, $\hat{\beta}$ and $\hat{\sigma}$, are usually called M-estimators. Note that by choosing particular $\rho$ functions, a few well-known parameter estimators can be obtained (*e.g.*, least squares estimators, $L_p$-norm estimators, and maximum likelihood estimators).

To assess the discrepancy between the function in Eq. (7.15) evaluated under the true model and under candidate models, Shi and Tsai proposed the following generalized Kullback–Leibler information:

$$\Delta(\beta, \sigma) = E_*[D(\beta, \sigma)], \tag{7.16}$$

where $E_*$ denotes the expectation evaluated under the true model. If $\rho = -2\log(f)$, then Eq. (7.16) gives the classical Kullback–Leibler information.

Next we assume that $\rho$ is a second-order differentiable function, and that there is a $k \times k_*$ matrix $M$ such that $M\beta_* = \tilde{\beta}_*$. Under these assumptions, the second-order Taylor expansion of $\rho\{(y_i - x_i'\beta)/\sigma\}$ for $\tilde{\beta}_*$ leads to the following approximation of the generalized Kullback–Leibler information in Eq. (7.16):

$$\Delta(\beta, \sigma) \simeq \Delta(\tilde{\beta}_*, \sigma) + \dot{\Delta}(\tilde{\beta}_*, \sigma)(\beta - \tilde{\beta}_*) + (\beta - \tilde{\beta}_*)'\ddot{\Delta}(\tilde{\beta}_*, \sigma)(\beta - \tilde{\beta}_*)/2, \tag{7.17}$$

where

$$\dot{\Delta}(\tilde{\beta}_*, \sigma) = E_*[\partial D(\beta, \sigma)/\partial\beta]_{\beta=\tilde{\beta}_*}$$

and

$$\ddot{\Delta}(\tilde{\beta}_*, \sigma) = E_*[\partial^2 D(\beta, \sigma)/\partial\beta\partial\beta']_{\beta=\tilde{\beta}_*}.$$

If we now adopt Huber's (1981, p. 165) assumption that $\dot{\Delta}(\tilde{\beta}_*, \sigma) = 0$ for a given $\sigma > 0$, Eq. (7.17) becomes

$$\Delta(\beta, \sigma) \simeq \Delta(\tilde{\beta}_*, \sigma) + (\beta - \tilde{\beta}_*)'\ddot{\Delta}(\tilde{\beta}_*, \sigma)(\beta - \tilde{\beta}_*)/2.$$

One criterion for assessing candidate models with respect to the data is $E_*[\Delta(\hat{\beta}, \hat{\sigma})]$, where $\hat{\beta}$ and $\hat{\sigma}$ are the M-estimators of $\beta$ and $\sigma$, respectively (Huber, 1981, p. 176).

Applying Silvapulle's (1985, p. 1494) results that $\sqrt{n}(\hat{\beta}-\tilde{\beta}_*)$ and $\ddot{\Delta}(\tilde{\beta}_*,\hat{\sigma})$ are asymptotically independent, Shi and Tsai (1998) replace $\mathrm{var}[\sqrt{n}(\hat{\beta}-\tilde{\beta}_*)]$ by its asymptotic covariance, and $E_*[\Delta(\hat{\beta},\hat{\sigma})]$ can be approximated as

$$E_*[\Delta(\hat{\beta},\hat{\sigma})] \simeq E_*[\Delta(\tilde{\beta}_*,\hat{\sigma})] + \frac{1}{2}tr\{A\}, \tag{7.18}$$

where

$$A = cE_*[\ddot{\Delta}(\tilde{\beta}_*,\hat{\sigma})](X'X)^{-1},$$
$$c = \mathrm{var}[\psi(\tilde{\varepsilon}_*)]/E_*^2[\dot{\psi}(\tilde{\varepsilon}_*)],$$
$$\tilde{\varepsilon}_* = (y_1 - x'_{*1}\beta_*)/\sigma_*,$$
$$\psi(t) = d\rho(t)/dt,$$
$$\dot{\psi}(t) = d\psi(t)/dt,$$

$X$ is an $n\times k$ matrix with the $i$th row $x_i'$, and $\mathrm{var}[\cdot]$ indicates the variance evaluated under the true model. By further applying Silvapulle's (1985) Theorem 1 and the asymptotic covariance of $\mathrm{var}[\sqrt{n}(\hat{\beta}-\tilde{\beta}_*)]$, Shi and Tsai obtained

$$E_*[\Delta(\tilde{\beta}_*,\hat{\sigma})] \simeq E_*\left[\frac{1}{n}\sum_{i=1}^{n} E_*\left[\rho\left(\frac{y_i - x_i'\hat{\beta}}{\sigma}\right)\right]_{\hat{\sigma}}\right] + \frac{1}{2}tr\{A\}.$$

Finally, substituting the above equation into Eq. (7.18) leads to

$$E_*[\Delta(\hat{\beta},\hat{\sigma})] \simeq E_*\left[\frac{1}{n}\sum_{i=1}^{n} E_*\left[\rho\left(\frac{y_i - x_i'\hat{\beta}}{\sigma}\right)\right]_{\hat{\sigma}}\right] + tr\{A\}. \tag{7.19}$$

Eq. (7.18) and Eq. (7.19) unify most existing Akaike information criteria, and will also allow us to derive generalized criteria that are more robust against outliers and other departures from the distributional assumptions of the model. In the next Section we will show how these equations can be useful by applying them to location–scale regression models.

### *7.6.2. Location–Scale Regression Models*

In this Section we use Eq. (7.18) and Eq. (7.19) to obtain the generalized AIC criteria AICR* and AICcR. Three examples are given to illustrate the relationship between AICR* and its parent criterion AIC, and the relationship of AICcR* to AICc and to AICi.

Assume that the data $y_i$ are generated from the true model in Eq. (7.3) and then fitted with candidate models of the form in Eq. (7.2), where $\varepsilon_{*i}$ and

$\varepsilon_i$ are independent and identically distributed. The true location–scale model has probability density function $f_*(\varepsilon_{*i}, \sigma_*) = \sigma_*^{-1} g(\varepsilon_{*i}/\sigma_*)$. Let the candidate model have density $f(\varepsilon_i, \sigma) = \sigma^{-1} g(\varepsilon_i/\sigma)$. In both densities, $g(\cdot)$ is a positive function, $\sigma_*$ denotes the true scale parameter, and $\sigma$ is the candidate model's scale parameter.

For location–scale regression,

$$\ddot{\Delta}(\tilde{\beta}_*, \sigma) = \frac{1}{n}\Big(\int \dot{\psi}\Big(t\frac{\sigma_*}{\sigma}\Big) g(t)dt\Big) X'X \frac{\sigma_*^2}{\sigma^2}.$$

Now, $tr\{A\}$ from Eq. (7.18) and Eq. (7.19) becomes

$$tr\{A\} = \frac{1}{n} E_* \left[\frac{\sigma_*^2}{\hat{\sigma}^2} \int \dot{\psi}\left(\frac{t\sigma_*}{\hat{\sigma}}\right) g(t)dt\right] \frac{k \int \psi^2(t) g(t)dt}{\left(\int \dot{\psi}(t) g(t)dt\right)^2}. \tag{7.20}$$

Substituting Eq. (7.20) into Eq. (7.18), Shi and Tsai (1998) obtained the generalized AICc model selection criterion

$$\text{AICcR} = E_*\left[\Delta(\tilde{\beta}_*, \hat{\sigma})\right] + \frac{1}{n} E_* \left[\frac{\sigma_*^2}{\hat{\sigma}^2} \int \dot{\psi}\left(\frac{t\sigma_*}{\hat{\sigma}}\right) g(t)dt\right] \frac{k \int \psi^2(t) g(t)dt}{2(\int \dot{\psi}(t) g(t)dt)^2}.$$

In practice, we may consider replacing the nonpivotal quantity in the first term of AICcR by its unbiased estimator.

Substituting Eq. (7.20) into Eq. (7.19) by replacing $\sigma_*$ with its consistent estimator $\hat{\sigma}$, and then using $\frac{1}{n}\sum_{i=1}^{n} \rho\{(y_i - x_i'\hat{\beta})/\hat{\sigma}\}$ to estimate the expectation

$$E_*\left[\frac{1}{n}\sum_{i=1}^{n} E_*\left[\rho\Big(\frac{y_i - x_i'\hat{\beta}}{\sigma}\Big)\right]_{\hat{\sigma}}\right],$$

Shi and Tsai obtained the generalized AIC selection criterion, scaled by sample size,

$$\text{AICR}^* = \frac{1}{n}\sum_{i=1}^{n} \rho\Big(\frac{y_i - x_i'\hat{\beta}}{\hat{\sigma}}\Big) + \frac{k}{n} \frac{\int \psi^2(t) g(t)dt}{\int \dot{\psi}(t) g(t)dt}. \tag{7.21}$$

AICR* given in Eq. (7.21) is the same as Ronchetti's (1985) AICR when his $\alpha$ is chosen to be the ratio in the last term of Eq. (7.21), $\int \psi^2(t) g(t)dt / \int \dot{\psi}(t) g(t)dt$. Note that Ronchetti obtained his penalty term by using Stone's (1977) results.

Shi and Tsai derived their corrected AICcR* by showing that the penalty term in AICR* can be inflated by the usual $n/(n-k-2)$. This yields

$$\text{AICcR}^* = \frac{1}{n}\sum_{i=1}^{n} \rho\Big(\frac{y_i - x_i'\hat{\beta}}{\hat{\sigma}}\Big) + \frac{k}{n-k-2} \frac{\int \psi^2(t) g(t)dt}{\int \dot{\psi}(t) g(t)dt}. \tag{7.22}$$

Next we adopt Shi and Tsai's (1998) three examples to illustrate the relationships between the generalized criteria and their predecessors.

**Example 1. AICR* and AIC**

Let

$$\rho\left(\frac{y_i - x_i'\beta}{\sigma}\right) = -2\log\big(f(y_i - x_i'\beta, \sigma)\big),$$

where $f(t,\sigma) = \sigma^{-1}g(t/\sigma)$ and $g(\cdot)$ is the density function of the standard normal distribution. The second term of Eq. (7.21) can be reduced to $2k/n$. Hence, if the constant term $\log(2\pi) + (1 - 2/n)$ is ignored, AICR* reduces to the usual AIC.

**Example 2. AICcR and AICc**

For this example as well, let

$$\rho\left(\frac{y_i - x_i'\beta}{\sigma}\right) = -2\log\big(f(y_i - x_i'\beta, \sigma)\big),$$

where $f(t,\sigma) = \sigma^{-1}g(t/\sigma)$ and $g(\cdot)$ is a normal density function with mean zero and variance $\sigma^2$. Also let $\hat{\beta}$ and $\hat{\sigma}^2$ be the maximum likelihood estimators of $\beta$ and $\sigma^2$, respectively. Given these conditions,

$$E_*[\Delta(\tilde{\beta}_*, \hat{\sigma})] = E_*[\log(\hat{\sigma}^2)] + E_*[\sigma_*^2/\hat{\sigma}^2] + 1 + \log(2\pi)$$

and

$$E_*[\sigma_*^2/\hat{\sigma}^2] = n/(n-k-2).$$

Furthermore, the second term of AICcR simplifies to $E_*[\sigma_*^2/\hat{\sigma}^2]k/n$. Then, replacing $E_*[\log(\hat{\sigma}^2)]$ by its unbiased estimator $\log(\hat{\sigma}^2)$ and ignoring $\log(2\pi)+1$, AICcR reduces to AICc.

**Example 3. AICcR and AICi**

In order to illustrate the relationship between AICcR and AICi, let

$$\rho\left(\frac{y_i - x_i'\beta}{\sigma}\right) = -2\log\left(\sigma^{-1}g\left(\frac{y_i - x_i'\beta}{\sigma}\right)\right),$$

where $g(t) = \exp(-t)\exp\left(-e^{-t}\right)$ is the density of the extreme value distribution (see Lawless, 1982, p. 298). It can be shown that

$$\frac{\int \psi^2(t)g(t)dt}{\left\{\int \dot{\psi}(t)g(t)dt\right\}^2} = 1$$

and

$$\int \dot{\psi}\left(\frac{t\sigma_*}{\hat{\sigma}}\right) g(t)dt = 2\int \exp\left(-\frac{t\sigma_*}{\hat{\sigma}}\right) g(t)dt$$
$$= 2\Gamma\left(1 + \frac{\sigma_*}{\hat{\sigma}}\right).$$

Also,

$$\Delta(\tilde{\beta}_*, \hat{\sigma}) = \log(\hat{\sigma}^2) + 2E_*[\varepsilon_{*1}]/\hat{\sigma} + 2E_*\left[\exp\left(-\frac{\varepsilon_{*1}}{\sigma}\right)\right]\Big|_{\hat{\sigma}}$$
$$= \log(\hat{\sigma}^2) + 2\gamma\frac{\sigma_*}{\hat{\sigma}} + 2\Gamma\left(1 + \frac{\sigma_*}{\hat{\sigma}}\right),$$

where $\gamma$ is Euler's constant. Using the above results and replacing $E_*[\log(\hat{\sigma}^2)]$ by its unbiased estimator $\log(\hat{\sigma}^2)$, we can obtain AICcR as

$$\log(\hat{\sigma}^2) + 2E_*\left[\gamma\frac{\sigma_*}{\hat{\sigma}} + \Gamma\left(1 + \frac{\sigma_*}{\hat{\sigma}}\right)\right] + \frac{k}{n}E_*\left[\frac{\sigma_*^2}{\hat{\sigma}^2}\Gamma\left(1 + \frac{\sigma_*}{\hat{\sigma}}\right)\right]. \tag{7.23}$$

According to Theorem G4 in Lawless (1982, p. 538), $\sigma_*/\hat{\sigma}$ is pivotal; *i.e.*, its distribution is independent of the parameters $\beta_*$ and $\sigma_*$. This means that Hurvich and Tsai's (1990) Monte Carlo approach is a suitable method for computing the two expectations in Eq. (7.23). The criterion in Eq. (7.23) calculated via this method was designated AICi by Hurvich, Shumway and Tsai (1990).

### 7.6.3. Monte Carlo Study

Next we will compare the relative performances of AIC, AICc, AICR*, AICcR, and AICcR*. One thousand realizations were generated from the true model in Eq. (7.3) with $k_* = 5$, $\beta_1 = 1, \ldots, \beta_5 = 1$, and $x_{i,j}$ *i.i.d.* $N(0,1)$ for $i = 1, \ldots, n$ and $j = 1, \ldots, 9$. The $\varepsilon_{*i}$ are *i.i.d.* random errors chosen from either the $N(0, 0.7^2)$, $t_4$ (Student's $t$ with four degrees of freedom), or $0.9N(0,1) + 0.1N(0,7^2)$ distributions, over sample sizes of 20, 50 and 80. Nine candidate variables were stored in an $n \times 9$ matrix of independent identically distributed normal random variables. The candidate models are linear, and are listed in columns in a sequential nested fashion; thus the set of candidate models includes the true model. In order to examine the impact of robust functions on model selection criteria, we used Huber's (1964) $\psi$-function with the bending point $c = 1.345$, as proposed by Ronchetti and Staudte (1994), to conduct our simulation studies. We also apply Huber's (1981, Section 7.8) algorithm to compute the M-estimates of $\beta$ and $\sigma$.

Table 7.6. Proportion of correct order selection.

| $N(0, 0.7^2)$ | | | |
|---|---|---|---|
| | $n = 20$ | $n = 50$ | $n = 80$ |
| AIC | 0.532 | 0.673 | 0.723 |
| AICc | 0.911 | 0.804 | 0.798 |
| AICR* | 0.747 | 0.749 | 0.758 |
| AICcR* | 0.945 | 0.858 | 0.835 |
| AICcR | 0.409 | 0.420 | 0.413 |
| $t_4$ | | | |
| | $n = 20$ | $n = 50$ | $n = 80$ |
| AIC | 0.472 | 0.668 | 0.691 |
| AICc | 0.719 | 0.802 | 0.773 |
| AICR* | 0.562 | 0.764 | 0.739 |
| AICcR* | 0.758 | 0.855 | 0.815 |
| AICcR | 0.142 | 0.322 | 0.320 |
| $0.9N(0, 1) + 0.1N(0, 7^2)$ | | | |
| | $n = 20$ | $n = 50$ | $n = 80$ |
| AIC | 0.370 | 0.584 | 0.664 |
| AICc | 0.485 | 0.698 | 0.757 |
| AICR* | 0.535 | 0.767 | 0.756 |
| AICcR* | 0.713 | 0.862 | 0.821 |
| AICcR | 0.106 | 0.278 | 0.330 |

We see from Table 7.6 that AICc, AICR*, and AICcR* outperform AIC across all error distributions and sample sizes. AICc selects the correct model order more often than AICR*, except when the error distribution is the contaminated normal. It is not surprising that AICcR* performs better than AICc when the error distributions depart from normality. Overall, the Monte Carlo studies show that AICcR* provides the best model selections across all error distributions and sample sizes. AICcR performed worst. Therefore, while AICcR is of theoretical interest, we do not recommend its use in practice.

In the previous sections we have discussed various robust model selection criteria that can take into account nonnormal distributions. However, if we know that our data can be fitted with some known extended quasi-likelihood model (Nelder and Pregibon, 1987), then we may consider obtaining model selection criteria for these models directly. This is the focus of the next Section.

## 7.7. Quasi-likelihood

Generalized linear models, such as those for logistic regression (binary outcomes), Poisson regression, and exponential regression, are useful when the errors are not additive. Quasi-likelihood is useful when the full distribution is unknown but the structure for the mean and variance is known. For nonnormal regression models, Pregibon (1979) and Hosmer *et al.* (1989) proposed two different versions of Mallows's Cp for generalized linear models. Pregibon's Cp is based on the deviance, while that of Hosmer *et al.* is based on a Pearson chi-square goodness-of-fit statistic. Efron (1986) also proposed a version of AIC for generalized linear models. The generalized linear models considered by these investigators mentioned above do not include a dispersion parameter. McCullagh and Nelder (1989, p. 90) suggested that it is often wise to assume a dispersion parameter is present in the model unless the data or prior information indicate otherwise, and they obtained a version of the Cp criterion for such models. Unfortunately, their Cp criterion can lead to gross overfitting when the sample size is small or when the number of fitted parameters is a moderate to large fraction of the sample size. This deficiency occurs in normal as well as nonnormal regression models.

Hurvich and Tsai (1995) subsequently generalized the model selection criterion AICc to an extended quasi-likelihood model (Nelder and Pregibon, 1987, and McCullagh and Nelder, 1989), which includes the generalized linear model with a dispersion parameter as a special case. In Section 7.7.1, we will review Hurvich and Tsai's generalized AICc for quasi-likelihood, and present some competing selection criteria that are potentially suitable for the extended quasi-likelihood model. In Section 7.7.2, we compare all of these criteria in a small-sample Monte Carlo study. We will see that under quasi-likelihood AICc exhibits the best performance, and examine the reasons why.

### *7.7.1. Selection Criteria for Extended Quasi-likelihood Models*

In studying generalized versions of AIC, AICc, and Cp for use with quasi-likelihood models, we will essentially follow the notation of Nelder and Pregibon (1987). Consider the true extended quasi-likelihood model

$$Q^+(Y;\mu_*,\sigma_*^2) = -\frac{n}{2}\log(\sigma_*^2) - \frac{1}{2\sigma_*^2}D(Y;\mu_*),$$

where $Y = (y_1, \ldots, y_n)'$, the $y_i$ are independent, $E_*[Y] = \mu_* = (\mu_{*1}, \ldots, \mu_{*n})'$, $E_*$ denotes the expectation evaluated under the true model, $D(Y;\mu_*) = 2\{Q(Y;Y) - Q(Y;\mu_*)\}$, and $Q(Y;\mu_*)$ is the quasi-likelihood model. Now consider the exponential family of the extended quasi-likelihood model proposed by McCullagh and Nelder (1989, p. 336),

$$Q(Y;\mu_*) = Y'\theta_* - b(\theta_*) + c(Y),$$

where $b(\cdot)$ and $c(\cdot)$ are suitably chosen functions and $\theta_* = (\theta_{*1}, \ldots, \theta_{*n})'$. Furthermore, the relationship between the mean of $y_i$ and its covariates is defined by the function $g(\mu_{*i}) = x'_{*i}\beta_*$, $x_{*i} = (x_{i,0}, \ldots, x_{i,k_*-1})'$ for $i = 1, \ldots, n$, and $\beta_*$ is a $k_* \times 1$ vector.

The family of candidate models is given by

$$Q^+(Y; \mu, \sigma^2) = -\frac{n}{2}\log(\sigma^2) - \frac{1}{2\sigma^2}D(Y; \mu),$$

where

$$D(Y; \mu) = 2\{Q(Y; Y) - Q(Y; \mu)\},$$

$$Q(Y; \mu) = Y'\theta - b(\theta) + c(Y),$$

$$E[Y] = \mu = (\mu_1, \ldots, \mu_n)' \text{ and } \theta = (\theta_1, \ldots, \theta_n)'.$$

In addition, we assume that $g(\mu_i) = x_i'\beta$, where $x_i = (x_{i,0}, \ldots, x_{i,k-1})'$ for $i = 1, \ldots, n$, $\beta$ is $k \times 1$ vector representing the number of parameters under the candidate model.

The Kullback–Leibler measure of the discrepancy between the true and candidate model is

$$\begin{aligned}\Delta(\hat{\beta}, \sigma^2) &= E_*[-2Q^+(Y; \mu, \sigma^2)] \\ &= n\log\sigma^2 + \frac{2}{\sigma^2}\left(E_*[Y'Y - b(Y)] - (\mu_*'\theta - b(\theta))\right),\end{aligned}$$

where $\mu_i = \partial b(\theta)/\partial\theta_i, \mu_{*i} = \partial b(\theta_*)/\partial\theta_{*i}, g(\mu_i) = x_i'\beta$ and $g(\mu_{*i}) = x'_{*i}\beta_*$ for $i = 1, \ldots, n$. Thus,

$$\Delta(\hat{\beta}, \hat{\sigma}^2) = n\log\hat{\sigma}^2 + \frac{2}{\hat{\sigma}^2}\left(E_*[Y'Y - b(Y)] - (\mu_*'\hat{\theta} - b(\hat{\theta}))\right),$$

where $\hat{\beta}$ and $\hat{\sigma}^2$ are the maximum quasi-likelihood estimators of $\beta$ and $\sigma^2$ obtained from the candidate model (see Nelder and Pregibon, 1987). Note that $\hat{\beta}$ minimizes $D(Y; \mu), \hat{\sigma}^2 = D(Y, \hat{\mu})/n = 2\{Y'Y - b(Y) - Y'\hat{\theta} + b(\hat{\theta})\}/n$, and $\hat{\mu}$ and $\hat{\theta}$ are $\mu$ and $\theta$ evaluated at $\hat{\beta}$.

A reasonable criterion for judging the quality of the candidate model is $\Delta(k) = E_*[\Delta(\hat{\beta}, \hat{\sigma}^2)]$. Given a collection of competing candidate models, the one that minimizes $\Delta(k)$ is preferred. The quantity $\Delta(k)$ cannot be used as a practical model selection criterion, however, since it depends on the unknown true model. The key motivation behind AICc, as well as AIC, is to provide an estimate of $\Delta(k)$ based on the observed data. We now assume, as did Akaike (1974) in his original derivation of AIC, that the set of candidate models includes the true model. Hurvich and Tsai (1995) show that (see Appendix 7A) an approximately unbiased estimator of $\Delta(k)$, scaled by $n$, is given by

$$\text{AICc} = \log(\hat{\sigma}^2) + 1 + \frac{2(k+1)}{n-k-2}. \tag{7.24}$$

The corresponding Akaike information criterion is

$$\text{AIC} = \log(\hat{\sigma}^2) + 1 + \frac{2(k+1)}{n}. \tag{7.25}$$

Pregibon's (1979) Cp* and Hosmer *et al.'s* (1989) Cp, both of which are generalized from Mallows's Cp, can be extended to the quasi-likelihood case. The generalization of Cp* is based on the deviance, and that of Hosmer *et al.* is based on the Pearson chi-square goodness-of-fit statistic. Cp* and Cp are defined as follows:

$$\text{Cp}^* = (n-K)D_k/D_K - n + 2k \tag{7.26}$$

and

$$\text{Cp} = (n-K)\mathbf{X}_k/\mathbf{X}_K - n + 2k, \tag{7.27}$$

where $K$ is the dimension of the full model, which includes all available candidate variables, and $D_k$, $D_K$ and $\mathbf{X}_k, \mathbf{X}_K$ are deviances and Pearson chi-square statistics, respectively, evaluated under the model indicated by their corresponding subscripts. In fact, Pregibon's Cp* is the same as McCullagh and Nelder's (1989) Q-statistic if we choose $\alpha = 2$ in their Equation (3.10).

### 7.7.2. Quasi-likelihood Monte Carlo Study

Here we compare the performance of quasi-likelihood model selection criteria using extended quasi-likelihood logistic models. Let $y_i$ be independent binomial distributions with the dispersion parameter $\sigma_*^2 = 1$ for $i = 1, \ldots, n$. The Kullback–Leibler discrepancy comparing binomials is used as the basis of K-L observed efficiency. Let the true model have probability distribution

$$f_*(y_i, \pi_{*i}) = \binom{m_i}{y_i} \pi_{*i}^{y_i}(1-\pi_{*i})^{m_i - y_i},$$

where

$$\pi_{*i} = \frac{\exp(x'_{*i}\beta_*)}{1+\exp(x'_{*i}\beta_*)}.$$

The candidate models have probability distribution

$$f(y_i, \pi_i) = \binom{m_i}{y_i} \pi_i^{y_i}(1-\pi_i)^{m_i - y_i}$$

where

$$\pi_i = \frac{\exp(x_i'\beta)}{1+\exp(x_i'\beta)}.$$

The maximum likelihood estimates for $\pi_i$ are then computed, and the Kullback–Leibler discrepancy (ignoring the scale parameter) is

$$\text{K-L} = m_i\pi_{*i}\log\left(\frac{\pi_{*i}}{\hat{\pi}_i}\right) + m_i(1-\pi_{*i})\log\left(\frac{1-\pi_{*i}}{1-\hat{\pi}_i}\right). \tag{7.28}$$

$L_2$ and $L_1$ distance can be defined as

$$L_2 = \frac{1}{n}\sum_{i=1}^{n}(m_i\pi_{*i} - m_i\hat{\pi}_i)^2 \tag{7.29}$$

and

$$L_1 = \frac{1}{n}\sum_{i=1}^{n}|m_i\pi_{*i} - m_i\hat{\pi}_i|. \tag{7.30}$$

Before discussing the simulation results, we introduce the selection criteria used as well as the four special case models. In addition to the quasi-likelihood criteria given in Eqs. (7.24)–(7.27), we include RTp, Eq. (7.12), the consistent criterion SIC $= \log(\hat{\sigma}^2) + \log(n)k/n$, another consistent criterion HQ $= \log(\hat{\sigma}^2) + 2\log\log(n)k/n$, its small-sample correction HQc $= \log(\hat{\sigma}^2) + 2\log\log(n)k/(n-k-2)$, and AICu $= \log(s^2) + (n+k)/(n-k-2)$. The dispersion parameter estimators, $\hat{\sigma}^2 = D(Y,\hat{\mu})/n$ and $s^2 = \mathbf{X}_k/(n-k)$, are the MLE and approximately unbiased estimator for $\sigma^2$, respectively. The AICu considered in this Monte Carlo study is obtained by analogy to AICu in Chapter 2, by substituting the unbiased estimate $s^2$ for the MLE $\hat{\sigma}^2$ in AICc.

The four models we use to compare the above criteria all have true order $k_* = 6$, $\sigma_* = 1$, and $m_i = 10$. With $\sigma_* = 1$, the true model is the usual logistic regression. Models 19 and 21 have true parameter structure $\beta_j = 1$ ($\beta_1 = \cdots = \beta_5 = 1$) and $\beta_0 = 1$. Models 20 and 22 have true parameter structure $\beta_j = 1/j$ ($\beta_1 = 1$, $\beta_2 = 1/2$, $\beta_3 = 1/3$, $\beta_4 = 1/4$, $\beta_5 = 1/5$) and $\beta_0 = 1$. These are the same two parameter structures used in Chapter 2. Models 19 and 20 have a sample size of 25 and Models 21 and 22 have a sample size of 50. One thousand realizations of each model are generated. For each realization, a new design matrix $X$ and observation matrix $Y$ are generated. Candidate models include all subsets.

For each candidate model, we compute the distance between the true model and the fitted candidate model as well as the observed efficiency of the selection.

The Kullback–Leibler observed efficiency (K-L) is computed using Eq. (7.28) and Eq. (1.2), $L_2$ observed efficiency is computed using Eq. (7.29) and Eq. (1.1), and $L_1$ observed efficiency is computed using Eq. (7.30) and Eq. (7.6).

Selection criterion performance is based on average observed efficiency over the one thousand realizations. In addition, counts of the model orders selected are presented. These count patterns are useful in studying overfitting and underfitting. Results for Models 19–22 are summarized in Tables 7.7–7.10.

The count and observed efficiency results are nearly identical for Model 19. The top performers, AICc and HQc, have intermediate penalty functions that

Table 7.7. Simulation results for Model 19. Counts and observed efficiency. $n = 25$, $m_i = 10$, $\beta_j = 1$, $k_* = 6$.

| | | | | | counts | | | | | | | |
|---|---|---|---|---|---|---|---|---|---|---|---|---|
| $k$ | AIC | AICc | AICu | SIC | HQ | HQc | Cp* | Cp | RTp | $L_1$ | $L_2$ | K-L |
| 1 | 0 | 0 | 0 | 0 | 0 | 0 | 0 | 2 | 0 | 0 | 0 | 0 |
| 2 | 0 | 0 | 1 | 0 | 0 | 0 | 0 | 0 | 0 | 0 | 0 | 0 |
| 3 | 0 | 0 | 3 | 0 | 0 | 0 | 0 | 0 | 10 | 0 | 0 | 0 |
| 4 | 0 | 2 | 17 | 1 | 0 | 2 | 1 | 5 | 70 | 0 | 0 | 0 |
| 5 | 2 | 21 | 73 | 12 | 3 | 30 | 7 | 22 | 190 | 0 | 0 | 0 |
| 6 | 242 | 630 | 661 | 458 | 312 | 690 | 415 | 371 | 296 | 351 | 582 | 795 |
| 7 | 307 | 277 | 205 | 335 | 321 | 228 | 349 | 363 | 276 | 402 | 310 | 176 |
| 8 | 258 | 63 | 36 | 140 | 234 | 44 | 162 | 157 | 124 | 175 | 89 | 27 |
| 9 | 144 | 7 | 3 | 44 | 99 | 6 | 54 | 66 | 30 | 66 | 16 | 2 |
| 10 | 42 | 0 | 1 | 10 | 28 | 0 | 12 | 14 | 4 | 6 | 2 | 0 |
| 11 | 5 | 0 | 0 | 0 | 3 | 0 | 0 | 0 | 0 | 0 | 1 | 0 |
| true | 235 | 609 | 620 | 438 | 302 | 669 | 401 | 349 | 13 | 350 | 582 | 794 |

| | | | | K-L observed efficiency | | | | | | | | |
|---|---|---|---|---|---|---|---|---|---|---|---|---|
| | AIC | AICc | AICu | SIC | HQ | HQc | Cp* | Cp | RTp | $L_1$ | $L_2$ | K-L |
| ave | 0.624 | 0.797 | 0.783 | 0.714 | 0.656 | 0.822 | 0.699 | 0.684 | 0.269 | 0.97 | 0.99 | 1.00 |
| med | 0.596 | 1.000 | 1.000 | 0.710 | 0.627 | 1.000 | 0.682 | 0.674 | 0.196 | 0.99 | 1.00 | 1.00 |
| sd | 0.266 | 0.275 | 0.298 | 0.285 | 0.276 | 0.271 | 0.283 | 0.282 | 0.226 | 0.06 | 0.03 | 0.00 |
| rank | 8 | 2 | 3 | 4 | 7 | 1 | 5 | 6 | 9 | | | |

| | | | | $L_2$ observed efficiency | | | | | | | | |
|---|---|---|---|---|---|---|---|---|---|---|---|---|
| | AIC | AICc | AICu | SIC | HQ | HQc | Cp* | Cp | RTp | $L_1$ | $L_2$ | K-L |
| ave | 0.642 | 0.803 | 0.786 | 0.725 | 0.670 | 0.827 | 0.711 | 0.693 | 0.260 | 0.97 | 1.00 | 0.99 |
| med | 0.626 | 0.991 | 0.992 | 0.747 | 0.658 | 0.995 | 0.717 | 0.698 | 0.185 | 1.00 | 1.00 | 1.00 |
| sd | 0.258 | 0.265 | 0.293 | 0.275 | 0.267 | 0.260 | 0.273 | 0.275 | 0.228 | 0.05 | 0.00 | 0.02 |
| rank | 8 | 2 | 3 | 4 | 7 | 1 | 5 | 6 | 9 | | | |

| | | | | $L_1$ observed efficiency | | | | | | | | |
|---|---|---|---|---|---|---|---|---|---|---|---|---|
| | AIC | AICc | AICu | SIC | HQ | HQc | Cp* | Cp | RTp | $L_1$ | $L_2$ | K-L |
| ave | 0.779 | 0.875 | 0.860 | 0.828 | 0.797 | 0.888 | 0.820 | 0.810 | 0.459 | 1.00 | 0.99 | 0.99 |
| med | 0.789 | 0.974 | 0.976 | 0.867 | 0.806 | 0.980 | 0.844 | 0.831 | 0.416 | 1.00 | 1.00 | 0.99 |
| sd | 0.168 | 0.164 | 0.189 | 0.174 | 0.171 | 0.163 | 0.173 | 0.178 | 0.195 | 0.00 | 0.02 | 0.02 |
| rank | 8 | 2 | 3 | 4 | 7 | 1 | 5 | 6 | 9 | | | |

balance underfitting and overfitting well. Both identify the correct model over 60% of the time, roughly double the performance of AIC and HQ, which due to their weak penalty functions overfit excessively. RTp does not perform well in Model 19. This may be due to the Wald test requiring a larger sample size in quasi-likelihood. More work is needed to deduce why RTp worked well in the additive error case earlier in Chapter 7, but does not work as well here. Both versions of Cp behave similarly, ranking in the middle. SIC also performs near the middle. Since its penalty function is larger, SIC overfits less than either AIC or HQ. Weak penalty functions result

Table 7.8. Simulation results for Model 20. Counts and observed efficiency. $n = 25$, $m_i = 10$, $\beta_j = 1/j$, $k_* = 6$.

| | | | | | | counts | | | | | | |
|---|---|---|---|---|---|---|---|---|---|---|---|---|
| $k$ | AIC | AICc | AICu | SIC | HQ | HQc | Cp* | Cp | RTp | $L_1$ | $L_2$ | K-L |
| 1 | 0 | 0 | 0 | 0 | 0 | 0 | 0 | 0 | 0 | 0 | 0 | 0 |
| 2 | 1 | 10 | 55 | 11 | 3 | 13 | 4 | 5 | 5 | 0 | 0 | 0 |
| 3 | 24 | 130 | 261 | 125 | 42 | 183 | 71 | 72 | 60 | 0 | 2 | 5 |
| 4 | 132 | 315 | 370 | 259 | 171 | 358 | 217 | 223 | 192 | 16 | 25 | 25 |
| 5 | 229 | 352 | 240 | 317 | 273 | 309 | 323 | 316 | 271 | 136 | 146 | 173 |
| 6 | 285 | 146 | 67 | 176 | 264 | 112 | 234 | 225 | 289 | 405 | 591 | 646 |
| 7 | 191 | 39 | 6 | 79 | 155 | 24 | 105 | 108 | 134 | 304 | 189 | 132 |
| 8 | 97 | 8 | 1 | 26 | 66 | 1 | 34 | 39 | 43 | 123 | 43 | 15 |
| 9 | 31 | 0 | 0 | 5 | 19 | 0 | 8 | 9 | 5 | 14 | 4 | 3 |
| 10 | 8 | 0 | 0 | 2 | 7 | 0 | 4 | 3 | 1 | 2 | 0 | 1 |
| 11 | 2 | 0 | 0 | 0 | 0 | 0 | 0 | 0 | 0 | 0 | 0 | 0 |
| true | 53 | 29 | 17 | 40 | 47 | 24 | 50 | 48 | 11 | 236 | 433 | 489 |

| | | | | K-L observed efficiency | | | | | | | | |
|---|---|---|---|---|---|---|---|---|---|---|---|---|
| | AIC | AICc | AICu | SIC | HQ | HQc | Cp* | Cp | RTp | $L_1$ | $L_2$ | K-L |
| ave | 0.511 | 0.512 | 0.493 | 0.509 | 0.510 | 0.506 | 0.516 | 0.517 | 0.479 | 0.96 | 0.99 | 1.00 |
| med | 0.485 | 0.483 | 0.450 | 0.463 | 0.481 | 0.480 | 0.478 | 0.495 | 0.454 | 0.99 | 1.00 | 1.00 |
| sd | 0.220 | 0.225 | 0.227 | 0.227 | 0.223 | 0.223 | 0.230 | 0.229 | 0.216 | 0.07 | 0.03 | 0.00 |
| rank | 4 | 3 | 8 | 6 | 5 | 7 | 2 | 1 | 9 | | | |

| | | | | $L_2$ observed efficiency | | | | | | | | |
|---|---|---|---|---|---|---|---|---|---|---|---|---|
| | AIC | AICc | AICu | SIC | HQ | HQc | Cp* | Cp | RTp | $L_1$ | $L_2$ | K-L |
| ave | 0.521 | 0.514 | 0.489 | 0.514 | 0.520 | 0.506 | 0.523 | 0.522 | 0.469 | 0.96 | 1.00 | 0.99 |
| med | 0.495 | 0.487 | 0.453 | 0.477 | 0.494 | 0.478 | 0.494 | 0.497 | 0.440 | 0.99 | 1.00 | 1.00 |
| sd | 0.222 | 0.226 | 0.230 | 0.229 | 0.225 | 0.224 | 0.233 | 0.231 | 0.217 | 0.06 | 0.00 | 0.03 |
| rank | 3 | 5 | 8 | 5 | 4 | 7 | 1 | 2 | 9 | | | |

| | | | | $L_1$ observed efficiency | | | | | | | | |
|---|---|---|---|---|---|---|---|---|---|---|---|---|
| | AIC | AICc | AICu | SIC | HQ | HQc | Cp* | Cp | RTp | $L_1$ | $L_2$ | K-L |
| ave | 0.701 | 0.694 | 0.676 | 0.695 | 0.698 | 0.689 | 0.701 | 0.700 | 0.663 | 1.00 | 0.98 | 0.98 |
| med | 0.705 | 0.698 | 0.675 | 0.690 | 0.700 | 0.691 | 0.698 | 0.701 | 0.668 | 1.00 | 1.00 | 1.00 |
| sd | 0.159 | 0.161 | 0.169 | 0.162 | 0.159 | 0.161 | 0.163 | 0.165 | 0.165 | 0.00 | 0.03 | 0.03 |
| rank | 1 | 6 | 8 | 5 | 4 | 7 | 1 | 3 | 9 | | | |

in excessive overfitting and reduced observed efficiency. However, if the penalty function is too large, underfitting results and observed efficiency is lost, as can be seen in AICu. While there is good agreement between K-L and $L_2$ observed efficiency, $L_1$ has a shallow minimum distance and as a result, all $L_1$ observed efficiencies are higher than either K-L or $L_2$. $L_1$ does not penalize either underfitting or overfitting severely. Next we look at Model 20, a weakly identifiable model with a small sample size.

Model 20 represents the weakly identifiable case where the correct model is difficult to detect; consequently, all the criteria underfit. None of the criteria

Table 7.9. Simulation results for Model 21. Counts and observed efficiency. $n = 50$, $m_i = 10$, $\beta_j = 1$, $k_* = 6$.

| | | | | | counts | | | | | | | |
|---|---|---|---|---|---|---|---|---|---|---|---|---|
| $k$ | AIC | AICc | AICu | SIC | HQ | HQc | Cp* | Cp | RTp | $L_1$ | $L_2$ | K-L |
| 1 | 0 | 0 | 0 | 0 | 0 | 0 | 0 | 0 | 0 | 0 | 0 | 0 |
| 2 | 0 | 0 | 0 | 0 | 0 | 0 | 0 | 0 | 0 | 0 | 0 | 0 |
| 3 | 0 | 0 | 0 | 0 | 0 | 0 | 0 | 0 | 0 | 0 | 0 | 0 |
| 4 | 0 | 0 | 0 | 0 | 0 | 0 | 0 | 1 | 5 | 0 | 0 | 0 |
| 5 | 0 | 0 | 8 | 0 | 0 | 0 | 0 | 6 | 88 | 0 | 0 | 0 |
| 6 | 324 | 516 | 641 | 707 | 491 | 694 | 414 | 396 | 229 | 474 | 679 | 889 |
| 7 | 392 | 363 | 281 | 239 | 367 | 254 | 393 | 377 | 323 | 353 | 266 | 105 |
| 8 | 214 | 109 | 63 | 47 | 120 | 49 | 157 | 169 | 239 | 136 | 49 | 5 |
| 9 | 60 | 9 | 7 | 6 | 19 | 3 | 31 | 45 | 99 | 35 | 5 | 1 |
| 10 | 10 | 3 | 0 | 1 | 3 | 0 | 5 | 6 | 14 | 2 | 1 | 0 |
| 11 | 0 | 0 | 0 | 0 | 0 | 0 | 0 | 0 | 3 | 0 | 0 | 0 |
| true | 324 | 516 | 638 | 706 | 491 | 693 | 414 | 394 | 75 | 474 | 679 | 889 |
| | | | | K-L observed efficiency | | | | | | | | |
| | AIC | AICc | AICu | SIC | HQ | HQc | Cp* | Cp | RTp | $L_1$ | $L_2$ | K-L |
| ave | 0.690 | 0.770 | 0.832 | 0.850 | 0.759 | 0.845 | 0.729 | 0.734 | 0.476 | 0.99 | 1.00 | 1.00 |
| med | 0.685 | 1.000 | 1.000 | 1.000 | 0.836 | 1.000 | 0.736 | 0.757 | 0.413 | 1.00 | 1.00 | 1.00 |
| sd | 0.263 | 0.267 | 0.253 | 0.250 | 0.268 | 0.251 | 0.266 | 0.264 | 0.376 | 0.03 | 0.01 | 0.00 |
| rank | 8 | 4 | 3 | 1 | 5 | 2 | 7 | 6 | 9 | | | |
| | | | | $L_2$ observed efficiency | | | | | | | | |
| | AIC | AICc | AICu | SIC | HQ | HQc | Cp* | Cp | RTp | $L_1$ | $L_2$ | K-L |
| ave | 0.688 | 0.767 | 0.829 | 0.848 | 0.756 | 0.843 | 0.726 | 0.730 | 0.471 | 0.99 | 1.00 | 1.00 |
| med | 0.678 | 0.986 | 0.998 | 1.000 | 0.835 | 0.999 | 0.734 | 0.756 | 0.406 | 1.00 | 1.00 | 1.00 |
| sd | 0.264 | 0.268 | 0.254 | 0.250 | 0.269 | 0.251 | 0.267 | 0.265 | 0.378 | 0.03 | 0.00 | 0.01 |
| rank | 8 | 4 | 3 | 1 | 5 | 2 | 7 | 6 | 9 | | | |
| | | | | $L_1$ observed efficiency | | | | | | | | |
| | AIC | AICc | AICu | SIC | HQ | HQc | Cp* | Cp | RTp | $L_1$ | $L_2$ | K-L |
| ave | 0.815 | 0.861 | 0.895 | 0.907 | 0.855 | 0.904 | 0.837 | 0.841 | 0.613 | 1.00 | 1.00 | 1.00 |
| med | 0.834 | 0.974 | 0.989 | 0.993 | 0.934 | 0.992 | 0.868 | 0.878 | 0.655 | 1.00 | 1.00 | 1.00 |
| sd | 0.169 | 0.166 | 0.162 | 0.154 | 0.168 | 0.154 | 0.169 | 0.170 | 0.310 | 0.00 | 0.01 | 0.01 |
| rank | 8 | 4 | 3 | 1 | 5 | 2 | 7 | 6 | 9 | | | |

identify the correct model much more than 5% of the time, thus counts are no longer a useful measure of performance and we concentrate on observed efficiency. The difficulty in identifying the correct model can be seen in the reduced observed efficiency of the criteria. Efficiencies for Model 20 tend to be lower than for Model 19. Here Cp and Cp* seem to balance underfitting and overfitting well, resulting in better performance. Our top performer in Model 19, HQc, now performs poorly in Model 20 because its strong penalty function overcompensates for overfitting and causes too much underfitting. The result is that HQc now ranks 7th under all three observed efficiency

Table 7.10. Simulation results for Model 22. Counts and observed efficiency. $n = 50$, $m_i = 10$, $\beta_j = 1/j$, $k_* = 6$.

| | | | | | counts | | | | | | | |
|---|---|---|---|---|---|---|---|---|---|---|---|---|
| $k$ | AIC | AICc | AICu | SIC | HQ | HQc | Cp* | Cp | RTp | $L_1$ | $L_2$ | K-L |
| 1 | 0 | 0 | 0 | 0 | 0 | 0 | 0 | 0 | 0 | 0 | 0 | 0 |
| 2 | 0 | 0 | 3 | 4 | 2 | 3 | 0 | 0 | 3 | 0 | 0 | 0 |
| 3 | 2 | 5 | 23 | 31 | 7 | 16 | 4 | 5 | 22 | 0 | 0 | 0 |
| 4 | 39 | 87 | 205 | 217 | 95 | 164 | 63 | 67 | 115 | 1 | 0 | 1 |
| 5 | 177 | 289 | 418 | 375 | 278 | 392 | 229 | 236 | 208 | 35 | 46 | 50 |
| 6 | 352 | 383 | 259 | 284 | 370 | 324 | 375 | 365 | 298 | 494 | 763 | 863 |
| 7 | 260 | 185 | 76 | 67 | 172 | 81 | 219 | 222 | 224 | 341 | 168 | 79 |
| 8 | 131 | 47 | 15 | 21 | 62 | 20 | 89 | 87 | 105 | 111 | 21 | 6 |
| 9 | 32 | 4 | 1 | 1 | 11 | 0 | 16 | 13 | 20 | 16 | 1 | 1 |
| 10 | 5 | 0 | 0 | 0 | 3 | 0 | 5 | 5 | 5 | 2 | 1 | 0 |
| 11 | 2 | 0 | 0 | 0 | 0 | 0 | 0 | 0 | 0 | 0 | 0 | 0 |
| true | 199 | 217 | 158 | 182 | 214 | 195 | 205 | 184 | 28 | 450 | 729 | 826 |

| | | | | | K-L observed efficiency | | | | | | | |
|---|---|---|---|---|---|---|---|---|---|---|---|---|
| | AIC | AICc | AICu | SIC | HQ | HQc | Cp* | Cp | RTp | $L_1$ | $L_2$ | K-L |
| ave | 0.601 | 0.604 | 0.560 | 0.565 | 0.600 | 0.583 | 0.600 | 0.590 | 0.449 | 0.99 | 1.00 | 1.00 |
| med | 0.573 | 0.563 | 0.521 | 0.521 | 0.561 | 0.538 | 0.562 | 0.557 | 0.407 | 1.00 | 1.00 | 1.00 |
| sd | 0.257 | 0.267 | 0.261 | 0.267 | 0.268 | 0.269 | 0.261 | 0.257 | 0.245 | 0.03 | 0.01 | 0.00 |
| rank | 2 | 1 | 8 | 7 | 3 | 6 | 3 | 5 | 9 | | | |

| | | | | | $L_2$ observed efficiency | | | | | | | |
|---|---|---|---|---|---|---|---|---|---|---|---|---|
| | AIC | AICc | AICu | SIC | HQ | HQc | Cp* | Cp | RTp | $L_1$ | $L_2$ | K-L |
| ave | 0.601 | 0.602 | 0.554 | 0.560 | 0.598 | 0.579 | 0.599 | 0.587 | 0.439 | 0.99 | 1.00 | 1.00 |
| med | 0.566 | 0.563 | 0.509 | 0.510 | 0.560 | 0.529 | 0.560 | 0.546 | 0.397 | 1.00 | 1.00 | 1.00 |
| sd | 0.258 | 0.268 | 0.263 | 0.270 | 0.270 | 0.271 | 0.262 | 0.259 | 0.245 | 0.03 | 0.00 | 0.01 |
| rank | 2 | 1 | 8 | 7 | 4 | 6 | 3 | 5 | 9 | | | |

| | | | | | $L_1$ observed efficiency | | | | | | | |
|---|---|---|---|---|---|---|---|---|---|---|---|---|
| | AIC | AICc | AICu | SIC | HQ | HQc | Cp* | Cp | RTp | $L_1$ | $L_2$ | K-L |
| ave | 0.758 | 0.758 | 0.726 | 0.729 | 0.754 | 0.742 | 0.756 | 0.750 | 0.640 | 1.00 | 1.00 | 1.00 |
| med | 0.759 | 0.755 | 0.719 | 0.719 | 0.752 | 0.734 | 0.753 | 0.752 | 0.640 | 1.00 | 1.00 | 1.00 |
| sd | 0.171 | 0.178 | 0.180 | 0.183 | 0.179 | 0.182 | 0.174 | 0.172 | 0.187 | 0.00 | 0.01 | 0.01 |
| rank | 1 | 1 | 8 | 7 | 4 | 6 | 3 | 5 | 9 | | | |

measures. However, AICc performs about the same in both Model 19 and Model 20, indicating that its penalty function strikes a good balance in terms of strength. The next two models we consider have increased sample size.

The combination of a strongly identifiable model with larger sample size nearly eliminates underfitting from the criteria. Now performance depends on overfitting properties, and the stronger the penalty function, the less overfitting and better the performance. SIC is a strongly consistent criterion with a very strong penalty in large samples. In Model 21, SIC overfits the least and has the highest observed efficiency. Due to the strongly identifiable model, observed efficiency is proportional to higher counts of selecting the correct model. HQc now has a much larger penalty function than AICc although both have similar structure. We can see in the count patterns that HQc overfits much less than AICc. Also, HQc has higher observed efficiency than AICc as well. The increased sample size helps prevent underfitting and this improves the performance of AICu, although AICu still underfits on occasion and overfits more than HQc. The large sample size shows a marked difference in performance between HQ and AIC. Now, HQ has a much larger penalty function than AIC and performs better, whereas in small samples, AIC and HQ perform similarly. In general, the larger sample size favors the consistent criteria, particularly since the true model is strongly identifiable. Both Cp versions perform near the bottom. RTp still has troubles and performs the worst, overfitting severely. All three observed efficiency measure give the same relative results, and $L_1$ observed efficiencies remain higher than either K-L or $L_2$ observed efficiency.

Even though the sample size is larger, Model 22 is weakly identifiable and the consistent criteria no longer perform as well as they did in Model 21. Consistency seems best applied in larger samples with strongly identifiable true models. AICc performs the best in Model 22. We can see from the count patterns that AICc does not overfit or underfit excessively. AIC also performs well. Although AIC and AICc have similar observed efficiencies, the count patterns reveal that AIC still overfits more than AICc, reducing its observed efficiency. The penalty function of HQ is the weakest of our three consistent criteria and in a sample size of 50, its performance remains good, although HQ underfits more than either AIC or AICc. SIC, AICu, and HQc have even stronger penalty functions and hence underfit more than HQ. RTp both underfits and overfits and it performs the worst of our criteria. The two versions of Cp now perform differently: Cp* now clearly outperforms Cp, but both still rank near the middle.

These four models illustrate four conditions a researcher might encounter:

a small sample/strong model, a small sample/weak model, a moderate sample/strong model, and a moderate sample/weak model. While no criterion is clearly the best performer overall, AICc never performs poorly and we feel that it is the best choice in this quasi-likelihood study. We also recommend HQc for the strong model case, since it performed well in both the small and moderate sample sizes. However, in order to know in practice whether the true model is weakly or strongly identifiable, more work needs to be done to devise a data driven measure of model identifiability and strength.

## 7.8. Summary

Once a model departs from the standard assumptions of normality, new model structures and assumptions are required. This is often where robust techniques come into play. One conclusion that is supported by the simulation results throughout this Chapter is that naive implementations of existing criteria still perform well. In fact, they often outperform criteria specially derived for robust conditions.

There are many possible exceptions to the assumption of normality. If we assume that the additive errors are double exponentially distributed, then the maximum likelihood estimates are the same as those for $L_1$ regression. Here, $L_1$ regression is related to maximum likelihood just as normal distributions are related to $L_2$ (least squares) regression. L1AICc estimates the Kullback–Leibler information for double exponential distributions. However, unlike the normal distribution case, no convenient calculations exist for computing L1AICc. Not only is L1AICc much more computationally intensive than the naive AICc, AICc also seems to perform as well as the specially adapted L1AICc in our simulations. Therefore we feel that the increased computational effort involved in computing L1AICc makes it of limited use.

Although the Kullback–Leibler discrepancy usually requires a specific distribution assumption, a generalized Kullback–Leibler information can be obtained for measuring the distance between a robust function evaluated under the true model and a fitted model. This generalized K-L information can be used to obtain robust model selection criteria that have several advantages. They not only fit the majority of the data, but also take into account nonnormal errors. This gives them a performance advantage over the other criteria derived under robust conditions or their naive counterparts. In addition, they unify most existing Akaike information criteria. Kullback–Leibler can also be used to extend the criteria from Chapter 2 to quasi-likelihood. Our simulations show that the relative performance of these criteria is not substantially changed under quasi-likelihood—criteria with weak penalty functions tend to overfit, criteria with penalty terms that excessively resist overfitting are prone to underfitting, particularly when the true model is weak.

## Chapter 7 Appendix

### Appendix 7A. Derivation of AICc under Quasi-likelihood

The following derivations are adapted from Hurvich and Tsai's (1995) paper, with permission from Biometrics. Assume that the family of candidate models contains the true model. Under this assumption, the covariate vectors in the candidate model can be rearranged so that $x_i = (x'_{*i}, x'_{1i})'$, where $x_{1i}$ is a $(k-k_*)\times 1$ vector. Hence, the mean of $y_i$ satisfies $\mu_{*i} = g^{-1}(x'_{*i}\beta_*) = g^{-1}(x'_i\tilde{\beta}_*)$, where $\tilde{\beta}_* = (\beta'_*, 0')'$ and 0 is a $(k - k_*) \times 1$ vector of zeros. The quadratic expansion of $\{\mu'_*\hat{\theta} - b(\hat{\theta})\}$ at $\beta = \tilde{\beta}_*$ is given by

$$(\mu'_*\theta_* - b(\theta_*)) - (\hat{\beta} - \tilde{\beta}_*)'V'_*V_*(\hat{\beta} - \tilde{\beta}_*)/2,$$

where

$$V_* = (\partial^2 b(\theta_*)/\partial\theta_*\partial\theta'_*)^{\frac{1}{2}}(\partial\mu_*/\partial\tilde{\eta}_*)X,$$

$\tilde{\eta}_* = X\tilde{\beta}_*, X$ is the $n \times k$ matrix and the $i$th row of X is $x'_i$.

Let

$$\sigma_*^2 = \frac{2}{n}E_*[Y'Y - b(Y) - \mu'_*\theta_* + b(\theta_*)].$$

Then, applying Jørgensen's (1987, Section 4) asymptotic results, $\hat{\beta} - \tilde{\beta}_*$ is approximately multivariate normal, $N(0, \sigma_*^2(V'_*V_*)^{-1})$, the quantity $n\,\hat{\sigma}^2/\sigma_*^2$ is approximately distributed as $\chi^2_{n-k}$ and independent of $\hat{\beta}$. Hence,

$$E_*\left[\frac{\sigma_*^2}{\hat{\sigma}^2}\right] \cong \frac{n}{n-k-2}$$

and

$$E_*\left[\frac{(\hat{\beta} - \tilde{\beta}_*)'V'_*V_*(\hat{\beta} - \tilde{\beta}_*)}{\hat{\sigma}^2}\right] \cong \frac{nk}{n-k-2}.$$

Thus,

$$\Delta(k) \cong E_*[n\log(\hat{\sigma}^2)] + \frac{n(n+k)}{n-k-2}.$$

Consequently, scaling by $n$,

$$\begin{aligned}\text{AICc} &= \log(\hat{\sigma}^2) + \frac{n+k}{n-k-2}\\ &= \log(\hat{\sigma}^2) + 1 + \frac{2(k+1)}{n-k-2}.\end{aligned}$$

## Chapter 8
# Nonparametric Regression and Wavelets

Many different methods have been proposed for constructing nonparametric estimates of smooth regression functions. Some of the most common are the local polynomial, (convolution) kernel, and smoothing spline estimators. These estimators use a smoothing parameter to control the amount of smoothing performed on a given data set, where the parameter is chosen using a selection criterion. However, due to great variability and a tendency to undersmooth, the "classical" criteria, such as generalized cross-validation, GCV, and AIC, are not ideal for this purpose. Hurvich, Simonoff and Tsai (1998) addressed these problems by proposing a nonparametric version of their AICc criterion. Unlike plug-in methods, AICc can be used to choose smoothing parameters for any linear smoother, including local quadratic and smoothing spline estimators. AICc is competitive with plug-in methods for choosing smoothing parameters, and also performs well when a plug-in approach fails or is unavailable. Also, because neither parametric nor nonparametric estimation may give a reasonable fit to the data in practice, Shi and Tsai (1997) and Simonoff and Tsai (1997) obtained AICc for semiparametric regression models (Speckman, 1988 and Wahba, 1990). When addressing this case we will focus on Shi and Tsai's approach.

Finally, we consider the selection of a hard wavelet threshold for recovery of a signal embedded in additive Gaussian white noise, a problem closely related to that of selecting a subset model in orthogonal normal linear regression. We start by discussing existing approaches, such as Donoho and Johnstone's (1994) universal method, and we give a computationally efficient algorithm for implementing a cross-validatory method proposed by Nason (1996). We then present a cross-validatory version of AICc, proposed by Hurvich and Tsai (1998), that, like universal thresholding and Nason's method, can be implemented in $O(n \log n)$ operations (where $n$ is the sample size). Simulation results show that both of the cross-validatory methods outperform universal thresholding.

## 8.1. Model Selection in Nonparametric Regression

For our discussion of nonparametric estimation of smooth regression functions we assume that we have data $Y = (y_1, \cdots, y_n)'$, generated by the model

$$y_i = \mu_*(x_i) + \varepsilon_{*i}, \quad i = 1, \cdots, n,$$

where $\mu_*(\cdot)$ is an unknown smooth function, the $x_i$ are given real numbers in the interval $[a, b]$, and the $\varepsilon_{*i}$ are independent random variables with mean zero and variance $\sigma_*^2$. The predictor vector $\mathbf{x} = (x_1, \ldots, x_n)'$ can be either random or nonrandom, but if it is random the analysis proceeds conditional on the observed values.

Many different estimators of $\mu_*$ have been proposed in the literature, but three of the most common are the local polynomial kernel, convolution kernel, and smoothing spline estimators. A $k$th-order local polynomial estimator is defined as the constant term $\hat{\beta}_0$ of the minimizer of

$$\sum_{i=1}^{n} \left(y_i - \beta_0 - \cdots - \beta_k (x - x_i)^k\right)^2 K\left(\frac{x - x_i}{h}\right),$$

where $K(\cdot)$ is the kernel function, generally taken to be a symmetric probability density function with finite second derivative, and $h$ is the bandwidth (also called the smoothing parameter). Typical choices of $k$ are 0, 1, 2, and 3, where the local linear ($k = 1$) and local cubic ($k = 3$) estimators have certain asymptotic and boundary bias-correction advantages over the local constant ($k = 0$) and local quadratic ($k = 2$), respectively. Higher values of $k$ (2 or 3) also can take advantage of increased smoothness in $\mu_*$ by yielding a faster convergence rate to zero of MSE $= E\left[\left(\hat{\mu}_h(x) - \mu_*(x)\right)^2\right]$, the mean squared error of the estimator.

The Gasser–Müller convolution kernel estimator takes the form

$$\hat{\mu}_h(x) = h^{-1} \sum_{i=1}^{n} \left( \int_{s_{i-1}}^{s_i} K\left(\frac{x - t}{h}\right) dt \right) y_i, \tag{8.1}$$

where $x_{i-1} \le s_{i-1} \le x_i$. A common choice is $s_{i-1} = (x_{i-1} + x_i)/2$, with $s_0$ and $s_n$ being the upper and lower limits of the range of $x$, respectively. Unlike the $k$th-order polynomial estimator, here the kernel function $K(\cdot)$ need not be a probability density function because higher-order kernels can yield improved MSE convergence rates for smoother $\mu_*$ (analogous to the local quadratic and cubic estimators). However, the kernel functions must be corrected for potential bias effects in the boundary regions of the data using boundary kernels.

A cubic smoothing spline estimator minimizes

$$\frac{1}{n}\sum_{i=1}^{n}\big(y_i-\mu_*(x_i)\big)^2+\alpha\int\mu_*''(t)^2dt$$

over the class of functions with $\mu_*$ and $\mu_*'$ absolutely continuous and $\mu_*''$ square-integrable. Note that $\mu_*'$ and $\mu_*''$ are the first and second derivatives of $\mu_*$ with respect to $t$.

Although these estimators are defined in different ways, there are connections between them. For example, for a fixed design of equidistant values of $\mathbf{x}$, local polynomial and convolution kernel estimators are asymptotically equivalent in the interior (and at the boundary if boundary kernels are used). Despite this, the finite sample properties of the estimators can be very different. The key property for all the derivations in this Section is that all of the estimators are linear, in the sense that $\hat{\boldsymbol{\mu}}_h=\hat{\mu}_h(\mathbf{x})=HY$, where the matrix $H$ is commonly called the hat matrix or smoother matrix. $H$ depends only on $\mathbf{x}$, not on Y.

Because it controls the smoothness of the resulting estimate, choosing the smoothing parameter ($h$ for the local polynomial and kernel estimators, $\alpha$ for the smoothing spline) is a crucial step in estimating $\boldsymbol{\mu}_*$. Automatic smoothing parameter selection criteria generally fall into two broad classes: classical and plug-in approaches. Classical methods are based on minimization of an approximately unbiased estimator of either the mean average squared error,

$$\text{MASE}=\frac{1}{n}E[(\hat{\boldsymbol{\mu}}_h-\boldsymbol{\mu}_*)'(\hat{\boldsymbol{\mu}}_h-\boldsymbol{\mu}_*)]$$

(*e.g.*, GCV, Craven and Wahba, 1979), or the expected Kullback–Leibler discrepancy given in Eq. (8.3) (*e.g.*, AIC, Akaike, 1973). Here $\boldsymbol{\mu}_*$ represents $\mu_*(\mathbf{x})$, and $h$ represents a generic smoothing parameter for any linear smoother (including the smoothing spline). The smoothing parameter is chosen to minimize $\log\hat{\sigma}^2+\psi(H)$, where

$$\hat{\sigma}^2=\frac{1}{n}\sum_{i=1}^{n}\big(y_i-\hat{\mu}_h(x_i)\big)^2=Y'(I-H)'(I-H)Y/n$$

and $\psi(\cdot)$ is a penalty function designed to decrease with increasing smoothness of $\hat{\boldsymbol{\mu}}_h$. Common choices of $\psi$ lead to $GCV[\psi(H)=-2\log(1-tr\{H\}/n)]$, AIC$[\psi(H)=2tr\{H\}/n]$, and $T[\psi(H)=-\log(1-2tr\{H\}/n]$ (Rice, 1984). Each of these selection criteria depends on $H$ though its trace, which can be interpreted as the effective number of parameters used in the smoothing fit (see Hastie and Tibshirani, 1990, Section 3.5).

Classical bandwidth selection criteria, particularly GCV and AIC, are no longer popular for use with local polynomial and kernel estimators because of two unfavorable properties: they lead to highly variable choices of smoothing parameters, and they possess a noticeable tendency towards undersmoothing (too large a value of $tr\{H\}$). These drawbacks inspired the formulation of plug-in methods.

The plug-in criterion of Ruppert, Sheather, and Wand (1995) for the local linear estimator is a typical example. It can be shown that the bandwidth that asymptotically minimizes the weighted conditional mean integrated squared error

$$\text{MISE}(\hat{\mu}_h|x_1, \cdots, x_n) = E\left[\int_a^b [\hat{\mu}_h(t) - \mu_*(t)]^2 f_X(t)dt|x_1, \ldots, x_n\right]$$

is

$$h_{*,a} = \left(\frac{S(K)\sigma_*^2}{n\mu_2(K)^2 \int \mu_*''(t)^2 f_X(t)dt}\right)^{1/5}, \tag{8.2}$$

where $S(K) = \int K(t)^2 dt, \mu_2(K) = \int t^2 K(t)dt$, and $f_X(t)$ is the density function for the predictors. Fixed designs take $x_i = F_X^{-1}(i/n)$, where $F_X$ is the cumulative function of the density $f_X$ of the design. The plug-in bandwidth is given by the right-hand side of Eq. (8.2), with estimates of $\sigma_*^2$ and $\int \mu_*''(t)^2 f_X(t)dt$ substituted (plugged in) for the actual values. The estimator that results is much less variable than that based on GCV, and does not undersmooth.

Despite these favorable properties, plug-in selection criteria have several theoretical and practical problems. First, they have been defined only for situations where the asymptotically optimal bandwidth $h_{*,a}$ has a simple form. This is not the case for the local quadratic estimator, where $h_{*,a}$ depends on $\mu_*'''$, $\mu_*^{(iv)}$, and $f_X'$. Similarly, no plug-in methods have been proposed for smoothing splines.

Plug-in selection criteria also have philosophical drawbacks. The main theoretical advantages of plug-in over classical selection criteria are related to estimation of $h_*$, the bandwidth that minimizes MISE for the given sample size and design. This estimation, which approaches $h_{*,a}$ as $n \to \infty$, is thus optimal with respect to average performance over all possible data sets for a given population, but not with respect to performance for the observed data set. While plug-in criteria are far better at estimating $h_*$ than are classical selection criteria, these advantages do not carry over to estimation of $\hat{h}_*$, the bandwidth that minimizes integrated squared error (ISE) or average squared error (ASE) for the observed data set. In our opinion, $\hat{h}_*$ is a more reasonable target from an applied point of view, and therefore many of the theoretical advantages of plug-in selection criteria are not relevant to the questions that are likely to interest the data analyst. Further discussion of issues related to estimating $h_*$ versus $\hat{h}_*$ can be found in Mammen (1990), Hall and Marron (1991), Jones (1991), Jones and Kappenman (1991), and Grund, Hall, and Marron (1994).

Note also that plug-in selection criteria target $h_{*,a}$ rather than $h_*$, so it is not obvious how well they will work for small-to-moderate sample sizes. The local linear plug-in selection criterion proceeds by estimating $\int \mu_*''(t)^2 f_X(t)dt$, which requires assuming the existence of roughly four continuous derivatives for $\mu_*$. That much smoothness renders the local linear estimator itself asymptotically inefficient, thus calling into question the entire operation. This point was noted by Terrell (1992) and Loader (1995), among others. Estimating this function typically requires choosing preliminary parameters in either a data-dependent or fixed fashion, and the properties of the final plug-in bandwidth can be sensitive to these choices.

To circumvent the high variability and undersmoothing tendencies of GCV, and the continuity assumptions of the plug-in method, Hurvich, Simonoff and Tsai (1998) proposed an adaptation of AIC within a nonparametric context. Hart and Yi (1996) have also proposed a variant of cross-validation with the same goal in mind. Improvements to AIC-based classical smoothing parameter selection criteria can be derived using the Kullback–Leibler information. As is true for all classical methods, the selection criteria are defined for all linear estimators. We will review the derivations in the next Section, Section 8.1.1, and in Section 8.1.2 we will compare the performance of each selection criterion via Monte Carlo simulations.

### 8.1.1. AIC for Smoothing Parameter Selection

The corrected Akaike Information Criterion, AICc, was designed by Hurvich and Tsai (1989) for parametric models. It is a less-biased estimate of the expected Kullback–Leibler information than AIC itself, especially for small samples. For nonparametric regression, Hurvich, Simonoff and Tsai (1998) developed two criteria, $\text{AICc}_0$ and $\text{AICc}_1$, which are specifically designed to be approximately unbiased estimates in the nonparametric context. $\text{AICc}_0$ is the more exact of the two, but requires numerical integration for its evaluation. $\text{AICc}_1$ is an approximation to $\text{AICc}_0$ that performs identically to $\text{AICc}_0$

in practice, but is simpler to evaluate (although it does require the calculation of all elements of an $n \times n$ matrix). A third criterion, nonparametric AICc, is an approximation to $\text{AICc}_1$. Because it is based on the smoother only through $tr\{H\}$, it is as simple to apply as GCV, AIC, and T.

*8.1.1.1. Derivation of $AICc_0$*

To derive $\text{AICc}_0$, we generate the data Y from the true model $Y = \boldsymbol{\mu}_* + \varepsilon_*$, $\varepsilon_* \sim N(0, \sigma_*^2 I_n)$, and we consider the candidate model $Y = \boldsymbol{\mu} + \varepsilon$, where $\varepsilon \sim N(0, \sigma^2 I_n)$. It should be stressed that in spite of the normality assumption imposed for $\text{AICc}_0$, and thus for its approximations $\text{AICc}_1$ and AICc, the resulting criteria exhibit good performance in simulations with both normal and nonnormal errors. If $f(Y)$ denotes the likelihood for $(\boldsymbol{\mu}, \sigma^2)$ and $E_*$ denotes expectation with respect to the true model, we can write the Kullback–Leibler discrepancy function (ignoring the constant $E_*[-2\log f_*(Y)]$, where $f_*(Y)$ is the likelihood with $\boldsymbol{\mu}_*$ and $\sigma_*^2$ from the true model) as

$$\begin{aligned}\Delta(\boldsymbol{\mu}, \sigma^2) &= E_*[-2\log f(Y)] \\ &= n\log(2\pi\sigma^2) + E_*[(\boldsymbol{\mu}_* + \varepsilon_* - \boldsymbol{\mu})'(\boldsymbol{\mu}_* + \varepsilon_* - \boldsymbol{\mu})/\sigma^2] \\ &= n\log(2\pi\sigma^2) + n\frac{\sigma_*^2}{\sigma^2} + (\boldsymbol{\mu}_* - \boldsymbol{\mu})'(\boldsymbol{\mu}_* - \boldsymbol{\mu})/\sigma^2.\end{aligned}$$

Thus,

$$\Delta(\hat{\boldsymbol{\mu}}_h, \hat{\sigma}^2) = n\log(2\pi\hat{\sigma}^2) + n\frac{\sigma_*^2}{\hat{\sigma}^2} + (\boldsymbol{\mu}_* - \hat{\boldsymbol{\mu}}_h)'(\boldsymbol{\mu}_* - \hat{\boldsymbol{\mu}}_h)/\hat{\sigma}^2.$$

A reasonable criterion for judging the quality of the estimator $\hat{\boldsymbol{\mu}}_h$ with respect to the data is $\Delta(h) = E_*[\Delta(\hat{\boldsymbol{\mu}}_h, \hat{\sigma}^2)]$. Ignoring the constant, $n\log(2\pi)$, we have

$$\Delta(h) = E_*[n\log\hat{\sigma}^2] + n\sigma_*^2 E_*[1/\hat{\sigma}^2] + E_*[(\boldsymbol{\mu}_* - \hat{\boldsymbol{\mu}}_h)'(\boldsymbol{\mu}_* - \hat{\boldsymbol{\mu}}_h)/\hat{\sigma}^2]. \qquad (8.3)$$

Unfortunately, since it depends on the true regression function, $\boldsymbol{\mu}_*$, $\Delta(h)$ will not be known in practice. Therefore, we need an approximately unbiased estimate of $\Delta(h)$ that depends only on the observed data $Y$. Although it will rarely hold exactly, at this stage it is helpful to make the simplifying assumption that $\hat{\boldsymbol{\mu}}_h$ is unbiased; that is, $E_*[\hat{\boldsymbol{\mu}}_h] = \boldsymbol{\mu}_*$, or equivalently, $H\boldsymbol{\mu}_* = \boldsymbol{\mu}_*$. Cleveland and Devlin (1988) make a similar assumption in their derivation of a nonparametric analog of Mallows's Cp. This assumption is needed only to facilitate the derivation of a feasible penalty function. It is analogous to the key simplifying assumption used in the derivation of AIC for parametric models-namely, that the candidate family of models includes the true model (see Akaike, 1974, and Linhart and Zucchini, 1986, p. 245). Although we make this assumption for the purposes of derivation, we will study the performance of $\mathrm{AICc}_0$ without regard to the assumptions underlying its derivation.

Assuming, then, that $H\boldsymbol{\mu}_* = \boldsymbol{\mu}_*$, $\Delta(h)$ reduces to

$$\tilde{\Delta}(h) = E_*[n\log\hat{\sigma}^2] + n^2 E_*\left[\frac{\sigma_*^2}{\varepsilon_*'(I-H)'(I-H)\varepsilon_*}\right] + nE_*\left[\frac{\varepsilon_*'H'H\varepsilon_*}{\varepsilon_*'(I-H)'(I-H)\varepsilon_*}\right]. \tag{8.4}$$

Even if $\hat{\boldsymbol{\mu}}_h$ is not unbiased, $\tilde{\Delta}(h)$ serves as an approximation to $\Delta(h)$. Let $B_1 = (I-H)'(I-H)$, and write $B_1 = \Gamma D \Gamma'$, where $D$ is a diagonal matrix of eigenvalues of $B_1$ and $\Gamma$ is an orthogonal matrix whose columns are the corresponding eigenvectors of $B_1$. Define $\tilde{\varepsilon} = \Gamma'\varepsilon_*/\sigma_*$. Then $\tilde{\varepsilon} \sim N(0, I_n)$, and

$$\varepsilon_*' B_1 \varepsilon_* / \sigma_*^2 = \tilde{\varepsilon}' D \tilde{\varepsilon}.$$

Let $A_1 = n^2 E_*[1/\tilde{\varepsilon}'D\tilde{\varepsilon}]$, so that $A_1$ is the second term on the right-hand side of Eq. (8.4). It follows from Jones (1986, Eq. 13) that

$$A_1 = n^2 \int_0^1 (1-t)^{r/2-2} \prod_{j=1}^{r} (1 - t + 2td_j)^{-1/2} dt, \tag{8.5}$$

where $r$ is the rank of $B_1$ and $d_j$ is the $j$th diagonal element of $D$.

Next, let $B_2 = H'H$ and $C = \Gamma' B_2 \Gamma$. Then

$$\frac{\varepsilon' H' H \varepsilon}{\varepsilon'(I-H)'(I-H)\varepsilon} = \frac{\tilde{\varepsilon}' C \tilde{\varepsilon}}{\tilde{\varepsilon}' D \tilde{\varepsilon}}.$$

Defining $A_2 = nE_*[\tilde{\varepsilon}'C\tilde{\varepsilon}/\tilde{\varepsilon}'D\tilde{\varepsilon}]$, so that $A_2$ is the final term in Eq. (8.4), it follows from Jones (1987, Eq. 3.6(a)) that

$$A_2 = n \int_0^\infty \sum_{i=1}^{n} \frac{c_{ii}}{1 + 2d_i t} \prod_{i=1}^{n} (1 + 2d_i t)^{-1/2} dt, \tag{8.6}$$

where $c_{ii}$ is the $i$th diagonal element of $C$.

Given the above, Hurvich, Simonoff and Tsai (1998) obtained the $\mathrm{AICc}_0$ criterion,

$$\mathrm{AICc}_0 = n \log \hat{\sigma}^2 + A_1 + A_2,$$

where $A_1$ and $A_2$ are given by (one-dimensional) numerical integration of Eq. (8.5) and Eq. (8.6). Note that $\mathrm{AICc}_0$ is exactly unbiased for $\tilde{\Delta}(h)$, regardless of whether $H\boldsymbol{\mu}_* = \boldsymbol{\mu}_*$ holds, but if $H\boldsymbol{\mu}_* \neq \boldsymbol{\mu}_*$, then $\tilde{\Delta}(h)$ will not coincide exactly with the true expected K-L information, $\Delta(h)$. This situation is similar to that for model selection in linear regression, where AIC is typically biased (even asymptotically) when the dimension of the candidate model is less than the dimension of the true model.

Although the terms $A_1$ and $A_2$ in $\text{AICc}_0$ are easily and accurately obtained, the necessity for using numerical integration (even in one dimension), as well as numerical eigensystem routines, may be considered a drawback. Hurvich, Simonoff and Tsai (1998) therefore derived an approximation to $\text{AICc}_0$ that can be evaluated without resorting to numerical integration.

*8.1.1.2. Derivation of $AICc_1$*

From Eq. (8.4) and the notation that follows it, we can write

$$\tilde{\Delta}(h) = E_*[n \log \hat{\sigma}^2] + n^2 E_* \left[ \frac{1}{\tilde{\varepsilon}' D \tilde{\varepsilon}} \right] + n E_* \left[ \frac{\tilde{\varepsilon}' C \tilde{\varepsilon}}{\tilde{\varepsilon}' D \tilde{\varepsilon}} \right].$$

Using the method described by Cleveland and Devlin (1988), the distributions of $\tilde{\varepsilon}' D \tilde{\varepsilon}$ and $\tilde{\varepsilon}' C \tilde{\varepsilon} / \tilde{\varepsilon}' D \tilde{\varepsilon}$ are approximated as follows:

$$\tilde{\varepsilon}' D \tilde{\varepsilon} \sim (\delta_2/\delta_1) \chi^2_{\delta_1^2/\delta_2},$$

and

$$\frac{\tilde{\varepsilon}' C \tilde{\varepsilon}}{\tilde{\varepsilon}' D \tilde{\varepsilon}} \sim (\nu_1/\delta_1) F_{\nu_1^2/\nu_2, \delta_1^2/\delta_2},$$

where $\delta_1 = tr\{B_1\}, \delta_2 = tr\{B_1^2\}, \nu_1 = tr\{B_2\}, \nu_2 = tr\{B_2^2\}$ and $B_1$, and $B_2$ are as defined above. Treating these distributional approximations as exact yields

$$E_* \left[ \frac{1}{\tilde{\varepsilon}' D \tilde{\varepsilon}} \right] = (\delta_1/\delta_2)/(\delta_1^2/\delta_2 - 2)$$

and

$$E_* \left[ \frac{\tilde{\varepsilon}' C \tilde{\varepsilon}}{\tilde{\varepsilon}' D \tilde{\varepsilon}} \right] = \frac{\nu_1(\delta_1/\delta_2)}{\delta_1^2/\delta_2 - 2}.$$

Substituting, Hurvich, Simonoff and Tsai (1998) obtained their approximately unbiased estimator of $\tilde{\Delta}(h)$, $\text{AICc}_1$:

$$\text{AICc}_1 = n \log \hat{\sigma}^2 + n \left\{ \frac{(\delta_1/\delta_2)(n + \nu_1)}{\delta_1^2/\delta_2 - 2} \right\}.$$

The accuracy of the approximation of $\text{AICc}_1$ to $\text{AICc}_0$ was examined in Monte Carlo simulations (not reported here), and was found to be excellent.

The criterion $\text{AICc}_1$, while simpler to apply than $\text{AICc}_0$, still requires calculating all elements of the $n \times n$ matrix $H$. Although these calculations can be facilitated using binning techniques, (see Turlach and Wand, 1996), we would like to simplify even further without sacrificing performance.

#### 8.1.1.3. Derivation of AICc for Nonparametric Selection

Hurvich and Tsai (1989) showed that for parametric linear regression and autoregressive time series, the bias-corrected AIC criterion AICc takes the form

$$\log \hat{\sigma}^2 + \frac{1 + k/n}{1 - (k+2)/n} = \log \hat{\sigma}^2 + 1 + \frac{2(k+1)}{n-k-2},$$

where $\hat{\sigma}^2$ is the estimated error (or innovations) variance, and $k$ is the number of regression parameters in the model. By analogy, Hurvich, Simonoff and Tsai (1998) obtained a version of AICc for smoothing parameter selection:

$$\begin{aligned} \text{AICc} &= \log \hat{\sigma}^2 + \frac{1 + tr\{H\}/n}{1 - (tr\{H\}+2)/n} \\ &= \log \hat{\sigma}^2 + 1 + \frac{2(tr\{H\}+1)}{n - tr\{H\} - 2}. \end{aligned} \tag{8.7}$$

Because it is a function of $H$ only through its trace, AICc is quite easy to apply. If $H$ is assumed to be symmetric and idempotent (an assumption we did not make in the derivation of $\text{AICc}_1$), then $\text{AICc}_1$ reduces to AICc. Since, in general, $H$ will not be symmetric and idempotent, we can think of AICc as an approximation to $\text{AICc}_1$ (which is, in turn, a very accurate approximation to $\text{AICc}_0$).

It follows from Härdle, Hall, and Marron (1988) that all of the classical selection criteria considered here are asymptotically equivalent to each other. Given this, one might wonder why they exhibit noticeably different performance in practice. The reason is that the asymptotic theory assumes that $tr\{H\}/n \to 0$, a situation that is not consistent with a small smoothing parameter. AIC and GCV have relatively weak penalties, and thus tend to undersmooth. $T$, on the other hand, has a very strong penalty, as it is effectively infinite for $tr\{H\}/n \geq 0.5$. This means that $T$ must lead to oversmoothing when a very small smoothing parameter is appropriate. The behavior of AICc is in between these two extremes; it is less susceptible to both the undersmoothing of AIC and GCV and the oversmoothing of $T$.

### *8.1.2. Nonparametric Monte Carlo Study*

We will follow Hurvich, Simonoff and Tsai's (1998) simulation format to investigate the properties of different selection criteria in practice. We will look at performance with respect to sample size, the pattern of predictor values, the true regression function, the true standard deviation of the errors, and the regression estimator used. The following levels of these factors were examined, with 500 simulation replications for each level: sample size of $n = 50$, 100, and 500, an equispaced fixed design and a random uniform design, both on [0, 1], and six regression functions, most of which have been used in other Monte Carlo studies (Ruppert, Sheather, and Wand, 1995; Hart and Yi, 1996; Herrmann, 1997). The six functions are:

(1) $\mu_*(x) = \sin(15\pi x)$ (a function with a good deal of fine structure);

(2) $\mu_*(x) = \sin(5\pi x)$ (a function with less fine structure);

(3) $\mu_*(x) = 1 - 48x + 218x^2 - 315x^3 + 145x^4$ (a function with less fine structure and a trend, typical of many regression situations);

(4) $\mu_*(x) = 0.3\exp(-64(x - 0.25)^2) + 0.7\exp(-256(x - 0.75)^2)$ (a function with noticeably different degrees of curvature for different values of the predictor);

(5) $\mu_*(x) = 10\exp(-10x)$ (a function with a trend, but no fine structure);

(6) $\mu_*(x) = \exp(x - 1/3)$ for $x < 1/3$ and $\exp(-2(x - 1/3))$ for $x \geq 1/3$ (a function with undefined first derivative at $x = 1/3$, which violates the standard assumptions for optimal performance of the estimators used here).

The regression functions given by (1) with $\sigma/r_y = 0.01$, given by (3) with $\sigma/r_y = 0.01$, and given by (5) with $\sigma/r_y = 0.05$ correspond to situations where either a relatively small, moderate, or large amount of smoothing, respectively, is appropriate (function (1) was deliberately chosen to represent an extreme low-smoothing situation). The models have error standard deviations of $\sigma_* = 0.01r_y$, $0.05r_y$, $0.25r_y$, and $0.5r_y$, where $r_y$ is the range of $\mu_*(x)$ over $x \in [0, 1]$. Regression estimators used are the local linear and quadratic estimators using a Gaussian kernel, second-order and fourth-order boundary-corrected Gasser–Müller convolution kernel estimators, as described in Herrmann (1997), and a cubic smoothing spline estimator. Tables 8.1–8.5 summarize Monte Carlo results for the six regression functions for an equispaced design only for $n = 100$, and Table 8.6 for a random design for $n = 100$. These tables give the strength of the regression relationship, $\sigma_*/r_y$, the averaged optimal average squared errors (based on the optimal smoothing parameter for the simulated data set), and the mean of the ratio of average squared error to the optimal value when using a particular selection criterion (therefore, the closer to 1, the better). The optimal and data based bandwidths were found using grid search routines. Boxplots of the ASEs for each simulation run (not presented) show that the summary measures given in the tables are representative of the actual relative behavior of the selection criteria. That is, the mean ratio of ASE to optimal ASE does not reflect unusual values, but rather the pattern of the entire distribution of ASE values. The median ratios, while generally smaller than the mean ratios, follow the same patterns as the mean values. If the difference in mean ASE ratios between two criteria is greater than 0.02, results from the signed rank test are generally statistically significant at the 0.05 level for all of the selection criteria except for GCV. When GCV is involved in the comparison, its high variability means that differences in mean ASE ratio up to 0.15 are sometimes not statistically significant.

Values for GCV, $T$, AICc, and $\text{AICc}_1$ are reported for all estimators, as are results for Herrmann's (1997) plug-in selection criterion and Ruppert, Sheather, and Wand's (1995) local linear plug-in selection criterion. Finally, a plug-in selection criterion for the local quadratic estimator has been calculated as $27\hat{h}_L/16$, where $\hat{h}_L$ is the local linear plug-in bandwidth. It yields a bandwidth for the local quadratic estimator that has the same asymptotic variance as the local linear estimator using $\hat{h}_L$, while having smaller asymptotic bias (Sheather, 1996). Because AIC almost always chose the smallest bandwidth available in the simulation run, results for the bandwidths chosen using AIC are not instructive and hence are not shown.

Table 8.1 gives results for the local linear estimator. The plug-in selection criterion has the best performance most often, with the exception of the function $\mu_*(x) = \sin(15\pi x)$, where it fails badly. $T$, AICc, and $\text{AICc}_1$ usually behave similarly, but with AICc noticeably better. Overall, even though it results in a slightly higher ASE, AICc is competitive with the plug-in selection criterion. GCV does well when a small bandwidth is appropriate (although not as well as the plug-in), but because of its tendency to undersmooth its performance deteriorates when a moderate or large bandwidth is best.

Table 8.2 gives results for the local quadratic estimator. The first important thing to note is that the values for optimal mean average squared errors are uniformly lower compared to those for the local linear estimator in Table 8.1; for strong regression relationships (small $\sigma_*/r_y$) it is often 40–50% smaller. This clear superiority of the local quadratic estimator is consistent with asymptotic properties, and can be contrasted with the situation in kernel density estimation, where higher-order kernel estimators are only an improvement on second-order kernels for sample sizes in the hundreds and even thousands (Marron and Wand, 1992). The advantage of the local quadratic estimator lessens for weaker regression relationships, and is small for regression function (6), where the kink in the function means that the local quadratic estimator is not asymptotically superior to the local linear. However, it is never worse. This superiority of higher order does not carry over to the local cubic estimator, where Monte Carlo simulations (not given here) indicate that increased variability outweighs any advantages in boundary bias correction.

The second important point regarding Table 8.2 is that the gain from using the local quadratic estimator is achievable in practice. AICc is clearly the best choice. As was the case for the local linear estimator, GCV has problems when a large bandwidth is appropriate, while AICc is similar to, but consistently better than, $T$ and $\mathrm{AICc}_1$. The local-linear-based local quadratic plug-in performs better than the plug-in applied to the local linear estimator, as it was designed to do, it is not competitive with the other selection criteria under strong relationships for most of the regression functions. Since it is not intended to target the actual optimal bandwidth for the local quadratic estimator, but only to improve on the local linear estimator, this is not surprising. Overall, Tables 8.1 and 8.2 clearly demonstrate that for $n = 100$ and equispaced fixed design, the best choice is the local quadratic estimator using AICc to select the bandwidth.

In Tables 8.3 and 8.4 we examine the results for the second- and fourth-order convolution kernel estimators. The optimal performance of the kernel estimators is comparable to that of the local linear and quadratic estimators, respectively. The fourth-order kernel's optimal mean ASE is consistently smaller than that of the second-order kernel. The second-order plug-in selection criterion is often better "tuned" for the estimator than the Ruppert *et al.* plug-in is for the local linear estimator, and the fourth-order plug-in selection criterion is much better than the local-linear-based local quadratic counterpart as well, since it does in fact target the true optimal bandwidth. AICc has the best overall performance of the classical methods, being generally a bit better than the plug-in for larger target bandwidths and somewhat worse for smaller target bandwidths. The best single choice would probably be the plug-in selection criteria, but AICc is a close second. We again see that GCV is effective for small bandwidths but less so when large bandwidths are appropriate.

Table 8.5 summarizes the results for the cubic smoothing spline estimator. Asymptotically, this estimator is comparable to the local quadratic (Table 8.2) and fourth-order kernel estimators (Table 8.4), and to a certain extent the results reflect this. While for most regression functions the optimal MASE is close to the values for the local quadratic and fourth-order kernel estimators,

Table 8.1. Simulation results for the local linear estimator.
$n = 100$.

| $\mu_*(x) = \sin(15\pi x)$ Very fine structure | | | | | | |
|---|---|---|---|---|---|---|
| $\sigma_*/r_y$ | *Optimal* | *GCV* | *T* | AICc | AICc$_1$ | *Plug-in* |
| 0.01 | $3.768\times10^{-4}$ | 1.088 | 16.995 | 8.108 | 14.382 | 19.551 |
| 0.05 | $5.444\times10^{-3}$ | 1.826 | 1.998 | 1.513 | 2.015 | 2.049 |
| 0.25 | $6.965\times10^{-2}$ | 1.074 | 1.156 | 1.112 | 1.240 | 1.190 |
| 0.5 | 0.214 | 1.173 | 1.185 | 1.133 | 1.239 | 1.705 |
| $\mu_*(x) = \sin(5\pi x)$ Fine structure | | | | | | |
| $\sigma_*/r_y$ | *Optimal* | *GCV* | *T* | AICc | AICc$_1$ | *Plug-in* |
| 0.01 | $1.746\times10^{-4}$ | 1.101 | 1.514 | 1.243 | 1.482 | 1.117 |
| 0.05 | $2.384\times10^{-3}$ | 1.121 | 1.106 | 1.086 | 1.109 | 1.003 |
| 0.25 | $3.153\times10^{-2}$ | 1.218 | 1.098 | 1.068 | 1.103 | 1.082 |
| 0.5 | $9.367\times10^{-2}$ | 1.238 | 1.144 | 1.134 | 1.164 | 1.137 |
| $\mu_*(x) = 1 - 48x + 218x^2 - 315x^3 + 145x^4$ Fine structure +trend | | | | | | |
| $\sigma_*/r_y$ | *Optimal* | *GCV* | *T* | AICc | AICc$_1$ | *Plug-in* |
| 0.01 | $4.495\times10^{-4}$ | 1.088 | 1.177 | 1.088 | 1.216 | 1.082 |
| 0.05 | $5.598\times10^{-3}$ | 1.139 | 1.074 | 1.080 | 1.107 | 1.064 |
| 0.25 | $7.332\times10^{-2}$ | 1.370 | 1.227 | 1.214 | 1.145 | 1.184 |
| 0.5 | 0.222 | 1.496 | 1.303 | 1.221 | 1.273 | 1.261 |
| $\mu_*(x) = 0.3\exp(-64(x-0.25)^2) + 0.7\exp(-256(x-0.75)^2)$ Different curvature for different values of the predictor | | | | | | |
| $\sigma_*/r_y$ | *Optimal* | *GCV* | *T* | AICc | AICc$_1$ | *Plug-in* |
| 0.01 | $2.562\times10^{-5}$ | 1.918 | 1.815 | 1.367 | 1.727 | 1.208 |
| 0.05 | $3.330\times10^{-4}$ | 1.098 | 1.129 | 1.076 | 1.164 | 1.122 |
| 0.25 | $4.170\times10^{-3}$ | 1.164 | 1.148 | 1.138 | 1.112 | 1.073 |
| 0.5 | $1.214\times10^{-2}$ | 1.229 | 1.157 | 1.110 | 1.142 | 1.158 |
| $\mu_*(x) = 10\exp(-10x)$ Trend only | | | | | | |
| $\sigma_*/r_y$ | *Optimal* | *GCV* | *T* | AICc | AICc$_1$ | *Plug-in* |
| 0.01 | $2.302\times10^{-3}$ | 1.101 | 1.089 | 1.112 | 1.167 | 1.046 |
| 0.05 | $2.779\times10^{-2}$ | 1.152 | 1.094 | 1.117 | 1.101 | 1.059 |
| 0.25 | 0.364 | 1.548 | 1.332 | 1.329 | 1.248 | 1.332 |
| 0.5 | 0.998 | 2.040 | 1.570 | 1.624 | 1.512 | 1.825 |
| $\mu_*(x) = \exp(x - 1/3), x < 1/3; \exp(-2(x - 1/3)), x \geq 1/3$ Undefined first derivative | | | | | | |
| $\sigma_*/r_y$ | *Optimal* | *GCV* | *T* | AICc | AICc$_1$ | *Plug-in* |
| 0.01 | $1.391\times10^{-5}$ | 1.132 | 1.119 | 1.081 | 1.174 | 1.226 |
| 0.05 | $1.666\times10^{-4}$ | 1.190 | 1.094 | 1.145 | 1.092 | 1.066 |
| 0.25 | $1.957\times10^{-3}$ | 1.404 | 1.297 | 1.267 | 1.274 | 1.264 |
| 0.5 | $5.816\times10^{-3}$ | 2.128 | 1.532 | 1.558 | 1.535 | 1.630 |

Table 8.2. Simulation results for the local quadratic estimator. $n = 100$.

| $\mu_*(x) = \sin(15\pi x)$ Very fine structure | | | | | | |
|---|---|---|---|---|---|---|
| $\sigma_*/r_y$ | *Optimal* | *GCV* | *T* | AICc | AICc$_1$ | *Plug-in* |
| 0.01 | $2.055\times10^{-4}$ | 1.210 | 2.141 | 1.183 | 1.645 | 6.470 |
| 0.05 | $3.541\times10^{-3}$ | 1.024 | 1.172 | 1.058 | 1.191 | 1.261 |
| 0.25 | $6.234\times10^{-2}$ | 1.069 | 1.063 | 1.089 | 1.102 | 1.148 |
| 0.5 | 0.204 | 1.089 | 1.138 | 1.139 | 1.243 | 1.943 |
| $\mu_*(x) = \sin(5\pi x)$ Fine structure | | | | | | |
| $\sigma_*/r_y$ | *Optimal* | *GCV* | *T* | AICc | AICc$_1$ | *Plug-in* |
| 0.01 | $8.354\times10^{-5}$ | 1.090 | 1.068 | 1.074 | 1.097 | 1.534 |
| 0.05 | $1.443\times10^{-3}$ | 1.093 | 1.104 | 1.085 | 1.066 | 1.373 |
| 0.25 | $2.350\times10^{-2}$ | 1.232 | 1.116 | 1.165 | 1.142 | 1.171 |
| 0.5 | $8.012\times10^{-2}$ | 1.298 | 1.215 | 1.155 | 1.197 | 1.150 |
| $\mu_*(x) = 1 - 48x + 218x^2 - 315x^3 + 145x^4$ Fine structure +trend | | | | | | |
| $\sigma_*/r_y$ | *Optimal* | *GCV* | *T* | AICc | AICc$_1$ | *Plug-in* |
| 0.01 | $2.105\times10^{-4}$ | 1.106 | 1.110 | 1.099 | 1.103 | 1.678 |
| 0.05 | $3.230\times10^{-3}$ | 1.247 | 1.140 | 1.150 | 1.131 | 1.537 |
| 0.25 | $5.874\times10^{-2}$ | 1.394 | 1.271 | 1.239 | 1.297 | 1.402 |
| 0.5 | 0.192 | 1.485 | 1.285 | 1.327 | 1.303 | 1.470 |
| $\mu_*(x) = 0.3\exp(-64(x-0.25)^2) + 0.7\exp(-256(x-0.75)^2)$ Different curvature for different values of the predictor | | | | | | |
| $\sigma_*/r_y$ | *Optimal* | *GCV* | *T* | AICc | AICc$_1$ | *Plug-in* |
| 0.01 | $1.518\times10^{-5}$ | 1.097 | 1.132 | 1.090 | 1.099 | 1.054 |
| 0.05 | $2.607\times10^{-4}$ | 1.095 | 1.068 | 1.082 | 1.039 | 1.040 |
| 0.25 | $4.014\times10^{-3}$ | 1.207 | 1.123 | 1.167 | 1.153 | 1.123 |
| 0.5 | $1.303\times10^{-2}$ | 1.311 | 1.229 | 1.208 | 1.184 | 1.171 |
| $\mu_*(x) = 10\exp(-10x)$ Trend only | | | | | | |
| $\sigma_*/r_y$ | *Optimal* | *GCV* | *T* | AICc | AICc$_1$ | *Plug-in* |
| 0.01 | $1.345\times10^{-3}$ | 1.126 | 1.084 | 1.095 | 1.086 | 1.339 |
| 0.05 | $1.938\times10^{-2}$ | 1.241 | 1.158 | 1.199 | 1.190 | 1.299 |
| 0.25 | 0.311 | 1.663 | 1.456 | 1.363 | 1.389 | 1.593 |
| 0.5 | 1.021 | 1.690 | 1.611 | 1.535 | 1.519 | 1.943 |
| $\mu_*(x) = \exp(x - 1/3), x < 1/3; \exp(-2(x-1/3)), x \geq 1/3$ Undefined first derivative | | | | | | |
| $\sigma_*/r_y$ | *Optimal* | *GCV* | *T* | AICc | AICc$_1$ | *Plug-in* |
| 0.01 | $1.269\times10^{-5}$ | 1.096 | 1.078 | 1.093 | 1.112 | 1.098 |
| 0.05 | $1.529\times10^{-4}$ | 1.210 | 1.116 | 1.194 | 1.111 | 1.084 |
| 0.25 | $1.924\times10^{-3}$ | 1.459 | 1.354 | 1.379 | 1.378 | 1.310 |
| 0.5 | $5.569\times10^{-3}$ | 1.513 | 1.507 | 1.502 | 1.468 | 1.694 |

Table 8.3. Simulation results for the second-order convolution kernel estimator. $n = 100$.

| | | | | | | |
|---|---|---|---|---|---|---|
| | | $\mu_*(x) = \sin(15\pi x)$ Very fine structure | | | | |
| $\sigma_*/r_y$ | *Optimal* | *GCV* | *T* | AICc | $\text{AICc}_1$ | *Plug-in* |
| 0.01 | $3.783\times10^{-4}$ | 4.767 | 9.119 | 5.404 | 7.243 | 1.050 |
| 0.05 | $5.274\times10^{-3}$ | 1.054 | 1.591 | 1.290 | 1.633 | 1.013 |
| 0.25 | $6.721\times10^{-2}$ | 1.215 | 1.125 | 1.115 | 1.096 | 1.020 |
| 0.5 | 0.202 | 1.164 | 1.156 | 1.151 | 1.211 | 1.069 |
| | | $\mu_*(x) = \sin(5\pi x)$ Fine structure | | | | |
| $\sigma_*/r_y$ | *Optimal* | *GCV* | *T* | AICc | $\text{AICc}_1$ | *Plug-in* |
| 0.01 | $1.631\times10^{-4}$ | 1.009 | 1.333 | 1.250 | 1.357 | 1.009 |
| 0.05 | $2.305\times10^{-3}$ | 1.128 | 1.074 | 1.073 | 1.124 | 1.080 |
| 0.25 | $3.001\times10^{-2}$ | 1.226 | 1.153 | 1.094 | 1.164 | 1.154 |
| 0.5 | $9.142\times10^{-2}$ | 1.420 | 1.199 | 1.190 | 1.139 | 1.141 |
| | | $\mu_*(x) = 1 - 48x + 218x^2 - 315x^3 + 145x^4$ Fine structure +trend | | | | |
| $\sigma_*/r_y$ | *Optimal* | *GCV* | *T* | AICc | $\text{AICc}_1$ | *Plug-in* |
| 0.01 | $4.603\times10^{-4}$ | 1.001 | 1.114 | 1.083 | 1.181 | 1.085 |
| 0.05 | $5.962\times10^{-3}$ | 1.162 | 1.081 | 1.107 | 1.107 | 1.084 |
| 0.25 | $7.749\times10^{-2}$ | 1.299 | 1.224 | 1.188 | 1.236 | 1.130 |
| 0.5 | 0.234 | 1.487 | 1.296 | 1.337 | 1.334 | 1.217 |
| | | $\mu_*(x) = 0.3\exp(-64(x-0.25)^2) + 0.7\exp(-256(x-0.75)^2)$ Different curvature for different values of the predictor | | | | |
| $\sigma_*/r_y$ | *Optimal* | *GCV* | *T* | AICc | $\text{AICc}_1$ | *Plug-in* |
| 0.01 | $2.289\times10^{-5}$ | 1.770 | 1.546 | 1.249 | 1.514 | 1.027 |
| 0.05 | $3.073\times10^{-4}$ | 1.153 | 1.097 | 1.039 | 1.109 | 1.086 |
| 0.25 | $4.004\times10^{-3}$ | 1.224 | 1.186 | 1.116 | 1.103 | 1.158 |
| 0.5 | $1.210\times10^{-2}$ | 1.435 | 1.188 | 1.202 | 1.228 | 1.228 |
| | | $\mu_*(x) = 10\exp(-10x)$ Trend only | | | | |
| $\sigma_*/r_y$ | *Optimal* | *GCV* | *T* | AICc | $\text{AICc}_1$ | *Plug-in* |
| 0.01 | $2.456\times10^{-3}$ | 1.084 | 1.111 | 1.080 | 1.127 | 1.204 |
| 0.05 | $3.075\times10^{-2}$ | 1.185 | 1.119 | 1.140 | 1.107 | 1.141 |
| 0.25 | 0.389 | 1.577 | 1.295 | 1.314 | 1.251 | 1.274 |
| 0.5 | 1.125 | 1.974 | 1.528 | 1.447 | 1.528 | 1.655 |
| | | $\mu_*(x) = \exp(x - 1/3), x < 1/3; \exp(-2(x-1/3)), x \geq 1/3$ Undefined first derivative | | | | |
| $\sigma_*/r_y$ | *Optimal* | *GCV* | *T* | AICc | $\text{AICc}_1$ | *Plug-in* |
| 0.01 | $1.348\times10^{-5}$ | 1.190 | 1.138 | 1.101 | 1.152 | 1.042 |
| 0.05 | $1.597\times10^{-4}$ | 1.079 | 1.176 | 1.160 | 1.160 | 1.129 |
| 0.25 | $1.791\times10^{-3}$ | 1.717 | 1.458 | 1.412 | 1.342 | 1.408 |
| 0.5 | $5.553\times10^{-3}$ | 2.310 | 1.663 | 1.634 | 1.622 | 2.009 |

Table 8.4. Simulation results for the fourth-order convolution kernel estimator. $n = 100$.

| $\mu_*(x) = \sin(15\pi x)$ Very fine structure | | | | | | |
|---|---|---|---|---|---|---|
| $\sigma_*/r_y$ | *Optimal* | *GCV* | *T* | AICc | AICc$_1$ | *Plug-in* |
| 0.01 | $2.753\times10^{-4}$ | 1.128 | 2.221 | 1.300 | 1.420 | 1.310 |
| 0.05 | $3.608\times10^{-3}$ | 1.032 | 1.156 | 1.054 | 1.185 | 1.080 |
| 0.25 | $5.960\times10^{-2}$ | 1.106 | 1.089 | 1.126 | 1.108 | 1.068 |
| 0.5 | 0.206 | 1.112 | 1.135 | 1.133 | 1.165 | 1.111 |
| $\mu_*(x) = \sin(5\pi x)$ Fine structure | | | | | | |
| $\sigma_*/r_y$ | *Optimal* | *GCV* | *T* | AICc | AICc$_1$ | *Plug-in* |
| 0.01 | $8.486\times10^{-5}$ | 1.099 | 1.057 | 1.071 | 1.045 | 1.025 |
| 0.05 | $1.418\times10^{-3}$ | 1.176 | 1.104 | 1.160 | 1.058 | 1.054 |
| 0.25 | $2.445\times10^{-2}$ | 1.300 | 1.185 | 1.246 | 1.221 | 1.171 |
| 0.5 | $7.978\times10^{-2}$ | 1.487 | 1.246 | 1.207 | 1.198 | 1.193 |
| $\mu_*(x) = 1 - 48x + 218x^2 - 315x^3 + 145x^4$ Fine structure +trend | | | | | | |
| $\sigma_*/r_y$ | *Optimal* | *GCV* | *T* | AICc | AICc$_1$ | *Plug-in* |
| 0.01 | $1.685\times10^{-4}$ | 1.143 | 1.145 | 1.131 | 1.118 | 1.060 |
| 0.05 | $2.648\times10^{-3}$ | 1.434 | 1.289 | 1.254 | 1.236 | 1.246 |
| 0.25 | $5.223\times10^{-2}$ | 1.532 | 1.414 | 1.414 | 1.380 | 1.687 |
| 0.5 | 0.186 | 1.813 | 1.490 | 1.449 | 1.387 | 1.961 |
| $\mu_*(x) = 0.3\exp(-64(x-0.25)^2) + 0.7\exp(-256(x-0.75)^2)$ Different curvature for different values of the predictor | | | | | | |
| $\sigma_*/r_y$ | *Optimal* | *GCV* | *T* | AICc | AICc$_1$ | *Plug-in* |
| 0.01 | $1.457\times10^{-5}$ | 1.106 | 1.072 | 1.037 | 1.084 | 1.009 |
| 0.05 | $2.585\times10^{-4}$ | 1.139 | 1.074 | 1.063 | 1.068 | 1.071 |
| 0.25 | $4.023\times10^{-3}$ | 1.254 | 1.118 | 1.161 | 1.162 | 1.192 |
| 0.5 | $1.310\times10^{-2}$ | 1.405 | 1.226 | 1.259 | 1.281 | 1.145 |
| $\mu_*(x) = 10\exp(-10x)$ Trend only | | | | | | |
| $\sigma_*/r_y$ | *Optimal* | *GCV* | *T* | AICc | AICc$_1$ | *Plug-in* |
| 0.01 | $1.129\times10^{-3}$ | 1.269 | 1.130 | 1.140 | 1.132 | 1.088 |
| 0.05 | $1.776\times10^{-2}$ | 1.406 | 1.270 | 1.296 | 1.272 | 1.227 |
| 0.25 | 0.295 | 1.932 | 1.512 | 1.482 | 1.409 | 1.977 |
| 0.5 | 1.163 | 1.897 | 1.544 | 1.520 | 1.423 | 2.344 |
| $\mu_*(x) = \exp(x - 1/3), x < 1/3; \exp(-2(x-1/3)), x \geq 1/3$ Undefined first derivative | | | | | | |
| $\sigma_*/r_y$ | *Optimal* | *GCV* | *T* | AICc | AICc$_1$ | *Plug-in* |
| 0.01 | $1.281\times10^{-5}$ | 1.160 | 1.144 | 1.129 | 1.093 | 1.196 |
| 0.05 | $1.594\times10^{-4}$ | 1.270 | 1.208 | 1.235 | 1.248 | 1.108 |
| 0.25 | $1.979\times10^{-3}$ | 1.521 | 1.405 | 1.381 | 1.379 | 1.434 |
| 0.5 | $6.574\times10^{-3}$ | 1.971 | 1.521 | 1.442 | 1.483 | 2.074 |

Table 8.5. Simulation results for the cubic smoothing spline estimator. $n = 100$.

| $\mu_*(x) = \sin(15\pi x)$ Very fine structure | | | | | |
|---|---|---|---|---|---|
| $\sigma_*/r_y$ | *Optimal* | *GCV* | $T$ | AICc | $\text{AICc}_1$ |
| 0.01 | $1.969\times10^{-4}$ | 1.031 | 1.689 | 1.145 | 1.185 |
| 0.05 | $3.504\times10^{-3}$ | 1.026 | 1.160 | 1.101 | 1.130 |
| 0.25 | $6.017\times10^{-2}$ | 1.051 | 1.069 | 1.105 | 1.119 |
| 0.5 | 0.204 | 1.117 | 1.235 | 1.234 | 1.421 |
| $\mu_*(x) = \sin(5\pi x)$ Fine structure | | | | | |
| $\sigma_*/r_y$ | *Optimal* | *GCV* | $T$ | AICc | $\text{AICc}_1$ |
| 0.01 | $8.459\times10^{-5}$ | 1.025 | 1.057 | 1.081 | 1.105 |
| 0.05 | $1.439\times10^{-3}$ | 1.123 | 1.063 | 1.077 | 1.070 |
| 0.25 | $2.418\times10^{-2}$ | 1.162 | 1.099 | 1.112 | 1.167 |
| 0.5 | $8.444\times10^{-2}$ | 1.258 | 1.132 | 1.139 | 1.165 |
| $\mu_*(x) = 1 - 48x + 218x^2 - 315x^3 + 145x^4$ Fine structure +trend | | | | | |
| $\sigma_*/r_y$ | *Optimal* | *GCV* | $T$ | AICc | $\text{AICc}_1$ |
| 0.01 | $3.188\times10^{-4}$ | 1.100 | 1.080 | 1.093 | 1.110 |
| 0.05 | $4.532\times10^{-3}$ | 1.108 | 1.046 | 1.097 | 1.119 |
| 0.25 | $6.603\times10^{-2}$ | 1.263 | 1.250 | 1.196 | 1.241 |
| 0.5 | 0.205 | 1.410 | 1.309 | 1.249 | 1.256 |
| $\mu_*(x) = 0.3\exp(-64(x-0.25)^2) + 0.7\exp(-256(x-0.75)^2)$ Different curvature for different values of the predictor | | | | | |
| $\sigma_*/r_y$ | *Optimal* | *GCV* | $T$ | AICc | $\text{AICc}_1$ |
| 0.01 | $1.466\times10^{-5}$ | 1.040 | 1.108 | 1.085 | 1.110 |
| 0.05 | $2.628\times10^{-4}$ | 1.068 | 1.100 | 1.071 | 1.125 |
| 0.25 | $3.926\times10^{-3}$ | 1.157 | 1.089 | 1.115 | 1.178 |
| 0.5 | $1.173\times10^{-2}$ | 1.206 | 1.228 | 1.205 | 1.201 |
| $\mu_*(x) = 10\exp(-10x)$ Trend only | | | | | |
| $\sigma_*/r_y$ | *Optimal* | *GCV* | $T$ | AICc | $\text{AICc}_1$ |
| 0.01 | $1.885\times10^{-3}$ | 1.147 | 1.075 | 1.108 | 1.131 |
| 0.05 | $2.616\times10^{-2}$ | 1.155 | 1.147 | 1.145 | 1.102 |
| 0.25 | 0.339 | 1.429 | 1.340 | 1.250 | 1.343 |
| 0.5 | 1.051 | 1.865 | 1.645 | 1.566 | 1.579 |
| $\mu_*(x) = \exp(x-1/3), x < 1/3; \exp(-2(x-1/3)), x \geq 1/3$ Undefined first derivative | | | | | |
| $\sigma_*/r_y$ | *Optimal* | *GCV* | $T$ | AICc | $\text{AICc}_1$ |
| 0.01 | $1.232\times10^{-5}$ | 1.141 | 1.074 | 1.085 | 1.130 |
| 0.05 | $1.447\times10^{-4}$ | 1.245 | 1.114 | 1.109 | 1.120 |
| 0.25 | $1.737\times10^{-3}$ | 1.430 | 1.355 | 1.372 | 1.334 |
| 0.5 | $5.318\times10^{-3}$ | 1.928 | 1.627 | 1.616 | 1.640 |

Table 8.6. Simulation results for uniform random design.
$n = 100$.

$\mu_*(x) = \sin(5\pi x)$
Fine structure

*Local quadratic estimator*

| $\sigma_*/r_y$ | *Optimal* | *GCV* | *T* | AICc | *Plug-in* |
|---|---|---|---|---|---|
| 0.01 | $9.105\times10^{-5}$ | 1.139 | 1.049 | 1.073 | 2.048 |
| 0.05 | $1.469\times10^{-3}$ | 1.499 | 1.178 | 1.221 | 2.089 |
| 0.25 | $2.683\times10^{-2}$ | 1.928 | 1.492 | 1.492 | 1.803 |
| 0.5 | $8.551\times10^{-2}$ | 2.202 | 1.624 | 1.561 | 1.730 |

*Fourth-order kernel estimator*

| $\sigma_*/r_y$ | *Optimal* | *GCV* | *T* | AICc | *Plug-in* |
|---|---|---|---|---|---|
| 0.01 | $2.889\times10^{-4}$ | 1.066 | 1.029 | 1.025 | 2.140 |
| 0.05 | $1.765\times10^{-3}$ | 1.108 | 1.079 | 1.077 | 1.050 |
| 0.25 | $2.524\times10^{-2}$ | 1.282 | 1.169 | 1.200 | 1.197 |
| 0.5 | $8.651\times10^{-2}$ | 1.346 | 1.249 | 1.280 | 1.226 |

*Cubic smoothing spline estimator*

| $\sigma_*/r_y$ | *Optimal* | *GCV* | *T* | AICc |
|---|---|---|---|---|
| 0.01 | $1.147\times10^{-4}$ | 1.045 | 1.067 | 1.038 |
| 0.05 | $1.524\times10^{-3}$ | 1.112 | 1.092 | 1.054 |
| 0.25 | $2.526\times10^{-2}$ | 1.148 | 1.110 | 1.129 |
| 0.5 | $8.555\times10^{-2}$ | 1.297 | 1.182 | 1.138 |

$\mu_*(x) = 0.3\exp(-64(x-0.25)^2) + 0.7\exp(-256(x-0.75)^2)$
Different curvature for different values of the predictor

*Local quadratic estimator*

| $\sigma_*/r_y$ | *Optimal* | *GCV* | *T* | AICc | *Plug-in* |
|---|---|---|---|---|---|
| 0.01 | $1.618\times10^{-5}$ | 1.051 | 1.070 | 1.084 | 1.069 |
| 0.05 | $2.597\times10^{-4}$ | 1.263 | 1.093 | 1.072 | 1.111 |
| 0.25 | $3.969\times10^{-3}$ | 1.553 | 1.246 | 1.214 | 1.163 |
| 0.5 | $1.227\times10^{-2}$ | 1.729 | 1.425 | 1.486 | 1.177 |

*Fourth-order kernel estimator*

| $\sigma_*/r_y$ | *Optimal* | *GCV* | *T* | AICc | *Plug-in* |
|---|---|---|---|---|---|
| 0.01 | $2.897\times10^{-5}$ | 1.102 | 1.177 | 1.141 | 1.399 |
| 0.05 | $2.872\times10^{-4}$ | 1.113 | 1.067 | 1.074 | 1.103 |
| 0.25 | $4.270\times10^{-3}$ | 1.230 | 1.156 | 1.197 | 1.203 |
| 0.5 | $1.289\times10^{-2}$ | 1.345 | 1.284 | 1.284 | 1.196 |

*Cubic smoothing spline estimator*

| $\sigma_*/r_y$ | *Optimal* | *GCV* | *T* | AICc |
|---|---|---|---|---|
| 0.01 | $2.728\times10^{-5}$ | 1.075 | 1.050 | 1.054 |
| 0.05 | $2.612\times10^{-4}$ | 1.083 | 1.054 | 1.046 |
| 0.25 | $4.020\times10^{-3}$ | 1.144 | 1.099 | 1.082 |
| 0.5 | $1.166\times10^{-2}$ | 1.314 | 1.213 | 1.252 |

it is as much as 50% worse (larger) for functions (3) and (5), functions without very fine structure. Once again $T$, AICc, and AICc$_1$ are generally very similar, with AICc performing best (except in the case of very fine structure). There is no plug-in selection criterion for the smoothing spline, but all of the classical selection criteria generally work at least as well (compared to the optimal performance) as they do for the other estimators. Still, the comparatively poor performance of the spline compared to the local quadratic and fourth-order kernel estimators casts doubt on whether choosing it over one of the other estimators is appropriate for an equispaced design.

Results for $n = 50$ and $n = 500$ are too voluminous to include here, but they can be summarized as follows. At $n = 50$ variability of the estimators is naturally a problem, and accordingly the low variability of the plug-in selection criteria works in their favor. For this reason the plug-in selection criteria are generally better than the other selection criteria in this case, although AICc is still competitive. At $n = 500$ the plug-in's properties are very similar to those of the classical selection criteria (except GCV).

Finally, we will look at the results for estimators under uniform random design. It is well-known that the convolution kernel, Eq. (8.1), is asymptotically inefficient for random designs; therefore, to study selection criterion performance under these conditions, we need an adapted version of our estimator. Herrmann (1996) discussed general versions of the Gasser–Müller estimator,

$$\hat{\mu}_h(x) = h^{-1} \sum_{i=1}^{n} c_i \left[ \int_{a_i}^{b_i} K\left(\frac{x-t}{h}\right) dt \right] y_i, \tag{8.8}$$

pointing out that it is the variability of the differences $b_i - a_i(x_{i+1} - x_{i-1})/2$ for Eq. (8.1) that potentially inflates the variance of the estimator Eq. (8.8) (see also Chu and Marron, 1991 and Jones, Davies and Park, 1994). To minimize this problem, Herrmann suggested taking $c_i = 1$ and using kernel quantile estimators to determine $a_i$ and $b_i$,

$$a_i = \tilde{g}^{-1} \sum_{j=1}^{n} x_j \int_{(j-0.5)/(n+1)}^{(j+0.5)/(n+1)} K_s\left(\frac{(i-0.5)/(n+1) - t}{\tilde{g}}\right) dt$$

and

$$b_i = \tilde{g}^{-1} \sum_{j=1}^{n} x_j \int_{(j-0.5)/(n+1)}^{(j+0.5)/(n+1)} K_s\left(\frac{(i+0.5)/(n+1) - t}{\tilde{g}}\right) dt,$$

where $K_s$ is a symmetric boundary-corrected kernel of order $k_s = 2$ or $4, \tilde{g} = 0.75(n+1)^{-(3k+1)/((2k+1)(k_s+1))}$, and $k$ is the order of the kernel $K(\cdot)$. The estimator constructed under these conditions does not suffer the asymptotic inefficiency of Eq. (8.1) under random designs.

Table 8.6 gives representative simulation results for the modified local quadratic, fourth-order kernel, and cubic smoothing spline estimators at high and low levels of fine structure when the predictor values in each simulation run are taken as a random sample from a uniform distribution. Results for $\text{AICc}_1$ are not given here due to calculation constraints.

The mean ASE values for the local quadratic and cubic smoothing spline estimators are close to those for the fixed uniform design in Tables 8.2 and 8.5, which is consistent with those estimators' asymptotic equivalence under fixed and random designs. The higher mean ASE value for the smoothing spline under regression function (4) and $\sigma_*/r_y = 0.01$ is a bit misleading, since the median value is very close to the value under the fixed design. The values for Herrmann's modified convolution kernel estimator are sometimes considerably higher, however, indicating that the corrective action for randomness has not taken hold yet at $n = 100$. The plug-in selection criterion for this estimator is not always as close to the optimal ASE as that for the fixed uniform design version. AICc generally performs well.

## 8.2. Semiparametric Regression Model Selection

In many applications the parametric model itself is at best an approximation of the true model, and the search for an adequate model from the parametric family is not easy. When the true mean response is assumed to be linearly related to one or more variables, but the relation to an additional variable or variables is not assumed to be easily parameterized, then the family of semiparametric regression models is an alternative. Semiparametric regression contains both parametric and nonparametric components, which provides a convenient way to include nonlinearities of an unspecified form in a regression model. Many investigators have explored this area; for example, Chen and Shiau (1994) studied the asymptotic behavior of two efficient estimators of the parametric component in a semiparametric model when the smoothing parameter is chosen by either generalized cross-validation (GCV) (Craven and Wahba, 1979) or by Mallows's Cp. Green and Silverman (1994) suggested using Allen's cross-validation CV and GCV to choose an appropriate value for the smoothing parameter. These studies do not consider selection of both components together—the explanatory variables from the parametric component and the smoothing parameters from the nonparametric component. We will investigate how we may do both. Simonoff and Tsai (1997) generalized the three AICc-based selection criteria from Section 8.1 to semiparametric regression models. Here, we only focus on obtaining AICc for semiparametric regression models when the nonparametric component is approximated by a B-spline function (see Shi and Tsai, 1997).

Assume the true model is generated by

$$\begin{aligned} Y &= \boldsymbol{\mu}_* + \varepsilon_* \\ &= X_*\beta_* + g_* + \varepsilon_*, \end{aligned} \tag{8.9}$$

where $Y$ and $\boldsymbol{\mu}_*$ are $n \times 1$ vectors, $X_*$ is a known $n \times k_*$ matrix, $\beta_*$ is the $k_* \times 1$ vector of unknown parameters, $g_* = (g_*(T_1), ..., g_*(T_n))'$, $g_*(\cdot)$ is an unknown function defined on $[0,1]$ with $\int_0^1 g_*(t)dt = 0$, and $\varepsilon_* \sim N(0, \sigma_*^2 I_{n\times n})$.

Let $M = j + k_n$, where $j$ is a given nonnegative integer and $k_n$ is a positive integer, and let $t_1, \cdots, t_{k_n}$ $(t_1 < \cdots < t_{k_n} = 1)$ be a $D_0$-quasi uniform sequence partitioned over $[0,1]$ (see Schumaker, 1981, p. 216). Further, let

$$s_1 = \cdots = s_{j+1} = 0,$$

$$s_{j+2} = t_1, \cdots, s_{j+k_n} = t_{k_n-1},$$

and

$$s_{j+k_n+1} = \cdots = s_{2j+k_n+1} = 1,$$

and let $B(t) = (B_1(t), .., B_M(t))'$ be a vector of normalized B-splines of order $j+1$ associated with the extended partition $\{s_1, \cdots, s_M\}$ of $[0,1]$ (see Schumaker, 1981, p. 224). Then, from Corollary 6.21 in Schumaker (1981), we can approximate $g_*(t)$ by the B-spline function $\pi(t)'\alpha$, where $\pi(t) = B(t) - \int_0^1 B(t)dt$, and $\alpha$ is an $M \times 1$ vector of unknown parameters.

In practical situations, a criterion such as AIC is usually used to choose spline knots from the sample points $T_1, \ldots, T_n$ based on the forward placement and backward deletion algorithms of He and Shi (1996), except that the spline knots are optimized at the deletion step. This results in a selection of knot sets, $\Gamma_2$, based on the order statistics of $T_1, \ldots, T_n$.

### *8.2.1. The Family of Candidate Models*

In order to approximate the true model in Eq. (8.9), consider the family of candidate models

$$Y = X(\gamma)\beta(\gamma) + \Pi(\gamma)\alpha(\gamma) + \varepsilon,$$

where $\beta(\gamma) \in R^k$, $\alpha(\gamma) \in R^N$, $\varepsilon \sim N(0, \sigma^2 I_n)$, $\gamma = \{\gamma_1, \gamma_2\}$, $\gamma_1 = (1, \ldots, k) \in \Gamma_1$ and $\gamma_2 \in \Gamma_2$. $\Gamma_1$ is a collection of all candidate models of the parametric part, $X(\gamma)$ and $\Pi(\gamma) = (\pi(T_1), \ldots, \pi(T_n))'$ are $n \times k$ and $n \times N$ matrices determined by the indices $\gamma_1$ and $\gamma_2$, respectively, $N = N(\gamma) = j + l$, and $l$ is the cardinal number of $\gamma_2$. Thus the true model can be expressed in the form

$$Y = X_* \beta_* + \Pi(\gamma)\alpha(\gamma) + R_n + \varepsilon \tag{8.10}$$

for the given $\gamma$, where $R_n$ is an $n \times 1$ vector such that $n^{-1} R_n' R_n \leq C_1 N^{-2C_2}$, $C_1$ is some constant, and $C_2 \geq j$ is an indicator of the smoothness of $g_*$. A detailed discussion of $R_n$ can be found in Shi and Li (1995).

### *8.2.2. AICc*

Let $f_*(y)$ and $f(y)$ be the densities of the true and candidate models, respectively. Then, the Kullback–Leibler information criterion is $\delta\{\theta(\gamma)\} = 2E_*[\log\{f_*(y)/f(y)\}]$. Given a collection of competing candidate models for $Y$, the one that minimizes $E_*[\Delta\{\hat{\theta}(\gamma)\}]$ has the smallest discrepancy and is therefore preferred, where $\hat{\theta}(\gamma) = (\hat{\beta}(\gamma), \hat{\alpha}(\gamma), \hat{\sigma}^2(\gamma))'$ and $\hat{\beta}(\gamma), \hat{\alpha}(\gamma), \hat{\sigma}^2(\gamma)$ are the maximum likelihood estimators of $\beta, \alpha$ and $\sigma^2$, respectively, obtained for the candidate model. Applying techniques similar to those used in Chapter 2, Shi and Tsai (1997) obtained the selection criteria

$$\text{AICc} = n \log(\hat{\sigma}^2(\gamma)) + n + \frac{2n}{n - N - k - 2}(N + k + 1), \tag{8.11}$$

and

$$\text{AIC} = n \log(\hat{\sigma}^2(\gamma)) + n + 2(N + k + 1). \tag{8.12}$$

The details of the derivations and the relevant properties can be found in Shi and Tsai (1997).

### *8.2.3. Semiparametric Monte Carlo Study*

In order to evaluate selection criterion performance under semiparametric conditions, we will briefly discuss results from Shi and Tsai's (1997) study. Five hundred realizations were generated from the model given by Eq. (8.9) with standard normal errors and $n$ = 15, 20, 30, 40, and 50. For each specified interval of $t$, $(T_1, \ldots, T_n)$ were chosen from the uniform distribution to be independent and identically distributed. The true order of the parametric component is three, and candidate models that include up to four extraneous variables were considered. The columns of $X$ were generated such that the rows of $X$ are *i.i.d.* $N(0, I_7)$, where the first $k_*$ columns contain the variables belonging to the true model with $k_* = 3$, $\beta_* = (3, 2, 1)'$. The values of $g_*$ have been chosen to represent each of the three most important classes of functions—polynomials, triangular functions, and exponential functions. They are $g_* = t^2/2 - 2/3$ where $t \in [-2, 2]$, $g_* = 2\sin(2\pi t)$, and $g_* = 0.1\exp(4t) - (\exp(4) - 1)/40$ where $t \in [0, 1]$.

Table 8.7 presents the estimated probability that the correct order of the true parametric component of Eq. (8.9) was chosen by each criterion. AICc clearly outperforms AIC.

Table 8.7. Estimated probability of choosing the correct order of the true parametric component.

| $g_*$ | $n$ | AICc | AIC |
|---|---|---|---|
| $t^2/2 - 2/3$ | 15 | 0.671 | 0.022 |
| | 20 | 0.835 | 0.213 |
| | 30 | 0.838 | 0.436 |
| | 40 | 0.847 | 0.531 |
| | 50 | 0.810 | 0.566 |
| $2\sin(2\pi t)$ | 15 | 0.551 | 0.014 |
| | 20 | 0.821 | 0.218 |
| | 30 | 0.831 | 0.453 |
| | 40 | 0.837 | 0.536 |
| | 50 | 0.812 | 0.569 |
| $0.1\exp(4t) - \frac{\exp(4)-1}{40}$ | 15 | 0.655 | 0.018 |
| | 20 | 0.841 | 0.224 |
| | 30 | 0.835 | 0.438 |
| | 40 | 0.845 | 0.545 |
| | 50 | 0.807 | 0.556 |

## 8.3. A Cross-validatory AIC for Hard Wavelet Thresholding

In this Section, we focus the selection of hard wavelet thresholds. A nice introduction to wavelet methods can be found in Hubbard (1996). Selective wavelet reconstruction, proposed by Donoho and Johnstone (1994), provides an automatic, spatially adaptive estimate of a function that has been embedded in additive noise. The method is appealing for several reasons. First, for a sample size of $n$, the estimate results in a mean integrated squared error (which can be computed without any knowledge of the function) that is within a factor of $O(\log^2 n)$ of the performance of piecewise polynomial and variable-knot spline methods, even when these methods are given the unfair advantage of an "oracle" that tells them how best to choose their tuning parameters. Without the oracle, if the function or any of its low-order derivatives is discontinuous, these methods would typically perform much worse. Second, for a given selection of the coefficients which are to be retained, the proposed estimate is obtained by orthogonal linear regression. This simplicity allows for a reasonably comprehensive understanding of the method's theoretical properties. Third, because the orthogonal basis consists of compactly supported wavelets, each coefficient in the estimate can be naturally attributed to a localized oscillation near a specific location and a specific frequency. Fourth, the estimate can be computed in $O(n)$ calculations, using the Fast Wavelet Transform of Mallat (1989).

In selective wavelet reconstruction, given a data set of noisy observations, a subset of the empirical wavelet coefficients is used to estimate the function. A natural question arises as to which subset should be used. One answer, proposed by Donoho and Johnstone (1994), is a computationally efficient technique called *hard wavelet thresholding*, in which only those wavelet coefficients whose absolute value exceeds some threshold are retained. Donoho and Johnstone (1994) suggested a specific choice of the threshold, $\sigma\lambda_n^u = \sigma\sqrt{2\log n}$, where $\sigma$ is the standard deviation of the noise. The quantity $\sigma\lambda_n^u$ is called the universal threshold. One drawback is that universal thresholding requires knowledge of $\sigma$, and in practice $\sigma$ will be unknown and must be estimated. Donoho and Johnstone suggested using an estimate of $\sigma$ based on the median absolute deviation of the highest-frequency wavelet coefficients, but some of their theoretical results still require that $\sigma$ be known. Another potential drawback is that universal thresholding may fail to come close to minimizing the mean integrated squared error for the signal actually at hand.

Alternatively, Hurvich and Tsai (1998) proposed a data-dependent method of hard threshold selection based on a cross-validatory version of AICc, and compared it to universal and other hard thresholding methods. This method requires no knowledge of either $\sigma$ or of the underlying function, and will be discussed later in this Section. Our discussion of wavelet thresholding is organized as follows: Section 8.3.1 contains a brief introduction to wavelet reconstruction and thresholding, and in Section 8.3.2 we discuss Nason's (1996) cross-validation method and give an $O(n \log n)$ algorithm for its implementation. In Section 8.3.3 we present Hurvich and Tsai's (1998) method of selecting the threshold and demonstrate its implementation in $O(n \log n)$ operations, and in Section 8.3.4, we present their results on the rate at which this method attains the optimal mean integrated squared error. In Section 8.3.5, we investigate the relative performance of various thresholding methods via a Monte Carlo study. The contents of Section 8.3.1 to Section 8.3.5 are directly or indirectly cited from Hurvich and Tsai's (1998) paper, which are used with permission from Oxford University Press.

### 8.3.1. Wavelet Reconstruction and Thresholding

Of the several specific implementations proposed in Donoho and Johnstone (1994), we focus on their universal hard wavelet thresholding. We will describe the essential elements of the method here, but further details can be found in Donoho and Johnstone (1994), Donoho, Johnstone, Kerkyacharian and Picard (1995), Strang (1993), Nason and Silverman (1994), Mallat (1989), and Daubechies (1992).

Suppose we have data $\{y_i\}_{i=1}^n$, where $n = 2^{J+1}$, and $J$ is a positive integer. As Donoho and Johnstone (1994) did, for simplicity we assume that $n$ is a power of two in order to facilitate calculations via the Fast Wavelet Transform.

The model is

$$y_i = \mu(t_i) + \varepsilon_i, \quad i = 1, \dots, n, \tag{8.13}$$

where $t_i = i/n$, the $\{\varepsilon_i\}$ are independent identically distributed $N(0, \sigma^2)$, and $\mu(\cdot)$ is the unknown function on [0, 1] we would like to estimate.

For a given choice of various parameters such as the number of vanishing moments, the support width, and the low-resolution cutoff ($j_o$) (see Donoho and Johnstone, 1994), we can construct the finite wavelet transform matrix (an $n \times n$ orthogonal matrix $\tilde{W}_n$). This yields the vector $w = \tilde{W}_n Y$ of wavelet coefficients of $Y = (y_1, \dots, y_n)'$. It is helpful to didactically index a certain set of $n - 1$ elements of $w$, obtaining $\{w_{j,l}\}$ for $j = 0, \dots, J$ and $l = 0, \dots, 2^j - 1$. The remaining element is denoted by $w_{-1,0}$. The inversion formula $Y = \tilde{W}_n' w$ may then be expressed as

$$y_i = \sum_{j,l} w_{j,l} W_{njl}(i), \tag{8.14}$$

where $W_{njl}$ is the $(j, l)$ column of $\tilde{W}_n'$. For $j$ neither too small nor too large, we have the approximation:

$$\sqrt{n} W_{njl}(i) \approx 2^{j/2} \Psi(2^j t - l), t = i/n,$$

where $\Psi$ is an oscillating function of compact support known as the mother wavelet. Thus $W_{njl}$ is essentially localized to spatial positions near $t = l2^{-j}$ and frequencies near $2^j$.

In selective wavelet reconstruction, a subset of the wavelet coefficients of $Y$ is used in Eq. (8.14), yielding

$$\hat{\boldsymbol{\mu}} = \sum_{(j,l)\in\delta} w_{j,l} W_{njl},$$

where $\delta$ is a list of the $(j,l)$ pairs to be used. An important question is how to choose $\delta$. A natural measure of risk for this choice is the mean integrated squared error,

$$R(\hat{\boldsymbol{\mu}}, \boldsymbol{\mu}) = n^{-1} E\left[\| \hat{\boldsymbol{\mu}} - \boldsymbol{\mu} \|^2\right] = n^{-1} \sum_{i=1}^{n} E[(\hat{\mu}(t_i) - \mu(t_i))^2],$$

where $\boldsymbol{\mu} = (\mu(t_1), \ldots, \mu(t_n))'$. The $\delta$ that minimizes $R$ is denoted by $\Delta(\boldsymbol{\mu})$, and the corresponding estimate $\hat{\boldsymbol{\mu}}$ is referred to as the ideal selective wavelet reconstruction. Donoho and Johnstone (1994) have derived the theoretical properties of this ideal reconstruction; however, in practice, ideal reconstruction is not feasible since $\Delta$ depends on both the unknown function $\boldsymbol{\mu}$ and on the error variance $\sigma^2$. Thus, Donoho and Johnstone (1994) proposed hard wavelet thresholding, whereby only the wavelet coefficients whose absolute value exceeds some threshold $\lambda$ are retained, with $\lambda$ not depending on $\boldsymbol{\mu}$. They also consider soft thresholding, whereby the coefficients exceeding the threshold are downweighted rather than being completely retained.

An important issue to be resolved is the choice of $\lambda$. One possibility is universal thresholding, $\lambda = \sigma\sqrt{2\log n}$, which is shown by Donoho and Johnstone (1994) to be asymptotically minimax with respect to a mean integrated squared error risk, and to come within a factor of $2\log n$ of the performance of ideal selective wavelet reconstruction. The asymptotic minimax risk bound is the same for both soft and hard thresholding. Since universal thresholding cannot be implemented without an estimate of $\sigma$, Donoho and Johnstone (1995) propose

$$\tilde{\sigma} = [\text{median absolute deviation } \{w_{J,l}\}_{l=0}^{2^J-1}]/0.6745,$$

but this choice seems somewhat *ad hoc*, and no theory is given regarding properties of the corresponding feasible universal wavelet threshold estimate $\hat{\boldsymbol{\mu}}$, which uses threshold $\tilde{\sigma}\sqrt{2\log n}$. In a subsequent paper, however, Donoho, Johnstone, Kerkyacharian and Picard (1995) do state that the feasible version comes very close to being asymptotically minimax.

### 8.3.2. Nason's Cross-validation Method

Nason (1996) has proposed a method of threshold selection based on half-sample cross-validation. Here we will describe the method and present a computationally efficient algorithm for its implementation.

First, the data vector $Y$ is split into two halves, $Y^{odd} = (y_1, y_3, \ldots, y_{n-1})'$, and $Y^{even} = (y_2, y_4, \ldots, y_n)'$. Each of these vectors has length $n^* = n/2$ which, like $n$ itself, is a power of 2. According to the model in Eq. (8.13), $Y^{odd}$ and $Y^{even}$ are independent. Thresholding each half of the data with a threshold $\lambda$ produces an estimate of the function $\boldsymbol{\mu}$, the quality of which can be assessed by its ability to predict an interpolated version of the other half. Denote the estimators based on the even and odd data by $\hat{\mu}_{\lambda,i}^{even}$ and $\hat{\mu}_{\lambda,i}^{odd}$, where in both cases $i$ has been reindexed to lie in the set $1, \ldots, n^*$. Then $\lambda$ is chosen to minimize the double cross-validation function

$$\hat{R}(\lambda) = \sum_{i=1}^{n^*} [(\hat{\mu}_{\lambda,i}^{even} - \tilde{y}_i^{odd})^2 + (\hat{\mu}_{\lambda,i}^{odd} - \tilde{y}_i^{even})^2], \tag{8.15}$$

where $\tilde{y}_i^{odd} = 0.5(y_{2i-1} + y_{2i+1})$ for $i = 1, \ldots, n^* - 1$, $\tilde{y}_{n^*}^{odd} = 0.5(y_1 + y_{n-1})$, and $\tilde{Y}^{even}$ is defined in an analogous fashion. Nason (1996) notes that the threshold $\lambda_{n^*}^{CV}$ that minimizes $\hat{R}(\cdot)$ would be appropriate for use based on a sample of size $n^*$ from the model in Eq. (8.13). He then proposed converting $\lambda_{n^*}^{CV}$ into a threshold $\lambda_n^{CV}$ appropriate for use on the full sample of size $n$ by the formula $\lambda_n^{CV} = \lambda_{n^*}^{CV}[1 - \log 2/\log n]^{-1/2}$. This suggestion, although somewhat heuristic, seems sensible provided that the optimal threshold for a sample of size $n$ is $\lambda_n = \lambda^* \sqrt{2 \log n}$ and that $\lambda^*$ is effectively determined as $\lambda^* = \lambda_{n^*}^{CV}/\sqrt{2 \log n^*}$.

The naive cost of computing $\hat{R}(\lambda)$ for all thresholds $\lambda$ is $O(n^2)$ operations. However, if hard thresholding is used the cost may be easily reduced to $O(n \log n)$ operations using the Fast Wavelet Transform. To verify this, we consider the wavelet decompositions of $Y^{even}$ and $Y^{odd}$ given by $w^{even} = \tilde{W}_{n^*} Y^{even}$, and $w^{odd} = \tilde{W}_{n^*} Y^{odd}$. The estimators $\hat{\boldsymbol{\mu}}_{\lambda}^{even}$ and $\hat{\boldsymbol{\mu}}_{\lambda}^{odd}$ are selective wavelet reconstructions of $Y^{even}$ and $Y^{odd}$, which include only those coefficients whose absolute value exceeds $\lambda$. The function $\hat{R}(\cdot)$ is therefore piecewise constant, with jumps at values of $\lambda$ which coincide with the absolute value of one of the entries of either $w^{even}$ or $w^{odd}$. Thus, $\hat{R}(\cdot)$ is completely determined once it has been evaluated at these $n$ values of $\lambda$. The complete evaluation of $\hat{R}(\cdot)$ directly from Eq. (8.15) would therefore require $O(n^2)$ computations, since the sum in Eq. (8.15) would be directly evaluated $n$ times, at a cost of $O(n)$ computations for each evaluation. Similarly, if it is only desired to evaluate $\hat{R}(\cdot)$ for a specific set of $O(n^{\alpha})$ thresholds, then the naive cost would be $O(n^{1+\alpha})$.

To further improve the cost of computing the function $\hat{R}(\cdot)$ to $O(n \log n)$ operations, we note that, by Parseval's relation,

$$\hat{R}(\lambda) = \sum_{j,l} (w_{j,l}^{even} I_{\{|w_{j,l}^{even}|>\lambda\}} - \tilde{w}_{j,l}^{odd})^2 + \sum_{j,l} (w_{j,l}^{odd} I_{\{|w_{j,l}^{odd}|>\lambda\}} - \tilde{w}_{j,l}^{even})^2, \quad (8.16)$$

where $\tilde{w}^{odd} = \tilde{W}_{n^*} \tilde{Y}^{odd}$, $\tilde{w}^{even} = \tilde{W}_{n^*} \tilde{Y}^{even}$, and $I_{\{\cdot\}}$ denotes an indicator function. If we form an $n$-dimensional vector consisting of all elements of $|w^{even}|$ and $|w^{odd}|$, and sort this vector from lowest to highest at a cost of $O(n \log n)$ operations, this yields the values $\lambda_1, \ldots, \lambda_n$. These are the sorted jump points of $\hat{M}(\cdot)$. For simplicity, assume that these values are all distinct. If $\lambda = 0$, all coefficients of both $w^{even}$ and $w^{odd}$ contribute to Eq. (8.16). If $\lambda = \lambda_1$, the nonzero coefficient of $w^{even}$ or $w^{odd}$ with the smallest absolute value is deleted from the sum. In general, if $\lambda = \lambda_m (m > 1)$, the sum in Eq. (8.16) is precisely as it was for $\lambda_{m-1}$ except for the deletion of one wavelet coefficient. That is,

$$\hat{R}(\lambda_m) = \hat{R}(\lambda_{m-1}) - (w_{j,l}^{even} - \tilde{w}_{j,l}^{odd})^2 + (0 - \tilde{w}_{j,l}^{odd})^2$$

if $\lambda_m = |w_{j,l}^{even}|$, and

$$\hat{R}(\lambda_m) = \hat{R}(\lambda_{m-1}) - (w_{j,l}^{odd} - \tilde{w}_{j,l}^{even})^2 + (0 - \tilde{w}_{j,l}^{even})^2$$

if $\lambda_m = |w_{j,l}^{odd}|$. Using these updating formulas, we can obtain $\hat{R}(\lambda_m)$ from $\hat{R}(\lambda_{m-1})$ in $O(1)$ operations, and obtain the entire set $\hat{R}(\lambda_1), \ldots, \hat{R}(\lambda_n)$ in $O(n) + O(n \log n) = O(n \log n)$ operations.

As we will see in the Monte Carlo study of Section 8.3.5, Nason's cross-validation method outperforms universal thresholding for all but one of the examples considered by Donoho and Johnstone (1994). This, in combination with its computational efficiency, makes cross-validation an attractive alternative to universal thresholding. One drawback, however, is that relatively little theory is currently available on its performance. Although Nason (1996) studied the concavity of the function $\hat{R}(\cdot)$, he did not establish any results on the mean integrated squared error of the thresholded estimator that minimizes $\hat{R}(\cdot)$. In the next Section we will discuss a method of threshold selection that is based on classical model selection ideas from univariate regression, and for which some theoretical results are available.

### *8.3.3. Cross-validatory AICc*

The use of any particular hard threshold $\lambda$ yields a specific data-determined subset model that, when fitted to the data $Y$ by least squares, produces the corresponding estimate $\hat{\mu}$. As the threshold is decreased, the set of wavelet coefficients used in the estimate increases in a hierarchically nested fashion. Thus, the set of all possible thresholds determines a nested sequence of $n$ candidate models. For a given data set we can therefore recast the problem of selecting the threshold as the problem of selecting the index, $k \in \{1, \ldots, n\}$, of this nested sequence of models. Here $k$ denotes both the number of parameters in the candidate linear regression model, and, implicitly, the particular data-determined set of variables which appear in this model.

Whereas the number of candidate models in selective wavelet reconstruction is $2^n - 1$, corresponding to all possible subsets, the number of candidate models in hard wavelet thresholding is just $n$. However, these $n$ subsets are determined by the data. Thus, wavelet thresholding does not rule out any of the $2^n - 1$ possible subsets *a priori*, but uses the data to reduce these to $n$ candidates.

Given the $n$ candidates, one may be tempted to choose $k$ using classical model selection techniques such as AIC (Akaike, 1973) or SIC (Schwarz, 1978). However, because the candidates were determined from the same set of data which is to be used to select $k$, we would not expect this to work well.

Hurvich and Tsai (1998) found a way around the problem by applying classical model selection criteria in a cross-validatory setting, as described in the following algorithm. The algorithm is symmetrical in its treatment of $Y^{even}$ and $Y^{odd}$, as will be seen in step 7.

1. Obtain the vectors $Y^{odd}$, $Y^{even}$, $w^{odd}$, and $w^{even}$, at a total cost of $O(n)$ operations.
2. Apply all possible thresholds to $w^{odd}$, thereby obtaining a nested sequence of $n^*$ models. Note that the candidate variables here are the columns of $\tilde{W}_{n^*}{}'$. The computational cost of this step is $O(n \log n)$, since the nested sequence can be readily obtained from a ranking of the absolute values of the elements of $w^{odd}$.
3. Compute the residual sums of squares $RSS^{even}(k)$ that would result from fitting the $n^*$ models determined above by $Y^{odd}$ to the independent replication $Y^{even}$. This fitting would yield $n^*$ selective wavelet reconstructions of $Y^{even}$, indexed by $k = 1, \ldots, n^*$, where $k$ is the number of elements of the wavelet coefficient vector $w^{even} = \tilde{W}_{n^*} Y^{even}$ retained in the reconstruction. Since the candidate models are nested and since the columns of $\tilde{W}_{n^*}{}'$ are orthonormal, we see that if we define $RSS^{even}(0) = (w^{even})'(w^{even})$, then for $k \geq 1$, $RSS^{even}(k)$ is equal to $RSS^{even}(k-1)$ minus the square of the element of $w^{even}$ for which the correspondingly indexed element of

$w^{odd}$ has the $k$'th largest absolute value. Thus, once the absolute values of the entries of $w^{odd}$ have been sorted, the sequence $\{RSS^{even}(k)\}_{k=1}^{n^*}$ can be computed in $O(n^*)$ operations. It is not necessary to actually compute the function estimates themselves, which would be considerably more expensive.

4. Use the residual sums of squares from step 3 to compute an information criterion for model $k = 1, \ldots, K$ where $K$ is the largest model dimension under consideration. Because $K$ may be an appreciable fraction of the sample size, we will use AICc (Hurvich and Tsai, 1989):

$$\text{AICc}^{even}(k) = n^* \log RSS^{even}(k) + 2(k+1)\frac{n^*}{n^* - k - 2}.$$

If AICc$^{even}$ is used, $K$ may be taken as large as $n^* - 3$, if desired. Let $\tilde{k}$ denote the model which minimizes AICc$^{even}$.

5. Fit the model $\tilde{k}$ to $Y^{even}$ by least squares. Denote the vector of fitted values by $\hat{\boldsymbol{\mu}}^{even}$. Although $\hat{\boldsymbol{\mu}}^{even}$ is a selective wavelet reconstruction of $Y^{even}$, this reconstruction is not necessarily thresholded, since the retained coefficients are not necessarily precisely those whose absolute value exceeds some threshold.

Using results from Hurvich and Tsai (1995) on model selection with non-stochastic candidate models, we can derive some theoretical properties of $\hat{\boldsymbol{\mu}}^{even}$ (details are given in Section 8.3.4). To obtain our results, we take advantage of the independence of $Y^{even}$ and $Y^{odd}$; that is, we exploit the cross-validatory nature of the method. The drawback in using $\hat{\boldsymbol{\mu}}^{even}$, however, is that the wavelet coefficients it uses were estimated on just half of the data, $Y^{even}$. The remaining steps of our algorithm yield a threshold suitable for use with all the data, $Y$.

6. Find the threshold $\hat{\lambda}^{even}$ that, when applied to $Y^{even}$, produces a function estimate that best approximates $\hat{\boldsymbol{\mu}}^{even}$ in the sense of the norm $\|\cdot\|$. This can be achieved in $O(n)$ operations as follows. Any threshold $\lambda$ applied to $Y^{even}$ produces the thresholded estimator

$$\hat{\boldsymbol{\mu}}_{\lambda}^{even} = \sum_{j,l} w_{j,l}^{even} I_{\{|w_{j,l}^{even}| > \lambda\}} W_{n^* jl}.$$

Furthermore, the estimator from step 5 can be expressed as

$$\hat{\boldsymbol{\mu}}^{even} = \sum_{j,l} w_{j,l}^{even} I_{\{(j,l)\in\delta\}} W_{n^* jl},$$

where $\hat{\delta}$ is a list of the $(j, l)$ pairs used in the selective wavelet reconstruction $\hat{\boldsymbol{\mu}}^{even}$. By Parseval's formula, we have

$$\|\hat{\boldsymbol{\mu}}_{\lambda}^{even} - \hat{\boldsymbol{\mu}}^{even}\|^2 = \sum_{j,l}[w_{j,l}^{even}(I_{\{|w_{j,l}^{even}|>\lambda\}} - I_{\{(j,l)\in\hat{\delta}\}})]^2. \tag{8.17}$$

By an argument similar to that given in Section 8.3.2 on the evaluation of Eq. (8.16), it follows that Eq. (8.17) can be evaluated for all $\lambda$, and the minimizing threshold $\lambda = \hat{\lambda}^{even}$ can thereby be determined in $O(n)$ operations, assuming that the absolute values of the entries of $w^{even}$ have been sorted.

7. Repeat steps 2-6 above, replacing "even" with "odd," to obtain $\hat{\lambda}^{odd}$.
8. Average the selected thresholds $\hat{\lambda}^{even}$ and $\hat{\lambda}^{odd}$, and then convert this into a threshold suitable for use on a sample of size $n$ using Nason's proposed extrapolation, $\hat{\lambda} = 0.5(\hat{\lambda}^{odd} + \hat{\lambda}^{even})(1 - \log 2/\log n)^{-1/2}$. Use this threshold on the full data vector $Y$ to compute the estimator

$$\hat{\boldsymbol{\mu}}_{\hat{\lambda}} = \sum_{j,l} w_{j,l} I_{\{|w_{j,l}|>\hat{\lambda}\}} W_{njl}. \tag{8.18}$$

We will examine the actual performance of $\hat{\boldsymbol{\mu}}_{\hat{\lambda}}$ in simulations in Section 8.3.5.

### 8.3.4. Properties of the AICc Selected Estimator

In order to evaluate the performance of a given selection criterion we need to measure how well it estimates $\boldsymbol{\mu}$. We will discuss several methods based on minimizing the mean integrated squared error.
Donoho
and Johnstone (1994) proposed

$$MISE_1(\lambda) = E[\|\hat{\boldsymbol{\mu}}_{\lambda} - \boldsymbol{\mu}\|^2]$$

to measure the quality of a thresholded estimator $\hat{\boldsymbol{\mu}}_{\lambda}$ based on $Y$, where the expectation is taken with respect to the full realization $Y$ and $\lambda$ is held fixed. Donoho and Johnstone obtained useful upper bounds for $MISE_1(\lambda_n)$ for certain deterministic threshold sequences, $\lambda_n$, and although these bounds have been very important for understanding the worst-case performance of universal and other methods of thresholding when $\sigma$ is known, they do not reveal how close $MISE_1(\lambda_n)$ comes to the best possible performance, $\min_{\lambda} MISE_1(\lambda)$.

We will begin by considering the universal threshold, $\lambda_n = \sigma\sqrt{2\log n} = \sigma\lambda_n^u$. However, the threshold used in practice will be random, since the unknown $\sigma$ must be estimated. Therefore we turn our attention to the feasible version of the universal threshold estimator, $\hat{\boldsymbol{\mu}}_{\tilde{\lambda}_n}$, where $\tilde{\lambda}_n = \tilde{\sigma}\sqrt{2\log n}$. Three reasonable measures of the quality of $\hat{\boldsymbol{\mu}}_{\tilde{\lambda}}$ are $MISE_1(\tilde{\lambda}_n)$, which is a random variable, $E[MISE_1(\tilde{\lambda}_n)]$, and $E[\|\hat{\boldsymbol{\mu}}_{\tilde{\lambda}_n} - \boldsymbol{\mu}\|^2]$.

Within the context of cross-validation there are other ways to measure the quality of a thresholded estimator, still based on mean integrated squared error. Let $\hat{\boldsymbol{\mu}}^{\lambda,even}$ be the selective wavelet reconstruction

$$\hat{\boldsymbol{\mu}}^{\lambda,even} = \sum_{j,l} w_{j,l}^{even} I_{\{|w_{j,l}^{odd}|>\lambda\}} W_{n^*jl},$$

using those entries of $w^{even}$ for which the corresponding entries of $|w^{odd}|$ exceed the threshold $\lambda$. The set of $n^*$ estimates of $\hat{\boldsymbol{\mu}}^{\lambda,even}$, obtained for a given sample by allowing $\lambda$ to vary, form the collection of selective wavelet reconstructions described in step 3 of the algorithm of the previous Section. Let $E^{even}$ and $E^{odd}$ denote expectations with respect to $Y^{even}$ and $Y^{odd}$, respectively. Thus Hurvich and Tsai (1998) defined the quality measure

$$MISE_2(\lambda) = E^{odd}\ E^{even}[\|\hat{\boldsymbol{\mu}}^{\lambda,even} - \boldsymbol{\mu}^{even}\|^2 \mid Y^{odd}],$$

where $\boldsymbol{\mu}^{even} = [\mu(t_2), \mu(t_4), \ldots, \mu(t_n)]'$. A convenient closed-form expression for $MISE_2(\lambda)$ can be obtained as follows. Noting that the wavelet coefficients of $Y^{even}$ and $Y^{odd}$ are given by

$$w_{j,l}^{even} = \theta_{j,l}^{even} + \eta_{j,l}^{even}, w_{j,l}^{odd} = \theta_{j,l}^{odd} + \eta_{j,l}^{odd},$$

where $\theta_{j,l}^{even}$ and $\theta_{j,l}^{odd}$ are the wavelet coefficients of $\boldsymbol{\mu}^{even}$ and $\boldsymbol{\mu}^{odd} = [\mu(t_1), \mu(t_3), \ldots, \mu(t_{n-1})]'$, respectively, and the $\eta_{j,l}^{even}$ and $\eta_{j,l}^{odd}$ are all independently identically distributed as $N(0, \sigma^2)$, Hurvich and Tsai (1998) used Parseval's rule to obtain

$$\begin{aligned}
MISE_2(\lambda) =& E^{odd}\sum_{j,l} E^{even}[(w_{j,l}^{even} I_{\{|w_{j,l}^{odd}|>\lambda\}} - \theta_{j,l}^{even})^2 \mid Y^{odd}] \\
=& E^{odd}\sum_{j,l}[\sigma^2 I_{\{|w_{j,l}^{odd}|>\lambda\}} + (\theta_{j,l}^{even})^2 I_{\{|w_{j,l}^{odd}|\le\lambda\}}] \\
=& \sum_{j,l}[\sigma^2 Pr\{|w_{j,l}^{odd}| > \lambda\} + (\theta_{j,l}^{even})^2 Pr\{|w_{j,l}^{odd}| \le \lambda\}] \\
=& \sum_{j,l}\{\sigma^2 - \sigma^2\Phi[(\lambda - \theta_{j,l}^{odd})/\sigma] + \sigma^2\Phi[(-\lambda - \theta_{j,l}^{odd})/\sigma] \\
&+ (\theta_{j,l}^{even})^2\Phi[(\lambda - \theta_{j,l}^{odd})/\sigma] - (\theta_{j,l}^{even})^2\Phi[(-\lambda - \theta_{j,l}^{odd})/\sigma]\},
\end{aligned}$$

where $\Phi(\cdot)$ is the cumulative distribution function of a standard normal random variable.

Still another quality measure is a function of $k$, the number of wavelet coefficients retained in the estimator. For any $k \in \{1, \ldots, n^*\}$, let $(j_1, l_1), \ldots, (j_k, l_k)$ be the indices corresponding to the $k$ largest entries of $|w^{odd}|$. Define $\hat{\boldsymbol{\mu}}^{even}(k) = \tilde{W}_{n^*}'(k) w^{even}(k)$, where $w^{even}(k) = (w^{even}_{j_1,l_1}, \ldots, w^{even}_{j_k,l_k})'$, and $\tilde{W}_{n^*}'(k)$ is an $n^* \times k$ matrix consisting of column $(j_1, l_1), \ldots, (j_k, l_k)$ of $\tilde{W}_{n^*}'$. Note that $\hat{\boldsymbol{\mu}}^{even}(k)$ is a selective wavelet reconstruction of $Y^{even}$, which retains $k$ elements of the wavelet coefficient vector $w^{even} = \tilde{W}_{n^*}' Y^{even}$, and the choice of the particular $k$ coefficients to be retained is determined by thresholding $w^{odd}$. Each such thresholding choice determines a candidate model, which can be identified with $k$ or with the columns of $\tilde{W}_{n^*}'(k)$. The candidate model is then fitted to the independent replication $Y^{even}$, as described in step 3 of Section 8.3.3. Let $J^{odd}_{n^*}$ denote the class of candidate models determined by $Y^{odd}$, as $k$ ranges from 1 to $K$. A quality measure for these candidate models is given by

$$L_{n^*}(k) = E^{even}[\|\hat{\boldsymbol{\mu}}^{even}(k) - \boldsymbol{\mu}^{even}\|^2 \mid Y^{odd}],$$

a random variable depending of $Y^{odd}$.

We can also use a model selection criterion whose performance is asymptotically equivalent to that of AICc:

$$S^{even}_{n^*}(k) = \{n^* + 2k\} RSS^{even}(k)/n^* = \{n^* + 2k\} \|Y^{even} - \hat{\boldsymbol{\mu}}^{even}(k)\|^2 / n^*.$$

Equivalence can be established using Shibata (1980), Theorem 4.2, and Shibata (1981), Section 5. In order to study the rate at which our proposed method attains the mean integrated squared error, we make the following assumptions:

**Assumption 1:** $\max_{k \in J^{odd}_{n^*}}(k) = o_p((n^*)^a)$ for some constant $a$ with $0 < a \leq 1$.

**Assumption 2:** There exists $b$ with $0 \leq b < 1/2$ such that for any $\delta \in (0, 1)$

$$E^{odd} \sum_{k \in J^{odd}_{n^*}} \delta^{L_{n^*}(k)/(n^*)^{2b}} \to 0.$$

Since Assumption 2 implies that

$$E^{odd} \min_{k \in J^{odd}_{n^*}} L_{n^*}(k)/(n^*)^{2b} \to \infty,$$

it is seen that the larger the value of $b$, the faster the expectation of the smallest $L_{n^*}(k)$ is guaranteed to diverge to infinity. Let $\hat{k}$ be the model selected from $J_{n^*}^{odd}$ by minimizing $S_{n^*}^{even}(k)$, and let $k^{*(n^*)}$ be the element of $J_{n^*}^{odd}$ which minimizes $L_{n^*}(k)$. The following Theorem provides the relative rate of convergence of the mean integrated squared error of the selected estimator compared to the best possible estimator in the class of candidates under consideration. Hurvich and Tsai (1998) apply the same techniques used in Shibata's (1981) Theorem 2.2 to prove the special case of the theorem with $a = 1, b = 0$. The proof for general $a$ and $b$ can then be obtained straightforwardly from the techniques used in the proof of the special case and the proof of Hurvich and Tsai's (1995) Theorem.

**Theorem 8.1:**

If Assumptions 1 and 2 are satisfied, then

$$L_{n^*}(\hat{k})/L_{n^*}(k^{*(n^*)}) - 1 = o_p(n^{-c}),$$

where $c = \min\{(1-a)/2, b\}$.

*8.3.5. Wavelet Monte Carlo Study*

In this Section we investigate the performance of various criteria for selecting a hard wavelet threshold. The criteria we will consider are Nason's cross-validation CV (see Section 8.3.2), SIC, Donoho and Johnstone's universal thresholding (UNIV), based on a threshold of $\tilde{\sigma}\sqrt{2\log n}$, and our proposed AICc. The AICc estimate of $\boldsymbol{\mu}$ is given by Eq. (8.18), and the SIC estimate is given by the modification of Eq. (8.18) that results from replacing AICc$^{even}$ by

$$\text{SIC}^{even}(k) = n^* \log RSS^{even}(k) + k \log n^*$$

in step 4 of the algorithm in Section 8.3.3. The criteria were tested against four signals: Blocks, Bumps, HeaviSine and Doppler (described in detail by Donoho and Johnstone, 1994), with a sample size of $n = 2048$. Simulations were conducted at two different signal-to-noise ratios (SNR), 7 and 3. For each pairing of signal and SNR, one hundred simulated realizations were generated by multiplicatively scaling the signal so that its sample standard deviation was equal to the SNR, and then adding 2048 simulated independent identically distributed standard normal random variables. On each simulated realization the four criteria were used to select a threshold, and for each criterion the integrated squared error was computed as $ISE = n^{-1}\|\hat{\boldsymbol{\mu}} - \boldsymbol{\mu}\|$, where $\hat{\boldsymbol{\mu}}$ is the thresholded estimate selected by the criterion. Wavelet thresholding was carried out in S-Plus using the Wavethresh software (Nason and Silverman, 1994), Distribution 2.2. We used the default settings, yielding the $n^* = 2$

Daubechies compactly-supported wavelet from the DaubExPhase family, with periodic boundary handling. Thresholds were set manually.

Table 8.8 gives averages of the one hundred $ISE$ values for each criterion and process. CV and AICc perform similarly, almost always outperforming SIC and UNIV, where the superiority is somewhat more noticeable at $SNR = 3$ than at $SNR = 7$. The HeaviSine case is the exception; here the UNIV method outperforms the others. For all other processes, however, AICc significantly outperforms UNIV, as measured by a two-sided Wilcoxon signed rank test on the differences of the $ISEs$ for each realization ($p - value < 10^{-8}$). The ratio of the average integrated squared error for UNIV to that for AICc is 1.23 for the Blocks and Bumps signals when $SNR = 3$. For the remaining situations except for the HeaviSine process, the ratio is at least 1.12.

Table 8.8. Average $ISE$ values of hard threshold estimators. $n = 2048$.

| Process | SNR | CV | AICc | SIC | UNIV |
|---|---|---|---|---|---|
| Blocks | 7 | 0.165 | 0.169 | 0.181 | 0.190 |
| Blocks | 3 | 0.189 | 0.189 | 0.252 | 0.232 |
| Bumps | 7 | 0.241 | 0.249 | 0.244 | 0.286 |
| Bumps | 3 | 0.234 | 0.235 | 0.291 | 0.289 |
| HeaviSine | 7 | 0.0765 | 0.0772 | 0.0801 | 0.0789 |
| HeaviSine | 3 | 0.0702 | 0.0688 | 0.0819 | 0.0684 |
| Doppler | 7 | 0.228 | 0.229 | 0.255 | 0.257 |
| Doppler | 3 | 0.164 | 0.165 | 0.188 | 0.186 |

## 8.4. Summary

Chapter 8 extends the theme of Chapter 7 by further considering the cases where standard regression assumptions may be violated. However, unlike earlier chapters, instead of selecting a model we select a smoothing parameter. In some cases, the function relating $x$ and $y$ may not be linear, or more importantly, unknown. The errors are still additive with unknown (or known) distribution. Nonparametric regression is one solution for studying the relationship between $x$ and $y$, where the scatter plot between $x$ and $y$ is smoothed, and the task is to choose the smoothing parameter. We have seen in this Chapter that AICc is competitive with existing smoothing parameter selection criteria in this regard. Further extensions of AICc to categorical data analysis and density estimation can be found in Simonoff (1998).

Semiparametric models are in-between parametric and nonparametric regression having components of both. Part of the model is linear (parametric) and part is some general function (nonparametric). Again the errors are additive with unknown (or known) distribution. AIC and AICc have been adapted for use in semiparametric modeling, and simulation studies show that AICc performs very well.

Sometimes it is more convenient to think of the model as a signal embedded in noise. The usual regression model can be thought of as a special case where the signal is $X\beta$ and the noise is the additive $\varepsilon$. A signal embedded in noise also encompasses semiparametric and nonparametric regression models. wavelets are a new and useful method for recovering this signal. A computationally efficient cross-validatory version of AICc has been devised for wavelets. Simulation studies indicate that the cross-validation algorithm outperforms universal thresholding.

## Chapter 9
## Simulations and Examples

### 9.1. Introduction

In this Chapter we will compare the relative performance of 16 model selection criteria for univariate regression, autoregressive models, and moving average models, and for 18 selection criteria for multivariate regression and vector autoregressive models. In each case we will start by focusing on the effect of parameter structure and ease of model identification, and then broaden the scope via large-scale simulation studies in which many factors are widely varied. Large sample and real data examples are also discussed. Since our regression models include the intercept and our time series models do not, we have adopted the following notation. For regression there are $k$ variables in the model, which includes the intercept $\beta_0$. In our time series models $p$ represents both the order of the model as well as the number of variables included in the model. Both regressive models and autoregressive time series models will be referred to by their orders.

As before, we will use K-L and $L_2$ observed efficiencies, average rank based on observed efficiencies, and in some cases, counts of correct model choice, to measure performance. We found in previous chapters that when the parameter structure of a given model is weak, counts are not a useful measure of performance. Therefore, in such cases for this Chapter, observed efficiency and rank results will be emphasized. One problem with using rank as a measure of performance is distinguishing true differences in performance between ranks. In order to identify whether differences in rank reflect true differences in performance, we have developed a nonparametric test of selection criterion performance based on rank that is quite simple to apply (see Section 9.1.3). Its results indicate that the selection criteria we consider often cluster into groups with nearly identical performance, which we shall be able to see when criteria are listed in rank order for each simulation. Unlike previous chapters, here rank values are based on the results of this test. The tables in Chapter 9 summarize results by rank as well by counts for the true model (where appropriate), given in the "true" column. We have observed that K-L tends to penalize underfitting much more severely than overfitting while $L_2$ does the reverse. A good selection criteria should neither overfit nor underfit excessively. In other words, a good criterion should perform well in both K-L and $L_2$. To reflect this idea, the criteria are sorted on the basis of the sum of their K-L and $L_2$ rankings. Where $L_2$ is not a scalar the criteria are sorted based on the sum of their K-L and $\mathrm{tr}\{L_2\}$ rankings. Counts for overfitting and underfitting are summarized, but the details can be found in Appendix 9A.

In the last few decades the stepwise regression procedure has been widely used in variable selections. However, since this procedure is not compatible with the model selection criteria discussed in this Chapter, we only present its performance in Appendix 9B with respect to the F-test at three different levels of significance.

### *9.1.1. Univariate Criteria List*

In this Chapter we consider not only the criteria we have covered earlier in this book, but also several new criteria not previously discussed in detail. We have previously discussed the classic efficient and consistent criteria: AIC, Eq. (2.13) and Eq. (3.9), AICc, Eq. (2.14) and Eq. (3.10), SIC, Eq. (2.15) and Eq. (3.15), HQ, Eq. (2.16) and Eq. (3.16), Mallows's Cp, Eq. (2.12) and Eq. (3.14), and FPE, Eq. (2.11) and Eq. (3.12). We have also discussed signal-to-noise adjusted variants AICu, Eq. (2.18) and Eq. (3.11), HQc, Eq. (2.21) and Eq. (3.17), and FPEu, Eq. (2.20) and Eq. (3.13), as well as the withhold-1 cross-validation CV, Eq. (6.3) and Eq. (6.6), and bootstrap criteria DCVB, Eq. (6.22) and Eq. (6.26) and BFPE, Eq. (6.21) and Eq. (6.25). The new criteria are listed below.

The efficient criterion Rp $= s_k^2(n-1)/(n-k)$ (Breiman and Freedman, 1983). Rp is derived under the assumption that the true model is of infinite order and that $X$ is random. Rp is similar to Shibata's (1980) Sp criterion (not included in these studies).

The criterion FPE4 $= \hat{\sigma}^2(1+4k/(n-k)) = \hat{\sigma}^2(n+3k)/(n-k)$. This criterion is a variant of Bhansali and Downham's (1977) FPE$\alpha$ with $\alpha = 4$, and thus has an asymptotically much smaller chance of overfitting than FPE (which is FPE$\alpha$ with $\alpha = 2$).

The consistent criterion GM (Geweke and Meese, 1981) is a variant of Mallows's Cp. GM $= \mathrm{SSE}/s_K^2 + \log(n)k$, and is asymptotically equivalent to SIC. If the signal-to-noise ratio for GM using techniques discussed in Chapter 2 were computed, we would see that its small-sample signal-to-noise ratio is larger than that for SIC. This means that GM should overfit less than SIC in small samples.

The criterion $R^2_{adj}$ in Eq. (2.17) chooses the best model when $R^2_{adj}$ attains a maximum (when $s^2$ is minimized). While $R^2_{adj}$ almost never underfits, it is prone to excessive overfitting.

### 9.1.2. Multivariate Criteria List

We need to make some changes to the list of selection criteria we will use for multivariate regression and vector autoregressive models. Some of the criteria used for univariate models are absent here, such as GM, FPE4, Rp, or $R^2_{adj}$. Since we cannot compute the true signal-to-noise ratio for FPE under multivariate regression, we cannot derive a signal-to-noise corrected variant. Also, since not all criteria remain a scalar under these circumstances, we present the determinant of FPE, Eq. (4.11) and Eq. (5.9), trFPE (the trace of MFPE), deCV Eq. (6.8) and Eq. (6.13), trCV Eq. (6.9) and Eq. (6.14), deBFPE, Eq. (6.31) and Eq. (6.36), trBFPE, Eq. (6.32) and Eq. (6.37), deDCVB, Eq. (6.33) and Eq. (6.38), trDCVB, Eq. (6.34) and Eq. (6.39), trCp Eq. (4.13), and the maximum eigenvalue of Cp Eq. (4.12), meCp. Cp can be defined for bivariate vector autoregressive models as

$$\mathrm{Cp} = (n - 3P)\hat{\Sigma}_P^{-1}\hat{\Sigma}_p + (5p - n).$$

TrCp is the trace of Cp and meCp is the maximum eigenvalue of this Cp.

However, not all criteria for multivariate models are matrices. Criteria such as AIC are scalars and there is no confusion as to what form to use. Thus we can include some classic criteria such as AIC, Eq. (4.14) and Eq. (5.10), AICc, Eq. (4.15) and Eq. (5.11), SIC, Eq. (4.16) and Eq. (5.13), and HQ, Eq. (4.17) and Eq. (5.14). Signal-to-noise variants AICu, Eq. (4.18) and Eq. (5.12), and HQc, Eq. (4.20) and Eq. (5.15), are also included.

Two additional criteria designed for multiple responses will be considered: ICOMP (Bozdogan, 1990) and the consistent FIC (Wei, 1992).

$$\begin{aligned}\mathrm{ICOMP} =& \frac{1}{2}(n+k)\left\{q * \log\left(\frac{tr\{\hat{\Sigma}_k\}}{q}\right) - \log|\hat{\Sigma}_k|\right\} \\ &+ \frac{1}{2}q\left\{k \log\left(\frac{tr\{(X'X)^{-1}\}}{k}\right) + \log|X'X|\right\},\end{aligned}$$

and FIC, scaled by the sample size, is

$$\mathrm{FIC} = |\hat{\Sigma}_k| + |\hat{\Sigma}_K|\left(-k\log|\hat{\Sigma}_k| + q\log|X'X|\right)/n.$$

ICOMP balances the variance $\hat{\Sigma}_k$ with the complexity of the model, $X'X$. ICOMP does not specialize to univariate regression. In univariate regression, the error variance cancels and ICOMP becomes a function of $X'X$ only. FIC is consistent and should behave similarly to SIC, at least in large samples.

### *9.1.3. Nonparametric Rank Test for Criteria Comparison*

Consider the case where two selection criteria are compared, criterion A and criterion B. For any realization, there are three outcomes: observed efficiency of A > observed efficiency B (rank of A = 1); observed efficiency A = observed efficiency B (rank A = 1.5); observed efficiency A < observed efficiency B (rank A = 2). If the two criteria are similar, then the average rank of A should be 1.5, and if A performs better than B then rank A < 1.5. Suppose that there are n independent realizations. Then for A, there are $c1$ rank = 1 cases, $c2$ rank = 1.5 cases and $c3$ rank = 2 cases such that $c1 + c2 + c3 = N$. The average rank of A is $\bar{r} = (c1+1.5c2+2c2)/N$. Assume that the two criteria perform the same, or that the null hypothesis is $H_0$: criterion A performs the same as criterion B. Then the distribution of the ranks $r$ is a multinomial distribution with $P\{r = 1\} = (1-\pi)/2$, $P\{r = 1.5\} = \pi$, and $P\{r = 2\} = (1-\pi)/2$ and $\pi$ unknown. Under $H_0$, the expected average rank is 1.5 with variance $(1-\pi)/4$. The probability $\pi$ can be estimated by $c2/N$, yielding the estimated variance $v = (1 - c2/N)/4$. Let

$$z = \sqrt{N}(\bar{r} - 1.5)/\sqrt{v}. \qquad (9.1)$$

If the number of realizations is large enough (there are $N = 54,000$ realizations in the univariate regression study), then $z \sim N(0,1)$ under $H_0$. A large, negative $z$ value indicates that A outperforms B. Clusters are formed from the following relationship: For selection criteria A, B and C, if A = B and B = C, then A = C regardless of the test results comparing A and C. Although this weakens the power of the test, it does cluster the criteria that have similar performance.

Each of the remaining sections will be structured as follows. First we present true and candidate model structures, then we evaluate special case models, large-scale small-sample simulations, and large-sample simulations. Finally, we give real data examples (except for Section 9.4). The number of replications for the special case models is 10,000, for the large-scale small-sample simulations it is 100, and for the large-sample simulations it is 1,000.

## 9.2. Univariate Regression Models

### 9.2.1. Model Structure

Consider regression models of the form

$$y_i = \beta_0 + \beta_1 x_{i,1} + \cdots + \beta_{k_*-1} x_{i,k_*-1} + \varepsilon_{*i}, \quad \varepsilon_{*i} \sim N(0, \sigma_*^2), \quad i = 1, \ldots, n, \quad (9.2)$$

where the $\varepsilon_{*i}$ are independent. Regression models in this Section will be created by varying number of observations $n$, error variance $\sigma_*^2$, parameter structure $\beta_j$, true model order $k_*$, level of overfitting $o$, and degree of correlation between the columns of $X$, $\rho_x$. Candidate models of the form Eq. (2.3) and Eq. (2.4) are fit to the data. K-L and $L_2$ observed efficiencies are as defined as Eq. (2.10) in Eq. (1.2) and Eq. (2.9) in Eq. (1.1), respectively. Higher observed efficiency denotes selection of a model closer to the true model and thus better performance.

### 9.2.2. Special Case Models

Special case Models 1 and 2 given in Eqs. (2.28)–(2.30) from Chapter 2 are again considered here for the sixteen univariate model selection criteria in this Section. We recall that for both models, $n = 25$, $\sigma_*^2 = 1$, $k_* = 6$, $\rho_x = 0$, and the structures of these two models are:

**Model 1**

$$y_i = 1 + x_{i,1} + x_{i,2} + x_{i,3} + x_{i,4} + x_{i,5} + \varepsilon_{*i}$$

and

**Model 2**

$$y_i = 1 + x_{i,1} + \frac{1}{2} x_{i,2} + \frac{1}{3} x_{i,3} + \frac{1}{4} x_{i,4} + \frac{1}{5} x_{i,5} + \varepsilon_{*i}.$$

Table 9.1 summarizes the counts of true model selection, underfitting and overfitting, K-L observed efficiency rank, and $L_2$ observed efficiency rank for each criterion. Detailed tables for counts by model order, distance measure, and observed efficiencies can be found Table 9A.1, Appendix 9A.

We see from Table 9.1 that the top five performers are AICu, HQc, AICc, GM, and FPEu, and that the observed efficiency results parallel the counts. AICu and HQc, the criteria with the highest counts, also have the highest Kullback–Leibler and $L_2$ observed efficiencies. We also note that even the bootstrap criteria with stronger penalty functions are prone to overfitting, but still perform in the top half. The choice of $\alpha$ for FPE4 is based on asymptotic probabilities of overfitting. In small samples, these arguments may not hold, and in fact it performs near the lower middle. FPE4 has a structure similar to that of FPE, but performs better than FPE due to its larger penalty function. Cross-validation gives a disappointing performance here as well, but none of the criteria perform as poorly as $R^2_{adj}$. $R^2_{adj}$ has the weakest penalty function of all the criteria in this list and consequently overfits the most.

For Model 2 the true model is only weakly identifiable, and thus we expect more underfitting with respect to the true model order to be present. Table 9.2

Table 9.1. Simulation results summary for Model 1.
K-L observed efficiency ranks, $L_2$ observed efficiency ranks and counts.

| criterion | K-L ranking | $L_2$ ranking | true | underfitting | overfitting |
|---|---|---|---|---|---|
| AICu | 1 | 1 | 6509 | 1317 | 1795 |
| HQc | 2 | 2 | 6243 | 830 | 2535 |
| AICc | 3 | 3 | 5875 | 595 | 3165 |
| GM | 4 | 4 | 5686 | 702 | 3267 |
| FPEu | 5 | 5 | 4748 | 378 | 4602 |
| DCVB | 6 | 6 | 4378 | 812 | 4338 |
| SIC | 7 | 7 | 4307 | 328 | 5124 |
| BFPE | 8 | 8 | 3905 | 368 | 5392 |
| Cp | 9 | 9 | 3925 | 224 | 5622 |
| FPE4 | 9 | 9 | 4033 | 289 | 5454 |
| Rp | 11 | 11 | 3553 | 175 | 6067 |
| CV | 12 | 12 | 3187 | 210 | 6335 |
| HQ | 13 | 13 | 2881 | 132 | 6828 |
| FPE | 14 | 14 | 2585 | 92 | 7185 |
| AIC | 15 | 15 | 2338 | 83 | 7454 |
| $R^2_{adj}$ | 16 | 16 | 1202 | 18 | 8720 |

Table 9.2. Simulation results summary for Model 2
K-L observed efficiency ranks, $L_2$ observed efficiency ranks and counts.

| criterion | K-L ranking | $L_2$ ranking | true | underfitting | overfitting |
|---|---|---|---|---|---|
| HQc | 2 | 1 | 90 | 9002 | 998 |
| AICu | 1 | 3 | 48 | 9490 | 510 |
| AICc | 4 | 1 | 120 | 8584 | 1416 |
| GM | 3 | 4 | 82 | 8779 | 1221 |
| DCVB | 4 | 4 | 103 | 8412 | 1588 |
| FPEu | 6 | 4 | 149 | 7880 | 2120 |
| SIC | 7 | 4 | 156 | 7636 | 2364 |
| BFPE | 8 | 4 | 153 | 7539 | 2461 |
| FPE4 | 9 | 4 | 155 | 7427 | 2573 |
| Cp | 10 | 4 | 180 | 6827 | 3173 |
| Rp | 11 | 4 | 203 | 6294 | 3706 |
| CV | 11 | 4 | 185 | 6045 | 3955 |
| HQ | 13 | 13 | 195 | 5672 | 4328 |
| FPE | 14 | 14 | 207 | 4965 | 5035 |
| AIC | 15 | 15 | 201 | 4666 | 5334 |
| $R^2_{adj}$ | 16 | 16 | 179 | 2408 | 7592 |

summarizes counts, the K-L and $L_2$ rankings for Model 2, and Table 9A.2 gives the detailed results.

In Table 9.2, we see that four of the top five performers in Model 2 are the same as those for Model 1—are HQc, AICu, AICc, GM, and that DCVB has replaced FPEu. With a weakly identifiable true model, none of the criteria select the true model much more than 2% of the time. Even with a weakly identifiable model, overfitting remains a strong concern, and we see that all the criteria both underfit and overfit. The criteria with weak penalty functions continue to overfit excessively; for example, $R^2_{adj}$.

### 9.2.3. Large-scale Small-sample Simulations

Consider models of the form given by Eq. (9.2). In this Section we vary $n$, $\sigma_*^2$, $\beta_j$, and $k_*$, as shown in Table 9.3. All combinations are considered, resulting in $5 \times 3 \times 3 \times 2 \times 2 \times 3 = 540$ models. Table 9.4 summarizes the relationship between parameter structure and true model order.

Table 9.3. Summary of the regression models in simulation study.

| Sample Size $n$ | Error Variance $\sigma_*^2$ | Parameter structure $\beta_j$ | True Order $k_*$ | Overfitting $o$ | $\rho_x$ |
|---|---|---|---|---|---|
| 15 | 0.1, 1, 10 | $1/j^2$, $1/j$, 1 | 3, 6 | 2, 5 | 0, 0.4, 0.9 |
| 25 | 0.1, 1, 10 | $1/j^2$, $1/j$, 1 | 3, 6 | 2, 5 | 0, 0.4, 0.9 |
| 35 | 0.1, 1, 10 | $1/j^2$, $1/j$, 1 | 3, 6 | 2, 5 | 0, 0.4, 0.9 |
| 50 | 0.1, 1, 10 | $1/j^2$, $1/j$, 1 | 3, 6 | 2, 5 | 0, 0.4, 0.9 |
| 100 | 0.1, 1, 10 | $1/j^2$, $1/j$, 1 | 3, 6 | 2, 5 | 0, 0.4, 0.9 |

Table 9.4. Relationship between parameter structure and true order.

Parameter structure 1: $\beta_j = 1/j^2$
- $k_* = 3$ $\beta_0 = 1, \beta_1 = 1, \beta_2 = 1/4$
- $k_* = 6$ $\beta_0 = 1, \beta_1 = 1, \beta_2 = 1/4, \beta_3 = 1/9, \beta_4 = 1/16, \beta_5 = 1/25$

Parameter structure 2: $\beta_j = 1/j$
- $k_* = 3$ $\beta_0 = 1, \beta_1 = 1, \beta_2 = 1/2$
- $k_* = 6$ $\beta_0 = 1, \beta_1 = 1, \beta_2 = 1/2, \beta_3 = 1/3, \beta_4 = 1/4, \beta_5 = 1/5$

Parameter structure 3: $\beta_j = 1$
- $k_* = 3$ $\beta_0 = 1, \beta_1 = 1, \beta_2 = 1$
- $k_* = 6$ $\beta_0 = 1, \beta_1 = 1, \beta_2 = 1, \beta_3 = 1, \beta_4 = 1, \beta_5 = 1$

Unlike the examples involving the special case models, which focus on the effect of model identifiability, we have chosen the parameter levels in Tables 9.3 and 9.4 to represent a wide range of values. This will allow us to observe the behavior of the criteria under a variety of conditions. For example, the sample sizes represent a range from small ($n = 15$) to moderate ($n = 100$).

The ease with which the correct model can be identified depends on the size of the smallest nonzero noncentrality parameter, which in turn is a function of $X_*\beta_*$ and $\sigma_*^2$. In general, the larger this noncentrality parameter, the easier it is to identify the correct model. Larger $X_*\beta_*$ and smaller $\sigma_*^2$ will increase model identifiability. Let $\mathrm{r}_*^2 = var[X_*\beta_*]/(var[X_*\beta_*] + \sigma_*^2)$ when $X$ is random. Typically, models with low $\mathrm{r}_*^2$ are more difficult to identify than models with high $\mathrm{r}_*^2$. Therefore, $\sigma_*^2 = 0.1$ represents easily identified models where the errors contribute little to the variability in $Y$ compared to $X_*\beta_*$, whereas $\sigma_*^2 = 10$ represents models that may be difficult to detect, where the variability in $Y$ is mostly due to the errors. The ease with which the true model can be identified also depends on the $\beta_j$ parameters; for example, $\beta_j = 1/j^2$ represents models where the relative strength of the $\|\beta_j x_j\|$ decreases rapidly. In a testing context, $\beta_{k_*-1} = 1/(k_* - 1)^2$ should be difficult to detect due to its small value (the true order of the model may appear to be less than $k_*$). Structure $\beta_j = 1/j$ represents a moderately weak model, and $\beta_j = 1$ represents models that should be easy to detect. For computational convenience, the $x_{i,j} \sim N(0,1)$ with $x_{i,0} = 1$. However, some correlation between the columns of $X$ is included. Correlated $x_{i,j}$ are generated by letting the pairs $(x_{i,j}, x_{i,j+1})$ be bivariate normals with correlation $\rho_x$ and $N(0,1)$ marginals. The columns of $X$ are generated by conditioning on the previous column. Let $x_{i,1} \sim N(0,1)$ and generate $x_{i,j+1}|x_{i,j}$ for $j = 1, \ldots, k_* - 2$. A value of $\rho_x = 0$ represents independent columns, $\rho_x = 0.4$ represents moderate collinearity, and $\rho_x = 0.9$ represents strong collinearity. Because overfitting also plays a role in model selection performance, we consider a small opportunity to overfit, $o = 2$, and a larger opportunity to overfit, $o = 5$. The total number of variables in the model including the intercept is $K = k_* + o$. $K$ is the dimension of the largest model.

One hundred realizations were generated for each of the 540 individual models. For each realization the criteria select one of the candidate models, and the $L_2$ and K-L distance are computed for the selected model. The observed efficiency for each chosen model is then computed from these distances and compared to those of all the other criteria, and rank 1 is awarded to the selection criterion with the highest observed efficiency. Ties are frequent, since the selection criteria often select the same model. Ties receive the average rank for that trial. An overall average rank from all models and realizations is then computed, and these ranks are summarized in Tables 9A.3 and 9A.4.

In many cases the average ranks are nearly the same. Due to the large number of realizations over all models (54,000), results from the test defined in Eq. (9.1) at the $\alpha = 0.05$ level are used to determine whether any true difference in performance exists, and forming clusters of criteria that perform similarly under K-L and $L_2$. Final overall performance is determined by pairwise comparisons. All pairwise comparisons involving AICu indicated that AICu had higher observed efficiency. Since none of the other criteria tested equivalent to or better than AICu, AICu received rank 1. All pairwise comparisons between GM and the other criteria indicated that only AICu beat or tied GM. Hence GM received rank 2. HQ, Rp, and CV formed a cluster in the K-L comparisons. Since ten criteria outperformed this cluster, HQ, Rp, and CV each receive rank 11 in the K-L observed efficiency rankings. Table 9.5 summarizes the rankings of the criteria over the 540 regression models, sorted on the basis of the sum of their K-L and $L_2$ ranks.

Table 9.5. Simulation results over 540 models.
Summary of overall rank by K-L and $L_2$ observed efficiency.

| criterion | K-L ranking | $L_2$ ranking |
|---|---|---|
| AICu | 1 | 1 |
| GM | 2 | 2 |
| HQc | 3 | 3 |
| FPEu | 5 | 4 |
| SIC | 5 | 5 |
| FPE4 | 5 | 5 |
| DCVB | 4 | 7 |
| AICc | 5 | 8 |
| BFPE | 9 | 9 |
| Cp | 10 | 10 |
| HQ | 11 | 11 |
| Rp | 11 | 12 |
| CV | 11 | 13 |
| FPE | 14 | 14 |
| AIC | 15 | 15 |
| $R^2_{adj}$ | 16 | 16 |

We see from Table 9.5 that AICc, SIC, FPEu and FPE4 form a cluster, all performing equivalently under K-L. Since the best any one of them could do is rank 5, each of the four is assigned rank 5. Due the combined K-L and $L_2$ rank sorting, AICc is presented further down the list than the other three in its cluster. We also see from Table 9.5 that the $L_2$ results are very similar to those for K-L. This may be due to the wide variety of models considered. The criteria ranked 1–5 (there are six of them due to a tie) are AICu, GM, HQc, FPEu, and SIC and FPE4 (tied). We see that the distinction between consistency and efficiency is less important than small-sample signal-to-noise ratios, and as such, the doubly cross-validated bootstrap performs in the top half, while $R^2_{adj}$ performs worst. While Mallows's Cp does not perform well strictly on the basis of model selection, its strength lies in its ability to select models with good predictive ability. Although a good predictive model may not be the closest possible to the true model, it is certainly worth further investigation. However, CV's comparative observed efficiency is necessarily lowered since selecting the model closest to the true model is not the purpose for which it is best suited.

In general, the efficient criteria with weak penalty functions in small samples, such as Rp and AIC, tend to perform poorly overall due to overfitting (for which the results are not presented here). The criterion with the worst tendency to overfit is $R^2_{adj}$. The consistent criteria tend to perform better due to their larger penalty functions. Even in small samples, consistent criteria tend to have larger penalty functions than the efficient criteria. Our signal-to-noise corrected variants AICu, HQc and FPEu all have large penalty functions and all perform well over the 540 models. The penalty weighted bootstraps perform near the middle. DCVB performs the best of the data resampling criteria.

It is important to keep in mind that the above results cover many models and realizations. For any given model or realization, any of the top five could perform poorly. Tables 9A.3 and 9A.4 summarize the observed efficiency rankings for each realization. Since many of the criteria select the same model, ties are common. These ties are assigned their average rank. It is difficult to see patterns in these tables due to the large number of middling ranks, but two columns are of particular interest—the counts of realizations when a criterion was assigned Rank 1 (best), and the count of realizations for Rank 16 (worst). We see that even the top performing criteria sometimes finish last, and even the worst overall performer, $R^2_{adj}$, can perform well. Indeed, $R^2_{adj}$ has the highest counts for rank 1! Unfortunately, this is offset by its highest counts for rank 16. Since we do not know when a particular criterion will perform well or poorly, this suggests that rather than selecting models on the basis of one criterion alone, it may be a more sound strategy to use criteria under both K-L and $L_2$ and compare the selected models carefully. If the criteria select different models, then more time should be spent investigating the candidates.

Some general trends can be seen over the 540 models with respect to the parameters we have varied. As $n$ increases, observed efficiency increases. As $\sigma^2_*$ increases, model identifiability and $r^2_*$ decrease, and thus observed efficiency decreases. Correlated $X$ decreases observed efficiency, as does the number of irrelevant variables $o$. As the true model order $k_*$ increases, observed efficiency decreases. This, along with the effect of $o$, leads to the sensible conclusion that experiments with small numbers of variables are easier to work with than complicated experiments. Observed efficiency increases as the $\beta_j$ increase. In actuality, observed efficiency should increase as the $\|\beta_j x_j\|$ increases. We expect that as model identifiability increases (large noncentrality parameters for all variables), observed efficiency increases, but we must keep in mind that observed efficiency also depends on the number of variables involved in the experiment.

### *9.2.4. Large-sample Simulations*

This simulation study demonstrates the effect of irrelevant variables on small-sample model selection when the true model belongs to the set of candidate models. The models used here, A1 and A2, differ only in the number of extraneous variables and hence in the opportunity for overfitting. Model A1 presents few opportunities for overfitting, $o = 2$, and Model A2 presents greater opportunities for overfitting, with $o = 5$. For both models $n = 25{,}000$, $k_* = 2$, $\beta_0 = 1$, $\beta_1 = 1$ and $\sigma_*^2 = 1$. Also, by virtue of the large sample size, we will be able to demonstrate the asymptotic equivalence of many of the criteria, particularly the efficient ones.

That the true model belongs to the set of candidate models is the key assumption behind consistency. However, if consistency holds, how do the efficient criteria perform? We will be able to examine this question here, since Models A1 and A2 represent some of the worst case scenarios for efficiency. We will see that the efficient criteria are no longer efficient, particularly if the true model is of finite order. We will also see that observed efficiency decreases as the order of the true model decreases. If a true, finite model belongs to the set of candidate models, then the consistent criteria are both consistent as well as efficient. As we saw in Chapter 2 (Section 2.4.1), efficient criteria asymptotically overfit by one variable 15% of the time. This percentage applies to only one variable; there is the same chance of overfitting for each irrelevant variable in the study. Thus overfitting can be an arbitrarily large problem.

Table 9.6 gives the summary results for counts, K-L and $L_2$ observed efficiencies for one thousand realizations. The detailed results are given in Table 9A.5. We can see from Table 9.6 how the efficient ($\alpha 2$), consistent ($\alpha\infty$), and the signal-to-noise corrected ($\alpha 3$) criteria differ. The efficient criteria AIC, AICc, Cp, FPE, Rp, and CV all perform about the same, overfitting more than 30% of the time. SIC and GM behave as we would expect for consistent criteria, correctly identifying the true model nearly every time, resulting in observed efficiencies of 99.5%. Although HQ and HQc are also consistent, even for $n$ as large as 25,000 their penalty functions are much smaller than that of SIC, and they overfit to some degree. However, their observed efficiency is still quite good, at 98% for both K-L and $L_2$. We recall from Chapter 2, Section 2.5, that the signal-to-noise corrected variants AICu and FPEu have an asymptotic probability of overfitting by one variable roughly halfway between AIC and SIC. We see from Tables 9.6 and 9A.5 that AICu and FPEu,

Table 9.6. Simulation results summary for Model A1.
K-L observed efficiency ranks, $L_2$ observed efficiency ranks and counts.

| criterion | K-L ranking | $L_2$ ranking | true | underfitting | overfitting |
|---|---|---|---|---|---|
| GM | 1 | 1 | 994 | 0 | 6 |
| SIC | 1 | 1 | 994 | 0 | 6 |
| HQ | 3 | 3 | 924 | 0 | 76 |
| HQc | 3 | 3 | 924 | 0 | 76 |
| FPE4 | 5 | 5 | 888 | 0 | 112 |
| AICu | 6 | 6 | 823 | 0 | 177 |
| BFPE | 6 | 6 | 818 | 0 | 182 |
| DCVB | 6 | 6 | 817 | 0 | 183 |
| FPEu | 6 | 6 | 823 | 0 | 177 |
| AIC | 10 | 10 | 691 | 0 | 309 |
| AICc | 10 | 10 | 691 | 0 | 309 |
| Cp | 10 | 10 | 691 | 0 | 309 |
| CV | 10 | 10 | 692 | 0 | 308 |
| FPE | 10 | 10 | 691 | 0 | 309 |
| Rp | 10 | 10 | 691 | 0 | 309 |
| $R^2_{adj}$ | 16 | 16 | 463 | 0 | 537 |

Table 9.7. Simulation results summary for Model A2.
K-L observed efficiency ranks, $L_2$ observed efficiency ranks and counts.

| criterion | K-L ranking | $L_2$ ranking | true | underfitting | overfitting |
|---|---|---|---|---|---|
| GM | 1 | 1 | 994 | 0 | 3 |
| SIC | 1 | 1 | 994 | 0 | 3 |
| HQ | 3 | 3 | 858 | 0 | 139 |
| HQc | 3 | 3 | 858 | 0 | 139 |
| FPE4 | 5 | 5 | 810 | 0 | 187 |
| AICu | 6 | 6 | 668 | 0 | 329 |
| BFPE | 6 | 6 | 666 | 0 | 333 |
| DCVB | 6 | 6 | 667 | 0 | 332 |
| FPEu | 6 | 6 | 668 | 0 | 329 |
| AIC | 10 | 10 | 454 | 0 | 543 |
| AICc | 10 | 10 | 454 | 0 | 543 |
| Cp | 10 | 10 | 454 | 0 | 543 |
| CV | 10 | 10 | 458 | 0 | 539 |
| FPE | 10 | 10 | 454 | 0 | 543 |
| Rp | 10 | 10 | 454 | 0 | 543 |
| $R^2_{adj}$ | 16 | 16 | 144 | 0 | 853 |

as well as the $\alpha3$ penalty weighted bootstrap criteria, are in fact more consistent and have higher observed efficiency than the $\alpha2$. The results show that when the true model belongs to the set of candidate models, $\alpha3$ criteria have higher efficiency than $\alpha2$ criteria, and $\alpha1$ criteria overfit much more than $\alpha2$ criteria. $R^2_{adj}$, the most likely to overfit, is $\alpha1$.

Table 9A.5 shows that overfitting by one variable is much more common than overfitting by two variables. This agrees with our asymptotic probability findings in Chapter 2. In large samples, the probability overfitting by $L$ variables decreases as $L$ increases. No underfitting is seen due to the very large sample size. Results for Model A2 are summarized in Table 9.7, with the details given in Table 9A.6.

With Model A2 there is more opportunity to overfit, and this is reflected by lower counts as well as decreased observed efficiencies overall (Table 9A.6). The more irrelevant variables included in the study, the more difficult is the task of selecting a good model. However, as long as the true model belongs to the set of candidate models and the sample size is large, the consistent criteria are unaffected by additional irrelevant variables and in fact we see from Table 9.7 that SIC and GM are not affected by the increase in $o$; they still have observed efficiencies of 99.7%. This is not true for the efficient criteria, which here overfit nearly half the time. However, the detailed Tables 9A.5 and 9A.6 show that overfitting by one variable is still much more common than overfitting by two variables. The results for these two models also illustrates the asymptotic performance of the criteria. We can see that the $\alpha3$ criteria, AICu, FPEu, BFPE, and DCVB, perform better than the efficient criteria, but worse than the consistent criteria.

So far, we have dealt with simulated data only. We next apply each model selection criterion to real data.

### 9.2.5. Real Data Example

Simulations give us a picture of what to expect by showing us general trends in the behavior of criteria when selected parameters are varied. However, to see how our expectations hold up under practical use, we will apply our selection criteria to an example by using real data. Consider the traffic safety data presented in Weisberg (1985), and found in Carl Hoffstedt's unpublished Master's thesis. Thirty nine sections of large Minnesota highways were selected and observed in 1973. The goal is to model accidents per million vehicle-miles ($Y$) by 13 independent variables, which are described below.

| Variable | Description | Units |
|---|---|---|
| x1 | Length of Highway Section | miles |
| x2 | Daily Traffic Count | 1000's |
| x3 | Truck Volume | % of Total |
| x4 | Speed Limit | Miles per Hour |
| x5 | Lane Width | Feet |
| x6 | Outer Shoulder Width | Feet |
| x7 | Freeway-Type Interchanges/Mile of Segment | Count |
| x8 | Signal Interchanges/Mile of Segment | Count |
| x9 | Access Points/Mile of Segment | Count |
| x10 | Lanes of Traffic in Both Directions | Count |
| x11 | Federal Aid Interstate Highway | 1 if Yes, 0 Otherwise |
| x12 | Principal Arterial Highway | 1 if Yes, 0 Otherwise |
| x13 | Major Arterial Highway | 1 if Yes, 0 Otherwise |

All subsets are considered. The results show that most of the selection criteria choose one of two models, as shown in Table 9.8.

Table 9.8. Model choices for highway data example.

| Model Selected | Selection Criteria |
|---|---|
| x1, x4, x8, x9, x12 | AIC, AICc, Cp, FPE, HQ, $R^2_{adj}$, Rp |
| x1, x4, x9 | AICu, BFPE, FPE4, FPEu, GM, HQc, SIC |
| x1, x3, x4, x9 | CV |
| x1, x4, x8, x12 | DCVB |

Table 9.9. Regression statistics for model x1, x4, x8, x9, x12.

| SOURCE | DF | SUM OF SQUARES | MEAN SQUARE | F | P-VALUE |
|---|---|---|---|---|---|
| MODEL | 5 | 111.671 | 22.334 | 19.29 | 0.0001 |
| ERROR | 33 | 38.215 | 1.158 | | |
| TOTAL | 38 | 149.886 | | | |

| VARIABLE | PARAMETER ESTIMATE | STAND ERROR | T | P-VALUE |
|---|---|---|---|---|
| INTERCEP | 9.944 | 2.582 | 3.85 | 0.001 |
| X1 | -0.074 | 0.025 | -3.02 | 0.005 |
| X4 | -0.105 | 0.041 | -2.54 | 0.016 |
| X8 | 0.797 | 0.369 | 2.16 | 0.038 |
| X9 | 0.064 | 0.030 | 2.12 | 0.041 |
| X12 | -0.774 | 0.411 | -1.89 | 0.068 |

The efficient criteria tend to select the model that includes (x1, x4, x8, x9, x12). Note that HQ also selected this model, supporting our observation that the small-sample behavior of HQ should be close to that of AIC. Criteria with larger penalty functions tended to select the second model, which contains two fewer variables: (x1, x4, x9). Both models exhibit similar residual characteristics, giving us no grounds to prefer one over the other. The regression statistics for these two models are shown in Tables 9.9 and 9.10. However, since the models are nested, a partial F-test can be used to further discriminate between them.

Table 9.10. Regression statistics for model x1, x4, x9.

| SOURCE | DF | SUM OF SQUARES | MEAN SQUARE | F | P-VALUE |
|---|---|---|---|---|---|
| MODEL | 3 | 105.040 | 35.013 | 27.33 | 0.0001 |
| ERROR | 35 | 44.847 | 1.281 | | |
| TOTAL | 38 | 149.886 | | | |

| VARIABLE | PARAMETER ESTIMATE | STAND ERROR | T | P-VALUE |
|---|---|---|---|---|
| INTERCEP | 9.326 | 2.617 | 3.56 | 0.0001 |
| X1 | -0.077 | 0.025 | -3.10 | 0.004 |
| X4 | -0.102 | 0.043 | -2.39 | 0.023 |
| X9 | 0.101 | 0.027 | 3.72 | 0.001 |

The partial F-test comparing model (x1, x4, x9) to model (x1, x4, x8, x9, x12) gives F=2.86, with a corresponding p-value of 0.0715. This leads us to conclude that model (x1, x4, x9) is the better model. Note that most of the criteria are functions of SSE, and that these cluster around the two models we have evaluated. Criteria that are not functions of SSE may choose different models. For example, CV and its bootstrapped version DCVB are not functions of SSE directly, and while they both choose models of order 5 (4 variables plus the intercept), the two models have different variables.

## 9.3. Autoregressive Models

### *9.3.1. Model Structure*

Recall the autoregressive model AR($p$) with true order $p_*$:

$$y_t = \phi_1 y_{t-1} + \cdots + \phi_{p_*} y_{t-p_*} + w_{*t}, \quad w_{*t} \sim N(0, \sigma_*^2), \quad t = p_* + 1, \ldots, n, \quad (9.3)$$

where the $w_{*t}$ are independent and $y_1, \ldots, y_n$ is the observed series. Candidate models of the form in Eq. (3.1) are fit to the data. Unlike regression models, the models are ordered and the effective sample size changes as the model order changes.

The data are generated as follows. Each time series $Y = (y_1, \ldots, y_n)'$ is generated starting at $y_{t-50}$ with $y_t = 0$ for all $t < -50$. Only observations

$y_1, \ldots, y_n$ are kept. K-L and $L_2$ observed efficiencies are as defined in Chapter 3 with K-L observed efficiency computed from Eq. (3.8) and Eq. (1.2) and $L_2$ observed efficiency computed using Eq. (3.7) and Eq. (1.1).

### *9.3.2. Two Special Case Models*

The two special case models described in Eqs. (3.18)–(3.20) from Chapter 3 are now reexamined using all 16 criteria. For both models, $n = 35$, $p_* = 5$, $\sigma_* = 1$, and $w_{*t} \sim N(0,1)$. The two special case models are:
**Model 3** (parameter structure 7 in Table 9.14)

$$y_t = y_{t-5} + w_{*t}$$

and
**Model 4** (parameter structure 3 in Table 9.14)

$$y_t = 0.434y_{t-1} + 0.217y_{t-2} + 0.145y_{t-3} + 0.108y_{t-4} + 0.087y_{t-5} + w_{*t}.$$

Model 3 is an example of a nonstationary time series with strongly identifiable parameters, a seasonal random walk with season = 5. By contrast, Model 4 is an AR model with weaker parameters that are much more difficult to identify at the true order $p_*$. The coefficient of $y_{t-j}$ in Model 4 is proportional to $1/j$ for $j = 1, \ldots, 5$. Each time series $Y = (y_1, \ldots, y_{35})'$ was generated. Ten thousand realizations were simulated. For each realization, a new time series $Y = (y_1, \ldots, y_35)$ was generated and for Cp the maximum order is $P = 10$. Count, K-L and $L_2$ observed efficiency results are given in Table 9.11. Detailed results are given in Table 9A.7.

Table 9.11 shows that AICu has the best observed efficiency and the highest count for selecting the true order, correctly identifying the true order 5 nearly 92% of the time. HQc also performs well, ranking second in observed efficiency and identifying the correct order 87% of the time, followed by AICc, the doubly cross-validated bootstrap (DCVB), and FPEu. In general, criteria with strong penalty functions do well, and those with weak penalty functions and thus weak signal-to-noise ratios, such as AIC, HQ, and FPE, tend to overfit.

By contrast, when we look at the results for a weakly identifiable time series with quickly decaying parameters, we expect that underfitting properties will have much more of an impact on performance than for Model 3. Results for Model 4 are given in Table 9.12, and details in Table 9A.8. Table 9.12 shows that three of the top performers for Model 3 reappear here. AICc, HQc, and DCVB are joined by BFPE and Rp. Rp is derived under the assumption that the true model is of infinite dimension and does not belong to the set of candidate models. Here, the parameters decrease fast enough that this assumption appears to hold, explaining why Rp performs well in Model 4 but not in Model 3. AICu overfits too much in this case due to its strong penalty function, and it drops in rank to 6th place. Due to their weak

Table 9.11. Simulation results summary for Model 3.
K-L observed efficiency ranks, $L_2$ observed efficiency ranks and counts.

| criterion | K-L ranking | $L_2$ ranking | true | underfitting | overfitting |
|---|---|---|---|---|---|
| AICu | 1 | 1 | 9188 | 23 | 789 |
| HQc | 2 | 2 | 8739 | 12 | 1249 |
| AICc | 3 | 3 | 8430 | 9 | 1561 |
| DCVB | 4 | 4 | 8061 | 16 | 1923 |
| FPEu | 5 | 5 | 7869 | 9 | 2122 |
| GM | 6 | 6 | 7725 | 16 | 2259 |
| BFPE | 7 | 7 | 7445 | 7 | 2548 |
| SIC | 7 | 7 | 7562 | 10 | 2428 |
| FPE4 | 9 | 9 | 7454 | 11 | 2535 |
| Rp | 10 | 10 | 6467 | 7 | 3526 |
| CV | 11 | 11 | 6340 | 7 | 3653 |
| Cp | 12 | 12 | 6044 | 7 | 3949 |
| HQ | 13 | 13 | 5912 | 7 | 4081 |
| FPE | 14 | 14 | 5436 | 5 | 4559 |
| AIC | 15 | 15 | 5213 | 5 | 4782 |
| $R^2_{adj}$ | 16 | 16 | 3071 | 2 | 6927 |

Table 9.12. Simulation results summary for Model 4.
K-L observed efficiency ranks, $L_2$ observed efficiency ranks and counts.

| criterion | K-L ranking | $L_2$ ranking | true | underfitting | overfitting |
|---|---|---|---|---|---|
| AICc | 1 | 1 | 530 | 9154 | 316 |
| DCVB | 1 | 2 | 418 | 9314 | 268 |
| HQc | 1 | 2 | 366 | 9449 | 185 |
| BFPE | 4 | 2 | 505 | 8984 | 511 |
| Rp | 9 | 2 | 848 | 7782 | 1370 |
| AICu | 4 | 8 | 191 | 9741 | 68 |
| CV | 11 | 2 | 881 | 7620 | 1499 |
| FPEu | 6 | 7 | 446 | 9130 | 424 |
| SIC | 7 | 9 | 363 | 9180 | 457 |
| FPE4 | 8 | 12 | 353 | 9172 | 475 |
| HQ | 12 | 9 | 681 | 7733 | 1586 |
| FPE | 14 | 9 | 858 | 6997 | 2145 |
| GM | 9 | 15 | 366 | 8976 | 658 |
| Cp | 13 | 12 | 696 | 7480 | 1824 |
| AIC | 15 | 12 | 822 | 6941 | 2237 |
| $R^2_{adj}$ | 16 | 16 | 1104 | 3911 | 4985 |

penalty functions, AIC, HQ, and FPE once again allow excessive overfitting, reducing their observed efficiencies.

### *9.3.3. Large-scale Small-sample Simulations*

We have observed the behavior of our criteria with respect to the two special case time series models that focused only on how the relative strength or weakness of the true model affects performance. We would now like to see how the criteria behave when applied to many models with widely varying

Table 9.13. Summary of the autoregressive models. All models have $\sigma_* = 1$.

| Sample Size $n$ | Parameter Structure $\phi_j$ | True Model Order $p_*$ | Overfitting $o$ | Largest Order Considered $P$ |
|---|---|---|---|---|
| 15 | 8 structures | 2, 5, 10 | 2, 5, 10 | $\min(p_* + o, 6)$ |
| 25 | 8 structures | 2, 5, 10 | 2, 5, 10 | $\min(p_* + o, 10)$ |
| 35 | 8 structures | 2, 5, 10 | 2, 5, 10 | $\min(p_* + o, 15)$ |
| 50 | 8 structures | 2, 5, 10 | 2, 5, 10 | $\min(p_* + o, 20)$ |
| 100 | 8 structures | 2, 5, 10 | 2, 5, 10 | $\min(p_* + o, 20)$ |

Table 9.14. Relationship between parameter structure and true model order.

Parameter structure 1: $\phi_j \propto 1/j^2$
- $p_* = 2$ $\phi_1 = 0.792, \phi_2 = 0.198$
- $p_* = 5$ $\phi_1 = 0.676, \phi_2 = 0.169, \phi_3 = 0.075, \phi_4 = 0.042, \phi_5 = 0.027$

Parameter structure 2: $\phi_j \propto 1/j^{1.5}$
- $p_* = 2$ $\phi_1 = 0.731, \phi_2 = 0.259$
- $p_* = 5$ $\phi_1 = 0.562, \phi_2 = 0.199, \phi_3 = 0.108, \phi_4 = 0.070, \phi_5 = 0.050$

Parameter structure 3: $\phi_j \propto 1/j$
- $p_* = 2$ $\phi_1 = 0.660, \phi_2 = 0.330$
- $p_* = 5$ $\phi_1 = 0.434, \phi_2 = 0.217, \phi_3 = 0.145, \phi_4 = 0.108, \phi_5 = 0.087$

Parameter structure 4: $\phi_j \propto 1/\sqrt{j}$
- $p_* = 2$ $\phi_1 = 0.580, \phi_2 = 0.410$
- $p_* = 5$ $\phi_1 = 0.306, \phi_2 = 0.217, \phi_3 = 0.177, \phi_4 = 0.153, \phi_5 = 0.137$

Parameter structure 5: $\phi_j \propto 1$
- $p_* = 2$ $\phi_1 = 0.495, \phi_2 = 0.495$
- $p_* = 5$ $\phi_1 = 0.198, \phi_2 = 0.198, \phi_3 = 0.198, \phi_4 = 0.198, \phi_5 = 0.198$

Parameter structure 6: SAR(1) (seasonal AR)
- $p_* = 2$ $\phi_2 = 0.5$
- $p_* = 5$ $\phi_5 = 0.5$

Parameter structure 7: seasonal random walk
- $p_* = 2$ $\phi_2 = 1$
- $p_* = 5$ $\phi_5 = 1$

Parameter structure 8
- $p_* = 2$ $\phi_1 = 0.5, \phi_2 = 0.5$
- $p_* = 5$ $\phi_2 = 0.5, \phi_5 = 0.5$

characteristics. The different AR models are formed by varying the components of Eq. (9.3), and Table 9.13 describes the model parameters.

There is some additional information to keep in mind when considering these AR models. The sample size ranges from very short ($n = 15$) to moderately large ($n = 100$). However, unlike regression models, our AR models will have variable effective sample size, which will impact the degrees of freedom and the largest possible candidate model order. Since $y_t$ is a function of the infinite past errors $w_j$, the variability of $y_t$ depends on the variability of $w_t$ alone. Any time series can be rescaled for arbitrary error variance, and all of these models have $\sigma_*^2 = 1$. The first five parameter structures in Table 9.14 parallel some of the model structures in our regression study. However, if we let $\phi_j = 1$ or $\phi_j \propto 1/j$, the result is a nonstationary model. Furthermore, these models are unstable in the sense that they are very difficult to simulate (the $y_t$ explode towards $\pm\infty$). For all but parameter structures 6–8, the $\phi_j$ parameters are rescaled to ensure stationarity. Structure 7, the seasonal random walk, is nonstationary, but is stable with respect to generating the data. Another consideration is that some combinations of true order $p_*$ and overfitting $o$ can yield models too large for the given sample size. For example, a criterion like AICc needs at least $n - 2P - 2 > 0$, or to have the largest model order $P = p_* + o$ less than $n/2 - 1$. We have noted that the largest model will depend on sample size, and the last column of Table 9.13 lists the largest order considered. This restriction causes some redundancy in the total

Table 9.15. Simulation results over 360 models.
Summary of overall rank by K-L and $L_2$ observed efficiency.

| criterion | K-L ranking | $L_2$ ranking |
|---|---|---|
| HQc | 2 | 1 |
| AICc | 3 | 1 |
| DCVB | 3 | 1 |
| AICu | 1 | 6 |
| FPEu | 5 | 4 |
| BFPE | 6 | 4 |
| FPE4 | 7 | 9 |
| Rp | 9 | 7 |
| CV | 10 | 8 |
| GM | 8 | 10 |
| SIC | 11 | 11 |
| Cp | 12 | 11 |
| FPE | 13 | 13 |
| HQ | 14 | 14 |
| AIC | 15 | 14 |
| $R^2_{adj}$ | 16 | 16 |

number of different AR models considered in this study, which for now we will ignore. The actual parameters involved depend on the structure as well as the order of the true model, and Table 9.14 summarizes the relationship between the eight parameter structures and true model order.

All together, all combinations of the five $n$, eight parameter structures, three true orders and three $o$ gives us $5 \times 8 \times 3 \times 3 = 360$ AR models. One hundred realizations are generated for each model, resulting in 36,000 realizations. For each realization, the criteria select a model and observed efficiency is computed, with the best observed efficiency given a rank of 1. Ranks for each criterion are averaged over all realizations, and performance is based on these average rankings as well as the results of the pairwise comparison test defined in Eq. (9.1) discussed earlier. Because it is not practical to summarize individual models, the results are summarized by rank in Table 9.15. Detailed results are given in Tables 9A.9 and 9A.10.

In this study we see a great deal of variability in performance between the K-L and $L_2$ observed efficiencies. However, we would like to be able to identify criteria that balance overfitting and underfitting tendencies, and thus perform well under both distance measures. Therefore we have sorted the criteria in Table 9.15 by the sum of their K-L and $L_2$ rankings. HQc ranks second in K-L and ties for first in $L_2$, for a sum of 3. Since HQc performs well with respect to both observed efficiency measures, it is listed first in Table 9.15. HQc thus should have the best balance between overfitting and underfitting over a wide range of model situations. On the other hand, while AICu has the highest K-L observed efficiency, under $L_2$ it has fallen to sixth due to its tendency to underfit, resulting in a sum of 7. AICc does better with a K-L rank of 3 and an $L_2$ rank of 1 for a sum of 4, and so it appears above AICu. Taking results for both $L_2$ and K-L performance into account, the top five selection criteria are HQc, AICc and DCVB (tie), AICu, and FPEu. Excessive overfitting is penalized by both observed efficiency measures and criteria with weak penalty functions are found at the bottom of Table 9.15.

### *9.3.4. Large-sample Simulations*

In order to demonstrate overfitting in AR models we will also consider two asymptotic AR models, Models A3 and A4. In both cases the true model is $y_t = 0.9y_{t-1} + w_{*t}$ with $w_{*t} \sim N(0, 1)$, and 1000 realizations were generated with sample size $n = 25{,}000$. When extra variables are included in the model the opportunity for overfitting increases slowly, in contrast to regression where the opportunity for overfitting increases rapidly. This is due to the sequential nature of fitting AR models versus the all subsets approach in regression. Models A3 and A4 differ only in the largest order considered, which is 3 for A3 (two extra variables) and 6 for A4 (five extra variables). The summary of results for Model A3 is given in Table 9.16, and detailed results can be found in Table 9A.11.

Since a true model of finite order belongs to the set of candidate models

Table 9.16. Simulation results summary for Model A3.
K-L observed efficiency ranks, $L_2$ observed efficiency ranks and counts.

| criterion | K-L ranking | $L_2$ ranking | true | underfitting | overfitting |
|---|---|---|---|---|---|
| GM | 1 | 1 | 993 | 0 | 7 |
| SIC | 1 | 1 | 993 | 0 | 7 |
| HQ | 3 | 3 | 924 | 0 | 76 |
| HQc | 3 | 3 | 924 | 0 | 76 |
| FPE4 | 5 | 5 | 890 | 0 | 110 |
| AICu | 6 | 6 | 812 | 0 | 188 |
| BFPE | 6 | 6 | 809 | 0 | 191 |
| DCVB | 6 | 6 | 810 | 0 | 190 |
| FPEu | 6 | 6 | 812 | 0 | 188 |
| AIC | 10 | 10 | 692 | 0 | 308 |
| AICc | 10 | 10 | 692 | 0 | 308 |
| Cp | 10 | 10 | 692 | 0 | 308 |
| CV | 10 | 10 | 691 | 0 | 309 |
| FPE | 10 | 10 | 692 | 0 | 308 |
| Rp | 10 | 10 | 692 | 0 | 308 |
| $R^2_{adj}$ | 16 | 16 | 468 | 0 | 532 |

Table 9.17. Simulation results summary for Model A4.
K-L observed efficiency ranks, $L_2$ observed efficiency ranks and counts.

| criterion | K-L ranking | $L_2$ ranking | true | underfitting | overfitting |
|---|---|---|---|---|---|
| GM | 1 | 1 | 994 | 0 | 6 |
| SIC | 1 | 1 | 994 | 0 | 6 |
| HQ | 3 | 3 | 931 | 0 | 69 |
| HQc | 3 | 3 | 931 | 0 | 69 |
| FPE4 | 5 | 5 | 900 | 0 | 100 |
| AICu | 6 | 6 | 798 | 0 | 202 |
| BFPE | 6 | 6 | 794 | 0 | 206 |
| DCVB | 6 | 6 | 798 | 0 | 202 |
| FPEu | 6 | 6 | 798 | 0 | 202 |
| AIC | 10 | 10 | 615 | 0 | 385 |
| AICc | 10 | 10 | 615 | 0 | 385 |
| Cp | 10 | 10 | 615 | 0 | 385 |
| CV | 10 | 10 | 612 | 0 | 388 |
| FPE | 10 | 10 | 615 | 0 | 385 |
| Rp | 10 | 10 | 615 | 0 | 385 |
| $R^2_{adj}$ | 16 | 16 | 333 | 0 | 667 |

and the sample size is large, consistency holds in this case. No underfitting is possible since the true model is AR(1), the smallest candidate model considered. The observed efficiency ranking results are the same for K-L and $L_2$, and as expected, the consistent criteria perform best, with SIC and GM tied for first, followed by HQ or HQc. The $\alpha 2$ efficient criteria all perform the same, overfitting nearly 25% of the time. The $\alpha 3$ AICu overfits less than the efficient criteria and performs better under consistent conditions, where the true model belongs to the set of candidate models.

Table 9A.11 presents the detailed count patterns for the three possible candidate models, and we see that even the strongly consistent criteria SIC and GM overfit on occasion and that none of the criteria identify the true model every time. In general, overfitting by one variable is much more common than overfitting by two variables. Overfitting depends on the asymptotic strength of the penalty function, and this explains the very similar count patterns between asymptotically equivalent criteria. The criteria form clear clusters depending on whether the penalty function is $\alpha 1$, $\alpha 2$, $\alpha 3$, or $\alpha\infty$. We will next look at the results for Model A4, for which the opportunity for overfitting is even greater. The summary is given in Table 9.17, and detailed results in Table 9A.12.

Results for Model A4 are similar to those for Model A3. As expected, the increased opportunity for overfitting results in more overfitting from some of the criteria. The consistent criteria SIC and GM are unaffected by the increased opportunity for overfitting, whereas the efficient criteria overfit even more severely, identifying the true model less than two thirds of the time.

Table 9A.12 details the count patterns. Both distance measures have a well-defined minimum at the correct order, and in general, the overfitting counts decrease as the order increases. Although criteria with stronger asymptotic penalty functions overfit less than criteria with weaker asymptotic penalty functions, there is not much difference in the count patterns in Tables 9A.11 and 9A.12. Although counts for selecting the correct model are high in both tables, the overfitting patterns in Table 9A.12 are spread out over higher orders. This results in lower average observed efficiencies in Table 9A.12.

### *9.3.5. Real Data Example*

Our real data example comes from series W2 in Wei (1990, pp. 446–447), which consists of the Wolf yearly sunspot numbers from 1700 to 1983. We choose this data since no moving average components are commonly added, and it can be modeled as a purely AR process. However, some transformations are required to stabilize the variance and to remove the mean, and we will carry out our analysis on the transformed data $y^*$ where $y_t^* = \sqrt{y_t} - 6.298$. All the selection criteria were applied to this data, and all selected the AR(9). Regression statistics for this model are given in Table 9.19.

Table 9.18. Model choices for Wolf sunspot data.

| Order $p$ | Selection Criteria |
|---|---|
| 9 | AIC, AICc, AICu, BFPE, Cp, CV, DCVB, FPE, FPE4, FPEu, GM, HQ, HQc, $R^2_{adj}$, Rp, SIC |

Table 9.19. AR(9) model statistics.

| VARIABLE | PARAMETER ESTIMATE | STAND ERROR | T | P-VALUE |
|---|---|---|---|---|
| $\phi_1$ | 1.1003 | 0.0582 | 18.91 | 0.000 |
| $\phi_2$ | -0.3703 | 0.0879 | -4.21 | 0.000 |
| $\phi_3$ | -0.1410 | 0.0896 | -1.57 | 0.115 |
| $\phi_4$ | 0.1957 | 0.0962 | 2.03 | 0.042 |
| $\phi_5$ | -0.1368 | 0.1026 | -1.33 | 0.183 |
| $\phi_6$ | -0.1340 | 0.1019 | -1.31 | 0.189 |
| $\phi_7$ | 0.2870 | 0.1024 | 2.80 | 0.005 |
| $\phi_8$ | -0.2916 | 0.1004 | -2.90 | 0.004 |
| $\phi_9$ | 0.3446 | 0.0638 | 5.40 | 0.000 |

Analysis of the AR(9) residuals show that they appear to be white noise. They do seem to be heavier-tailed than the normal distribution, but this non-normality apparently does not affect the selection process, since all criteria agreed on the model choice. The AR(9) model choice agrees with earlier analyses (see Wei, 1990, p. 152).

In the beginning of this Chapter we observed that autoregressive models are of finite order; however, time series models do allow us a convenient way to generate models of infinite order by using moving averages. In the next Section we will examine the behavior of such models, once again by using Monte Carlo simulations.

## 9.4. Moving Average MA(1) Misspecified as Autoregressive Models

### *9.4.1. Model Structure*

The moving average model can be used to illustrate models of infinite order. Consider the moving average MA(1) model of the form

$$y_t = \theta_1 w_{*t-1} + w_{*t}, \quad w_{*t} \sim N(0, \sigma_*^2), \quad t = 1, \ldots n, \tag{9.4}$$

where the $w_{*t}$ are independent. We can form different MA(1) models by varying sample size, $n$, and $\theta_1$. Candidate AR models of order 1 through 15 are fit to the data. Since only autoregressive AR($p$) models are considered as candidates, the true MA(1) model does not belong to the set of candidate models and consistency does not apply. The observed efficiency measures used to evaluate criterion performance, K-L and $L_2$, are as defined in Chapter 3. Here, K-L observed efficiency is defined by Eq. (3.25) and Eq. (1.2) and $L_2$ observed efficiency is defined by Eq. (3.24) and Eq. (1.1).

### *9.4.2. Two Special Case Models*

Once again we will first revisit special case models from an earlier chapter, in this case two misspecified MA models, originally discussed in Chapter 3 in Eqs. (3.21)–(3.23). They are:

**Model 5**

$$y_t = 0.5w_{*t-1} + w_{*t}$$

and

**Model 6**

$$y_t = 0.9w_{*t-1} + w_{*t}.$$

Both MA models are stationary and can be written in terms of an infinite order AR model. The $\phi_j = \theta_1^j$ parameters decay much more quickly for Model 5 than for Model 6. In small samples, Model 5 may be approximated by a finite order AR model, but because Model 6 has AR parameters that decay much more slowly, in finite samples no good approximation may exist. Both models have sample size $n = 35$, error variance $\sigma_*^2 = 1$, and 10,000 realizations. Each time series $Y$ is formed by generating white noise $w_0, \ldots, w_{35}$. Since the true model does not belong to the set of candidate models, no counts are presented in Tables 9.20 and 9.21. However, the detailed counts of chosen model orders are given in Tables 9A.13 and 9A.14. Model 5 details can be found in Table 9A.13. Observed efficiency ranks are summarized in Table 9.20.

Table 9.20 shows that the top five ranked criteria are AICu, DCVB, HQc, AICc, and BFPE and FPE4 (tie). The lack of a true model belonging to the set of candidate models does not seem to affect the rankings, which are similar to those seen in earlier sections. With no true finite AR order, underfitting and overfitting are not easily defined and performance in terms of $L_2$ and K-L are similarly poor. In this case underfitting and overfitting must be defined in terms of the order that is closest to the true model as chosen by K-L or $L_2$. Table 9A.13 shows that 60% observed efficiency is typical.

We next consider Model 6 with $\theta_1 = 0.9$. Counts and observed efficiency details are given in Table 9.A14, and the summary of observed efficiency ranks is given in Table 9.21.

We can see the impact of $\theta_1$ on the performance of the criteria, where the larger $\theta_1$ value mimics a longer order AR model. We see this in the count patterns in Table 9A.14, where the order of the model closest to the true model is larger in Model 6 than in Model 5. The $L_2$ and K-L distances for Model 6 typically find the closest model to be of order 3, 4, or 5. These larger orders allow us to evaluate underfitting as well as overfitting. Large penalty functions

Table 9.20. Simulation results summary for Model 5.
K-L observed efficiency ranks and $L_2$ observed efficiency ranks.

| criterion | K-L ranking | $L_2$ ranking |
|---|---|---|
| AICu | 1 | 1 |
| DCVB | 2 | 1 |
| HQc | 3 | 1 |
| AICc | 4 | 4 |
| BFPE | 4 | 5 |
| FPEu | 4 | 5 |
| FPE4 | 4 | 7 |
| GM | 4 | 7 |
| SIC | 9 | 9 |
| Rp | 10 | 10 |
| CV | 10 | 11 |
| Cp | 10 | 12 |
| FPE | 13 | 13 |
| HQ | 14 | 14 |
| AIC | 15 | 15 |
| $R^2_{adj}$ | 16 | 16 |

Table 9.21. Simulation results summary for Model 6.
K-L observed efficiency ranks and $L_2$ observed efficiency ranks.

| criterion | K-L ranking | $L_2$ ranking |
|---|---|---|
| AICc | 1 | 1 |
| HQc | 2 | 4 |
| CV | 7 | 2 |
| DCVB | 3 | 5 |
| Rp | 7 | 2 |
| BFPE | 5 | 5 |
| AICu | 3 | 8 |
| FPEu | 5 | 7 |
| FPE4 | 9 | 10 |
| SIC | 10 | 11 |
| FPE | 13 | 9 |
| Cp | 12 | 12 |
| GM | 11 | 13 |
| HQ | 14 | 13 |
| AIC | 15 | 15 |
| $R^2_{adj}$ | 16 | 16 |

result in too much underfitting in this model, as is seen for AICu, SIC, and GM. However, unlike AICu, SIC and GM also show excessive overfitting due to their weaker penalty functions. Since AICu underfits more than it overfits, AICu has much better K-L performance than $L_2$ performance (3rd and 8th respectively). AICc is the best performer in both K-L and $L_2$. HQc, CV, DCVB, and Rp round out the top five. In the next Section, we consider performance across a wider range of MA(1) models.

### *9.4.3. Large-scale Small-sample Simulations*

In our large-scale study involving 50 MA models, only two components in Eq. (9.4) are allowed to vary, the sample size $n$ and $\theta_1$. We will vary sample size from small to moderately large, and $\theta_1$ from 0.1 to 1.0 (see Table 9.22). Sample sizes $n = 15, 25, 35, 50, 100$ are used, with ten values of $\theta_1$, for a total of 50 models. Models with small $\theta_1$ values can be approximated by short AR($p$) models, and as $\theta_1$ increases so does the order of the approximating AR model. Since there is no true finite AR($p$) model order, the maximum order considered is based on sample size.

Table 9.22. Summary of the misspecified MA(1) models.
All models have $\sigma_* = 1$.

| Sample Size $n$ | Parameter $\theta_1$ | Largest Order Considered $P$ |
|---|---|---|
| 15 | 0.1, 0.2, 0.3, 0.4, 0.5, 0.6, 0.7, 0.8, 0.9, 1.0 | 6 |
| 25 | 0.1, 0.2, 0.3, 0.4, 0.5, 0.6, 0.7, 0.8, 0.9, 1.0 | 10 |
| 35 | 0.1, 0.2, 0.3, 0.4, 0.5, 0.6, 0.7, 0.8, 0.9, 1.0 | 15 |
| 50 | 0.1, 0.2, 0.3, 0.4, 0.5, 0.6, 0.7, 0.8, 0.9, 1.0 | 23 |
| 100 | 0.1, 0.2, 0.3, 0.4, 0.5, 0.6, 0.7, 0.8, 0.9, 1.0 | 48 |

Observed efficiency rankings over the 50 models are summarized in Table 9.23, and the detailed results are given in Tables 9A.15 and 9A.16.

All fifty MA models have a similar structure in that the approximating AR parameters decay smoothly as $\phi_j = \theta_1^j$. The top five performers are AICu, HQc/DCVB (tie), AICc, and BFPE/FPEu (tie). Once again the signal-to-noise corrected variants outperform the $\alpha 2$ efficient criteria. However, although it is consistent, HQc's penalty function and signal-to-noise ratio seems to balance underfitting and overfitting even when the true model does not belong to the set of candidate models. We saw this pattern for the special case models as well. SIC and GM still perform more poorly than they did when a true model belonged to the set of candidate models and consistency held.

Table 9.23. Simulation results over 50 models.
Summary of overall rank by K-L and $L_2$ observed efficiency.

| criterion | K-L ranking | $L_2$ ranking |
|---|---|---|
| AICu | 1 | 1 |
| DCVB | 2 | 1 |
| HQc | 2 | 1 |
| AICc | 4 | 1 |
| BFPE | 5 | 5 |
| FPEu | 5 | 5 |
| FPE4 | 7 | 7 |
| CV | 9 | 8 |
| Rp | 9 | 8 |
| GM | 8 | 10 |
| SIC | 9 | 11 |
| Cp | 12 | 12 |
| FPE | 13 | 13 |
| HQ | 14 | 14 |
| AIC | 15 | 15 |
| $R^2_{adj}$ | 16 | 16 |

### *9.4.4. Large-sample Simulations*

We here observe the behavior of our model selection criteria when the true model does not belong to the set of candidate models by applying them to a model of infinite order. Model A5 is defined as $y_t = 0.7w_{*t-1} + w_{*t}$ with $w_{*t} \sim N(0,1)$, where $P = 15$ is the maximum order considered and the sample size

Table 9.24. Simulation results summary for Model A5.
K-L observed efficiency ranks and $L_2$ observed efficiency ranks.

| criterion | K-L ranking | $L_2$ ranking |
|---|---|---|
| AIC | 1 | 1 |
| AICc | 1 | 1 |
| Cp | 1 | 1 |
| CV | 1 | 1 |
| FPE | 1 | 1 |
| $R^2_{adj}$ | 1 | 1 |
| Rp | 1 | 1 |
| AICu | 8 | 8 |
| BFPE | 8 | 8 |
| DCVB | 8 | 8 |
| FPEu | 8 | 8 |
| FPE4 | 12 | 12 |
| HQ | 13 | 13 |
| HQc | 13 | 13 |
| GM | 15 | 15 |
| SIC | 15 | 15 |

$n = 25{,}000$. With a large sample size and no true finite AR order we would expect the efficient criteria to perform best. Table 9.24 summarizes the observed efficiency results for the selection criteria, and details are given in Table 9A.17.

Since no true finite AR order exists, counts of model order choice are useful only for demonstrating general trends in behavior by the different groups of criteria. None of the criteria chose orders 6 or below, but we can see from Table 9A.17 that the consistent criteria tend to choose shorter order AR models than the efficient criteria. The orders selected by AICu fall in between those chosen by the efficient and consistent criteria.

The efficient criteria indeed have the highest observed efficiency in both the $L_2$ and the K-L sense, and not surprisingly the consistent SIC and GM have the lowest observed efficiency. In a large sample size of 25,000 the asymptotic properties of observed efficiency and consistency become important. Since no true model belongs to the set of candidate models, observed efficiency should be the desired property and consistency should be meaningless. This is borne out by the good performance from the efficient criteria in Table 9.24, and the poor performance from the consistent criteria. The detailed Table 9A.17 shows that the consistent criteria underfit, resulting in a loss of observed efficiency. The $\alpha 3$ criteria, like AICu, fall in the middle. This is because if the true model does not belong to the set of candidate models, AICu, while not efficient, has higher efficiency than the consistent criteria. Conversely, as we saw in the previous Section, if a true model does belong to the set of candidate models, AICu, while not asymptotically consistent, is more consistent than the efficient criteria.

## 9.5. Multivariate Regression Models

### 9.5.1. Model Structure

In this Section we revisit the multivariate regression model from Chapter 4, described as

$$y_i = B_0 + B_1 x_{i,1} + \cdots B_{k_*-1} x_{i,k_*-1} + \varepsilon_{*i}, \quad \varepsilon_{*i} \sim N(0, \Sigma_*), \quad i = 1, \ldots, n, \quad (9.5)$$

where the $\varepsilon_{*i}$ are independent and $y_i$ is a $q \times 1$ vector of responses. $\Sigma_*$ is the covariance matrix of the $q \times 1$ error vector, $\varepsilon_{*i}$. Candidate models of the form of Eq. (4.3) are fit to the data. For the simulation in this Section we generate different multivariate regression models by varying $k_*$, $n$, $\Sigma_*$, parameter matrices $B$, and the amount of overfitting $o$.

Observed efficiencies for K-L, and the trace of $L_2$ and the determinant of $L_2$, are as defined as follows: K-L observed efficiency is computed using

Eq. (4.10) and Eq. (1.2); $\text{tr}\{L_2\}$ observed efficiency is computed using the trace of Eq. (4.9) and Eq. (1.1); $\det(L_2)$ observed efficiency is computed using the determinant of Eq. (4.9) and Eq. (1.1). Higher observed efficiency denotes selection of a model closer to the true model and better performance.

### 9.5.2. Two Special Case Models

Here we reexamine the two special case bivariate regression models from Chapter 4, Models 7 and 8 from Eqs. (4.24)–(4.26). In each case the true model has $n = 25$, $k_* = 5$, independent columns of $X$, and $cov[\varepsilon_{*i}] = \Sigma_* = \begin{pmatrix} 1 & 0.7 \\ 0.7 & 1 \end{pmatrix}$. Model 7 has strongly identifiable parameters relative to the error, and Model 8 has much more weakly identifiable parameters. They are:

**Model 7**

$$y_i = \begin{pmatrix} 1 \\ 1 \end{pmatrix} + \begin{pmatrix} 1 \\ 1 \end{pmatrix} x_{i,1} + \begin{pmatrix} 1/4 \\ 1 \end{pmatrix} x_{i,2} + \begin{pmatrix} 0 \\ 1 \end{pmatrix} x_{i,3} + \begin{pmatrix} 0 \\ 1 \end{pmatrix} x_{i,4} + \varepsilon_{*i}$$

and

**Model 8**

$$y_i = \begin{pmatrix} 1 \\ 1 \end{pmatrix} + \begin{pmatrix} 1 \\ 1 \end{pmatrix} x_{i,1} + \begin{pmatrix} 1/4 \\ 1/4 \end{pmatrix} x_{i,2} + \begin{pmatrix} 1/9 \\ 1/9 \end{pmatrix} x_{i,3} + \begin{pmatrix} 1/16 \\ 1/16 \end{pmatrix} x_{i,4} + \varepsilon_{*i}.$$

Table 9A.18 gives details on the count patterns for the selection criteria as well as the closest models for the distance measures. The summary in Table 9.25 shows that AICu ranks first under all observed efficiencies, and that HQc ranks second, outperforming the other consistent criteria. In general there is good agreement across the three measures as to relative performance. To avoid redundancy, we will confine our subsequent presentation of results to $\text{tr}\{L_2\}$. For the nonscalar criteria, the maximum eigenvalue of Cp (meCp) performs better than the trace. Determinants of the bootstraps outperform traces, probably because the determinant takes into account correlation of $\varepsilon_{i,1}$ and $\varepsilon_{i,2}$, whereas the trace focuses on the individual variances. ICOMP has the worst performance overall due to excessive underfitting, illustrating that too large a penalty function is as bad as too weak a penalty function. AIC overfits excessively and is also heavily penalized by low observed efficiency.

Next we consider the results for Model 8, which has a weaker parameter structure than Model 7. Summary results are given in Table 9.26, and details in Table 9A.19. In Model 8 underfitting is expected due to the weak parameters. Nevertheless, we see that the top four criteria for Model 8 are identical to those for Model 7—AICu, HQc, and deDCVB, and AICc. Since the candidate models closest to that identified by the distance measures tend to have fewer variables as a result of some underfitting, we note that AIC and trFPE and

Table 9.25. Simulation results summary for Model 7.
K-L observed efficiency ranks, $L_2$ observed efficiency ranks and counts.

| criterion | K-L ranking | tr$\{L_2\}$ ranking | det$(L_2)$ ranking | true | underfitting | overfitting |
|---|---|---|---|---|---|---|
| AICu | 1 | 1 | 1 | 8100 | 1176 | 598 |
| HQc | 2 | 1 | 2 | 8038 | 610 | 1198 |
| AICc | 3 | 3 | 3 | 7838 | 425 | 1580 |
| deDCVB | 4 | 4 | 4 | 6496 | 624 | 2683 |
| SIC | 5 | 5 | 5 | 6247 | 182 | 3460 |
| deBFPE | 6 | 6 | 6 | 5847 | 233 | 3787 |
| FIC | 6 | 6 | 6 | 5718 | 45 | 4157 |
| meCp | 6 | 6 | 6 | 5338 | 51 | 4443 |
| trDCVB | 6 | 9 | 9 | 5390 | 1385 | 2812 |
| trBFPE | 10 | 9 | 10 | 5218 | 658 | 3864 |
| deCV | 12 | 11 | 12 | 4555 | 100 | 5232 |
| trCp | 12 | 11 | 12 | 4637 | 44 | 5245 |
| trCV | 11 | 13 | 11 | 4471 | 302 | 4987 |
| HQ | 14 | 14 | 14 | 4177 | 39 | 5725 |
| trFPE | 14 | 15 | 14 | 3822 | 103 | 5970 |
| FPE | 16 | 16 | 16 | 3491 | 23 | 6436 |
| AIC | 17 | 17 | 17 | 3209 | 21 | 6724 |
| ICOMP | 18 | 18 | 18 | 140 | 5607 | 1456 |

Table 9.26. Simulation results summary for Model 8.
K-L observed efficiency ranks, $L_2$ observed efficiency ranks and counts.

| criterion | K-L ranking | tr$\{L_2\}$ ranking | det$(L_2)$ ranking | true | underfitting | overfitting |
|---|---|---|---|---|---|---|
| AICu | 1 | 1 | 1 | 1 | 9960 | 2 |
| HQc | 2 | 2 | 2 | 6 | 9851 | 13 |
| deDCVB | 3 | 3 | 3 | 16 | 9656 | 54 |
| AICc | 4 | 3 | 4 | 14 | 9729 | 28 |
| trDCVB | 5 | 5 | 4 | 31 | 9462 | 108 |
| SIC | 6 | 5 | 6 | 25 | 9361 | 170 |
| deBFPE | 7 | 7 | 7 | 31 | 9237 | 176 |
| trBFPE | 7 | 8 | 7 | 44 | 9045 | 245 |
| ICOMP | 9 | 10 | 9 | 106 | 7234 | 929 |
| FIC | 10 | 9 | 11 | 75 | 8450 | 365 |
| trCV | 10 | 10 | 10 | 82 | 7841 | 631 |
| trCp | 12 | 10 | 12 | 75 | 7849 | 757 |
| deCV | 13 | 10 | 13 | 66 | 7823 | 666 |
| HQ | 13 | 10 | 13 | 75 | 7755 | 875 |
| trFPE | 15 | 15 | 13 | 112 | 7106 | 1122 |
| FPE | 16 | 16 | 16 | 87 | 6838 | 1335 |
| AIC | 16 | 17 | 17 | 91 | 6579 | 1570 |
| meCp | 18 | 18 | 18 | 170 | 6158 | 579 |

FPE do not overfit as excessively here as for Model 7. MeCp has the lowest K-L observed efficiency, at 40.3%. ICOMP performs much better here than in Model 7 because its underfitted model choices match the models selected by the three distances much better, resulting in higher observed efficiency and better performance.

### 9.5.3. Large-scale Small-sample Simulations

Our goal for this simulation study is to see how the criteria behave over a wide range of models by varying the model characteristics likely to have an impact on performance. Table 9.27 summarizes the characteristics and values to be used in generating the 504 models to study.

Using models of the form in Eq. (9.5), different multivariate models can be created by varying its components. The sample sizes in this study range from small (15) to moderate (100), and six values of $\Sigma_*$ are chosen. The diagonal elements of $\Sigma_*$ affect the identifiability of the model, and the off-diagonals represent the covariance between the errors. This may lead to greater differences between traces and determinants, since the trace ignores off-diagonal elements of the residual error matrix. The error correlation is either 0.2 or 0.7, and since overfitting has an impact on model selection, two levels of overfitting ($o$) are used, $o = 2$ and $o = 4$. Seven parameter structures including true order $k_*$ are considered, where the relationship between parameter structure and true order is explained in Table 9.28. Lastly, the columns of $X$ may be correlated.

Table 9.27. Summary of multivariate regression models.

| Error Covariance $\Sigma_*$ | Sample Size $n$ | Parameter Structure and True Order $k_*$ | Overfitting $o$ | $\rho_x$ |
|---|---|---|---|---|
| $\begin{pmatrix} 0.1 & 0.02 \\ 0.02 & 0.1 \end{pmatrix}$ | 15, 35, 100 | 7 structures | 2, 4 | 0, 0.8 |
| $\begin{pmatrix} 1 & 0.2 \\ 0.2 & 1 \end{pmatrix}$ | 15, 35, 100 | 7 structures | 2, 4 | 0, 0.8 |
| $\begin{pmatrix} 10 & 2 \\ 2 & 10 \end{pmatrix}$ | 15, 35, 100 | 7 structures | 2, 4 | 0, 0.8 |
| $\begin{pmatrix} 0.1 & 0.07 \\ 0.07 & 0.1 \end{pmatrix}$ | 15, 35, 100 | 7 structures | 2, 4 | 0, 0.8 |
| $\begin{pmatrix} 1 & 0.7 \\ 0.7 & 1 \end{pmatrix}$ | 15, 35, 100 | 7 structures | 2, 4 | 0, 0.8 |
| $\begin{pmatrix} 10 & 7 \\ 7 & 10 \end{pmatrix}$ | 15, 35, 100 | 7 structures | 2, 4 | 0, 0.8 |

Table 9.28. Relationship between parameter structure and true order.

| |
|---|
| Structure 1 |
| $k_* = 3 \quad B_0 = \binom{1}{1}, B_1 = \binom{1}{1}, B_2 = \binom{1/4}{1/4}$ |
| Structure 2 |
| $k_* = 5 \quad B_0 = \binom{1}{1}, B_1 = \binom{1}{1}, B_2 = \binom{1/4}{1/4}, B_3 = \binom{1/9}{0}, B_4 = \binom{1/16}{0}$ |
| Structure 3 |
| $k_* = 5 \quad B_0 = \binom{1}{1}, B_1 = \binom{1}{1}, B_2 = \binom{1/4}{1/4}, B_3 = \binom{1/9}{1/9}, B_4 = \binom{1/16}{1/16}$ |
| Structure 4 |
| $k_* = 3 \quad B_0 = \binom{1}{1}, B_1 = \binom{1}{1}, B_2 = \binom{1}{1}$ |
| Structure 5 |
| $k_* = 5 \quad B_0 = \binom{1}{1}, B_1 = \binom{1}{1}, B_2 = \binom{1/4}{1}, B_3 = \binom{1/9}{0}, B_4 = \binom{1/16}{0}$ |
| Structure 6 |
| $k_* = 5 \quad B_0 = \binom{1}{1}, B_1 = \binom{1}{1}, B_2 = \binom{1/4}{1}, B_3 = \binom{0}{1}, B_4 = \binom{0}{1}$ |
| Structure 7 |
| $k_* = 5 \quad B_0 = \binom{1}{1}, B_1 = \binom{1}{1}, B_2 = \binom{1/4}{1}, B_3 = \binom{1/9}{1}, B_4 = \binom{1/16}{1}$ |

Let the correlation be of the form $\rho_x = corr(x_{i,j}, x_{i,j+1})$. We use either $\rho_x = 0$, independent regressors (low collinearity), or $\rho_x = 0.8$.

As in univariate regression, the columns of $X$ are generated by conditioning on the previous column. Let $x_{i,0} = 1$ and $x_{i,1} \sim N(0,1)$, and generate $x_{i,j+1}|x_{i,j}$ for $j = 1, \ldots, k_* - 2$. With true order $k_*$ and overfitting $o$, the total number of variables in the model, including the intercept, yields $K = k_* + o$ as the dimension of the largest model. All subsets are considered. Summary results of overall rank are given in Table 9.29, and details in Tables 9A.20, 9A.21 and 9A.22.

We have noted before that, in small samples, the asymptotic properties of selection criteria are less important than their small-sample signal-to-noise ratios. Therefore it is not surprising that the top performers from the multivariate special case models reappear here, AICu, HQc, AICc, and deDCVB. However, it is important to remember that these criteria may not have the highest observed efficiency for each realization and model, but that the above rankings reflect overall performance tendencies over all realizations. This is clearly seen in the detail tables (9A.20–9A.22), where AICu is the best overall performer but finishes last in K-L in 158 of the 50,400 realizations. Although ICOMP finishes last overall, it has the highest K-L observed efficiency in 3101 of the realizations. For any particular model or realization any of the criteria may perform poorly. Other trends reappear as well; the trace of Cp outperforms maximum eigenvalue of Cp, but in general, the determinants outperform their trace counterparts. The cross-validation criteria and the bootstraps (with the exception of deDCVB and deBFPE) perform near the middle.

Table 9.29. Simulation results over 504 Models.
Summary of overall rank by K-L and $L_2$ observed efficiency.

| criterion | K-L ranking | tr$\{L_2\}$ ranking | det$(L_2)$ ranking |
|---|---|---|---|
| AICu | 1 | 1 | 1 |
| HQc | 2 | 1 | 2 |
| AICc | 3 | 3 | 4 |
| deDCVB | 4 | 3 | 3 |
| deBFPE | 5 | 5 | 5 |
| SIC | 6 | 6 | 6 |
| trDCVB | 7 | 7 | 11 |
| trBFPE | 8 | 8 | 12 |
| deCV | 9 | 9 | 7 |
| trCp | 10 | 10 | 8 |
| HQ | 11 | 11 | 9 |
| meCp | 13 | 12 | 13 |
| trCV | 12 | 13 | 16 |
| FIC | 14 | 14 | 10 |
| FPE | 16 | 15 | 14 |
| trFPE | 15 | 16 | 17 |
| AIC | 17 | 16 | 15 |
| ICOMP | 18 | 18 | 18 |

## 9.5.4. Large-sample Simulations

Our last model simulation for multivariate regression uses a very large sample size of 25,000 in order to evaluate large-sample behavior of the selection criteria as well as their asymptotic properties. We will also be able to observe the effect of varying the number of extraneous variables by using two models that differ only in the amount of overfitting possible. Model A6 represents a model with small opportunity for overfitting, $o = 2$. Model A7 represents a model with larger opportunity for overfitting, $o = 5$. In both models $k_* = 2$, one thousand realizations were generated, and

$$y_i = \begin{pmatrix} 1 \\ 1 \end{pmatrix} + \begin{pmatrix} 1 \\ 1 \end{pmatrix} x_{i,1} + \varepsilon_{*i} \text{ with } cov[\varepsilon_{*i}] = \Sigma_* = \begin{pmatrix} 1 & 0.7 \\ 0.7 & 1 \end{pmatrix}.$$

Summary results for Model A6 are given in Table 9.30, and details in Table 9A.23.

SIC and FIC are strongly consistent, identifying the true model every time for observed efficiencies of 1. While HQ and HQc are asymptotically consistent, even for a large sample size of $n = 25{,}000$ they are less so, identifying the true model in only 973 of the 1000 realizations. The efficient criteria AIC, AICc,

Table 9.30. Simulation results summary for Model A6.
K-L observed efficiency ranks, $L_2$ observed efficiency ranks and counts.

| criterion | K-L ranking | tr$\{L_2\}$ ranking | det$(L_2)$ ranking | true | underfitting | overfitting |
|---|---|---|---|---|---|---|
| FIC | 1 | 1 | 1 | 1000 | 0 | 0 |
| SIC | 1 | 1 | 1 | 1000 | 0 | 0 |
| HQ | 3 | 3 | 3 | 973 | 0 | 27 |
| HQc | 3 | 3 | 3 | 973 | 0 | 27 |
| AICu | 5 | 5 | 5 | 889 | 0 | 111 |
| deBFPE | 5 | 5 | 5 | 885 | 0 | 115 |
| deDCVB | 5 | 5 | 5 | 885 | 0 | 115 |
| trBFPE | 8 | 8 | 8 | 859 | 0 | 141 |
| trDCVB | 8 | 8 | 8 | 860 | 0 | 140 |
| trFPE | 10 | 10 | 10 | 745 | 0 | 255 |
| trCV | 11 | 10 | 10 | 742 | 0 | 258 |
| AIC | 12 | 10 | 10 | 727 | 0 | 273 |
| AICc | 12 | 10 | 10 | 727 | 0 | 273 |
| deCV | 12 | 10 | 10 | 727 | 0 | 273 |
| FPE | 12 | 10 | 10 | 727 | 0 | 273 |
| trCp | 12 | 10 | 10 | 726 | 0 | 274 |
| meCp | 17 | 10 | 17 | 689 | 0 | 311 |
| ICOMP | 18 | 18 | 18 | 626 | 0 | 374 |

Table 9.31. Simulation results summary for Model A7.
K-L observed efficiency ranks, $L_2$ observed efficiency ranks and counts.

| criterion | K-L ranking | tr$\{L_2\}$ ranking | det$(L_2)$ ranking | true | underfitting | overfitting |
|---|---|---|---|---|---|---|
| FIC | 1 | 1 | 1 | 1000 | 0 | 0 |
| SIC | 1 | 1 | 1 | 1000 | 0 | 0 |
| HQ | 3 | 3 | 3 | 960 | 0 | 40 |
| HQc | 3 | 3 | 3 | 960 | 0 | 40 |
| AICu | 5 | 5 | 5 | 817 | 0 | 183 |
| deBFPE | 5 | 5 | 5 | 811 | 0 | 189 |
| deDCVB | 5 | 5 | 5 | 813 | 0 | 187 |
| trBFPE | 8 | 8 | 8 | 733 | 0 | 267 |
| trDCVB | 8 | 8 | 8 | 734 | 0 | 266 |
| AIC | 10 | 10 | 10 | 566 | 0 | 434 |
| AICc | 10 | 10 | 10 | 566 | 0 | 434 |
| deCV | 10 | 10 | 10 | 567 | 0 | 433 |
| FPE | 10 | 10 | 10 | 566 | 0 | 434 |
| trCp | 10 | 10 | 10 | 566 | 0 | 434 |
| trCV | 10 | 15 | 15 | 519 | 0 | 481 |
| trFPE | 10 | 15 | 15 | 521 | 0 | 479 |
| ICOMP | 17 | 17 | 17 | 373 | 0 | 627 |
| meCp | 18 | 18 | 18 | 132 | 0 | 868 |

trCp, and FPE, all perform about the same; that is, poorly. The efficient criteria identify the true model less than 73% of the time, overfitting the rest of the time. We also see differences between the performance of trace- and determinant-based criteria. For the bootstrapped criteria, the determinant outperforms the trace. On the other hand, while the determinant FPE (FPE) and deCV are both efficient, their traces perform slightly better. The trace of Cp (trCp) is efficient and outperforms meCp. MeCp and ICOMP all perform worse than the efficient criteria. As we would expect, AICu ranks in between the consistent and the efficient criteria. The determinants of the bootstraps behave similarly to AICu, overfitting by one variable with roughly the same probability.

Results for Model A7, with $o$ increased to 5, are given in Table 9.31. Detailed results are given in Table 9A.24. As consistent criteria, SIC and FIC are not affected by the increased opportunity for overfitting. However, larger $o$ translates into lower correct model counts and decreased observed efficiencies for the efficient criteria, which identify the true model less than 57% of the time. AICu is not affected as strongly by the increased opportunity to overfit, again ranking in between the consistent and efficient criteria. AICu, deBFPE, and deDCVB all behave similarly with the determinants outperform the traces. ICOMP and the maximum eigenvalue of Cp still overfit excessively, resulting in poor performance.

We can see in Table 9A.23 and 9A.24 that increasing the number of extraneous variables increases results in more actual overfitting for Model A7 (Table 9A.23) than for Model A6 (Table 9A.24). Not only does total overfitting increase in Model A7, but the degree of overfitting increases also. This suggests that care should be taken when compiling the list of possible variables to include in a study.

### *9.5.5. Real Data Example*

To examine performance under practical conditions for multivariate regression, the selection criteria are applied to a real data example. Consider the tobacco leaf data from Anderson and Bancroft (1952, p. 205) and presented in Bedrick and Tsai (1994). We use the first two columns of the data as $y_1$ and $y_2$, and ignore $y_3$ in this analysis. The data relate $y_1$, rate of cigarette burn in inches per 1,000 seconds, and $y_2$, percent of sugar in leaf, to the following independent variables: x1, percent nitrogen; x2, percent chlorine; x3, percent potassium; x4, percent phosphorus; x5, percent calcium; x6, percent magnesium. There are $n = 25$ observations and all subsets are considered.

The various models chosen by the criteria are given in Table 9.32.

Table 9.32. Multivariate real data selected models.

| Variables Selected | Selection Criteria |
|---|---|
| x1 | ICOMP |
| x1 x2 x6 | AICc, AICu, HQc, SIC, trCp |
| x2 x4 x6 | trDCVB |
| x1 x2 x3 x4 | trCV |
| x1 x2 x4 x6 | AIC, deCV, FIC, FPE, HQ, meCp, trFPE |
| x2 x3 x4 x6 | trBFPE, deDCVB |
| x2 x3 x4 x5 x6 | deBFPE |

Table 9.33. Multivariate regression results for x1, x2, x6.

| VARIABLE | PARAMETER ESTIMATE | COVARIANCE $\hat{B}$ | P-VALUE |
|---|---|---|---|
| INTERCEPT | $\begin{pmatrix} 2.078 \\ 25.446 \end{pmatrix}$ | $\begin{pmatrix} 0.047 & -0.046 \\ -0.046 & 6.007 \end{pmatrix}$ | 0.0001 |
| X1 | $\begin{pmatrix} 0.317 \\ -3.760 \end{pmatrix}$ | $\begin{pmatrix} 0.011 & -0.011 \\ -0.011 & 1.401 \end{pmatrix}$ | 0.0023 |
| X2 | $\begin{pmatrix} -0.140 \\ 1.414 \end{pmatrix}$ | $\begin{pmatrix} 0.0015 & -0.0014 \\ -0.0014 & 0.1874 \end{pmatrix}$ | 0.0008 |
| X6 | $\begin{pmatrix} -0.754 \\ -4.386 \end{pmatrix}$ | $\begin{pmatrix} 0.053 & -0.051 \\ -0.051 & 6.372 \end{pmatrix}$ | 0.0050 |

$$\hat{\Sigma} = \begin{pmatrix} 0.0103 & -0.0100 \\ -0.0100 & 1.3120 \end{pmatrix} \quad S^2 = \begin{pmatrix} 0.0123 & -0.0119 \\ -0.0119 & 1.5619 \end{pmatrix}$$

Table 9.34. Multivariate regression results for x1, x2, x4, x6.

| VARIABLE | PARAMETER ESTIMATE | COVARIANCE $\hat{B}$ | P-VALUE |
|---|---|---|---|
| INTERCEPT | $\begin{pmatrix} 2.4464 \\ 18.823 \end{pmatrix}$ | $\begin{pmatrix} 0.1692 & -0.0055 \\ -0.0055 & 19.6958 \end{pmatrix}$ | 0.0001 |
| x1 | $\begin{pmatrix} 0.2882 \\ -3.236 \end{pmatrix}$ | $\begin{pmatrix} 0.0117 & -0.0004 \\ -0.0004 & 1.3642 \end{pmatrix}$ | 0.0053 |
| x2 | $\begin{pmatrix} -0.1401 \\ 1.411 \end{pmatrix}$ | $\begin{pmatrix} .0015 & -0.0000 \\ -0.0000 & 0.1706 \end{pmatrix}$ | 0.0005 |
| x4 | $\begin{pmatrix} -0.7592 \\ 13.664 \end{pmatrix}$ | $\begin{pmatrix} 0.5206 & -0.0171 \\ -0.0171 & 60.5907 \end{pmatrix}$ | 0.1648 |
| x6 | $\begin{pmatrix} -0.6849 \\ -5.627 \end{pmatrix}$ | $\begin{pmatrix} .0570 & -0.0019 \\ -0.0019 & 6.6358 \end{pmatrix}$ | 0.0085 |

$$\hat{\Sigma} = \begin{pmatrix} 0.0098 & -0.0003 \\ -0.0003 & 1.1370 \end{pmatrix} \quad S^2 = \begin{pmatrix} 0.0122 & -0.0004 \\ -0.0004 & 1.4214 \end{pmatrix}$$

The two most popular models, (x1, x2, x6) and (x1, x2, x4, x6), warrant further investigation. Table 9.33 summarizes the regression results from the (x1, x2, x6) model. We can see that this model contains variables that are important to both $y_1$ and $y_2$, with the possible exception of x6, which is doubtful for $y_2$. An examination of residual plots for this model revealed no problems. Regression results for the second most popular model, which contains the additional variable x4, are given in Table 9.34.

The regression statistics reveal that the x4 variable is of borderline significance to both $y_1$ and $y_2$, and thus may be the result of overfitting. The residual plots do not indicate any problems for this model either. In the interest of simplicity and the absence of any overwhelming evidence that x4 is important, we conclude that the (x1, x2, x6) model is the more appropriate.

## 9.6. Vector Autoregressive Models

### *9.6.1. Model Structure*

The last set of simulations in this Chapter covers vector autoregressive (VAR) models. We recall the true vector autoregressive VAR($p_*$) model given by

$$y_t = \Phi_1 y_{t-1} + \cdots + \Phi_{p_*} y_{t-p_*} + w_{*t}, \quad w_{*t} \sim N_q(0, \Sigma_*), \quad t = p_* + 1, \ldots, n, \quad (9.6)$$

where the $w_{*t}$ are independent, $y_t$ is a $q \times 1$ vector observed at $t = 1, \ldots, n$, and the $\Phi_j$ are $q \times q$ matrices of unknown parameters. For the simulations in this Section we will generate different VAR models by choosing varying $p_*$, $n$, $\Sigma_*$, dimension $q$ and parameter matrices $\Phi$.

For each realization a new $w_*$ error matrix was generated, and hence a new $Y = (y_1, \ldots, y_n)'$. Then candidate models of the form in Eq. (5.1) are fit to the data. Three observed efficiency measures are computed for each candidate model. K-L observed efficiency is computed using Eq. (5.8) and Eq. (1.2); $\mathrm{tr}\{L_2\}$ observed efficiency is computed using Eq. (5.6) and Eq. (1.1); $\det(L_2)$ observed efficiency is computed using Eq. (5.7) and Eq. (1.1).

### *9.6.2. Two Special Case Models*

Our two special case VAR models are two-dimensional models that have either strongly identifiable or weakly identifiable parameters. In both cases the true model has order $n = 35$, $p_* = 4$, and $cov[w_{*t}] = \Sigma_* = \begin{pmatrix} 1 & 0.7 \\ 0.7 & 1 \end{pmatrix}$. The largest model order considered is $P = 8$. The two models are:

**Model 9** (parameter structure 22 in Table 9.38)

$$y_t = \begin{pmatrix} 0.090 & 0 \\ 0 & 0.090 \end{pmatrix} y_{t-1} + \begin{pmatrix} 0 & 0 \\ 0 & 0 \end{pmatrix} y_{t-2} + \begin{pmatrix} 0 & 0 \\ 0 & 0 \end{pmatrix} y_{t-3} + \begin{pmatrix} 0 & 0.900 \\ 0.900 & 0 \end{pmatrix} y_{t-4} + w_{*t}$$

and

**Model 10** (parameter structure 4 in Table 9.38)

$$y_t = \begin{pmatrix} 0.024 & 0.241 \\ 0.024 & 0.241 \end{pmatrix} y_{t-1} + \begin{pmatrix} 0 & 0.241 \\ 0 & 0.241 \end{pmatrix} y_{t-2} + \begin{pmatrix} 0 & 0.241 \\ 0 & 0.241 \end{pmatrix} y_{t-3} + \begin{pmatrix} 0 & 0.241 \\ 0 & 0.241 \end{pmatrix} y_{t-4} + w_{*t}.$$

These are the same models described in Eqs. (5.18)–(5.20). Tables 9A.25 and 9A.26 present average observed efficiency over the 10,000 realizations. Unlike all subsets regression, the candidate orders are nested. Table 9.35 gives the summary results for Model 9, and details can be found in Table 9A.25.

For the strongly identifiable Model 9, Table 9.35 shows that HQc not only outperforms the other consistent criteria SIC, HQ and FIC, it ties with AICc for first over all measures. Both criteria identify the correct model over 99%

Table 9.35. Simulation results summary for Model 9.
K-L observed efficiency ranks, $L_2$ observed efficiency ranks and counts.

| criterion | K-L ranking | tr$\{L_2\}$ ranking | det$(L_2)$ ranking | true | underfitting | overfitting |
|---|---|---|---|---|---|---|
| AICc | 1 | 1 | 1 | 9906 | 26 | 68 |
| HQc | 1 | 1 | 1 | 9918 | 55 | 27 |
| AICu | 3 | 3 | 3 | 9864 | 124 | 12 |
| deDCVB | 4 | 4 | 4 | 9510 | 20 | 470 |
| trDCVB | 5 | 5 | 5 | 9239 | 56 | 705 |
| deBFPE | 6 | 6 | 6 | 9192 | 15 | 793 |
| trBFPE | 7 | 7 | 7 | 8897 | 36 | 1067 |
| SIC | 8 | 8 | 8 | 8743 | 14 | 1243 |
| deCV | 9 | 9 | 9 | 8354 | 6 | 1640 |
| trCV | 10 | 10 | 10 | 8094 | 11 | 1895 |
| trCp | 11 | 11 | 11 | 6585 | 4 | 3411 |
| trFPE | 12 | 12 | 12 | 6117 | 3 | 3880 |
| FPE | 13 | 13 | 13 | 6041 | 1 | 3958 |
| meCp | 13 | 13 | 13 | 4531 | 1 | 5468 |
| HQ | 15 | 15 | 15 | 6059 | 2 | 3939 |
| FIC | 16 | 16 | 16 | 3895 | 0 | 6105 |
| AIC | 17 | 17 | 17 | 4464 | 0 | 5536 |
| ICOMP | 17 | 18 | 18 | 1601 | 8361 | 38 |

Table 9.36. Simulation results summary for Model 10.
K-L observed efficiency ranks, $L_2$ observed efficiency ranks and counts.

| criterion | K-L ranking | tr$\{L_2\}$ ranking | det$(L_2)$ ranking | true | underfitting | overfitting |
|---|---|---|---|---|---|---|
| trDCVB | 7 | 2 | 7 | 754 | 9117 | 129 |
| trBFPE | 8 | 3 | 8 | 837 | 8913 | 250 |
| trCV | 11 | 1 | 11 | 1627 | 7542 | 831 |
| ICOMP | 9 | 3 | 9 | 1308 | 7866 | 826 |
| AICc | 3 | 12 | 3 | 79 | 9920 | 1 |
| deDCVB | 5 | 10 | 5 | 250 | 9727 | 23 |
| deCV | 10 | 5 | 10 | 995 | 8542 | 463 |
| HQc | 2 | 14 | 2 | 23 | 9977 | 0 |
| deBFPE | 6 | 11 | 6 | 290 | 9653 | 57 |
| SIC | 4 | 14 | 3 | 161 | 9726 | 113 |
| AICu | 1 | 18 | 1 | 4 | 9996 | 0 |
| meCp | 16 | 5 | 16 | 2605 | 2866 | 4529 |
| trCp | 13 | 8 | 13 | 1253 | 6454 | 2293 |
| trFPE | 15 | 7 | 15 | 1815 | 5632 | 2553 |
| FPE | 14 | 8 | 14 | 1287 | 6525 | 2188 |
| HQ | 12 | 12 | 12 | 761 | 7513 | 1726 |
| AIC | 17 | 14 | 17 | 1044 | 5260 | 3696 |
| FIC | 18 | 14 | 18 | 1783 | 2586 | 5631 |

of the time. AICu tends to underfit (see Table 9A.25), but still performs well overall. Both forms of the doubly cross-validated bootstrap rank in the top five, and in general, the determinant forms outperform the trace forms. AIC severely overfits, as does ICOMP, which is worst overall. Next we look at the weakly identifiable VAR Model 10.

Table 9.36 summarizes the observed efficiency rankings, for which details can be found in Table 9A.26. The top five in Table 9.36 are trDCVB, trBFPE, trCV, ICOMP, and AICc. Model 10 is an example of a case where heavy penalty functions may hinder performance. Because special case Models 9 and 10 yield very different results, we need a wider range of VAR models to attempt to judge overall performance. Our large-scale small-sample simulation study, given in the next Section, covers 864 VAR models.

### *9.6.3. Large-scale Small-sample Simulations*

Here we expand our consideration of VAR models to 864 models that cover six covariance matrices $\Sigma_*$, three sample sizes $n$ = 15, 35, 100, two levels of overfitting $o = 2, 4$ and 24 parameter structures, as summarized in Table 9.37. The maximum order considered is $P = p_* + o$. Each true VAR model is of the form in Eq. (9.6), and the 24 parameter structures are summarized in Table 9.38.

Table 9.37. Summary of vector autoregressive (VAR) models.

| Error Covariance $\Sigma_*$ | Sample Size $n$ | Parameter Structure and True Order $p_*$ | Overfitting $o$ |
|---|---|---|---|
| $\begin{pmatrix} 1 & 0.7 \\ 0.7 & 1 \end{pmatrix}$ | 15, 35, 100 | 24 structures | 2, 4 |
| $\begin{pmatrix} 1 & 0.99 \\ 0.99 & 1 \end{pmatrix}$ | 15, 35, 100 | 24 structures | 2, 4 |
| $\begin{pmatrix} 2 & 0.99 \\ 0.99 & 2 \end{pmatrix}$ | 15, 35, 100 | 24 structures | 2, 4 |
| $\begin{pmatrix} 2 & 1.4 \\ 1.4 & 1 \end{pmatrix}$ | 15, 35, 100 | 24 structures | 2, 4 |
| $\begin{pmatrix} 1 & 0.99 \\ 0.99 & 2 \end{pmatrix}$ | 15, 35, 100 | 24 structures | 2, 4 |
| $\begin{pmatrix} 1 & 1.4 \\ 1.4 & 2 \end{pmatrix}$ | 15, 35, 100 | 24 structures | 2, 4 |

Table 9.38. Relationship between parameter structure and true model order.

Structure 1: $p_* = 2$

$$\Phi_1 = \begin{pmatrix} 0.471 & 0.047 \\ 0.047 & 0.471 \end{pmatrix}, \Phi_2 = \begin{pmatrix} 0.471 & 0 \\ 0 & 0.471 \end{pmatrix}$$

Structure 2: $p_* = 4$

$$\Phi_1 = \begin{pmatrix} 0.241 & 0.024 \\ 0.024 & 0.241 \end{pmatrix}, \Phi_2 = \begin{pmatrix} 0.241 & 0 \\ 0 & 0.241 \end{pmatrix}, \Phi_3 = \begin{pmatrix} 0.241 & 0 \\ 0 & 0.241 \end{pmatrix}, \Phi_4 = \begin{pmatrix} 0.241 & 0 \\ 0 & 0.241 \end{pmatrix}$$

Structure 3: $p_* = 2$

$$\Phi_1 = \begin{pmatrix} 0.047 & 0.471 \\ 0.047 & 0.471 \end{pmatrix}, \Phi_2 = \begin{pmatrix} 0 & 0.471 \\ 0 & 0.471 \end{pmatrix}$$

Structure 4: $p_* = 4$

$$\Phi_1 = \begin{pmatrix} 0.024 & 0.241 \\ 0.024 & 0.241 \end{pmatrix}, \Phi_2 = \begin{pmatrix} 0 & 0.241 \\ 0 & 0.241 \end{pmatrix}, \Phi_3 = \begin{pmatrix} 0 & 0.241 \\ 0 & 0.241 \end{pmatrix}, \Phi_4 = \begin{pmatrix} 0 & 0.241 \\ 0 & 0.241 \end{pmatrix}$$

Structure 5: $p_* = 2$

$$\Phi_1 = \begin{pmatrix} 0.047 & 0.471 \\ 0.471 & 0.047 \end{pmatrix}, \Phi_2 = \begin{pmatrix} 0 & 0.471 \\ 0.471 & 0 \end{pmatrix}$$

Structure 6: $p_* = 4$

$$\Phi_1 = \begin{pmatrix} 0.024 & 0.241 \\ 0.241 & 0.024 \end{pmatrix}, \Phi_2 = \begin{pmatrix} 0 & 0.241 \\ 0.241 & 0 \end{pmatrix}, \Phi_3 = \begin{pmatrix} 0 & 0.241 \\ 0.241 & 0 \end{pmatrix}, \Phi_4 = \begin{pmatrix} 0 & 0.241 \\ 0.241 & 0 \end{pmatrix}$$

Structure 7: $p_* = 2$

$$\Phi_1 = \begin{pmatrix} 0.248 & 0.248 \\ 0.248 & 0.248 \end{pmatrix}, \Phi_2 = \begin{pmatrix} 0.248 & 0.248 \\ 0.248 & 0.248 \end{pmatrix}$$

Structure 8: $p_* = 4$

$$\Phi_1 = \begin{pmatrix} 0.124 & 0.124 \\ 0.124 & 0.124 \end{pmatrix}, \Phi_2 = \begin{pmatrix} 0.124 & 0.124 \\ 0.124 & 0.124 \end{pmatrix}, \Phi_3 = \begin{pmatrix} 0.124 & 0.124 \\ 0.124 & 0.124 \end{pmatrix}, \Phi_4 = \begin{pmatrix} 0.124 & 0.124 \\ 0.124 & 0.124 \end{pmatrix}$$

Structure 9: $p_* = 2$

$$\Phi_1 = \begin{pmatrix} 0.471 & 0.047 \\ 0.090 & 0 \end{pmatrix}, \Phi_2 = \begin{pmatrix} 0.471 & 0 \\ 0 & 0.090 \end{pmatrix}$$

Structure 10: $p_* = 4$

$$\Phi_1 = \begin{pmatrix} 0.241 & 0.024 \\ 0.090 & 0 \end{pmatrix}, \Phi_2 = \begin{pmatrix} 0.241 & 0 \\ 0 & 0 \end{pmatrix}, \Phi_3 = \begin{pmatrix} 0.241 & 0 \\ 0 & 0 \end{pmatrix}, \Phi_4 = \begin{pmatrix} 0.241 & 0 \\ 0 & 0.900 \end{pmatrix}$$

Table 9.38. Continued

Structure 11: $p_* = 2$

$$\Phi_1 = \begin{pmatrix} 0.047 & 0.471 \\ 0.090 & 0 \end{pmatrix}, \Phi_2 = \begin{pmatrix} 0 & 0.471 \\ 0 & 0.900 \end{pmatrix}$$

Structure 12: $p_* = 4$

$$\Phi_1 = \begin{pmatrix} 0.024 & 0.241 \\ 0.090 & 0 \end{pmatrix}, \Phi_2 = \begin{pmatrix} 0 & 0.241 \\ 0 & 0 \end{pmatrix}, \Phi_3 = \begin{pmatrix} 0 & 0.241 \\ 0 & 0 \end{pmatrix}, \Phi_4 = \begin{pmatrix} 0 & 0.241 \\ 0 & 0.900 \end{pmatrix}$$

Structure 13: $p_* = 2$

$$\Phi_1 = \begin{pmatrix} 0.090 & 0 \\ 0.471 & 0.047 \end{pmatrix}, \Phi_2 = \begin{pmatrix} 0 & 0.900 \\ 0.471 & 0 \end{pmatrix}$$

Structure 14: $p_* = 4$

$$\Phi_1 = \begin{pmatrix} 0.090 & 0 \\ 0.241 & 0.024 \end{pmatrix}, \Phi_2 = \begin{pmatrix} 0 & 0 \\ 0.241 & 0 \end{pmatrix}, \Phi_3 = \begin{pmatrix} 0 & 0 \\ 0.241 & 0 \end{pmatrix}, \Phi_4 = \begin{pmatrix} 0 & 0.900 \\ 0.241 & 0 \end{pmatrix}$$

Structure 15: $p_* = 2$

$$\Phi_1 = \begin{pmatrix} 0.330 & 0 \\ 0.330 & 0 \end{pmatrix}, \Phi_2 = \begin{pmatrix} 0.330 & 0.330 \\ 0.330 & 0.330 \end{pmatrix}$$

Structure 16: $p_* = 4$

$$\Phi_1 = \begin{pmatrix} 0.198 & 0 \\ 0.198 & 0 \end{pmatrix}, \Phi_2 = \begin{pmatrix} 0.198 & 0 \\ 0.198 & 0 \end{pmatrix}, \Phi_3 = \begin{pmatrix} 0.198 & 0 \\ 0.198 & 0 \end{pmatrix}, \Phi_4 = \begin{pmatrix} 0.198 & 0.198 \\ 0.198 & 0.198 \end{pmatrix}$$

Structure 17: $p_* = 2$

$$\Phi_1 = \begin{pmatrix} 0 & 0.090 \\ 0.090 & 0 \end{pmatrix}, \Phi_2 = \begin{pmatrix} 0.900 & 0 \\ 0 & 0.900 \end{pmatrix}$$

Structure 18: $p_* = 4$

$$\Phi_1 = \begin{pmatrix} 0 & 0.090 \\ 0.090 & 0 \end{pmatrix}, \Phi_2 = \begin{pmatrix} 0 & 0 \\ 0 & 0 \end{pmatrix}, \Phi_3 = \begin{pmatrix} 0 & 0 \\ 0 & 0 \end{pmatrix}, \Phi_4 = \begin{pmatrix} 0.900 & 0 \\ 0 & 0.900 \end{pmatrix}$$

Structure 19: $p_* = 2$

$$\Phi_1 = \begin{pmatrix} 0.090 & 0 \\ 0.090 & 0 \end{pmatrix}, \Phi_2 = \begin{pmatrix} 0 & 0.900 \\ 0 & 0.900 \end{pmatrix}$$

Structure 20: $p_* = 4$

$$\Phi_1 = \begin{pmatrix} 0.090 & 0 \\ 0.090 & 0 \end{pmatrix}, \Phi_2 = \begin{pmatrix} 0 & 0 \\ 0 & 0 \end{pmatrix}, \Phi_3 = \begin{pmatrix} 0 & 0 \\ 0 & 0 \end{pmatrix}, \Phi_4 = \begin{pmatrix} 0 & 0.900 \\ 0 & 0.900 \end{pmatrix}$$

Structure 21: $p_* = 2$

$$\Phi_1 = \begin{pmatrix} 0.090 & 0 \\ 0 & 0.090 \end{pmatrix}, \Phi_2 = \begin{pmatrix} 0 & 0.900 \\ 0.900 & 0 \end{pmatrix}$$

Structure 22: $p_* = 4$

$$\Phi_1 = \begin{pmatrix} 0.090 & 0 \\ 0 & 0.090 \end{pmatrix}, \Phi_2 = \begin{pmatrix} 0 & 0 \\ 0 & 0 \end{pmatrix}, \Phi_3 = \begin{pmatrix} 0 & 0 \\ 0 & 0 \end{pmatrix}, \Phi_4 = \begin{pmatrix} 0 & 0.900 \\ 0.900 & 0 \end{pmatrix}$$

Structure 23: $p_* = 2$

$$\Phi_1 = \begin{pmatrix} 0 & 0 \\ 0 & 0 \end{pmatrix}, \Phi_2 = \begin{pmatrix} 0.495 & 0.495 \\ 0.495 & 0.495 \end{pmatrix}$$

Structure 24: $p_* = 4$

$$\Phi_1 = \begin{pmatrix} 0 & 0 \\ 0 & 0 \end{pmatrix}, \Phi_2 = \begin{pmatrix} 0 & 0 \\ 0 & 0 \end{pmatrix}, \Phi_3 = \begin{pmatrix} 0 & 0 \\ 0 & 0 \end{pmatrix}, \Phi_4 = \begin{pmatrix} 0.495 & 0.495 \\ 0.495 & 0.495 \end{pmatrix}$$

Table 9.39 summarizes the overall observed efficiency rankings for the small-sample VAR model simulation. Details are given in Tables 9A.27, 9A.28, and 9A.29.

The top performers over the 864 small-sample models are AICc, deDCVB, HQc, deBFPE, and AICu. Also, all the bootstrap criteria, with their strong penalty functions to prevent overfitting, perform quite well. It is interesting to note that in addition to the expected effects of $n$, $\det(\Sigma_*)$, $o$, and $\Phi$ on observed efficiency, for this simulation the seasonal models tended to be easier to identify and had higher observed efficiency. Less underfitting was observed for these models, and even when criteria chose an overfitted model their choices often matched those of the distance measures, resulting in better observed efficiency. Other parameter structures such as the one used in Model 10 (structure 4) had less agreement between the distance measures. The true order was difficult to detect, and as a result the criteria had smaller observed efficiencies.

Some general trends can be seen in Table 9.40. Because determinants take into account more elements of the residual error matrix (traces only include the diagonal elements), they outperform their trace counterparts. When the off-diagonals are important, the determinant may be a more reasonable basis for assessing models. trCp outperforms meCp, since the maximum eigenvalue focuses only on one eigenvalue of the residual error matrix and may give too little detail of the residual error matrix structure. For these reasons, we prefer using determinant variants. Also, our scalar criteria that estimate K-L (such as AICc) use determinants.

Overall, those criteria with stronger penalty functions and good signal-to-noise ratios performed well over all 864 models. Criteria with weak penalty functions and signal-to-noise ratios tended to overfit more and performed

Table 9.39. Simulation results over 864 Models.
Summary of overall rank by K-L and $L_2$ observed efficiency.

| criterion | K-L ranking | $\text{tr}\{L_2\}$ ranking | $\det(L_2)$ ranking |
|---|---|---|---|
| AICc | 2 | 3 | 3 |
| deDCVB | 4 | 1 | 1 |
| HQc | 1 | 4 | 1 |
| deBFPE | 5 | 2 | 3 |
| AICu | 2 | 6 | 5 |
| deCV | 8 | 5 | 6 |
| trDCVB | 6 | 7 | 8 |
| trBFPE | 7 | 8 | 8 |
| SIC | 8 | 9 | 7 |
| trCV | 10 | 10 | 10 |
| trCp | 11 | 11 | 11 |
| FPE | 12 | 12 | 12 |
| HQ | 12 | 13 | 13 |
| trFPE | 14 | 14 | 15 |
| meCp | 15 | 14 | 15 |
| FIC | 16 | 16 | 14 |
| AIC | 17 | 17 | 17 |
| ICOMP | 18 | 18 | 18 |

worse. Of all our models in this Chapter, VAR models showed the least amount of overfitting in terms of counts. At the other extreme, ICOMP does not perform well in our VAR models due to underfitting. The trends seen in Table 9.39 are similar to the trends seen for multivariate regression models in Table 9.29.

### *9.6.4. Large-sample Simulations*

For our large-sample simulation we use two models that differ only in the number of irrelevant variables in order to study the effect of overfitting $o$ on VAR model selection. Models A8 and A9 are described as

$$y_t = \begin{pmatrix} 0.95 & 0 \\ 0 & 0.95 \end{pmatrix} y_{t-1} + w_{*t} \text{ with } cov[w_{*t}] = \Sigma_* = \begin{pmatrix} 1 & 0.7 \\ 0.7 & 1 \end{pmatrix},$$

where $n = 25{,}000$, $p_* = 1$. For Model A8 $o = 2$, and for Model A9 $o = 4$. One thousand realizations were generated.

Table 9.40 gives the summary of count and observed efficiency results and the details can be found in Table 9A.30. The results here are in good agreement with those of the large-sample studies we have conducted in previous sections.

Table 9.40. Simulation results summary for Model A8.
K-L observed efficiency ranks, $L_2$ observed efficiency ranks and counts.

| criterion | K-L ranking | $\text{tr}\{L_2\}$ ranking | $\det(L_2)$ ranking | true | underfitting | overfitting |
|---|---|---|---|---|---|---|
| FIC | 1 | 1 | 1 | 1000 | 0 | 0 |
| HQ | 1 | 1 | 1 | 998 | 0 | 2 |
| HQc | 1 | 1 | 1 | 998 | 0 | 2 |
| SIC | 1 | 1 | 1 | 1000 | 0 | 0 |
| AICu | 5 | 5 | 5 | 971 | 0 | 29 |
| deBFPE | 5 | 5 | 5 | 970 | 0 | 30 |
| deDCVB | 7 | 7 | 7 | 929 | 0 | 71 |
| trBFPE | 7 | 7 | 7 | 938 | 0 | 62 |
| ICOMP | 9 | 9 | 9 | 868 | 0 | 132 |
| trDCVB | 9 | 9 | 9 | 877 | 0 | 123 |
| AIC | 11 | 11 | 11 | 839 | 0 | 161 |
| AICc | 11 | 11 | 11 | 839 | 0 | 161 |
| deCV | 11 | 11 | 11 | 839 | 0 | 161 |
| FPE | 11 | 11 | 11 | 839 | 0 | 161 |
| trCp | 11 | 11 | 11 | 839 | 0 | 161 |
| trCV | 16 | 16 | 16 | 802 | 0 | 198 |
| trFPE | 16 | 16 | 16 | 801 | 0 | 199 |
| meCp | 18 | 18 | 18 | 443 | 0 | 557 |

The consistent criteria SIC and FIC identify the true model every time. HQ and HQc are not quite as strongly consistent as SIC and FIC, thus, even for this large sample size, they do not identify the true model 100% of the time. When the true model belongs to the set of candidate models the efficient criteria (such as AIC) are no longer efficient. Because trCp and deCV are asymptotically equivalent to the efficient criterion FPE, they behave at the same poor level as the efficient criteria. Once again, AICu performs in between the efficient and consistent groups, as do the determinants of the bootstraps.

Summary results for Model A9 are given in Table 9.41, and details are given in Table 9A.31. We expect that the increased opportunity to overfit will result in lower observed efficiencies and hence lower counts, but that the overall trends for the criteria will hold.

We can see from Table 9.41 that the counts do in fact decrease, but not by much. The performance trends we observed in Table 9.40 for observed efficiency, consistency, and for AICu to behave somewhere in the middle also appear here. Once again the efficient criteria overfit more with increased $o$, but the strongly consistent criteria SIC and FIC are unaffected. The maximum eigenvalue of Cp, meCp, is prone to overfitting in large samples, and thus performs worst overall.

Table 9.41. Simulation results summary for Model A9.
K-L observed efficiency ranks, $L_2$ observed efficiency ranks and counts.

| criterion | K-L ranking | tr$\{L_2\}$ ranking | det$(L_2)$ ranking | true | underfitting | overfitting |
|---|---|---|---|---|---|---|
| FIC | 1 | 1 | 1 | 1000 | 0 | 0 |
| HQ | 1 | 1 | 1 | 998 | 0 | 2 |
| HQc | 1 | 1 | 1 | 998 | 0 | 2 |
| SIC | 1 | 1 | 1 | 1000 | 0 | 0 |
| AICu | 5 | 5 | 5 | 959 | 0 | 41 |
| deBFPE | 5 | 5 | 5 | 957 | 0 | 43 |
| deDCVB | 7 | 7 | 7 | 920 | 0 | 80 |
| trBFPE | 7 | 7 | 7 | 928 | 0 | 72 |
| trDCVB | 9 | 9 | 9 | 874 | 0 | 126 |
| AIC | 10 | 10 | 10 | 839 | 0 | 161 |
| AICc | 10 | 10 | 10 | 839 | 0 | 161 |
| deCV | 10 | 10 | 10 | 836 | 0 | 164 |
| FPE | 10 | 10 | 10 | 839 | 0 | 161 |
| ICOMP | 10 | 10 | 10 | 855 | 0 | 145 |
| trCp | 10 | 10 | 10 | 839 | 0 | 161 |
| trCV | 16 | 16 | 16 | 775 | 0 | 225 |
| trFPE | 16 | 16 | 16 | 775 | 0 | 225 |
| meCp | 18 | 18 | 18 | 295 | 0 | 705 |

### 9.6.5. Real Data Example

Our practical example for VAR data will make use of Wei (1990, p. 330, exercise 13.5). We will model a two-dimensional relationship between house sales (in thousands) and housing starts (in thousands) for the time period January 1965 and December 1975. The data are first centered by removing $\bar{Y}_{sales} = 79.255$ and $\bar{Y}_{starts} = 45.356$, and our criteria are used to select a model order. Table 9.42 summarizes model orders selected.

Table 9.42. VAR real data selected models.

| Order $p$ | Selection Criteria |
|---|---|
| 2 | HQC, SIC |
| 3 | ICOMP |
| 7 | FIC |
| 9 | meCp, trBFPE, trDCVB |
| 11 | AIC, AICc, AICu, deCV, deBFPE, deDCVB, FPE, HQ, trCp, trCV, trFPE |

We see that most of the criteria selected order 11, whereas SIC and HQc selected order 2. VAR modeling results for order 2 and order 11 models are shown in Tables 9.43 and 9.44, respectively. We will examine these two candidate models in more detail.

Table 9.43. Summary of VAR(2) model.

| Order | $\hat{\Phi}$ | s.e.$(\hat{\Phi}_{p,i,j})$ | P-VALUE |
|---|---|---|---|
| 1 | $\begin{pmatrix} 0.9776 & 1.2389 \\ -0.1159 & 0.4090 \end{pmatrix}$ | $\begin{pmatrix} 0.0966 & 0.1720 \\ 0.0517 & 0.0921 \end{pmatrix}$ | 0.000 |
| 2 | $\begin{pmatrix} 0.3277 & 0.6047 \\ -0.1205 & -0.3578 \end{pmatrix}$ | $\begin{pmatrix} 0.1308 & 0.2328 \\ 0.0410 & 0.0730 \end{pmatrix}$ | 0.000 |

Residuals from the VAR(2) model show significant peaks in their individual autocorrelation functions (ACF), indicating that the VAR(2) model is underfitted and that important relationships with past time periods have been omitted. On the other hand, the VAR(11) model residuals appear to be a white noise. The cross-correlation function (CCF) indicates that the residuals of house sales and house starts are correlated at lag 0 only, and no past dependencies are observed. The individual residual series are only weakly dependent on each other, but each is strongly correlated with its own past values. The residuals appear to be a vector of white noise and no underfitting is present from VAR(11). P-values in Table 9.44 for lag 11 elements show that at least one element of the $\Phi_{11}$ parameter is nonzero, and thus the VAR(11) model does not seem to represent a case of overfitting. Since there is no evidence of underfitting or overfitting for this model, and it is the most popular choice among the selection criteria, we feel that VAR(11) is the best model.

Table 9.44. Summary of VAR(11) model.

| Order | $\hat{\Phi}$ | s.e.$(\hat{\Phi}_{p,i,j})$ | P-VALUE |
|---|---|---|---|
| 1 | $\begin{pmatrix} 0.9434 & 1.1631 \\ -0.0586 & 0.3078 \end{pmatrix}$ | $\begin{pmatrix} 0.1068 & 0.1950 \\ 0.0579 & 0.1057 \end{pmatrix}$ | 0.000 |
| 2 | $\begin{pmatrix} 0.2798 & 0.3790 \\ -0.1091 & -0.0385 \end{pmatrix}$ | $\begin{pmatrix} 0.1440 & 0.2629 \\ 0.0615 & 0.1122 \end{pmatrix}$ | 0.055 |
| 3 | $\begin{pmatrix} -0.1957 & 0.1162 \\ 0.1049 & 0.0454 \end{pmatrix}$ | $\begin{pmatrix} 0.1453 & 0.2653 \\ 0.0626 & 0.1143 \end{pmatrix}$ | 0.097 |
| 4 | $\begin{pmatrix} -0.0145 & -0.7758 \\ -0.0357 & -0.1038 \end{pmatrix}$ | $\begin{pmatrix} 0.1365 & 0.2491 \\ 0.0624 & 0.1138 \end{pmatrix}$ | 0.002 |
| 5 | $\begin{pmatrix} 0.0680 & 0.4361 \\ 0.0212 & 0.1342 \end{pmatrix}$ | $\begin{pmatrix} 0.1362 & 0.2486 \\ 0.0620 & 0.1132 \end{pmatrix}$ | 0.082 |
| 6 | $\begin{pmatrix} -0.0725 & -0.3129 \\ -0.0557 & -0.0359 \end{pmatrix}$ | $\begin{pmatrix} 0.1319 & 0.2407 \\ 0.0613 & 0.1120 \end{pmatrix}$ | 0.196 |
| 7 | $\begin{pmatrix} 0.0211 & 0.0977 \\ 0.1527 & 0.0016 \end{pmatrix}$ | $\begin{pmatrix} 0.1327 & 0.2422 \\ 0.0607 & 0.1109 \end{pmatrix}$ | 0.014 |
| 8 | $\begin{pmatrix} -0.2280 & -0.0592 \\ -0.1257 & -0.2471 \end{pmatrix}$ | $\begin{pmatrix} 0.1311 & 0.2393 \\ 0.0620 & 0.1132 \end{pmatrix}$ | 0.045 |
| 9 | $\begin{pmatrix} 0.1431 & 0.5134 \\ 0.0911 & 0.1816 \end{pmatrix}$ | $\begin{pmatrix} 0.1376 & 0.2512 \\ 0.0632 & 0.1153 \end{pmatrix}$ | 0.044 |
| 10 | $\begin{pmatrix} -0.0326 & -0.1430 \\ 0.1894 & -0.0931 \end{pmatrix}$ | $\begin{pmatrix} 0.1355 & 0.2472 \\ 0.0624 & 0.1139 \end{pmatrix}$ | 0.003 |
| 11 | $\begin{pmatrix} -0.2194 & -0.4079 \\ -0.0539 & 0.3191 \end{pmatrix}$ | $\begin{pmatrix} 0.1338 & 0.2443 \\ 0.0476 & 0.0868 \end{pmatrix}$ | 0.000 |

## 9.7. Summary

We have compared a large number of selection criteria over a wide variety of possible models. Our hope is that these large-scale studies give a better idea of selection criteria performance over a wide range of models and situations. No one criterion is uniformly better than the others; for any given situation, any of the selection criteria may perform poorly. However, some general patterns emerge.

Selection criteria with superlinear penalty functions that increase quickly as model order (and hence overfitting) increases, such as AICc and AICu, perform well in all our studies. One theme throughout this book is that selection criteria with superlinear penalty functions do not overfit excessively, in contrast to AIC which has a penalty function that is linear with respect to model order. Such linear penalty functions are too weak with respect to $\log(\hat{\sigma}^2)$ and result in overfitting in small samples. Similar small-sample problems can be seen in HQ. Signal-to-noise corrections such as HQc perform much better than their parent criteria in small samples. The bootstrapped criteria with penalty functions also performed well, however, we feel that the computational increase for bootstrapping is not worth the gains in performance. Other criteria perform better and are easier to apply.

We have not attempted to answer the question of whether efficiency or consistency is the better asymptotic property. Although efficiency and consistency each have properties that limit their applicability, because of the circumstances of everyday application, small-sample performance is much more important than large-sample properties.

Efficient criteria have penalty functions of the form $\alpha 2$, and consistent criteria have asymptotic penalty functions similar to $\alpha\infty$. We have proposed a class of selection criteria with asymptotic penalty functions of the form $\alpha 3$. Although neither efficient or consistent, $\alpha 3$ criteria have higher observed efficiency than the efficient criteria when the true model belongs to the set of candidate models, and have higher observed efficiency than the consistent criteria when the true model has infinite order and does not belong to the set of candidate models. This makes them a good choice when the nature of the true model is unknown.

The results we have seen in this Chapter have practical implications for the choice of model selection criteria. Some of the classical selection criteria consistently performed poorly, and are not recommended for use in practice. Results for AIC, FPE, and Mallows's Cp were disappointing, and $R^2_{adj}$ performed so poorly in the univariate case that we did not include it in the multivariate models.

On the other hand, some criteria consistently performed well and should be considered routinely for practical data analysis. An overall pattern has emerged for the better selection criteria. HQc, AICu and AICc performed well when the true model belonged to the set of candidate models, and also when the true model did not. Since the three criteria have different asymptotic properties, comparing the models selected by each should give different insights into the problem. Using this set of criteria together is a more well-rounded approach to choosing a model than to accept a model selected on the basis of just one criterion.

## Chapter 9 Appendices

## Appendix 9A. Details of Simulation Results

Table 9A.1. Counts and observed efficiencies for Model 1.

| | order $k$ | | | | | | | | | | | | K-L | $L_2$ |
|---|---|---|---|---|---|---|---|---|---|---|---|---|---|---|
| | 1 | 2 | 3 | 4 | 5 | 6 | 7 | 8 | 9 | 10 | 11 | true | ave | ave |
| AIC | 0 | 0 | 0 | 7 | 76 | 2463 | 3284 | 2542 | 1179 | 397 | 52 | 2338 | 0.498 | 0.652 |
| AICc | 0 | 1 | 12 | 84 | 498 | 6240 | 2533 | 568 | 59 | 4 | 1 | 5875 | 0.679 | 0.789 |
| AICu | 3 | 11 | 71 | 254 | 978 | 6888 | 1541 | 228 | 25 | 1 | 0 | 6509 | 0.714 | 0.802 |
| BFPE | 0 | 1 | 12 | 51 | 304 | 4240 | 3456 | 1474 | 398 | 60 | 4 | 3905 | 0.597 | 0.724 |
| Cp | 0 | 0 | 2 | 22 | 200 | 4154 | 3418 | 1561 | 520 | 111 | 12 | 3925 | 0.577 | 0.714 |
| CV | 0 | 0 | 1 | 20 | 189 | 3455 | 3607 | 1986 | 619 | 112 | 11 | 3187 | 0.557 | 0.694 |
| DCVB | 1 | 2 | 26 | 128 | 655 | 4850 | 3166 | 981 | 172 | 17 | 2 | 4378 | 0.631 | 0.738 |
| FPE | 0 | 0 | 1 | 8 | 83 | 2723 | 3470 | 2435 | 971 | 268 | 41 | 2585 | 0.511 | 0.663 |
| FPE4 | 0 | 1 | 8 | 40 | 240 | 4257 | 3246 | 1542 | 536 | 115 | 15 | 4033 | 0.581 | 0.716 |
| FPEu | 0 | 1 | 7 | 56 | 314 | 5020 | 3107 | 1165 | 283 | 43 | 4 | 4748 | 0.619 | 0.745 |
| GM | 0 | 0 | 20 | 106 | 576 | 6031 | 2365 | 735 | 146 | 18 | 3 | 5686 | 0.666 | 0.776 |
| HQ | 0 | 0 | 1 | 12 | 119 | 3040 | 3417 | 2231 | 880 | 257 | 43 | 2881 | 0.525 | 0.673 |
| HQc | 0 | 2 | 24 | 140 | 664 | 6635 | 2106 | 383 | 43 | 3 | 0 | 6243 | 0.699 | 0.800 |
| $R^2_{adj}$ | 0 | 0 | 0 | 2 | 16 | 1262 | 2905 | 3232 | 1837 | 659 | 87 | 1202 | 0.448 | 0.613 |
| Rp | 0 | 0 | 1 | 14 | 160 | 3758 | 3599 | 1791 | 566 | 101 | 10 | 3553 | 0.561 | 0.702 |
| SIC | 0 | 1 | 7 | 50 | 270 | 4548 | 3209 | 1371 | 444 | 89 | 11 | 4307 | 0.595 | 0.727 |
| K-L | 0 | 0 | 0 | 4 | 171 | 7818 | 941 | 747 | 274 | 45 | 0 | 7169 | 1.000 | 0.832 |
| $L_2$ | 0 | 0 | 0 | 0 | 1 | 9979 | 12 | 7 | 1 | 0 | 0 | 9970 | 0.916 | 1.000 |

Table 9A.2. Counts and observed efficiencies for Model 2.

| | order $k$ | | | | | | | | | | | | K-L | $L_2$ |
|---|---|---|---|---|---|---|---|---|---|---|---|---|---|---|
| | 1 | 2 | 3 | 4 | 5 | 6 | 7 | 8 | 9 | 10 | 11 | true | ave | ave |
| AIC | 2 | 84 | 598 | 1631 | 2351 | 2395 | 1659 | 830 | 353 | 89 | 8 | 201 | 0.332 | 0.474 |
| AICc | 9 | 405 | 2133 | 3427 | 2610 | 1066 | 292 | 49 | 9 | 0 | 0 | 120 | 0.411 | 0.492 |
| AICu | 30 | 1229 | 3414 | 3274 | 1543 | 410 | 84 | 14 | 2 | 0 | 0 | 48 | 0.449 | 0.486 |
| BFPE | 12 | 441 | 1784 | 2787 | 2515 | 1494 | 706 | 203 | 50 | 7 | 1 | 153 | 0.393 | 0.485 |
| Cp | 4 | 234 | 1374 | 2515 | 2700 | 1840 | 899 | 327 | 91 | 15 | 1 | 180 | 0.371 | 0.484 |
| CV | 2 | 134 | 889 | 2205 | 2815 | 2237 | 1137 | 438 | 121 | 21 | 1 | 185 | 0.363 | 0.483 |
| DCVB | 21 | 699 | 2248 | 3104 | 2340 | 1083 | 392 | 89 | 22 | 2 | 0 | 103 | 0.419 | 0.485 |
| FPE | 2 | 86 | 644 | 1763 | 2470 | 2448 | 1552 | 713 | 262 | 57 | 3 | 207 | 0.337 | 0.476 |
| FPE4 | 22 | 653 | 1958 | 2530 | 2264 | 1447 | 746 | 280 | 80 | 19 | 1 | 155 | 0.388 | 0.481 |
| FPEu | 14 | 531 | 2006 | 2906 | 2423 | 1355 | 562 | 164 | 33 | 5 | 1 | 149 | 0.397 | 0.486 |
| GM | 19 | 937 | 2816 | 3045 | 1962 | 824 | 297 | 79 | 18 | 2 | 1 | 82 | 0.423 | 0.482 |
| HQ | 4 | 165 | 932 | 2034 | 2537 | 2160 | 1308 | 580 | 224 | 53 | 3 | 195 | 0.349 | 0.479 |
| HQc | 14 | 620 | 2642 | 3480 | 2246 | 775 | 188 | 32 | 3 | 0 | 0 | 90 | 0.425 | 0.492 |
| $R^2_{adj}$ | 0 | 10 | 122 | 626 | 1650 | 2537 | 2579 | 1587 | 687 | 182 | 20 | 179 | 0.300 | 0.464 |
| Rp | 2 | 147 | 983 | 2297 | 2865 | 2132 | 1092 | 372 | 95 | 14 | 1 | 203 | 0.360 | 0.484 |
| SIC | 16 | 561 | 1973 | 2738 | 2348 | 1398 | 667 | 223 | 61 | 14 | 1 | 156 | 0.391 | 0.484 |
| K-L | 1 | 19 | 310 | 1668 | 3752 | 3501 | 612 | 118 | 16 | 3 | 0 | 1459 | 1.000 | 0.884 |
| $L_2$ | 0 | 2 | 101 | 746 | 2998 | 5402 | 647 | 100 | 4 | 0 | 0 | 3459 | 0.887 | 1.000 |

Table 9A.3. Simulation results for all 540 univariate regression models—K-L observed efficiency.

| | best 1 | 2,3 | 4,5 | 6,7 | rank 8,9 | 10,11 | 12,13 | 14,15 | worst 16 | ave. rank |
|---|---|---|---|---|---|---|---|---|---|---|
| AIC | 8 | 1105 | 2157 | 2917 | 24042 | 5833 | 8277 | 9098 | 563 | 9.82 |
| AICc | 50 | 2679 | 5955 | 8661 | 25322 | 5970 | 4724 | 592 | 47 | 7.85 |
| AICu | 1736 | 5390 | 8337 | 7702 | 23674 | 3435 | 2184 | 1192 | 350 | 7.00 |
| BFPE | 426 | 2944 | 5697 | 7217 | 24624 | 5375 | 4167 | 2362 | 1188 | 8.14 |
| Cp | 30 | 617 | 3109 | 7534 | 25762 | 7481 | 7597 | 1846 | 24 | 8.74 |
| CV | 785 | 1872 | 2961 | 5481 | 23380 | 6066 | 6918 | 4321 | 2216 | 9.10 |
| DCVB | 1543 | 4918 | 6367 | 6395 | 23251 | 3948 | 3333 | 2622 | 1623 | 7.72 |
| FPE | 12 | 1075 | 2192 | 3315 | 24467 | 6252 | 8769 | 7829 | 89 | 9.64 |
| FPE4 | 10 | 1770 | 7416 | 8328 | 25429 | 5936 | 3795 | 1293 | 23 | 7.87 |
| FPEu | 13 | 454 | 7723 | 9599 | 25901 | 6760 | 3184 | 344 | 22 | 7.82 |
| GM | 568 | 5190 | 8656 | 7817 | 23884 | 3205 | 2150 | 1662 | 868 | 7.26 |
| HQ | 6 | 494 | 3646 | 6375 | 24925 | 6910 | 6923 | 4523 | 198 | 9.02 |
| HQc | 20 | 3295 | 8943 | 8867 | 25036 | 4798 | 2283 | 730 | 28 | 7.35 |
| $R^2_{adj}$ | 1905 | 1324 | 1457 | 1407 | 13026 | 3727 | 5549 | 8236 | 17369 | 11.71 |
| Rp | 9 | 742 | 2387 | 6226 | 25517 | 7401 | 8593 | 3107 | 18 | 9.04 |
| SIC | 18 | 2965 | 6855 | 7674 | 24715 | 5250 | 4109 | 1937 | 477 | 7.92 |

Table 9A.4. Simulation results for all 540 regression models—$L_2$ observed efficiency.

| | best 1 | 2,3 | 4,5 | 6,7 | rank 8,9 | 10,11 | 12,13 | 14,15 | worst 16 | ave. rank |
|---|---|---|---|---|---|---|---|---|---|---|
| AIC | 13 | 1711 | 2961 | 3686 | 24308 | 5687 | 7472 | 7677 | 485 | 9.43 |
| AICc | 58 | 2170 | 5556 | 8279 | 25161 | 6207 | 5193 | 1251 | 125 | 8.08 |
| AICu | 1314 | 4382 | 7626 | 7211 | 23583 | 3872 | 2936 | 2094 | 982 | 7.51 |
| BFPE | 474 | 2913 | 5548 | 6963 | 24368 | 5374 | 4413 | 2548 | 1399 | 8.23 |
| Cp | 44 | 790 | 3530 | 7533 | 25536 | 7303 | 7234 | 1969 | 61 | 8.68 |
| CV | 980 | 2156 | 3306 | 5609 | 23312 | 5758 | 6474 | 4077 | 2328 | 8.96 |
| DCVB | 1373 | 4437 | 5797 | 6021 | 23029 | 4148 | 3858 | 3039 | 2298 | 8.04 |
| FPE | 21 | 1585 | 3016 | 4074 | 24659 | 6006 | 7764 | 6738 | 137 | 9.30 |
| FPE4 | 11 | 1669 | 7136 | 8169 | 25319 | 5938 | 4133 | 1583 | 42 | 7.97 |
| FPEu | 15 | 547 | 7281 | 9226 | 25639 | 6767 | 3878 | 608 | 39 | 7.95 |
| GM | 530 | 4639 | 7873 | 7196 | 23678 | 3583 | 3002 | 2377 | 1122 | 7.61 |
| HQ | 8 | 874 | 4252 | 6873 | 25059 | 6819 | 6368 | 3591 | 156 | 8.77 |
| HQc | 29 | 2756 | 8121 | 8323 | 24898 | 5144 | 3147 | 1514 | 68 | 7.66 |
| $R^2_{adj}$ | 2735 | 2072 | 2155 | 2179 | 13599 | 3713 | 5246 | 7239 | 15062 | 10.97 |
| Rp | 19 | 1006 | 3161 | 6577 | 25408 | 7106 | 7706 | 2969 | 48 | 8.86 |
| SIC | 17 | 2842 | 6742 | 7611 | 24665 | 5268 | 4250 | 2059 | 546 | 7.99 |

Table 9A.5. Counts and observed efficiencies for Model A1.

| | order $k$ | | | | | K-L | $L_2$ |
|---|---|---|---|---|---|---|---|
| | 1 | 2 | 3 | 4 | true | ave | ave |
| AIC | 0 | 691 | 268 | 41 | 691 | 0.815 | 0.785 |
| AICc | 0 | 691 | 268 | 41 | 691 | 0.815 | 0.785 |
| AICu | 0 | 823 | 159 | 18 | 823 | 0.884 | 0.868 |
| BFPE | 0 | 818 | 165 | 17 | 818 | 0.883 | 0.865 |
| Cp | 0 | 691 | 268 | 41 | 691 | 0.815 | 0.785 |
| CV | 0 | 692 | 268 | 40 | 692 | 0.816 | 0.786 |
| DCVB | 0 | 817 | 166 | 17 | 817 | 0.882 | 0.865 |
| FPE | 0 | 691 | 268 | 41 | 691 | 0.815 | 0.785 |
| FPE4 | 0 | 888 | 106 | 6 | 888 | 0.924 | 0.913 |
| FPEu | 0 | 823 | 159 | 18 | 823 | 0.884 | 0.868 |
| GM | 0 | 994 | 6 | 0 | 994 | 0.995 | 0.995 |
| HQ | 0 | 924 | 74 | 2 | 924 | 0.947 | 0.940 |
| HQc | 0 | 924 | 74 | 2 | 924 | 0.947 | 0.940 |
| $R^2_{adj}$ | 0 | 463 | 426 | 111 | 463 | 0.715 | 0.663 |
| Rp | 0 | 691 | 268 | 41 | 691 | 0.815 | 0.785 |
| SIC | 0 | 994 | 6 | 0 | 994 | 0.995 | 0.995 |
| K-L | 0 | 984 | 16 | 0 | 984 | 1.000 | 1.000 |
| $L_2$ | 0 | 998 | 2 | 0 | 998 | 1.000 | 1.000 |

Table 9A.6. Counts and observed efficiencies for Model A2.

| | order $k$ | | | | | | | | K-L | $L_2$ |
|---|---|---|---|---|---|---|---|---|---|---|
| | 1 | 2 | 3 | 4 | 5 | 6 | 7 | true | ave | ave |
| AIC | 0 | 457 | 387 | 129 | 23 | 4 | 0 | 454 | 0.659 | 0.611 |
| AICc | 0 | 457 | 388 | 128 | 23 | 4 | 0 | 454 | 0.659 | 0.611 |
| AICu | 0 | 671 | 282 | 37 | 10 | 0 | 0 | 668 | 0.783 | 0.754 |
| BFPE | 0 | 667 | 285 | 38 | 10 | 0 | 0 | 666 | 0.781 | 0.751 |
| Cp | 0 | 457 | 388 | 128 | 23 | 4 | 0 | 454 | 0.659 | 0.611 |
| CV | 0 | 461 | 383 | 128 | 24 | 4 | 0 | 458 | 0.661 | 0.614 |
| DCVB | 0 | 668 | 284 | 37 | 11 | 0 | 0 | 667 | 0.781 | 0.751 |
| FPE | 0 | 457 | 387 | 129 | 23 | 4 | 0 | 454 | 0.659 | 0.611 |
| FPE4 | 0 | 813 | 170 | 17 | 0 | 0 | 0 | 810 | 0.873 | 0.857 |
| FPEu | 0 | 671 | 282 | 37 | 10 | 0 | 0 | 668 | 0.783 | 0.754 |
| GM | 0 | 997 | 3 | 0 | 0 | 0 | 0 | 994 | 0.997 | 0.997 |
| HQ | 0 | 861 | 132 | 7 | 0 | 0 | 0 | 858 | 0.904 | 0.893 |
| HQc | 0 | 861 | 132 | 7 | 0 | 0 | 0 | 858 | 0.904 | 0.893 |
| $R^2_{adj}$ | 0 | 147 | 383 | 289 | 136 | 40 | 5 | 144 | 0.488 | 0.413 |
| Rp | 0 | 457 | 388 | 128 | 23 | 4 | 0 | 454 | 0.659 | 0.611 |
| SIC | 0 | 997 | 3 | 0 | 0 | 0 | 0 | 994 | 0.997 | 0.997 |
| K-L | 0 | 969 | 30 | 1 | 0 | 0 | 0 | 966 | 1.000 | 1.000 |
| $L_2$ | 0 | 998 | 2 | 0 | 0 | 0 | 0 | 995 | 1.000 | 1.000 |

Table 9A.7. Counts and observed efficiencies for Model 3.

| | order $p$ | | | | | | | | | | K-L | $L_2$ |
|---|---|---|---|---|---|---|---|---|---|---|---|---|
| | 1 | 2 | 3 | 4 | 5 | 6 | 7 | 8 | 9 | 10 | ave | ave |
| AIC | 0 | 0 | 1 | 4 | 5213 | 1342 | 894 | 779 | 660 | 1107 | 0.663 | 0.731 |
| AICc | 1 | 0 | 1 | 7 | 8430 | 1037 | 349 | 134 | 29 | 12 | 0.884 | 0.902 |
| AICu | 11 | 1 | 2 | 9 | 9188 | 615 | 130 | 38 | 5 | 1 | 0.927 | 0.933 |
| BFPE | 1 | 0 | 1 | 5 | 7445 | 1348 | 550 | 325 | 167 | 158 | 0.826 | 0.858 |
| Cp | 0 | 0 | 1 | 6 | 6044 | 1247 | 769 | 605 | 501 | 827 | 0.715 | 0.772 |
| CV | 0 | 0 | 2 | 5 | 6340 | 1427 | 830 | 583 | 388 | 425 | 0.749 | 0.798 |
| DCVB | 7 | 0 | 2 | 7 | 8061 | 1225 | 405 | 171 | 82 | 40 | 0.869 | 0.891 |
| FPE | 0 | 0 | 1 | 4 | 5436 | 1394 | 917 | 757 | 587 | 904 | 0.681 | 0.745 |
| FPE4 | 2 | 1 | 1 | 7 | 7454 | 1104 | 514 | 355 | 249 | 313 | 0.810 | 0.846 |
| FPEu | 1 | 0 | 1 | 7 | 7869 | 1104 | 463 | 265 | 148 | 142 | 0.843 | 0.871 |
| GM | 4 | 2 | 2 | 8 | 7725 | 962 | 423 | 304 | 222 | 348 | 0.826 | 0.858 |
| HQ | 0 | 0 | 1 | 6 | 5912 | 1269 | 792 | 633 | 519 | 868 | 0.707 | 0.766 |
| HQc | 1 | 1 | 2 | 8 | 8739 | 894 | 255 | 77 | 18 | 5 | 0.902 | 0.915 |
| $R^2_{adj}$ | 0 | 0 | 0 | 2 | 3071 | 1325 | 1109 | 1131 | 1215 | 2147 | 0.528 | 0.621 |
| Rp | 0 | 0 | 1 | 6 | 6467 | 1413 | 766 | 565 | 357 | 425 | 0.753 | 0.802 |
| SIC | 1 | 1 | 1 | 7 | 7562 | 1074 | 487 | 334 | 231 | 302 | 0.817 | 0.851 |
| K-L | 1 | 5 | 8 | 22 | 7726 | 1471 | 471 | 191 | 76 | 29 | 1.000 | 0.984 |
| $L_2$ | 0 | 0 | 0 | 0 | 7868 | 1226 | 496 | 222 | 126 | 62 | 0.984 | 1.000 |

Table 9A.8. Counts and observed efficiencies for Model 4.

| | order $p$ | | | | | | | | | | K-L | $L_2$ |
|---|---|---|---|---|---|---|---|---|---|---|---|---|
| | 1 | 2 | 3 | 4 | 5 | 6 | 7 | 8 | 9 | 10 | ave | ave |
| AIC | 1149 | 2434 | 2142 | 1216 | 822 | 514 | 421 | 384 | 381 | 537 | 0.520 | 0.564 |
| AICc | 1901 | 3634 | 2511 | 1108 | 530 | 174 | 82 | 35 | 22 | 3 | 0.610 | 0.609 |
| AICu | 3124 | 4098 | 1949 | 570 | 191 | 46 | 15 | 4 | 3 | 0 | 0.606 | 0.572 |
| BFPE | 2117 | 3560 | 2306 | 1001 | 505 | 198 | 126 | 89 | 58 | 40 | 0.599 | 0.596 |
| Cp | 1518 | 2775 | 2066 | 1121 | 696 | 414 | 319 | 326 | 339 | 426 | 0.532 | 0.561 |
| CV | 1229 | 2729 | 2323 | 1339 | 881 | 488 | 364 | 272 | 201 | 174 | 0.560 | 0.594 |
| DCVB | 2391 | 3724 | 2263 | 936 | 418 | 145 | 68 | 30 | 19 | 6 | 0.612 | 0.598 |
| FPE | 1111 | 2437 | 2186 | 1263 | 858 | 541 | 428 | 378 | 373 | 425 | 0.526 | 0.570 |
| FPE4 | 3050 | 3602 | 1824 | 696 | 353 | 147 | 100 | 81 | 80 | 67 | 0.575 | 0.555 |
| FPEu | 2398 | 3671 | 2177 | 884 | 446 | 172 | 105 | 63 | 50 | 34 | 0.592 | 0.584 |
| GM | 2982 | 3545 | 1765 | 684 | 366 | 174 | 131 | 102 | 119 | 132 | 0.568 | 0.550 |
| HQ | 1609 | 2905 | 2130 | 1089 | 681 | 391 | 296 | 265 | 264 | 370 | 0.543 | 0.569 |
| HQc | 2393 | 3878 | 2313 | 865 | 366 | 112 | 47 | 15 | 10 | 1 | 0.611 | 0.596 |
| $R^2_{adj}$ | 274 | 1081 | 1422 | 1134 | 1104 | 874 | 865 | 874 | 985 | 1387 | 0.410 | 0.495 |
| Rp | 1255 | 2804 | 2414 | 1309 | 848 | 462 | 308 | 248 | 197 | 155 | 0.563 | 0.594 |
| SIC | 2829 | 3672 | 1945 | 734 | 363 | 148 | 98 | 72 | 73 | 66 | 0.582 | 0.565 |
| K-L | 522 | 2421 | 3311 | 2399 | 953 | 259 | 82 | 35 | 13 | 5 | 1.000 | 0.938 |
| $L_2$ | 111 | 1538 | 3447 | 2961 | 1314 | 343 | 126 | 93 | 38 | 29 | 0.946 | 1.000 |

Table 9A.9. Simulation results for all 360 autoregressive models—K-L observed efficiency.

| | best | | | | rank | | | | worst | ave. |
|---|---|---|---|---|---|---|---|---|---|---|
| | 1 | 2,3 | 4,5 | 6,7 | 8,9 | 10,11 | 12,13 | 14,15 | 16 | rank |
| AIC | 2 | 849 | 1849 | 1984 | 17240 | 3406 | 4788 | 5387 | 495 | 9.57 |
| AICc | 65 | 2380 | 4687 | 5681 | 17816 | 2832 | 2114 | 359 | 66 | 7.50 |
| AICu | 829 | 3301 | 4803 | 4755 | 16945 | 2002 | 1913 | 1126 | 326 | 7.34 |
| BFPE | 91 | 1064 | 4421 | 5616 | 18000 | 3951 | 2154 | 589 | 114 | 7.86 |
| Cp | 94 | 804 | 2118 | 3761 | 17683 | 4259 | 5085 | 2104 | 92 | 8.86 |
| CV | 336 | 1187 | 2869 | 4243 | 17408 | 3850 | 3674 | 1961 | 472 | 8.52 |
| DCVB | 325 | 2529 | 4791 | 5130 | 17580 | 2705 | 1835 | 890 | 215 | 7.52 |
| FPE | 16 | 982 | 1929 | 2458 | 17496 | 3699 | 5130 | 4245 | 45 | 9.26 |
| FPE4 | 10 | 889 | 3923 | 5151 | 17998 | 3879 | 3035 | 1080 | 35 | 8.16 |
| FPEu | 5 | 529 | 4926 | 6154 | 18101 | 4141 | 1941 | 201 | 2 | 7.79 |
| GM | 312 | 1690 | 3172 | 4548 | 17025 | 3530 | 3449 | 1473 | 801 | 8.34 |
| HQ | 21 | 458 | 2039 | 3457 | 16782 | 3828 | 4307 | 3474 | 1634 | 9.42 |
| HQc | 41 | 2226 | 5294 | 5710 | 17796 | 2672 | 1749 | 492 | 20 | 7.43 |
| $R^2_{adj}$ | 1456 | 883 | 1113 | 1027 | 9550 | 2557 | 3960 | 5607 | 9847 | 11.33 |
| Rp | 46 | 907 | 2685 | 4464 | 18125 | 4258 | 4087 | 1423 | 5 | 8.50 |
| SIC | 77 | 1094 | 2906 | 4410 | 17426 | 3644 | 3824 | 2179 | 440 | 8.61 |

Table 9A.10. Simulation results for all 360 autoregressive models—$L_2$ observed efficiency.

| | best | | | | rank | | | | worst | ave. |
|---|---|---|---|---|---|---|---|---|---|---|
| | 1 | 2,3 | 4,5 | 6,7 | 8,9 | 10,11 | 12,13 | 14,15 | 16 | rank |
| AIC | 2 | 1329 | 2751 | 2796 | 17341 | 3035 | 3905 | 4437 | 404 | 9.05 |
| AICc | 67 | 1776 | 4120 | 5234 | 17682 | 3169 | 2674 | 1051 | 227 | 7.92 |
| AICu | 499 | 2272 | 3874 | 4159 | 16841 | 2549 | 2781 | 2082 | 943 | 8.15 |
| BFPE | 96 | 1089 | 3990 | 5530 | 17841 | 3876 | 2667 | 774 | 137 | 8.00 |
| Cp | 126 | 944 | 2628 | 4099 | 17551 | 3923 | 4490 | 2079 | 160 | 8.69 |
| CV | 380 | 1491 | 3310 | 4506 | 17333 | 3482 | 3325 | 1724 | 449 | 8.29 |
| DCVB | 283 | 2078 | 4094 | 4753 | 17447 | 2990 | 2525 | 1458 | 372 | 7.93 |
| FPE | 27 | 1424 | 2866 | 3243 | 17507 | 3201 | 4124 | 3532 | 76 | 8.79 |
| FPE4 | 5 | 687 | 3438 | 5034 | 17872 | 3902 | 3562 | 1431 | 69 | 8.39 |
| FPEu | 4 | 514 | 4193 | 5951 | 17967 | 4106 | 2809 | 443 | 13 | 8.02 |
| GM | 185 | 1446 | 2970 | 4414 | 16806 | 3487 | 3733 | 1871 | 1088 | 8.60 |
| HQ | 81 | 884 | 2694 | 3993 | 16949 | 3515 | 3816 | 2870 | 1198 | 8.97 |
| HQc | 36 | 1719 | 4343 | 5178 | 17660 | 3076 | 2664 | 1251 | 73 | 7.92 |
| $R^2_{adj}$ | 2173 | 1539 | 1868 | 1872 | 9908 | 2123 | 3403 | 4750 | 8364 | 10.39 |
| Rp | 60 | 1176 | 3403 | 4859 | 17937 | 3828 | 3416 | 1286 | 35 | 8.25 |
| SIC | 55 | 1009 | 2883 | 4494 | 17409 | 3608 | 3885 | 2133 | 524 | 8.64 |

Table 9A.11. Counts and observed efficiencies for Model A3.

| | order $p$ | | | K-L | $L_2$ |
|---|---|---|---|---|---|
| | 1 | 2 | 3 | ave | ave |
| AIC | 692 | 186 | 122 | 0.813 | 0.778 |
| AICc | 692 | 186 | 122 | 0.813 | 0.778 |
| AICu | 812 | 130 | 58 | 0.881 | 0.861 |
| BFPE | 809 | 131 | 60 | 0.879 | 0.860 |
| Cp | 692 | 186 | 122 | 0.813 | 0.778 |
| CV | 691 | 184 | 125 | 0.811 | 0.777 |
| DCVB | 810 | 131 | 59 | 0.880 | 0.860 |
| FPE | 692 | 186 | 122 | 0.813 | 0.778 |
| FPE4 | 890 | 83 | 27 | 0.931 | 0.920 |
| FPEu | 812 | 130 | 58 | 0.881 | 0.861 |
| GM | 993 | 7 | 0 | 0.995 | 0.994 |
| HQ | 924 | 63 | 13 | 0.952 | 0.945 |
| HQc | 924 | 63 | 13 | 0.952 | 0.945 |
| $R^2_{adj}$ | 468 | 240 | 292 | 0.703 | 0.636 |
| Rp | 692 | 186 | 122 | 0.813 | 0.778 |
| SIC | 993 | 7 | 0 | 0.995 | 0.994 |
| K-L | 978 | 22 | 0 | 1.000 | 1.000 |
| $L_2$ | 993 | 7 | 0 | 1.000 | 1.000 |

Table 9A.12. Counts and observed efficiencies for Model A4.

| | order $p$ | | | | | | K-L | $L_2$ |
|---|---|---|---|---|---|---|---|---|
| | 1 | 2 | 3 | 4 | 5 | 6 | ave | ave |
| AIC | 615 | 160 | 76 | 62 | 43 | 44 | 0.756 | 0.706 |
| AICc | 615 | 160 | 76 | 62 | 43 | 44 | 0.756 | 0.706 |
| AICu | 798 | 123 | 40 | 21 | 9 | 9 | 0.878 | 0.855 |
| BFPE | 794 | 128 | 36 | 21 | 10 | 11 | 0.874 | 0.850 |
| Cp | 615 | 160 | 76 | 62 | 43 | 44 | 0.756 | 0.706 |
| CV | 612 | 164 | 77 | 62 | 43 | 42 | 0.757 | 0.706 |
| DCVB | 798 | 124 | 37 | 20 | 10 | 11 | 0.876 | 0.853 |
| FPE | 615 | 160 | 76 | 62 | 43 | 44 | 0.756 | 0.706 |
| FPE4 | 900 | 77 | 16 | 6 | 0 | 1 | 0.939 | 0.928 |
| FPEu | 798 | 123 | 40 | 21 | 9 | 9 | 0.878 | 0.855 |
| GM | 994 | 6 | 0 | 0 | 0 | 0 | 0.995 | 0.995 |
| HQ | 931 | 57 | 10 | 2 | 0 | 0 | 0.956 | 0.949 |
| HQc | 931 | 57 | 10 | 2 | 0 | 0 | 0.956 | 0.949 |
| $R^2_{adj}$ | 333 | 139 | 107 | 126 | 122 | 173 | 0.562 | 0.478 |
| Rp | 615 | 160 | 76 | 62 | 43 | 44 | 0.756 | 0.706 |
| SIC | 994 | 6 | 0 | 0 | 0 | 0 | 0.995 | 0.995 |
| K-L | 973 | 25 | 2 | 0 | 0 | 0 | 1.000 | 0.999 |
| $L_2$ | 987 | 13 | 0 | 0 | 0 | 0 | 1.000 | 1.000 |

Table 9A.13. Counts and observed efficiencies for Model 5.

| | order $p$ | | | | | | | | | | | | | | | K-L | $L_2$ |
|---|---|---|---|---|---|---|---|---|---|---|---|---|---|---|---|---|---|
| | 1 | 2 | 3 | 4 | 5 | 6 | 7 | 8 | 9 | 10 | 11 | 12 | 13 | 14 | 15 | ave | ave |
| AIC | 3012 | 1817 | 831 | 544 | 322 | 237 | 196 | 178 | 145 | 129 | 122 | 161 | 258 | 478 | 1570 | 0.412 | 0.424 |
| AICc | 5241 | 2761 | 1067 | 490 | 232 | 111 | 66 | 27 | 5 | 0 | 0 | 0 | 0 | 0 | 0 | 0.626 | 0.598 |
| AICu | 6894 | 2271 | 573 | 179 | 57 | 14 | 10 | 2 | 0 | 0 | 0 | 0 | 0 | 0 | 0 | 0.664 | 0.610 |
| BFPE | 5771 | 2412 | 838 | 387 | 202 | 109 | 85 | 52 | 31 | 18 | 14 | 10 | 16 | 19 | 36 | 0.624 | 0.588 |
| Cp | 5311 | 1717 | 605 | 342 | 208 | 165 | 118 | 129 | 105 | 101 | 91 | 114 | 140 | 252 | 602 | 0.540 | 0.513 |
| CV | 3839 | 2482 | 1187 | 747 | 469 | 307 | 250 | 195 | 130 | 104 | 68 | 62 | 50 | 45 | 65 | 0.546 | 0.541 |
| DCVB | 6106 | 2445 | 804 | 347 | 155 | 69 | 41 | 22 | 8 | 0 | 2 | 0 | 1 | 0 | 0 | 0.649 | 0.607 |
| FPE | 3347 | 2089 | 976 | 646 | 412 | 303 | 254 | 225 | 192 | 165 | 145 | 161 | 220 | 287 | 578 | 0.468 | 0.472 |
| FPE4 | 6567 | 1961 | 535 | 204 | 130 | 66 | 57 | 44 | 29 | 28 | 22 | 34 | 46 | 75 | 202 | 0.621 | 0.574 |
| FPEu | 5993 | 2398 | 768 | 336 | 183 | 92 | 73 | 35 | 24 | 19 | 10 | 6 | 13 | 16 | 34 | 0.628 | 0.589 |
| GM | 7043 | 1309 | 379 | 174 | 110 | 63 | 70 | 59 | 47 | 56 | 45 | 59 | 70 | 126 | 390 | 0.619 | 0.564 |
| HQ | 3820 | 1917 | 777 | 435 | 233 | 176 | 145 | 112 | 101 | 73 | 80 | 118 | 202 | 396 | 1415 | 0.460 | 0.459 |
| HQc | 6004 | 2595 | 821 | 332 | 152 | 54 | 33 | 9 | 0 | 0 | 0 | 0 | 0 | 0 | 0 | 0.644 | 0.604 |
| $R^2_{adj}$ | 1083 | 1057 | 696 | 641 | 544 | 517 | 490 | 504 | 435 | 403 | 451 | 463 | 591 | 785 | 1340 | 0.264 | 0.301 |
| Rp | 3993 | 2503 | 1144 | 717 | 436 | 299 | 220 | 171 | 119 | 90 | 51 | 52 | 56 | 51 | 98 | 0.548 | 0.540 |
| SIC | 6173 | 2028 | 587 | 217 | 130 | 67 | 53 | 41 | 26 | 21 | 17 | 29 | 44 | 108 | 459 | 0.602 | 0.561 |
| K-L | 3215 | 4115 | 1895 | 513 | 177 | 49 | 24 | 7 | 1 | 3 | 0 | 1 | 0 | 0 | 0 | 1.000 | 0.950 |
| $L_2$ | 2395 | 4541 | 2187 | 568 | 181 | 68 | 31 | 17 | 2 | 4 | 3 | 3 | 0 | 0 | 0 | 0.961 | 1.000 |

Table 9A.14. Counts and observed efficiencies for Model 6.

| | order $p$ | | | | | | | | | | | | | | | K-L | $L_2$ |
|---|---|---|---|---|---|---|---|---|---|---|---|---|---|---|---|---|---|
| | 1 | 2 | 3 | 4 | 5 | 6 | 7 | 8 | 9 | 10 | 11 | 12 | 13 | 14 | 15 | ave | ave |
| AIC | 565 | 1404 | 1227 | 1089 | 625 | 595 | 348 | 313 | 218 | 212 | 222 | 228 | 335 | 660 | 1959 | 0.404 | 0.533 |
| AICc | 1592 | 3236 | 2268 | 1595 | 689 | 391 | 150 | 65 | 12 | 2 | 0 | 0 | 0 | 0 | 0 | 0.657 | 0.671 |
| AICu | 3136 | 3766 | 1788 | 888 | 282 | 98 | 34 | 8 | 0 | 0 | 0 | 0 | 0 | 0 | 0 | 0.656 | 0.625 |
| BFPE | 2041 | 3104 | 1963 | 1267 | 606 | 400 | 201 | 125 | 72 | 43 | 36 | 26 | 35 | 29 | 52 | 0.627 | 0.643 |
| Cp | 1925 | 2427 | 1398 | 981 | 510 | 399 | 266 | 221 | 156 | 159 | 150 | 130 | 172 | 348 | 758 | 0.521 | 0.575 |
| CV | 907 | 2132 | 1871 | 1653 | 955 | 774 | 477 | 381 | 240 | 170 | 122 | 84 | 83 | 72 | 79 | 0.584 | 0.659 |
| DCVB | 2406 | 3379 | 1959 | 1222 | 473 | 315 | 138 | 72 | 19 | 8 | 4 | 4 | 0 | 1 | 0 | 0.650 | 0.645 |
| FPE | 649 | 1631 | 1488 | 1326 | 817 | 728 | 461 | 408 | 269 | 268 | 258 | 205 | 269 | 418 | 805 | 0.480 | 0.592 |
| FPE4 | 2738 | 3010 | 1535 | 952 | 422 | 281 | 151 | 104 | 66 | 48 | 62 | 56 | 90 | 143 | 342 | 0.584 | 0.595 |
| FPEu | 2156 | 3198 | 1930 | 1276 | 562 | 358 | 182 | 101 | 53 | 35 | 31 | 17 | 30 | 22 | 49 | 0.629 | 0.640 |
| GM | 3632 | 2636 | 1122 | 663 | 309 | 227 | 155 | 112 | 98 | 79 | 66 | 70 | 103 | 203 | 525 | 0.560 | 0.558 |
| HQ | 931 | 1820 | 1313 | 1073 | 582 | 491 | 273 | 229 | 157 | 148 | 162 | 163 | 262 | 553 | 1843 | 0.436 | 0.543 |
| HQc | 2215 | 3502 | 2105 | 1320 | 476 | 251 | 90 | 35 | 6 | 0 | 0 | 0 | 0 | 0 | 0 | 0.657 | 0.652 |
| $R^2_{adj}$ | 85 | 392 | 615 | 801 | 667 | 821 | 634 | 631 | 534 | 567 | 527 | 593 | 664 | 913 | 1556 | 0.320 | 0.503 |
| Rp | 874 | 2166 | 1898 | 1633 | 966 | 801 | 464 | 340 | 204 | 153 | 118 | 73 | 93 | 90 | 127 | 0.580 | 0.655 |
| SIC | 2438 | 2941 | 1583 | 957 | 404 | 264 | 134 | 88 | 55 | 36 | 46 | 40 | 92 | 191 | 731 | 0.566 | 0.587 |
| K-L | 663 | 1403 | 2224 | 2041 | 1676 | 971 | 551 | 277 | 127 | 41 | 18 | 7 | 1 | 0 | 0 | 1.000 | 0.926 |
| $L_2$ | 48 | 601 | 1635 | 2120 | 2078 | 1434 | 916 | 513 | 326 | 144 | 84 | 50 | 26 | 18 | 7 | 0.914 | 1.000 |

Table 9A.15. Simulation results for all 50 misspecified MA(1) models—K-L observed efficiency.

| | best 1 | 2,3 | 4,5 | 6,7 | rank 8,9 | 10,11 | 12,13 | 14,15 | worst 16 | ave. rank |
|---|---|---|---|---|---|---|---|---|---|---|
| AIC | 0 | 115 | 202 | 200 | 2200 | 415 | 690 | 1017 | 161 | 10.08 |
| AICc | 13 | 296 | 702 | 914 | 2371 | 411 | 270 | 22 | 1 | 7.35 |
| AICu | 98 | 450 | 772 | 846 | 2246 | 307 | 207 | 67 | 7 | 6.99 |
| BFPE | 10 | 143 | 712 | 942 | 2376 | 535 | 240 | 39 | 3 | 7.58 |
| Cp | 18 | 165 | 325 | 592 | 2311 | 545 | 722 | 302 | 20 | 8.72 |
| CV | 58 | 166 | 379 | 638 | 2274 | 539 | 544 | 338 | 64 | 8.57 |
| DCVB | 42 | 325 | 742 | 875 | 2338 | 384 | 229 | 59 | 6 | 7.27 |
| FPE | 6 | 138 | 226 | 317 | 2264 | 483 | 782 | 776 | 8 | 9.54 |
| FPE4 | 1 | 175 | 631 | 866 | 2342 | 505 | 361 | 118 | 1 | 7.83 |
| FPEu | 1 | 75 | 769 | 993 | 2404 | 554 | 197 | 7 | 0 | 7.54 |
| GM | 113 | 346 | 450 | 699 | 2124 | 454 | 472 | 221 | 121 | 8.05 |
| HQ | 0 | 51 | 249 | 473 | 2178 | 518 | 582 | 684 | 265 | 9.74 |
| HQc | 4 | 267 | 808 | 971 | 2347 | 402 | 183 | 18 | 0 | 7.22 |
| $R^2_{adj}$ | 155 | 73 | 79 | 75 | 935 | 277 | 505 | 1007 | 1894 | 12.57 |
| Rp | 4 | 163 | 366 | 639 | 2396 | 567 | 580 | 285 | 0 | 8.53 |
| SIC | 15 | 201 | 428 | 734 | 2266 | 471 | 521 | 318 | 46 | 8.42 |

Table 9A.16. Simulation results for all 50 misspecified MA(1) models—$L_2$ observed efficiency.

| | best 1 | 2,3 | 4,5 | 6,7 | rank 8,9 | 10,11 | 12,13 | 14,15 | worst 16 | ave. rank |
|---|---|---|---|---|---|---|---|---|---|---|
| AIC | 0 | 172 | 308 | 261 | 2201 | 408 | 610 | 887 | 153 | 9.66 |
| AICc | 18 | 280 | 685 | 863 | 2358 | 454 | 273 | 62 | 7 | 7.47 |
| AICu | 68 | 369 | 661 | 787 | 2258 | 384 | 313 | 136 | 24 | 7.47 |
| BFPE | 8 | 147 | 658 | 898 | 2357 | 577 | 294 | 55 | 6 | 7.72 |
| Cp | 23 | 177 | 342 | 572 | 2283 | 548 | 690 | 325 | 40 | 8.73 |
| CV | 66 | 214 | 463 | 645 | 2251 | 526 | 472 | 295 | 68 | 8.32 |
| DCVB | 35 | 295 | 666 | 828 | 2321 | 438 | 301 | 100 | 16 | 7.54 |
| FPE | 5 | 199 | 332 | 377 | 2263 | 452 | 674 | 680 | 18 | 9.16 |
| FPE4 | 1 | 156 | 542 | 803 | 2332 | 554 | 449 | 161 | 2 | 8.09 |
| FPEu | 1 | 72 | 682 | 941 | 2390 | 595 | 292 | 25 | 2 | 7.75 |
| GM | 92 | 301 | 395 | 630 | 2095 | 484 | 537 | 286 | 180 | 8.44 |
| HQ | 3 | 95 | 311 | 505 | 2186 | 518 | 553 | 597 | 232 | 9.44 |
| HQc | 7 | 246 | 706 | 905 | 2332 | 467 | 283 | 49 | 5 | 7.50 |
| $R^2_{adj}$ | 235 | 137 | 160 | 134 | 963 | 266 | 475 | 904 | 1726 | 11.87 |
| Rp | 6 | 213 | 459 | 661 | 2371 | 549 | 487 | 252 | 2 | 8.27 |
| SIC | 11 | 185 | 384 | 704 | 2252 | 520 | 562 | 315 | 67 | 8.57 |

Table 9A.17. Counts and observed efficiencies for Model A5.

| | order $p$ | | | | | | | | | | K-L | $L_2$ |
|---|---|---|---|---|---|---|---|---|---|---|---|---|
| | 1-6 | 7 | 8 | 9 | 10 | 11 | 12 | 13 | 14 | 15 | ave | ave |
| AIC | 0 | 0 | 7 | 72 | 181 | 249 | 175 | 121 | 109 | 86 | 0.840 | 0.830 |
| AICc | 0 | 0 | 7 | 72 | 181 | 249 | 177 | 121 | 108 | 85 | 0.840 | 0.830 |
| AICu | 0 | 0 | 17 | 166 | 264 | 265 | 128 | 84 | 50 | 26 | 0.813 | 0.801 |
| BFPE | 0 | 0 | 17 | 158 | 259 | 272 | 129 | 85 | 57 | 23 | 0.814 | 0.803 |
| Cp | 0 | 0 | 7 | 72 | 181 | 249 | 176 | 121 | 109 | 85 | 0.840 | 0.830 |
| CV | 0 | 0 | 7 | 69 | 180 | 250 | 178 | 121 | 109 | 86 | 0.841 | 0.831 |
| DCVB | 0 | 0 | 17 | 158 | 258 | 273 | 128 | 85 | 57 | 24 | 0.814 | 0.803 |
| FPE | 0 | 0 | 7 | 72 | 181 | 249 | 175 | 121 | 109 | 86 | 0.840 | 0.830 |
| FPE4 | 0 | 0 | 34 | 263 | 316 | 237 | 87 | 37 | 19 | 7 | 0.780 | 0.767 |
| FPEu | 0 | 0 | 17 | 166 | 264 | 265 | 128 | 84 | 50 | 26 | 0.813 | 0.801 |
| GM | 0 | 35 | 377 | 430 | 131 | 26 | 1 | 0 | 0 | 0 | 0.580 | 0.562 |
| HQ | 0 | 1 | 52 | 324 | 329 | 199 | 60 | 22 | 8 | 5 | 0.756 | 0.742 |
| HQc | 0 | 1 | 52 | 324 | 330 | 198 | 60 | 22 | 8 | 5 | 0.756 | 0.742 |
| $R^2_{adj}$ | 0 | 0 | 0 | 23 | 69 | 143 | 150 | 158 | 209 | 248 | 0.851 | 0.842 |
| Rp | 0 | 0 | 7 | 72 | 181 | 249 | 177 | 121 | 108 | 85 | 0.840 | 0.830 |
| SIC | 0 | 35 | 377 | 430 | 131 | 26 | 1 | 0 | 0 | 0 | 0.580 | 0.562 |
| KL | 0 | 0 | 0 | 8 | 73 | 200 | 257 | 257 | 124 | 81 | 1.000 | 1.000 |
| $L_2$ | 0 | 0 | 0 | 6 | 74 | 195 | 259 | 259 | 125 | 82 | 1.000 | 1.000 |

Table 9A.18. Counts and observed efficiencies for Model 7.

| | order $k$ | | | | | | | | | | K-L | tr$\{L_2\}$ | det$(L_2)$ |
|---|---|---|---|---|---|---|---|---|---|---|---|---|---|
| | 1 | 2 | 3 | 4 | 5 | 6 | 7 | 8 | 9 | true | ave | ave | ave |
| AIC | 0 | 0 | 0 | 21 | 3255 | 3605 | 2155 | 819 | 145 | 3209 | 0.606 | 0.723 | 0.541 |
| AICc | 0 | 0 | 24 | 401 | 7995 | 1453 | 123 | 4 | 0 | 7838 | 0.827 | 0.897 | 0.839 |
| AICu | 1 | 26 | 132 | 1017 | 8226 | 568 | 28 | 2 | 0 | 8100 | 0.850 | 0.892 | 0.851 |
| deBFPE | 0 | 1 | 15 | 217 | 5980 | 2963 | 701 | 114 | 9 | 5847 | 0.740 | 0.830 | 0.719 |
| deCV | 0 | 0 | 1 | 99 | 4668 | 3600 | 1350 | 259 | 23 | 4555 | 0.682 | 0.784 | 0.639 |
| deDCVB | 0 | 6 | 51 | 567 | 6693 | 2309 | 346 | 25 | 3 | 6496 | 0.777 | 0.846 | 0.759 |
| FIC | 0 | 0 | 1 | 44 | 5798 | 3324 | 754 | 77 | 2 | 5718 | 0.725 | 0.825 | 0.707 |
| FPE | 0 | 0 | 1 | 22 | 3541 | 3769 | 1949 | 637 | 81 | 3491 | 0.622 | 0.737 | 0.562 |
| HQ | 0 | 0 | 1 | 38 | 4236 | 3558 | 1601 | 502 | 64 | 4177 | 0.651 | 0.761 | 0.605 |
| HQc | 0 | 2 | 41 | 567 | 8192 | 1116 | 79 | 3 | 0 | 8038 | 0.839 | 0.901 | 0.850 |
| ICOMP | 107 | 534 | 1542 | 3424 | 2937 | 1140 | 285 | 31 | 0 | 140 | 0.465 | 0.285 | 0.111 |
| meCp | 0 | 0 | 0 | 51 | 5506 | 3955 | 476 | 12 | 0 | 5338 | 0.740 | 0.830 | 0.709 |
| SIC | 0 | 0 | 14 | 168 | 6358 | 2657 | 665 | 130 | 8 | 6247 | 0.746 | 0.840 | 0.737 |
| trBFPE | 0 | 6 | 96 | 556 | 5478 | 2986 | 748 | 118 | 12 | 5218 | 0.737 | 0.778 | 0.680 |
| trCp | 0 | 0 | 1 | 43 | 4711 | 3541 | 1351 | 322 | 31 | 4637 | 0.674 | 0.781 | 0.637 |
| trCV | 0 | 1 | 27 | 274 | 4711 | 3468 | 1251 | 247 | 21 | 4471 | 0.704 | 0.751 | 0.638 |
| trDCVB | 1 | 29 | 267 | 1088 | 5803 | 2364 | 412 | 35 | 1 | 5390 | 0.747 | 0.773 | 0.677 |
| trFPE | 0 | 0 | 8 | 95 | 3927 | 3692 | 1775 | 444 | 59 | 3822 | 0.671 | 0.723 | 0.600 |
| K-L | 0 | 0 | 41 | 1111 | 8441 | 345 | 59 | 3 | 0 | 7459 | 1.000 | 0.839 | 0.803 |
| tr$\{L_2\}$ | 0 | 0 | 1 | 24 | 9962 | 12 | 1 | 0 | 0 | 9937 | 0.940 | 1.000 | 0.982 |
| det$(L_2)$ | 0 | 2 | 73 | 283 | 9640 | 2 | 0 | 0 | 0 | 9553 | 0.932 | 0.974 | 1.000 |

Table 9A.19. Counts and observed efficiencies for Model 8.

| | order $k$ | | | | | | | | | | K-L | $tr\{L_2\}$ | $det(L_2)$ |
|---|---|---|---|---|---|---|---|---|---|---|---|---|---|
| | 1 | 2 | 3 | 4 | 5 | 6 | 7 | 8 | 9 | true | ave | ave | ave |
| AIC | 4 | 1362 | 2625 | 2588 | 1851 | 1021 | 405 | 124 | 20 | 91 | 0.435 | 0.490 | 0.226 |
| AICc | 45 | 4470 | 3899 | 1315 | 243 | 27 | 1 | 0 | 0 | 14 | 0.644 | 0.608 | 0.414 |
| AICu | 213 | 6969 | 2410 | 368 | 38 | 2 | 0 | 0 | 0 | 1 | 0.770 | 0.661 | 0.529 |
| deBFPE | 65 | 4033 | 3473 | 1666 | 587 | 149 | 24 | 3 | 0 | 31 | 0.615 | 0.589 | 0.386 |
| deCV | 10 | 1830 | 3230 | 2753 | 1511 | 530 | 110 | 24 | 2 | 66 | 0.491 | 0.525 | 0.268 |
| deDCVB | 121 | 4957 | 3332 | 1246 | 290 | 50 | 4 | 0 | 0 | 16 | 0.672 | 0.615 | 0.434 |
| FIC | 3 | 1973 | 3641 | 2833 | 1185 | 311 | 50 | 4 | 0 | 75 | 0.501 | 0.531 | 0.279 |
| FPE | 4 | 1418 | 2766 | 2650 | 1827 | 926 | 322 | 77 | 10 | 87 | 0.443 | 0.496 | 0.233 |
| HQ | 11 | 2214 | 3125 | 2405 | 1370 | 598 | 218 | 50 | 9 | 75 | 0.495 | 0.525 | 0.279 |
| HQc | 74 | 5278 | 3550 | 949 | 136 | 12 | 1 | 0 | 0 | 6 | 0.685 | 0.627 | 0.450 |
| ICOMP | 3 | 1277 | 2914 | 3040 | 1837 | 718 | 184 | 24 | 3 | 106 | 0.576 | 0.520 | 0.329 |
| meCp | 2 | 232 | 1547 | 4377 | 3263 | 551 | 28 | 0 | 0 | 170 | 0.403 | 0.470 | 0.173 |
| SIC | 93 | 4670 | 3255 | 1343 | 469 | 143 | 23 | 4 | 0 | 25 | 0.641 | 0.601 | 0.414 |
| trBFPE | 8 | 3144 | 3805 | 2088 | 710 | 201 | 39 | 4 | 1 | 44 | 0.626 | 0.573 | 0.394 |
| trCp | 9 | 2032 | 3233 | 2575 | 1394 | 555 | 169 | 29 | 4 | 75 | 0.490 | 0.523 | 0.273 |
| trCV | 0 | 1586 | 3349 | 2906 | 1528 | 500 | 115 | 14 | 2 | 82 | 0.537 | 0.518 | 0.302 |
| trDCVB | 22 | 3850 | 3847 | 1743 | 430 | 99 | 9 | 0 | 0 | 31 | 0.670 | 0.599 | 0.430 |
| trFPE | 2 | 1359 | 2918 | 2827 | 1772 | 795 | 265 | 59 | 3 | 112 | 0.501 | 0.494 | 0.277 |
| K-L | 58 | 4666 | 4591 | 642 | 39 | 4 | 0 | 0 | 0 | 12 | 1.000 | 0.870 | 0.747 |
| $tr\{L_2\}$ | 0 | 1166 | 5374 | 3097 | 356 | 6 | 0 | 1 | 0 | 205 | 0.898 | 1.000 | 0.713 |
| $det(L_2)$ | 1590 | 3151 | 3742 | 1355 | 162 | 0 | 0 | 0 | 0 | 115 | 0.855 | 0.788 | 1.000 |

Table 9A.20. Simulation results for all 504 multivariate regression models—K-L observed efficiency.

| | best | | | rank | | | | | worst | ave. |
|---|---|---|---|---|---|---|---|---|---|---|
| | 1 | 2,3 | 4,5 | 6,7 | 8–10 | 11–13 | 14,15 | 16,17 | 18 | rank |
| AIC | 4 | 624 | 1489 | 5243 | 16401 | 9182 | 8042 | 8135 | 1280 | 11.18 |
| AICc | 42 | 4634 | 7072 | 10004 | 18748 | 6506 | 2584 | 759 | 51 | 7.82 |
| AICu | 1254 | 7088 | 8774 | 9831 | 16872 | 4098 | 1358 | 945 | 180 | 6.94 |
| deBFPE | 97 | 1834 | 6374 | 9908 | 19695 | 8107 | 3223 | 1007 | 155 | 8.47 |
| deCV | 194 | 956 | 2811 | 7356 | 18624 | 10340 | 6233 | 3263 | 623 | 9.92 |
| deDCVB | 357 | 3829 | 7880 | 10025 | 18797 | 6077 | 2190 | 1027 | 218 | 7.81 |
| FIC | 463 | 2456 | 4618 | 6984 | 14871 | 6154 | 4039 | 3692 | 7123 | 10.31 |
| FPE | 3 | 635 | 1604 | 5580 | 16988 | 9931 | 8619 | 6820 | 220 | 10.88 |
| HQ | 10 | 780 | 3652 | 7714 | 17688 | 8063 | 6048 | 5425 | 1020 | 10.10 |
| HQc | 55 | 5235 | 9070 | 10650 | 18423 | 4939 | 1319 | 683 | 26 | 7.28 |
| ICOMP | 3203 | 4330 | 4125 | 4001 | 7983 | 2868 | 1903 | 3567 | 18420 | 11.46 |
| meCp | 996 | 1676 | 3200 | 6098 | 14006 | 9132 | 5545 | 6120 | 3627 | 10.62 |
| SIC | 354 | 3024 | 6157 | 9027 | 17592 | 6982 | 4396 | 2417 | 451 | 8.67 |
| trBFPE | 176 | 4011 | 4917 | 6455 | 16860 | 5935 | 4022 | 7573 | 451 | 9.45 |
| trCp | 24 | 681 | 2322 | 7102 | 18681 | 10944 | 7486 | 3124 | 36 | 10.07 |
| trCV | 290 | 3126 | 3706 | 5147 | 15911 | 6782 | 5662 | 8441 | 1335 | 10.26 |
| trDCVB | 446 | 4836 | 5817 | 6716 | 16280 | 4958 | 3324 | 7459 | 564 | 9.07 |
| trFPE | 228 | 2638 | 3249 | 4533 | 15353 | 6927 | 6466 | 9259 | 1747 | 10.70 |

Table 9A.21. Simulation results for all 504 multivariate regression models—tr$\{L_2\}$ observed efficiency.

| | best 1 | 2,3 | 4,5 | 6,7 | rank 8–10 | 11–13 | 14,15 | 16,17 | worst 18 | ave. rank |
|---|---|---|---|---|---|---|---|---|---|---|
| AIC | 5 | 994 | 2410 | 5640 | 17795 | 10125 | 7339 | 5549 | 543 | 10.49 |
| AICc | 49 | 3899 | 6308 | 8894 | 18295 | 7433 | 3341 | 1904 | 277 | 8.39 |
| AICu | 811 | 5472 | 7444 | 8419 | 16483 | 5143 | 2557 | 2770 | 1301 | 8.08 |
| deBFPE | 154 | 1956 | 6177 | 9251 | 19319 | 8166 | 3513 | 1603 | 261 | 8.63 |
| deCV | 304 | 1449 | 3470 | 7398 | 19104 | 9936 | 5370 | 2784 | 585 | 9.57 |
| deDCVB | 349 | 3538 | 6975 | 8938 | 18003 | 6710 | 3132 | 2216 | 539 | 8.36 |
| FIC | 547 | 2697 | 4909 | 6888 | 15403 | 6276 | 3904 | 4169 | 5607 | 10.03 |
| FPE | 9 | 991 | 2568 | 6023 | 18385 | 10618 | 7297 | 4348 | 161 | 10.23 |
| HQ | 14 | 1054 | 4113 | 7735 | 18653 | 9259 | 5587 | 3605 | 380 | 9.65 |
| HQc | 84 | 4656 | 7978 | 9342 | 17666 | 5853 | 2598 | 2130 | 93 | 7.94 |
| ICOMP | 2388 | 3704 | 3176 | 3306 | 7535 | 3372 | 2481 | 4089 | 20349 | 12.32 |
| meCp | 1686 | 2464 | 3984 | 6341 | 14360 | 8037 | 4849 | 5575 | 3104 | 9.99 |
| SIC | 293 | 2813 | 5971 | 8556 | 17946 | 7553 | 4036 | 2404 | 828 | 8.80 |
| trBFPE | 293 | 4880 | 4521 | 5990 | 16721 | 6797 | 4363 | 6178 | 657 | 9.30 |
| trCp | 40 | 1023 | 3186 | 7370 | 19610 | 10818 | 6021 | 2240 | 92 | 9.64 |
| trCV | 426 | 4206 | 3623 | 4848 | 15541 | 7015 | 5295 | 7759 | 1687 | 10.07 |
| trDCVB | 513 | 5428 | 5159 | 6156 | 16247 | 5967 | 3875 | 6162 | 893 | 9.07 |
| trFPE | 389 | 3824 | 3276 | 4313 | 14987 | 7098 | 5984 | 8477 | 2052 | 10.44 |

Table 9A.22. Simulation results for all 504 multivariate regression models—det$(L_2)$ observed efficiency.

| | best 1 | 2,3 | 4,5 | 6,7 | rank 8–10 | 11–13 | 14,15 | 16,17 | worst 18 | ave. rank |
|---|---|---|---|---|---|---|---|---|---|---|
| AIC | 7 | 987 | 2255 | 6148 | 17580 | 9825 | 7049 | 5927 | 622 | 10.50 |
| AICc | 49 | 4179 | 6647 | 9839 | 18463 | 6793 | 2969 | 1305 | 156 | 8.09 |
| AICu | 844 | 6077 | 8060 | 9558 | 16778 | 4771 | 1949 | 1779 | 584 | 7.53 |
| deBFPE | 142 | 2090 | 6696 | 10297 | 19565 | 7413 | 2830 | 1156 | 211 | 8.33 |
| deCV | 292 | 1395 | 3459 | 7987 | 19032 | 9386 | 5245 | 2928 | 676 | 9.55 |
| deDCVB | 336 | 3800 | 7554 | 10005 | 18293 | 6093 | 2471 | 1514 | 334 | 7.97 |
| FIC | 633 | 2831 | 5047 | 7651 | 15626 | 6024 | 3275 | 3147 | 6166 | 9.82 |
| FPE | 12 | 972 | 2402 | 6549 | 18280 | 10066 | 7073 | 4849 | 197 | 10.25 |
| HQ | 13 | 1052 | 4327 | 8539 | 18773 | 8579 | 5138 | 3577 | 402 | 9.50 |
| HQc | 83 | 5023 | 8490 | 10462 | 18018 | 5279 | 1796 | 1184 | 65 | 7.52 |
| ICOMP | 2043 | 3469 | 3327 | 3450 | 7419 | 2798 | 2362 | 4692 | 20840 | 12.51 |
| meCp | 1388 | 2229 | 3824 | 6743 | 13969 | 7971 | 4757 | 5749 | 3770 | 10.21 |
| SIC | 372 | 3154 | 6499 | 9616 | 18278 | 6811 | 3406 | 1842 | 422 | 8.37 |
| trBFPE | 160 | 3355 | 4357 | 5894 | 16258 | 5886 | 4738 | 9027 | 725 | 9.95 |
| trCp | 48 | 997 | 3156 | 8026 | 19494 | 10077 | 5947 | 2580 | 75 | 9.61 |
| trCV | 310 | 2840 | 3320 | 4744 | 15247 | 6468 | 5991 | 9865 | 1615 | 10.62 |
| trDCVB | 302 | 3846 | 5012 | 6039 | 15722 | 5049 | 4212 | 9169 | 1049 | 9.78 |
| trFPE | 272 | 2513 | 3014 | 4347 | 14923 | 6750 | 6538 | 10209 | 1834 | 10.89 |

Table 9A.23. Counts and observed efficiencies for Model A6.

| | order $k$ 1 | 2 | 3 | 4 | true | K-L ave | $\text{tr}\{L_2\}$ ave | $\det(L_2)$ ave |
|---|---|---|---|---|---|---|---|---|
| AIC | 0 | 727 | 254 | 19 | 727 | 0.860 | 0.830 | 0.760 |
| AICc | 0 | 727 | 254 | 19 | 727 | 0.860 | 0.830 | 0.760 |
| AICu | 0 | 889 | 109 | 2 | 889 | 0.935 | 0.924 | 0.902 |
| deBFPE | 0 | 885 | 112 | 3 | 885 | 0.932 | 0.921 | 0.898 |
| deCV | 0 | 727 | 254 | 19 | 727 | 0.860 | 0.830 | 0.760 |
| deDCVB | 0 | 885 | 112 | 3 | 885 | 0.932 | 0.921 | 0.898 |
| FIC | 0 | 1000 | 0 | 0 | 1000 | 1.000 | 1.000 | 1.000 |
| FPE | 0 | 727 | 254 | 19 | 727 | 0.860 | 0.830 | 0.760 |
| HQ | 0 | 973 | 27 | 0 | 973 | 0.982 | 0.979 | 0.976 |
| HQc | 0 | 973 | 27 | 0 | 973 | 0.982 | 0.979 | 0.976 |
| ICOMP | 0 | 626 | 334 | 40 | 626 | 0.860 | 0.769 | 0.694 |
| meCp | 0 | 689 | 311 | 0 | 689 | 0.852 | 0.816 | 0.732 |
| SIC | 0 | 1000 | 0 | 0 | 1000 | 1.000 | 1.000 | 1.000 |
| trBFPE | 0 | 859 | 136 | 5 | 859 | 0.924 | 0.896 | 0.874 |
| trCp | 0 | 726 | 255 | 19 | 726 | 0.859 | 0.829 | 0.759 |
| trCV | 0 | 742 | 239 | 19 | 742 | 0.876 | 0.822 | 0.775 |
| trDCVB | 0 | 860 | 135 | 5 | 860 | 0.924 | 0.896 | 0.875 |
| trFPE | 0 | 745 | 236 | 19 | 745 | 0.877 | 0.824 | 0.777 |
| K-L | 0 | 1000 | 0 | 0 | 1000 | 1.000 | 1.000 | 1.000 |
| $\text{tr}\{L_2\}$ | 0 | 1000 | 0 | 0 | 1000 | 1.000 | 1.000 | 1.000 |
| $\det(L_2)$ | 0 | 1000 | 0 | 0 | 1000 | 1.000 | 1.000 | 1.000 |

Table 9A.24. Counts and observed efficiencies for Model A7.

| | order $k$ 1 | 2 | 3 | 4 | 5 | 6 | true | K-L ave | $\text{tr}\{L_2\}$ ave | $\det(L_2)$ ave |
|---|---|---|---|---|---|---|---|---|---|---|
| AIC | 0 | 566 | 333 | 88 | 12 | 1 | 566 | 0.775 | 0.720 | 0.612 |
| AICc | 0 | 566 | 333 | 88 | 12 | 1 | 566 | 0.775 | 0.720 | 0.612 |
| AICu | 0 | 817 | 170 | 12 | 1 | 0 | 817 | 0.897 | 0.876 | 0.836 |
| deBFPE | 0 | 811 | 179 | 9 | 1 | 0 | 811 | 0.893 | 0.870 | 0.830 |
| deCV | 0 | 567 | 333 | 87 | 12 | 1 | 567 | 0.775 | 0.721 | 0.614 |
| deDCVB | 0 | 813 | 178 | 8 | 1 | 0 | 813 | 0.894 | 0.872 | 0.832 |
| FIC | 0 | 1000 | 0 | 0 | 0 | 0 | 1000 | 1.000 | 1.000 | 1.000 |
| FPE | 0 | 566 | 333 | 88 | 12 | 1 | 566 | 0.775 | 0.720 | 0.612 |
| HQ | 0 | 960 | 40 | 0 | 0 | 0 | 960 | 0.973 | 0.972 | 0.963 |
| HQc | 0 | 960 | 40 | 0 | 0 | 0 | 960 | 0.973 | 0.972 | 0.963 |
| ICOMP | 0 | 373 | 419 | 170 | 33 | 5 | 373 | 0.749 | 0.598 | 0.479 |
| meCp | 0 | 132 | 732 | 135 | 1 | 0 | 132 | 0.653 | 0.533 | 0.281 |
| SIC | 0 | 1000 | 0 | 0 | 0 | 0 | 1000 | 1.000 | 1.000 | 1.000 |
| trBFPE | 0 | 733 | 243 | 21 | 3 | 0 | 733 | 0.865 | 0.811 | 0.766 |
| trCp | 0 | 566 | 333 | 88 | 12 | 1 | 566 | 0.775 | 0.720 | 0.612 |
| trCV | 0 | 519 | 378 | 90 | 13 | 0 | 519 | 0.771 | 0.673 | 0.577 |
| trDCVB | 0 | 734 | 242 | 21 | 3 | 0 | 734 | 0.866 | 0.811 | 0.767 |
| trFPE | 0 | 521 | 378 | 89 | 12 | 0 | 521 | 0.772 | 0.674 | 0.580 |
| K-L | 0 | 1000 | 0 | 0 | 0 | 0 | 1000 | 1.000 | 1.000 | 1.000 |
| $\text{tr}\{L_2\}$ | 0 | 1000 | 0 | 0 | 0 | 0 | 1000 | 1.000 | 1.000 | 1.000 |
| $\det(L_2)$ | 0 | 1000 | 0 | 0 | 0 | 0 | 1000 | 1.000 | 1.000 | 1.000 |

Table 9A.25. Counts and observed efficiencies for Model 9.

| | order $p$ | | | | | | | | K-L | $\text{tr}\{L_2\}$ | $\det(L_2)$ |
|---|---|---|---|---|---|---|---|---|---|---|---|
| | 1 | 2 | 3 | 4 | 5 | 6 | 7 | 8 | ave | ave | ave |
| AIC | 0 | 0 | 0 | 4464 | 1062 | 796 | 950 | 2728 | 0.576 | 0.706 | 0.571 |
| AICc | 10 | 11 | 5 | 9906 | 67 | 1 | 0 | 0 | 0.989 | 0.989 | 0.988 |
| AICu | 104 | 15 | 5 | 9864 | 12 | 0 | 0 | 0 | 0.988 | 0.984 | 0.982 |
| deBFPE | 6 | 6 | 3 | 9192 | 601 | 125 | 40 | 27 | 0.944 | 0.961 | 0.942 |
| deCV | 1 | 3 | 2 | 8354 | 1038 | 366 | 147 | 89 | 0.890 | 0.927 | 0.887 |
| deDCVB | 11 | 6 | 3 | 9510 | 421 | 44 | 4 | 1 | 0.965 | 0.975 | 0.963 |
| FIC | 0 | 0 | 0 | 3895 | 1786 | 1670 | 1504 | 1145 | 0.585 | 0.713 | 0.566 |
| FPE | 0 | 1 | 0 | 6041 | 1322 | 845 | 715 | 1076 | 0.712 | 0.805 | 0.708 |
| HQ | 1 | 1 | 0 | 6059 | 987 | 622 | 635 | 1695 | 0.696 | 0.792 | 0.696 |
| HQc | 38 | 13 | 4 | 9918 | 27 | 0 | 0 | 0 | 0.990 | 0.989 | 0.987 |
| ICOMP | 4399 | 2160 | 1802 | 1601 | 23 | 10 | 5 | 0 | 0.480 | 0.243 | 0.167 |
| meCp | 0 | 0 | 1 | 4531 | 3145 | 1655 | 591 | 77 | 0.734 | 0.812 | 0.692 |
| SIC | 7 | 6 | 1 | 8743 | 551 | 206 | 166 | 320 | 0.901 | 0.933 | 0.902 |
| trBFPE | 20 | 13 | 3 | 8897 | 766 | 191 | 68 | 42 | 0.931 | 0.944 | 0.925 |
| trCp | 1 | 3 | 0 | 6585 | 1148 | 708 | 600 | 955 | 0.749 | 0.830 | 0.747 |
| trCV | 6 | 4 | 1 | 8094 | 1108 | 472 | 194 | 121 | 0.881 | 0.907 | 0.872 |
| trDCVB | 29 | 19 | 8 | 9239 | 600 | 88 | 14 | 3 | 0.953 | 0.960 | 0.947 |
| trFPE | 1 | 2 | 0 | 6117 | 1284 | 914 | 726 | 956 | 0.734 | 0.798 | 0.720 |
| K-L | 55 | 78 | 166 | 9559 | 137 | 5 | 0 | 0 | 1.000 | 0.971 | 0.966 |
| $\text{tr}\{L_2\}$ | 0 | 0 | 0 | 9326 | 528 | 119 | 14 | 13 | 0.982 | 1.000 | 0.989 |
| $\det(L_2)$ | 0 | 0 | 0 | 9570 | 370 | 54 | 4 | 2 | 0.989 | 0.996 | 1.000 |

Table 9A.26. Counts and observed efficiencies for Model 10.

| | order $p$ | | | | | | | | K-L | $\text{tr}\{L_2\}$ | $\det(L_2)$ |
|---|---|---|---|---|---|---|---|---|---|---|---|
| | 1 | 2 | 3 | 4 | 5 | 6 | 7 | 8 | ave | ave | ave |
| AIC | 1641 | 2054 | 1565 | 1044 | 539 | 499 | 640 | 2018 | 0.485 | 0.608 | 0.398 |
| AICc | 5782 | 3452 | 686 | 79 | 1 | 0 | 0 | 0 | 0.812 | 0.631 | 0.715 |
| AICu | 8120 | 1777 | 99 | 4 | 0 | 0 | 0 | 0 | 0.834 | 0.552 | 0.773 |
| deBFPE | 5594 | 3129 | 930 | 290 | 42 | 10 | 5 | 0 | 0.791 | 0.635 | 0.699 |
| deCV | 3105 | 3423 | 2014 | 995 | 293 | 91 | 49 | 30 | 0.714 | 0.704 | 0.599 |
| deDCVB | 5249 | 3439 | 1039 | 250 | 19 | 4 | 0 | 0 | 0.797 | 0.649 | 0.699 |
| FIC | 256 | 840 | 1490 | 1783 | 1492 | 1498 | 1483 | 1158 | 0.363 | 0.607 | 0.271 |
| FPE | 1974 | 2586 | 1965 | 1287 | 610 | 436 | 431 | 711 | 0.580 | 0.669 | 0.477 |
| HQ | 3378 | 2718 | 1417 | 761 | 315 | 223 | 289 | 899 | 0.634 | 0.634 | 0.545 |
| HQc | 6994 | 2677 | 306 | 23 | 0 | 0 | 0 | 0 | 0.825 | 0.590 | 0.748 |
| ICOMP | 2591 | 3050 | 2225 | 1308 | 481 | 214 | 88 | 43 | 0.757 | 0.720 | 0.626 |
| meCp | 446 | 872 | 1548 | 2605 | 2589 | 1427 | 459 | 54 | 0.479 | 0.703 | 0.350 |
| SIC | 6915 | 2324 | 487 | 161 | 30 | 15 | 20 | 48 | 0.800 | 0.585 | 0.729 |
| trBFPE | 3056 | 3800 | 2057 | 837 | 176 | 44 | 18 | 12 | 0.768 | 0.717 | 0.645 |
| trCp | 1970 | 2581 | 1903 | 1253 | 624 | 414 | 454 | 801 | 0.580 | 0.666 | 0.478 |
| trCV | 1618 | 3174 | 2750 | 1627 | 498 | 192 | 97 | 44 | 0.688 | 0.747 | 0.559 |
| trDCVB | 2982 | 3916 | 2219 | 754 | 109 | 16 | 4 | 0 | 0.780 | 0.727 | 0.653 |
| trFPE | 1003 | 2304 | 2325 | 1815 | 837 | 513 | 507 | 696 | 0.565 | 0.692 | 0.449 |
| K-L | 3582 | 4747 | 1484 | 183 | 4 | 0 | 0 | 0 | 1.000 | 0.763 | 0.799 |
| $\text{tr}\{L_2\}$ | 159 | 2102 | 4273 | 3200 | 225 | 28 | 9 | 4 | 0.772 | 1.000 | 0.622 |
| $\det(L_2)$ | 5196 | 2571 | 1481 | 714 | 32 | 4 | 2 | 0 | 0.897 | 0.720 | 1.000 |

Table 9A.27. Simulation results for all 864 VAR models—K-L observed efficiency.

| | best 1 | 2,3 | 4,5 | 6,7 | rank 8–10 | 11–13 | 14,15 | 16,17 | worst 18 | ave. rank |
|---|---|---|---|---|---|---|---|---|---|---|
| AIC | 2 | 560 | 1782 | 5553 | 41046 | 7269 | 10402 | 14951 | 4835 | 11.26 |
| AICc | 60 | 4101 | 10650 | 17026 | 46368 | 5415 | 2017 | 754 | 9 | 7.81 |
| AICu | 892 | 6013 | 10670 | 15283 | 44032 | 4155 | 2355 | 2157 | 843 | 7.83 |
| deBFPE | 19 | 1278 | 9403 | 17382 | 47823 | 7516 | 2528 | 440 | 11 | 8.16 |
| deCV | 141 | 1177 | 5486 | 14228 | 47514 | 10081 | 5888 | 1786 | 99 | 8.89 |
| deDCVB | 99 | 2161 | 10011 | 17401 | 47339 | 6720 | 2105 | 522 | 42 | 8.01 |
| FIC | 139 | 1378 | 3393 | 7816 | 39870 | 6756 | 9822 | 13092 | 4134 | 10.76 |
| FPE | 16 | 766 | 2335 | 7984 | 44086 | 9611 | 11828 | 9541 | 233 | 10.34 |
| HQ | 7 | 407 | 3302 | 9779 | 44284 | 7703 | 8544 | 10476 | 1898 | 10.25 |
| HQc | 20 | 5084 | 11348 | 16501 | 45662 | 4172 | 2025 | 1513 | 75 | 7.75 |
| ICOMP | 2713 | 4352 | 5278 | 7109 | 19748 | 6136 | 3759 | 6273 | 31032 | 12.09 |
| meCp | 1283 | 1498 | 3098 | 6989 | 32015 | 12907 | 10098 | 11886 | 6626 | 11.16 |
| SIC | 103 | 1671 | 7034 | 14616 | 45236 | 6645 | 6056 | 4105 | 934 | 8.95 |
| trBFPE | 46 | 2567 | 6852 | 13426 | 46297 | 8022 | 3998 | 4801 | 391 | 8.85 |
| trCp | 72 | 882 | 2594 | 8420 | 44374 | 10373 | 11625 | 7875 | 185 | 10.16 |
| trCV | 172 | 2325 | 4862 | 11141 | 44597 | 9200 | 6423 | 6793 | 887 | 9.45 |
| trDCVB | 159 | 3056 | 7620 | 13491 | 45689 | 7268 | 3513 | 5278 | 326 | 8.74 |
| trFPE | 147 | 1806 | 3095 | 7650 | 40745 | 9150 | 9549 | 11280 | 2978 | 10.53 |

Table 9A.28. Simulation results for all 864 VAR models—tr$\{L_2\}$ observed efficiency.

| | best 1 | 2,3 | 4,5 | 6,7 | rank 8–10 | 11–13 | 14,15 | 16,17 | worst 18 | ave. rank |
|---|---|---|---|---|---|---|---|---|---|---|
| AIC | 31 | 1174 | 2996 | 6854 | 42456 | 8878 | 10209 | 11391 | 2411 | 10.58 |
| AICc | 56 | 2778 | 8830 | 15313 | 45584 | 6611 | 3905 | 3314 | 9 | 8.45 |
| AICu | 199 | 3201 | 8324 | 13413 | 42946 | 5433 | 4177 | 5549 | 3158 | 8.99 |
| deBFPE | 29 | 1535 | 8815 | 16382 | 46693 | 8076 | 3743 | 1110 | 17 | 8.35 |
| deCV | 245 | 2045 | 6944 | 14587 | 46584 | 9321 | 5120 | 1471 | 83 | 8.62 |
| deDCVB | 126 | 2230 | 9066 | 16080 | 46203 | 7483 | 3705 | 1432 | 75 | 8.30 |
| FIC | 410 | 2305 | 4351 | 8897 | 41606 | 8169 | 9126 | 9009 | 2527 | 10.05 |
| FPE | 45 | 1582 | 3941 | 9504 | 44993 | 9870 | 10135 | 6175 | 155 | 9.74 |
| HQ | 7 | 722 | 4040 | 10511 | 45089 | 9491 | 8782 | 7022 | 736 | 9.81 |
| HQc | 17 | 3294 | 9266 | 14600 | 44716 | 5532 | 4073 | 4717 | 185 | 8.53 |
| ICOMP | 873 | 2610 | 3746 | 5808 | 17789 | 4826 | 4046 | 6988 | 39714 | 13.46 |
| meCp | 2983 | 3047 | 4670 | 8475 | 33195 | 10765 | 7956 | 10583 | 4726 | 10.22 |
| SIC | 76 | 1375 | 6291 | 14044 | 44987 | 8443 | 6557 | 3452 | 1175 | 9.09 |
| trBFPE | 59 | 2627 | 6915 | 12964 | 45246 | 8166 | 4548 | 5447 | 428 | 8.95 |
| trCp | 138 | 1662 | 4004 | 9806 | 45058 | 10378 | 9879 | 5349 | 126 | 9.64 |
| trCV | 257 | 3171 | 5957 | 11450 | 44001 | 8410 | 5779 | 6578 | 797 | 9.22 |
| trDCVB | 150 | 2996 | 7347 | 12677 | 44566 | 7476 | 4475 | 6211 | 502 | 8.95 |
| trFPE | 299 | 2747 | 4342 | 8659 | 41258 | 8694 | 8203 | 9944 | 2254 | 10.04 |

Table 9A.29. Simulation results for all 864 VAR models—det($L_2$) observed efficiency.

| | best 1 | 2,3 | 4,5 | 6,7 | rank 8–10 | 11–13 | 14,15 | 16,17 | worst 18 | ave. rank |
|---|---|---|---|---|---|---|---|---|---|---|
| AIC | 15 | 844 | 2521 | 6577 | 42274 | 9103 | 10394 | 11957 | 2715 | 10.74 |
| AICc | 44 | 3223 | 9687 | 16049 | 45787 | 5851 | 3300 | 2407 | 52 | 8.20 |
| AICu | 486 | 4145 | 9360 | 14211 | 43317 | 4569 | 3498 | 4464 | 2350 | 8.56 |
| deBFPE | 27 | 1687 | 9758 | 17166 | 46919 | 7219 | 2885 | 717 | 22 | 8.15 |
| deCV | 209 | 1844 | 6821 | 14870 | 46825 | 9155 | 4993 | 1584 | 99 | 8.63 |
| deDCVB | 105 | 2377 | 9893 | 16798 | 46433 | 6707 | 2991 | 1014 | 82 | 8.11 |
| FIC | 271 | 1900 | 4128 | 8810 | 41341 | 8304 | 9161 | 9526 | 2959 | 10.21 |
| FPE | 39 | 1240 | 3388 | 9374 | 44965 | 10059 | 10410 | 6747 | 178 | 9.87 |
| HQ | 3 | 565 | 3973 | 10690 | 45180 | 9341 | 8492 | 7321 | 835 | 9.83 |
| HQc | 10 | 4024 | 10295 | 15458 | 45015 | 4662 | 3351 | 3424 | 161 | 8.19 |
| ICOMP | 1178 | 2763 | 3873 | 5927 | 17892 | 4505 | 3616 | 7295 | 39351 | 13.36 |
| meCp | 1901 | 2478 | 4249 | 8356 | 33095 | 11313 | 8342 | 11102 | 5564 | 10.57 |
| SIC | 204 | 1765 | 7197 | 14866 | 45368 | 7586 | 5687 | 2942 | 785 | 8.80 |
| trBFPE | 44 | 2052 | 6524 | 12872 | 45336 | 7920 | 4382 | 6715 | 555 | 9.12 |
| trCp | 109 | 1336 | 3573 | 9694 | 45093 | 10519 | 10111 | 5785 | 180 | 9.75 |
| trCV | 164 | 2407 | 5217 | 11127 | 43887 | 8841 | 6278 | 7591 | 888 | 9.47 |
| trDCVB | 98 | 2385 | 6991 | 12598 | 44610 | 7183 | 4213 | 7618 | 704 | 9.13 |
| trFPE | 221 | 2046 | 3598 | 8220 | 41068 | 9386 | 8850 | 10721 | 2290 | 10.29 |

Table 9A.30. Counts and observed efficiencies for Model A8.

| | order $p$ 1 | 2 | 3 | K-L ave | tr{$L_2$} ave | det($L_2$) ave |
|---|---|---|---|---|---|---|
| AIC | 839 | 114 | 47 | 0.904 | 0.887 | 0.848 |
| AICc | 839 | 114 | 47 | 0.904 | 0.887 | 0.848 |
| AICu | 971 | 29 | 0 | 0.982 | 0.978 | 0.973 |
| deBFPE | 970 | 30 | 0 | 0.980 | 0.978 | 0.972 |
| deCV | 839 | 114 | 47 | 0.904 | 0.887 | 0.848 |
| deDCVB | 929 | 57 | 14 | 0.957 | 0.949 | 0.933 |
| FIC | 1000 | 0 | 0 | 1.000 | 1.000 | 1.000 |
| FPE | 839 | 114 | 47 | 0.904 | 0.887 | 0.848 |
| HQ | 998 | 2 | 0 | 0.998 | 0.999 | 0.998 |
| HQc | 998 | 2 | 0 | 0.998 | 0.999 | 0.998 |
| ICOMP | 868 | 97 | 35 | 0.929 | 0.902 | 0.878 |
| meCp | 443 | 545 | 12 | 0.775 | 0.698 | 0.523 |
| SIC | 1000 | 0 | 0 | 1.000 | 1.000 | 1.000 |
| trBFPE | 938 | 51 | 11 | 0.964 | 0.953 | 0.941 |
| trCp | 839 | 114 | 47 | 0.904 | 0.887 | 0.848 |
| trCV | 802 | 136 | 62 | 0.888 | 0.853 | 0.813 |
| trDCVB | 877 | 96 | 27 | 0.931 | 0.910 | 0.885 |
| trFPE | 801 | 137 | 62 | 0.887 | 0.852 | 0.812 |
| K-L | 1000 | 0 | 0 | 1.000 | 1.000 | 1.000 |
| tr{$L_2$} | 1000 | 0 | 0 | 1.000 | 1.000 | 1.000 |
| det($L_2$) | 1000 | 0 | 0 | 1.000 | 1.000 | 1.000 |

Table 9A.31. Counts and observed efficiencies for Model A9.

| | order $p$ 1 | 2 | 3 | 4 | 5 | K-L ave | $\text{tr}\{L_2\}$ ave | $\det(L_2)$ ave |
|---|---|---|---|---|---|---|---|---|
| AIC | 839 | 118 | 28 | 10 | 5 | 0.904 | 0.882 | 0.848 |
| AICc | 839 | 118 | 28 | 10 | 5 | 0.904 | 0.882 | 0.848 |
| AICu | 959 | 39 | 1 | 0 | 1 | 0.974 | 0.967 | 0.961 |
| deBFPE | 957 | 40 | 2 | 0 | 1 | 0.972 | 0.966 | 0.959 |
| deCV | 836 | 120 | 29 | 10 | 5 | 0.902 | 0.880 | 0.846 |
| deDCVB | 920 | 70 | 6 | 2 | 2 | 0.952 | 0.941 | 0.925 |
| FIC | 1000 | 0 | 0 | 0 | 0 | 1.000 | 1.000 | 1.000 |
| FPE | 839 | 118 | 28 | 10 | 5 | 0.904 | 0.882 | 0.848 |
| HQ | 998 | 2 | 0 | 0 | 0 | 0.999 | 0.999 | 0.998 |
| HQc | 998 | 2 | 0 | 0 | 0 | 0.999 | 0.999 | 0.998 |
| ICOMP | 855 | 109 | 23 | 6 | 7 | 0.921 | 0.892 | 0.868 |
| meCp | 295 | 328 | 337 | 39 | 1 | 0.659 | 0.565 | 0.365 |
| SIC | 1000 | 0 | 0 | 0 | 0 | 1.000 | 1.000 | 1.000 |
| trBFPE | 928 | 64 | 5 | 2 | 1 | 0.956 | 0.944 | 0.933 |
| trCp | 839 | 118 | 28 | 10 | 5 | 0.904 | 0.882 | 0.848 |
| trCV | 775 | 157 | 36 | 13 | 19 | 0.868 | 0.829 | 0.791 |
| trDCVB | 874 | 102 | 15 | 3 | 6 | 0.926 | 0.905 | 0.884 |
| trFPE | 775 | 156 | 37 | 13 | 19 | 0.868 | 0.829 | 0.791 |
| K-L | 999 | 1 | 0 | 0 | 0 | 1.000 | 1.000 | 1.000 |
| $\text{tr}\{L_2\}$ | 1000 | 0 | 0 | 0 | 0 | 1.000 | 1.000 | 1.000 |
| $\det(L_2)$ | 1000 | 0 | 0 | 0 | 0 | 1.000 | 1.000 | 1.000 |

## Appendix 9B. Stepwise Regression

Since stepwise regression procedures are applied much differently than all subsets regression, no direct comparison can be accurately made. In general, stepwise procedures examine far fewer models than all the possible subsets. For many years, this made them much faster than all subsets regression. This has changed with the introduction of the leaps and bounds algorithm (Furnival and Wilson, 1974). Now best subsets can be computed almost as quickly as stepwise regression.

Here, we present results for 3 stepwise selection criteria: stepwise F-test procedures at $\alpha$ levels 0.05, 0.10, and 0.15, denoted by F05, F10, and F15, respectively. Since the models are built up sequentially by adding or removing one variable at a time, for a test of full model of order $k$ versus reduced model of order $k-1$, then $F_{obs} = (\text{SSE}_{red} - \text{SSE}_{full})/s^2_{full} \sim F_{1,n-k}$. We use the stepwise F-test procedure discussed in Rawlings (1988, p. 178) to test $H_0 : \beta_k = 0$ versus $H_1 : \beta_k \neq 0$ for $k > 1$. We begin with the intercept only model (order $k = 1$) then try adding one variable to the model. The variable added is the one with the largest $F_{obs}$. Of course this $F_{obs}$ need be significant at the $\alpha$ level to be included in the model. In general, the procedure is to try to add one variable in the forward step, and then try to remove as many variables as possible in backward steps. The procedure stops when no more variables can be added or removed. We assume that $\alpha_{enter} = \alpha_{remove}$. Now, we repeat the small-sample and large-scale simulations studies seen in Section 9.2. Tables 9B.1 and 9B.2 summarize the results for Model 1 and Model 2, respectively.

Table 9B.1. Stepwise counts and observed efficiencies for model 1.

| | order $k$ | | | | | | | | | | | | K-L | $L_2$ |
|---|---|---|---|---|---|---|---|---|---|---|---|---|---|---|
| | 1 | 2 | 3 | 4 | 5 | 6 | 7 | 8 | 9 | 10 | 11 | true | ave | ave |
| F05 | 58 | 307 | 673 | 899 | 1072 | 5449 | 1331 | 194 | 16 | 1 | 0 | 5112 | 0.630 | 0.676 |
| F10 | 7 | 40 | 134 | 306 | 591 | 5378 | 2717 | 708 | 108 | 9 | 2 | 4990 | 0.631 | 0.734 |
| F15 | 1 | 4 | 30 | 107 | 324 | 4352 | 3499 | 1332 | 310 | 38 | 3 | 4044 | 0.588 | 0.715 |
| K-L | 0 | 0 | 0 | 4 | 171 | 7818 | 941 | 747 | 274 | 45 | 0 | 7169 | 1.000 | 0.832 |
| $L_2$ | 0 | 0 | 0 | 0 | 1 | 9979 | 12 | 7 | 1 | 0 | 0 | 9970 | 0.916 | 1.000 |

Table 9B.2. Stepwise counts and observed efficiencies for Model 2.

| | order $k$ | | | | | | | | | | | | K-L | $L_2$ |
|---|---|---|---|---|---|---|---|---|---|---|---|---|---|---|
| | 1 | 2 | 3 | 4 | 5 | 6 | 7 | 8 | 9 | 10 | 11 | true | ave | ave |
| F05 | 122 | 2012 | 3817 | 2743 | 1026 | 234 | 42 | 4 | 0 | 0 | 0 | 35 | 0.467 | 0.476 |
| F10 | 22 | 741 | 2616 | 3270 | 2187 | 872 | 235 | 50 | 6 | 1 | 0 | 106 | 0.426 | 0.491 |
| F15 | 9 | 293 | 1552 | 2890 | 2797 | 1656 | 617 | 153 | 30 | 3 | 0 | 174 | 0.392 | 0.493 |
| K-L | 1 | 19 | 310 | 1668 | 3752 | 3501 | 612 | 118 | 16 | 3 | 0 | 1459 | 1.000 | 0.884 |
| $L_2$ | 0 | 2 | 101 | 746 | 2998 | 5402 | 647 | 100 | 4 | 0 | 0 | 3459 | 0.887 | 1.000 |

We can see from Table 9B.1 that F05 underfits excessively, far more than F10 or F15. On the other hand, the lower $\alpha$ value leads to less overfitting from F05. The choice of $\alpha$ balances overfitting and underfitting. High $\alpha$ favors overfitting, low $\alpha$ favors underfitting. In the strongly identifiable Model 1, we see that $\alpha = 0.05$ may be too low, causing too much underfitting. F10 has the highest observed efficiency of the three $\alpha$ levels. F15 overfits too much, resulting in lower K-L observed efficiency than the other two F-tests. In general, small $\alpha$ seem to be better than large $\alpha$ but, if $\alpha$ is too small, observed efficiency is lost due to excessive underfitting.

In Model 2, even more underfitting is seen from F05. K-L and $L_2$ find that the closest model tends to have order 5 or 6, thus orders 1 and 2 represent excessive underfitting. None of the procedures overfit excessively (low counts for orders 10 and 11). F05 has the highest K-L observed efficiency but the lowest $L_2$ observed efficiency. Once again the small $\alpha$ value in F$\alpha$ causes a loss of $L_2$ observed efficiency due to underfitting. The higher $\alpha$ values in F$\alpha$ causes a loss of K-L observed efficiency due to overfitting. Overall, Model 2 observed efficiencies are lower than Model 1 observed efficiencies due to the weak identifiability of Model 2.

We next examine the 540 regression model simulation study. Table 9B.3 summarizes the K-L observed efficiency rankings from each realization and

Table 9B.4 summarizes the $L_2$ observed efficiency results. From Tables 9B.3 and 9B.4 we can see that F05 is clearly the best of the three F-tests. F05 has the highest rank 1 counts and the lowest rank 3 counts. An application of the test defined in Eq. (9.1) indicates that the criteria rank as follows: F05, F10, followed by F15. The smaller $\alpha$ mimics a stronger penalty function (if such a thing existed for stepwise regression) and leads to better performance. However, notice that F15 can perform well, depending on the model structure. As noted above, F05 sometimes underfits excessively to the point where no variables are included in the model. This problem rarely occurs with F15. On the other hand, F15 overfits more than F05. A better strategy would be to use both F05 and F15 and compare the models. If the two agree, the resulting model probably has high observed efficiency. If the two disagree, more care should be taken to examine the differences.

In practice, the data analyst may be interested to know whether the stepwise or the all subsets selection procedure performs better. Unfortunately, theoretical justification for the routine use of either is lacking, and more work needs to be done in this area. We give empirical results here, repeating the small sample and large-scale simulation studies by comparing the selection criteria from Section 9.2 and Table 9B.1 using stepwise selection. Based on this limited study, we found that when using the stepwise approach, F05 performs best with respect to both K-L and $L_2$ observed efficiency.

Table 9B.3. Stepwise results for all 540 univariate regression models—K-L observed efficiency.

| | best 1 | 2 | worst 3 | ave. rank |
|---|---|---|---|---|
| F05 | 8890 | 39105 | 6005 | 1.86 |
| F10 | 686 | 43569 | 9745 | 1.99 |
| F15 | 2469 | 35632 | 15899 | 2.15 |

Table 9B.4. Stepwise results for all 540 univariate regression models—$L_2$ observed efficiency.

| | best 1 | 2 | worst 3 | ave. rank |
|---|---|---|---|---|
| F05 | 7413 | 38329 | 8258 | 1.93 |
| F10 | 827 | 43701 | 9472 | 1.98 |
| F15 | 3434 | 37018 | 13548 | 2.09 |

# References


Akaike, H. (1969). Statistical predictor identification. *Annals of the Institute of Statistical Mathematics* **22**, 203–217.

Akaike, H. (1973). Information theory and an extension of the maximum likelihood principle. In B.N. Petrov and F. Csaki ed. *2nd International Symposium on Information Theory* 267–281. Akademia Kiado, Budapest.

Akaike, H. (1978). A Bayesian analysis of the minimum AIC procedure. *Annals of the Institute of Statistical Mathematics* **30**, Part A, 9–14.

Allen, D.M. (1974). The relationship between variable selection and data augmentation and a method for prediction. *Technometrics* **16**, 125–127.

Anderson, R.L. and Bancroft, T.A. (1952). *Statistical Theory in Research.* McGraw Hill, New York.

Anderson, T.W. (1984). *An Introduction to Multivariate statistical Analysis.* Wiley, New York.

Antle, C.E. and Bain, L.J. (1969). A property of maximum likelihood estimators of location and scale parameters. *SIAM Review* **11**, 251–253.

Bates, D.M. and Watts, D.G. (1988). *Nonlinear Regression Analysis and Its Applications.* Wiley, New York.

Bedrick, E.J. and Tsai, C.L. (1994). Model selection for multivariate regression in small samples. *Biometrics* **50**, 226–231.

Bhansali, R.J. (1996). Asymptotically efficient autoregressive model selection for multistep prediction. *Annals of the Institute of Statistical Mathematics* **48**, 577-602.

Bhansali, R.J. and Downham, D.Y. (1977). Some properties of the order of an autoregressive model selected by a generalization of Akaike's EPF criterion. *Biometrika* **64**, 547–551.

Bloomfield, P. and Steiger, W.L. (1983). *Least Absolute Deviations Theory, Applications and Algorithms.* Birkhauser, Boston.

Box, G.E.P. and Jenkins, G.M. (1976). *Time Series Analysis, Forecasting and Control (Revised Edition)*, Holden-Day, San Francisco.

Bozdogan, H. (1990). On the information-based measure of covariance complexity and its application to the evaluation of multivariate linear models. *Communications in Statistics — Theory and Methods* **19**, 221–278.

Brieman, L. and Freedman, D. (1983). How many variables should be entered in a regression equation? *Journal of the American Statistical Association* **78**, 131–136.

Brockwell, P.J. and Davis, R.A. (1991). *Time Series: Theory and Methods*, 2nd edition. Springer-Verlag, New York.

Broersen, P.M.T. and Wensink, H.E. (1996). On the penalty for autoregressive order selection in finite samples. *IEEE Transactions On Signal Processing* **44**, 748–752.

Bunke, O. and Droge, B. (1984). Bootstrap and cross-validation estimates of the prediction error for linear regression models. *Annals of Statistics* **12**, 1400–1424.

Burg, J.P. (1978). A new analysis technique for time series data. In *Modern Spectrum Analysis* (Edited by D. G. Childers), 42-48. IEEE Press, New York.

Burman, P. (1989). A comparative study of ordinary cross-validation, $v$-hold cross-validation, and repeated learning-testing methods. *Biometrika* **76**, 503–514.

Burman, P. and Nolan, D. (1995). A general Akaike-type criterion for model selection in robust regression. *Biometrika* **82**, 877–886.

Burnham, K.P. and Anderson, D.R. (1998). *Model Selection and Inference: A Practical Information Theoretic Approach.* Springer-Verlag, New York.

Carlin, B.P. and Chib, S. (1995). Bayesian model choice via Markov Chain Monte Carlo methods. *Journal of the Royal Statistical Society, B* **57**, 473–484.

Carroll, R.J. and Ruppert, D. (1988). *Transformation and Weighting in Regression.* Chapman and Hall, London.

Carroll, R.J., Fan, J., Gijbels, I. and Wand, M.P. (1997). Generalized partially linear single-index models. *Journal of the American Statistical Association* **92**, 477–489.

Cavanaugh, J.E. and Shumway, R.H. (1997). A bootstrap variant of AIC for state-space model selection. *Statistica Sinica* **7**, 473–496.

Chen, H. and Shiau, J.H. (1994). Data-driven efficient estimators for a partly linear model. *Annals of Statistics* **22**, 211–237.

Chipman, H., Hamada, M. and Wu, C.F.J. (1997). A Bayesian variable-selection approach for analyzing designed experiments with complex aliasing. *Technometrics* **39**, 372–381.

Choi, B. (1992). *ARMA Model Identification.* Springer-Verlag, New York.

Chu, C.K. and Marron, J.S. (1991). Choosing a kernel regression estimator (with discussion). *Statistical Science* **6**, 404–436.

Cleveland, W.S and Devlin, S.J. (1988). Locally weighted regression: an approach to regression analysis by local fitting. *Journal of the American*

*Statistical Association* **83**, 596–610.

Craven, P. and Wahba, G. (1979). Smoothing noisy data with spline functions. *Numerische Mathematik* **31**, 375–382.

Daubechies, I. (1992). *Ten Lectures on Wavelets.* SIAM, Philadelphia.

Davisson, L.D. (1965). The prediction error of stationary Gaussian time series of unknown covariance, *IEEE Trans. Information Theory* **IT–11**, 527–532.

Diggle, P.J., Liang, K.Y., and Zeger, S.L. (1994). *Analysis of Longitudinal Data.* Oxford, New York.

Donoho, D.L. and Johnstone, I.M. (1994). Ideal spatial adaptation by wavelet shrinkage. *Biometrika* **81**, 425–455.

Donoho, D.L. and Johnstone, I.M. (1995). Adapting to unknown smoothness via wavelet shrinkage. *Journal of the American Statistical Association* **90**, 1200–1224.

Donoho, D.L., Johnstone, I.M., Kerkyacharian, G. and Picard, D. (1995). Wavelet shrinkage: asymptopia? *Journal of the Royal Statistical Society, B* **57**, 301–369.

Efron, B. (1979). Bootstrap methods: Another look at the jackknife. *Annals of Statistics* **7**, 1–26.

Efron, B. (1986). How biased is the apparent error rate of a prediction rule? *Journal of the American Statistical Association* **81**, 461–470.

Efron, B. and Tibshirani, R.J. (1993). *An Introduction to the Bootstrap.* Chapman and Hall, New York.

Fujikoshi, Y. and Satoh, K. (1997). Modified AIC and Cp in multivariate linear regression. *Biometrika* **84**, 707–716.

Fuller, W.A. (1987). *Measurement Error Models.* Wiley, New York.

Furnival, G.M. and Wilson, W. (1974). Regression by leaps and bounds. *Technometrics* **16**, 499–511.

Gasser, T. and Müller, H.G. (1979). Kernel estimation of regression functions, in *Smoothing Techniques in Curve Estimation* (Lecture Notes in Mathematics 757). Springer-Verlag, New York.

George, E.I. and McCulloch, R.E. (1993). Variable selection via Gibbs sampling. *Journal of the American Statistical Association* **88**, 881–889.

George, E.I. and McCulloch, R.E. (1997). Approaches for Bayesian variable selection. *Statistica Sinica* **7**, 339–373.

Geweke, J. and Meese, R. (1981). Estimating regression models of finite but unknown order. *International Economic Review* **22**, 55–70.

Gilmour, S.G. (1996). The interpretation of Mallows's Cp-statistic. *The Statistician* **45**, 49–56.

Gouriéroux, C. (1997). *ARCH Models and Financial Applications.* Springer-Verlag, New York.

Gradshteyn, I.S. and Ryzhik, I.M. (1965). *Table of Integrals, Series, and Products.* Academic Press, New York.

Green, P.J. and Silverman, B.W. (1994). *Nonparametric Regression and Generalized Linear Models: A Roughness Penalty Approach.* Chapman and Hall, London.

Grund, B., Hall, P. and Marron, J.S. (1994). Loss and risk in smoothing parameter selection. *Journal of Nonparametric Statistics* **4**, 107–132.

Hainz, G. (1995). The asymptotic properties of Burg estimators. Preprint Univ. of Heidelberg.

Haldane, J.B.S. (1951). A class of efficient estimates of a parameter. *Bulletin of the International Statistics Institute* **33**, 231–248.

Hall, P. and Marron, J.S. (1991). Lower bounds for bandwidth selection in density estimation. *Probability Theory and Related Fields* **90**, 149–173.

Hampel, F.R., Ronchetti, E.M., Rousseeuw, P.J. and Stahel, W.A. (1986). *Robust Statistics: The Approach Based on Influence Functions.* Wiley, New York.

Hannan, E.J. (1980). Estimation of the order of an ARMA process. *Annals of Statistics* **8**, 1071–1081.

Hannan, E.J. and Quinn, B.G. (1979). The determination of the order of an autoregression. *Journal of the Royal Statistical Society, B* **41**, 190–195.

Härdle, W., Hall, P. and Marron, J.S. (1988). How far are automatically chosen regression smoothing parameters from their optimum? *Journal of the American Statistical Association* **83**, 86–101.

Hart, J.D. and Yi, S. (1996). One-sided cross-validation. Unpublished manuscript.

Harvey, A.C. (1989). *Forecasting, Structural Time Series Models and the Kalman Filter.* Cambridge University Press, New York.

Hastie, T. and Tibshirani, R.J. (1990). *Generalized Additive Models.* Chapman and Hall, London.

He, X. and Shi, P. (1996). Bivariate tensor-product B-splines in a partly linear model. *Journal of Multivariate Analysis* **58**, 162–181.

Herrmann, E. (1997). Local bandwidth choice in kernel regression estimation. *Journal of Computational and Graphical Statistics* **6**, 35-54.

Herrmann, E. (1996). On the convolution type kernel regression estimator. Unpublished manuscript.

Hosmer, D.W., Jovanovic, B. and Lemeshow, S. (1989). Best subsets logistic regression. *Biometrics* **45**, 1265–1270.

Hubbard, B.B. (1996). *The World According To Wavelets.* A. K. Peters, MA.

Huber, P.J. (1964). Robust estimation of a location parameter. *Annals of Mathematical Statistics* **35**, 73–101.

Huber, P.J. (1981). *Robust Statistics.* Wiley, New York.

Hurvich, C.M., Shumway, R.H. and Tsai, C.L. (1990). Improved estimators of Kullback-Leibler information for autoregressive model selection in small samples. *Biometrika* **77**, 709–719.

Hurvich, C.M., Simonoff, J.S., and Tsai, C.L. (1998). Smoothing parameter selection in nonparametric regression using an improved Akaike information criterion. *Journal of the Royal Statistical Society, B*, to appear.

Hurvich, C.M. and Tsai, C.L. (1989). Regression and time series model selection in small samples. *Biometrika* **76**, 297–307.

Hurvich, C.M. and Tsai, C.L. (1990). Model selection for least absolute deviations regression in small samples. *Statistics and Probability Letters* **9**, 259–265.

Hurvich, C.M. and Tsai, C.L. (1990). The impact of model selection on inference in linear regression. *The American Statistician* **44**, 214–217.

Hurvich, C.M. and Tsai, C.L. (1991). Bias of the corrected AIC criterion for underfitted regression and time series models. *Biometrika* **78**, 499–509.

Hurvich, C.M. and Tsai, C.L. (1993). A corrected Akaike information criterion for vector autoregressive model selection. *Journal of Time Series* **14**, 271–279.

Hurvich, C.M. and Tsai, C.L. (1995). Model selection for extended quasi-likelihood in small samples. *Biometrics* **51**, 1077–1084.

Hurvich, C.M. and Tsai, C.L. (1995). Relative rates of convergence for efficient model selection criteria in linear regression. *Biometrika* **82**, 418–425.

Hurvich, C.M. and Tsai, C.L. (1996). The impact of unsuspected serial correlations on model selection in linear regression. *Statistics and Probability Letters* **33**, 115–126.

Hurvich, C.M. and Tsai, C.L. (1997). Selection of a multistep linear predictor for short time series. *Statistica Sinica* **7**, 395–406.

Hurvich, C.M. and Tsai, C.L. (1998). A cross-validatory AIC for hard wavelet thresholding in spatially adaptive function estimation. *Biometrika*, to appear.

Jones, M.C. (1986). Expressions for inverse moments of positive quadratic forms in normal variables. *Australian Journal of Statistics* **28**, 242–250.

Jones, M.C. (1987). On moments of ratios of quadratic forms in normal variables. *Statistics and Probability Letters* **6**, 129–136.

Jones, M.C. (1991). The roles of ISE and MISE in density estimation. *Statistics and Probability Letters* **12**, 51–56.

Jones, M.C. and Kappenman, R.F. (1991). On a class of kernel density estimate bandwidth selectors. *Scandinavian Journal of Statistics* **19**, 337–349.

Jones, M.C., Davies, S.J. and Park, B.U. (1994). Versions of kernel-type regression estimators. *Journal of the American Statistical Association* **89**, 825–832.

Jørgensen, B. (1987). Exponential dispersion models (with discussion). *Journal of the Royal Statistical Society, B* **49**, 127–162.

Konoshi, S. and Kitagawa, G. (1996). Generalized information criteria in model selection. *Biometrika* **83**, 875–890.

Kullback, S. and Leibler, R.A. (1951). On information and sufficiency. *Annals of Mathematical Statistics* **22**, 79–86.

Lai, T.L. and Lee, C.P. (1997). Information and prediction criteria for model selection in stochastic regression and ARMA models. *Statistica Sinica* **7**, 285–309.

Lawless, J.F. (1982). *Statistical Models And Methods for Lifetime Data.* Wiley, New York.

Léger, C. and Altman, N. (1993). Assessing influence in variable selection problems. *The Journal of the American Statistical Association* **88**, 547–556.

Li, K.C. (1991). Sliced inverse regression for dimension reduction. *Journal of the American Statistical Association* **86**, 316–342.

Linhart, H. and Zucchini, W. (1986). *Model Selection.* Wiley, New York.

Liu, S.I. (1996). Model selection for multiperiod forecasts. *Biometrika* **83**, 861–873.

Loader, C.R. (1995). Old Faithful erupts: bandwidth selection revisited. Unpublished manuscript.

Lütkepohl, H. (1985). Comparison of criteria for estimating the order of a vector autoregressive process. *Journal of Time Series* **6**, 35–52.

Lütkepohl, H. (1991). *Introduction to Multiple Time Series Analysis.* Springer-Verlag, New York.

Mallat, S.G. (1989). A theory for multiresolution signal decomposition: the wavelet representation. *IEEE Trans. on Pattern Analysis and Machine Intelligence* **11**, 674–693.

Mallows, C.L. (1973). Some comments on Cp. *Technometrics* **15**, 661–675.

Mallows, C.L. (1995). More comments on Cp. *Technometrics* **37**, 362–372.

Mammen, E. (1990). A short note on optimal bandwidth selection for kernel estimators. *Statistics and Probability Letters* **9**, 23–25.

Marron, J.S. and Wand, M.P. (1992). Exact mean integrated squared error. *Annals of Statistics* **20**, 712–736.

McCullagh, P. and Nelder, J.A. (1989). *Generalized Linear Models*, 2nd Edition. Chapman and Hall, New York.

McQuarrie, A.D.R. (1995). Small-sample model selection in regressive and autoregressive models: A signal-to-noise approach. Ph.D. Dissertation, Graduate Division, University of California at Davis.

McQuarrie, A.D.R., Shumway, R.H., and Tsai, C.L. (1997). The model selection criterion AICu. *Statistics and Probability Letters* **34**, 285–292.

Muirhead, R.J. (1982). *Aspects of Multivariate Statistical Theory*. Wiley, New York.

Nason, G.P. (1996). Wavelet regression by cross-validation. *Journal of the Royal Statistical Society, B* **58**, 463-479.

Nason, G.P. and Silverman, B.W. (1994). The discrete wavelet transform in S. *Journal of Computational and Graphical Statistics* **3**, 163–191.

Nelder, J.A. and Pregibon, D. (1987). An extended quasi-likelihood function. *Biometrika* **74**, 221–232.

Nishii, R. (1984). Asymptotic properties of criteria for selection of variables in multiple regression. *Annals of Statistics* **12**, 758–765.

Pregibon, D. (1979). Data analytic methods for generalized linear models. Ph.D. thesis, University of Toronto, Canada.

Press, W.H., Flannery, B.P., Teukolsky, S.A. and Vetterling, W.T. (1986). *Numerical Recipes*. Cambridge University Press, Cambridge.

Priestley, M.B. (1981). *Spectral Analysis and Time Series,* Vols. 1 and 2. Academic Press, New York.

Priestley, M.B. (1988). *Non-linear and Non-stationary Time Series Analysis*, Academic Press, London.

Pukkila, T., Koreisha, S., and Kallinen, A. (1990). The identification of ARMA models. *Biometrika* **77**, 537–548.

Rao, C.R. (1973). *Linear Statistical Inference and Its Applications*, 2nd edition. Wiley, New York.

Rao, C.R. and Wu, Y. (1989). A strongly consistent procedure for model selection in a regression problem. *Biometrika* **76**, 369–374.

Rawlings, J.O. (1988). *Applied Regression Analysis*. Wadsworth, Belmont.

Rice, J. (1984). Bandwidth choice for nonparametric regression. *Annals of Statistics* **12**, 1215–1230.

Ronchetti, E. (1985). Robust model selection in regression. *Statistics and Probability Letters* **3**, 21–23.

Ronchetti, E. (1997). Robustness aspects of model choice. *Statistica Sinica* **7**, 327-338.

Ronchetti, E., Field, C. and Blanchard W. (1997). Robust linear model selection by cross-validation. *Journal of the American Statistical Association* **92**, 1017–1032.

Ronchetti, E. and Staudte, G. (1994). A robust version of Mallows's Cp. *Journal of the American Statistical Association* **89**, 550-559.

Ruppert, D., Sheather, S.J. and Wand, M.P. (1995). An effective bandwidth selector for local least squares regression. *Journal of the American Statistical Association* **90**, 1257–1270.

Schumaker, L.L. (1981). *Spline Functions.* Wiley, New York.

Schwarz, G. (1978). Estimating the dimension of a model. *Annals of Statistics* **6**, 461–464.

Shao, J. (1993). Linear model selection by cross-validation. *Journal of the American Statistical Association* **88**, 486–494.

Shao, J. (1996). Bootstrap model selection. *Journal of the American Statistical Association* **91**, 655–665.

Shao, J. (1997). An asymptotic theory for linear model selection. *Statistica Sinica* **7**, 221–264

Sheather, J.S. (1996). Bandwidth selection: plug-in methods versus classical methods. Paper presented at Joint Statistical Meetings, Chicago, IL.

Shi, P. and Li, G.Y. (1995). Global rates of convergence of B-spline M-estimates for nonparametric regression. *Statistica Sinica* **5**, 303–318.

Shi, P. and Tsai, C.L. (1997). Semiparametric regression model selection. Technical Report, Graduate School of Management, University of California at Davis.

Shi, P. and Tsai, C.L. (1998). A note on the unification of the Akaike information criterion. *Journal of the Royal Statistical Society, B*, to appear.

Shi, P. and Tsai, C.L. (1998). On the use of marginal likelihood in model selection. Technical Report, Graduate School of Management, University of California at Davis.

Shibata, R. (1980). Asymptotic efficient selection of the order of the model for estimating parameters of a linear process. *Annals of Statistics* **8**, 147–164.

Shibata, R. (1981). An optimal selection of regression variables. *Biometrika* **68**, 45–54.

Shibata, R. (1984). Approximate efficiency of a selection procedure for the number of regression variables. *Biometrika* **71**, 43–49.

Shibata, R. (1997). Bootstrap estimate of Kullback-Leibler information for model selection. *Statistica Sinica* **7**, 375–394.

Silvapulle, M.J. (1985). Asymptotic behaviour of robust estimators of regression and scale parameters with fixed carriers. *Annals of Statistics* **13**, 1490–1497.

Simonoff, J.S. (1996). *Smoothing Methods in Statistics.* Springer-Verlag, New York.

Simonoff, J.S. (1998). Three sides of smoothing: categorical data smoothing, nonparametric regression, and density estimation. *International Statistical Review* to appear.

Simonoff, J.S. and Tsai, C.L. (1997). Semiparametric and additive model selection using an improved AIC criterion. Technical Report, Graduate School of Management, University of California at Davis.

Sommer, S. and Huggins, R.M. (1996). Variables selection using the Wald test and a robust Cp. *Applied Statistics* **45**, 15–29.

Sparks, R.S., Coutsourides, D. and Troskie, L. (1983). The Multivariate Cp. *Communications in Statistics — Theory and Methods.* **12**, 1775–1793.

Speckman, P.L. (1988). Kernel smoothing in partial linear models. *Journal of the Royal Statistical Society, B* **50**, 413–436.

Strang, G. (1993). Wavelet transforms versus Fourier transforms. *Bulletin (New Series) of the American Mathematical Society* **28**, 288–305.

Sugiura, N. (1978). Further analysis of the data by Akaike's information criterion and the finite corrections. *Communications in Statistics — Theory and Methods* **7**, 13–26.

Terrell, G.R. (1992). Discussion of "The performance of six popular bandwidth selection methods on some real data sets" and "Practical performance of several data driven bandwidth selectors." *Computational Statistics* **7**, 275–277.

Thall, P.F., Russell, K.E. and Simon, R.M. (1997). Variable selection in regression via repeated data splitting. *Journal of Computational and Graphical Statistics* **6**, 416–434.

Tibshirani, R.J. and Hastie, T. (1987). Local likelihood estimation. *Journal of the American Statistical Association* **82**, 559–568.

Tsay, R.S. (1984). Regression models with time series errors. *Journal of the American Statistical Association* **79**, 118–124.

Turlach, B.A. and Wand, M.P. (1996). Fast computation of auxiliary quantities in local polynomial regression. *Journal of Computational and*

*Graphical Statistics* **5**, 337-350.

Wahba, G. (1990). *Spline Models for Observational Data.* SIAM, Philadelphia, PA.

Wei, C.Z. (1992). On predictive least squares principles. *Annals of Statistics* **20**, 1–42.

Wei, W.S. (1990). *Time Series Analysis.* Addison-Wesley, New York.

Weisberg, S. (1981). A statistic for allocating Cp to individual cases. *Technometrics* **23**, 27–31.

Weisberg, S. (1985). *Applied Linear Regression*, 2nd edition. Wiley, New York.

Wu, C.F.J. (1986). Jackknife, bootstrap and other resampling methods in regression analysis (with discussions). *Annals of Statistics* **14**, 1261–1350.

Zeger, S.L. and Qaqish, B. (1988). Markov regression models for time series: A quasi-likelihood approach. *Biometrics* **44**, 1019–1031.

Zhang, P. (1993). On the convergence rate of model selection criteria. *Communications in Statistics — Theory and Methods* **22**, 2765–2775.

Zheng, X. and Loh, W.Y. (1995). Consistent variable selection in linear models. *Journal of the American Statistical Association* **90**, 151–156.

## Author Index

Akaike, H., 2–4, 15, 19, 20, 22, 89, 94–96, 146, 203, 252, 307, 318, 331, 335, 357
Allen, D. M., 3, 11, 252, 348
Altman, N., 13
Anderson, D. R., 13
Anderson, R. L., 399
Anderson, T. W., 158
Antle, C. E., 297
Bain, L. J., 297
Bancroft, T. A., 399
Bates, D. M., 13
Bedrick, E. J., 147–149, 399
Bhansali, R. J., 2, 24, 127, 219, 366
Blanchard, W., 291
Bloomfield, P., 295
Box, G. E. P., 90
Bozdogan, H., 367
Breiman, L., 366
Brockwell, P. J., 13, 235
Broersen, P. M. T., 13
Burg, J. P., 127
Burman, P., 254, 307–309
Burnham, K. P., 13
Carlin, B. P., 13
Carroll, R. J., 13
Cavanaugh, J. E., 267
Chen, H., 348
Chib, S., 13
Chipman, H., 13
Choi, B., 13
Chu, C. K., 347
Cleveland, W. S., 334, 336
Coutsourides, D., 141, 146
Craven, P., 331, 348
Daubechies, I., 353

Davies, S. J., 347
Davis, R. A., 13, 235
Davisson, L. D., 19
Devlin, S. J., 334, 336
Diggle, P. J., 13
Donoho, D. L., 12, 329, 351–354, 356, 359, 362
Downham, D. Y., 2, 24, 219, 366
Efron, B., 261, 263, 264, 272, 317
Fan, J., 13
Field, C., 291
Freedman, D., 366
Fujikoshi, Y., 180
Fuller, W. A., 13
Furnival, G. M., 427
Geisser, S., 254
George, E. I., 13
Geweke, J., 2, 366
Gijbels, I., 13
Gouriéroux, C., 13
Gradshteyn, I. S., 66–68
Green, P. J., 348
Grund, B., 333
Hainz, G., 127
Haldane, J. B. S., 6
Hall, P., 333, 337
Hamada, M., 13
Hampel, F. R., 293
Hannan, E. J., 2, 15, 23, 89, 96, 149, 206
Hart, J. D., 333, 338
Harvey, A. C., 13
Hastie, T., 332
He, X., 349
Herrmann, E., 338, 339, 347, 348
Hoffstedt, C., 377
Hosmer, D. W., 316, 317, 319
Hubbard, B. B., 351
Huber, P. J., 311, 315
Huggins, R. M., 306

Hurvich, C. M., 2, 3, 13, 15, 21, 22, 45, 93, 127–129, 205, 295–297, 304, 309, 310, 315, 317, 319, 327, 329, 333, 335–338, 352, 357, 358, 360, 362
Härdle, W., 337
Jenkins, G. M., 90
Johnstone, I. M., 12, 329, 351–354, 356, 359, 362
Jones, M. C., 333, 335, 347
Jørgensen, B., 327
Kallinen, A., 13
Kappenman, R. F., 333
Kerkyacharian, G., 353, 354
Kitagawa, G., 13
Konishi, S., 13
Koreisha, S., 13
Kullback, S., 6, 15
Lai, T. L., 13
Lawless, J. F., 13, 314, 315
Lee, C. P., 13
Leibler, R. A., 6, 15
Li, G. Y., 350
Li, K. C., 13
Liang, K. Y., 13
Linhart, H., 1, 6, 21, 261, 335
Liu, S. I., 129
Loh, W. Y., 13
Lütkepohl, H., 235
Léger, C., 13
Mallat, S. G., 352, 353
Mallows, C. L., 2, 15, 20, 95, 305
Mammen, E., 333
Marron, J. S., 333, 337, 339, 347
McCullagh, P., 293, 317, 319
McCulloch, R. E., 13
McQuarrie, A. D. R., 32
Meese, R., 2, 366
Muirhead, R. J., 181
Nason, G. P., 329, 352, 353, 355, 356, 359, 362
Nelder, J. A., 293, 316–319
Nishii, R., 4, 13, 22
Nolan, D., 307–309

Park, B. U., 347
Picard, D., 353, 354
Pregibon, D., 316–319
Press, W. H., *et al.*, 57
Priestley, M. B., 90, 96
Pukkila, T., 13
Qaqish, B., 13
Quinn, B. G., 2, 15, 23, 89, 96, 149, 206
Rao, C. R., 13, 46
Rawlings, J. O., 31, 427
Rice, J., 331
Ronchetti, E., 291, 293, 304, 305, 310, 313, 315
Rousseeuw, P. J., 293
Ruppert, D., 12, 13, 332, 338, 339
Russell, K. E., 291
Satoh, K., 180
Schumaker, L. L., 349
Schwarz, G., 2, 15, 22, 23, 96, 206, 357
Shao, J., 13, 254, 255, 266, 269, 273, 291
Sheather, S. J., 12, 332, 338, 339
Shi, P., 13, 310–314, 329, 349, 350
Shiau, J. H., 348
Shibata, R., 2, 3, 7, 13, 22, 24, 268, 361, 362, 366
Shumway, R. H., 32, 267, 310, 315
Silvapulle, M. J., 312
Silverman, B. W., 348, 353, 362
Simon, R. M., 291
Simonoff, J. S., 329, 333, 335–338, 348, 363
Sommer, S., 306
Sparks, R. S., 141, 146
Speckman, P. L., 329
Stahel, W. A., 293
Staudte, R. G., 304, 305, 315
Steiger, W. L., 295
Strang, G., 353
Sugiura, N., 2, 3, 15, 21
Terrell, G. R., 333
Thall, P. F., 291
Tibshirani, R. J., 263, 264, 272, 332

Troskie, L., 141, 146
Tsai, C. L., 2, 3, 13, 15, 21, 22, 32, 45, 93, 127–129, 147–149, 205, 295–297, 304, 309–315, 317, 319, 327, 329, 333, 335–338, 348–350, 352, 357, 358, 360, 362, 399
Tsay, R. S., 13
Turlach, B. A., 337
Wahba, G., 329, 331, 348
Wand, M. P., 12, 13, 332, 337–339
Watts, D. G., 13
Wei, C. Z., 367
Wei, W. S., 386, 387, 409
Weisberg, S., 13, 377
Wensink, H. E., 13
Wilson, W., 427
Wu, C. F. J., 13, 266
Wu, Y., 13
Yi, S., 333, 338
Zeger, S. L., 13
Zheng, X., 13
Zucchini, W., 1, 6, 21, 262, 335

# Index

AIC, 268, 309, 364, 366, 411
defined, 21, 93, 147, 204, 319, 350
$L_1$ regression, 295
misspecified MA(1) models, 124
multivariate regression, 147, 149, 154, 157, 167, 176, 393
nonparametric regression, 331
quasi-likelihood, 317
semiparametric regression, 349
univariate autoregressive models, 93, 97, 101, 116, 118, 130, 380
univariate regression, 18, 20, 25, 32, 36, 52, 374
vector autoregressive models, 204, 207, 213, 223, 232, 234, 403
wavelets, 357
AIC$\alpha$, 43
defined, 24
AICb, 268
AICc, 267, 274, 276, 280, 309, 363, 366, 410
defined, 22, 94, 148, 205, 319, 337, 350
$L_1$ regression, 295
misspecified MA(1) models, 124, 388
multivariate regression, 147, 149, 154, 157, 167, 176, 393
nonparametric regression, 333
quasi-likelihood, 317
semiparametric regression, 349
univariate autoregressive models, 93, 97, 101, 115, 118, 130, 380
univariate multistep autoregressive models, 128
univariate regression, 18, 20, 25, 32, 37, 45, 64, 369
vector autoregressive models, 205, 207, 214, 227, 231, 402
wavelets, 352
AICc$_0$, defined, 335
nonparametric regression, 333
AICc$_1$, defined, 336
nonparametric regression, 333
AICcm, defined, 128
univariate multistep autoregressive models, 128

AICcR, 310
    defined, 313
AICcR*, 312
    defined, 313
AICi, 310
AICm, defined, 129
    univariate multistep autoregressive models, 128
AICR, 310
AICR*, 310
    defined, 313
AICu, 93, 267, 280, 366, 410
    defined, 32, 94, 154, 205, 320
    misspecified MA(1) models, 124, 388
    multivariate regression, 154, 157, 167, 176, 180, 393
    quasi-likelihood, 320
    univariate autoregressive models, 97, 102, 115, 118, 130, 380
    univariate regression, 32, 37, 45, 62, 369
    vector autoregressive models, 205, 207, 215, 225, 231, 234, 403
Akaike Information Criterion *see* AIC
$\alpha 1$, 43, 282, 288, 386
$\alpha 2$, 43, 282, 288, 375, 386, 390, 411
$\alpha 3$, 43, 282, 288, 375, 386, 392, 411
$\alpha\infty$, 43, 288, 375, 386, 411
AR model *see* univariate autoregressive model
Asymptotic efficiency, 3
    defined, 7
Bayesian Information Criterion *see* BIC
Beta distribution, 47
BFPE, 267, 274, 276, 279, 366
    defined, 267, 270
    misspecified MA(1) models, 388
    univariate autoregressive models, 380
    univariate regression, 377
BFPE *see also* TrBFPE and DeBFPE
BIC, 22, 96
Binomial distribution, 319
Bootstrap, 261
    univariate regression, 262
Bootstrap *see also* naive bootstrap and refined bootstrap

BP, 279
BR, 279
Candidate model, extended quasi-likelihood, 318
  $L_1$ regression, 294
  multivariate regression, 142
  semiparametric regression, 349
  univariate autoregressive models, 89
  univariate regression, 16
  vector autoregressive models, 199
$\chi^2$ distribution, 28, 41, 106, 158, 213
Consistency, 3
Cp, 304, 306, 366, 411
  defined, 20, 95, 146, 319
  misspecified MA(1) models, 124
  multivariate regression, 146, 149, 154, 167, 393
  nonparametric regression, 334
  quasi-likelihood, 317
  semiparametric regression, 348
  univariate autoregressive models, 95, 97, 101, 111, 118, 380
  univariate regression, 19, 27, 37, 45, 373
  vector autoregressive models, 408
Cp *see also* TrCp and MeCp
Cp*, defined, 319
  quasi-likelihood, 324
Cross-validation *see* CV(1) and CV($d$)
Cubic smoothing spline estimator, 331
CV *see* CV(1)
CV *see* Nason's cross-validation
CV(1), 11, 290, 366
  defined, 253, 256
  misspecified MA(1) models, 390
  multivariate regression, 257
  semiparametric regression, 348
  univariate autoregressive models, 256, 387
  univariate regression, 252, 373
  vector autoregressive models, 260
CV(1) *see also* TrCV and DeCV
CV($d$), defined, 255, 256
  multivariate regression, 259

univariate regression, 254, 276, 278, 280
vector autoregressive models, 261
CV($d$) *see also* TrCV($d$) and DeCV($d$)
CVd *see* CV($d$)
DCVB, 267, 270, 274, 276, 279, 291, 366
defined, 267, 270
misspecified MA(1) models, 388
univariate autoregressive models, 380
univariate regression, 371
DCVB *see also* TrDCVB and DeDCVB
DeBFPE, 274, 276, 367
defined, 274, 276
multivariate regression, 397
vector autoregressive models, 405
DeCV, 258, 367
defined, 258, 261
multivariate regression, 399
vector autoregressive models, 408
DeCV($d$), defined, 259, 261
DeDCVB, 367
defined, 274, 276
multivariate regression, 393
vector autoregressive models, 405
Det($L_2$) distance, defined, 144, 201
multivariate regression, 179
vector autoregressive models, 233
Det($L_2$) expected distance, multivariate regression, 169, 171
vector autoregressive models, 223
Det($L_2$) expected efficiency, defined, 168, 223
multivariate regression, 169
vector autoregressive models, 223
Det($L_2$) observed efficiency, defined, 176, 230
multivariate regression, 177
vector autoregressive models, 234, 405
Distributions *see* Beta, binomial, $\chi^2$, double exponential, F, log-Beta, log-$\chi^2$, multivariate normal, noncentral Beta, noncentral $\chi^2$, noncentral log-Beta, noncentral log-$\chi^2$, normal, U, Wishart
Double exponential distribution, 295
Efficiency *see also* asymptotic efficiency, K-L, $L_2$, det($L_2$), and tr{$L_2$}

Exponential regression, 316
Extended quasi-likelihood, candidate model, 318
    true model, 317
F distribution, 36, 106, 158, 213
FIC, 367
    defined, 367
    multivariate regression, 397
    vector autoregressive models, 402
FPE, 145, 252, 256, 258, 263, 271, 290, 307, 366, 411
    defined, 19, 95, 146, 204
    misspecified MA(1) models, 124
    multivariate regression, 149, 154, 157, 167, 176, 399
    univariate autoregressive models, 93, 97, 101, 116, 118, 382
    univariate multistep autoregressive models, 128
    univariate regression, 19, 27, 33, 37, 45, 370
    vector autoregressive models, 203, 206, 215, 224, 233, 408
FPE4, 24, 367
    defined, 366
    misspecified MA(1) models, 388
    univariate regression, 369
FPE$\alpha$, 366
    defined, 24
FPEm, defined, 129
    univariate multistep autoregressive models, 128
FPEu, 93, 290, 366
    defined, 33, 95
    misspecified MA(1) models, 124, 390
    multivariate regression, 154
    univariate autoregressive models, 97, 102, 118, 380
    univariate regression, 32, 37, 45, 369
Gasser–Müller convolution kernel estimator, 330
Gasser–Müller estimator, 347
GCV, nonparametric regression, 331, 348
Generalized linear models, 316
Generating model *see* candidate model
Geweke and Meese Criterion *see* GM
GM, 304, 366
    defined, 366
    misspecified MA(1) models, 390

univariate autoregressive models, 386
univariate regression, 369
Hannan and Quinn Criterion *see* HQ
Hard wavelet thresholding, 352, 354
HQ, 23, 366, 411
defined, 23, 97, 149, 206, 320
$L_1$ regression, 297
misspecified MA(1) models, 124
multivariate regression, 149, 156, 157, 168, 176, 397
quasi-likelihood, 322
univariate autoregressive models, 93, 97, 101, 116, 118, 130, 380
univariate regression, 29, 34, 40, 45, 373
vector autoregressive models, 203, 207, 216, 223, 232, 402
HQc, 93, 366, 411
defined, 35, 97, 156, 206, 320
$L_1$ regression, 297
misspecified MA(1) models, 124, 388
multivariate regression, 154, 157, 168, 176, 180, 393
quasi-likelihood, 321
univariate autoregressive models, 97, 103, 115, 118, 130, 380
univariate regression, 32, 40, 45, 62, 369
vector autoregressive models, 206, 207, 216, 227, 231, 402
ICOMP, 367
defined, 367
multivariate regression, 393
vector autoregressive models, 403
K-L distance, 6
defined, 6, 19, 93, 121, 145, 203, 296, 320, 334
$L_1$ regression, 294
misspecified MA(1) models, 121
multivariate regression, 144, 179
nonparametric regression, 331, 336
quasi-likelihood, 319
univariate autoregressive models, 117
univariate regression, 17, 48, 280
vector autoregressive models, 233
K-L expected distance, misspecified MA(1) models, 123, 124
multivariate regression, 169, 171
univariate autoregressive models, 111

univariate regression, 49, 52
vector autoregressive models, 223
K-L expected efficiency, defined, 8, 52, 111, 124, 168, 223
misspecified MA(1) models, 124
multivariate regression, 169
univariate autoregressive models, 111
univariate regression, 54
vector autoregressive models, 223
K-L information *see* K-L distance
K-L observed efficiency, defined, 8, 61, 117, 121, 124, 176, 230, 278, 285, 298, 321
$L_1$ regression, 300
misspecified MA(1) models, 125, 388
multivariate regression, 177, 395
quasi-likelihood, 319, 323
univariate autoregressive models, 287, 384
univariate regression, 278, 369, 373
vector autoregressive models, 234, 405
Kullback–Leibler discrepancy *see* K-L distance
Kullback–Leibler information *see* K-L distance
Kullback–Leibler observed efficiency *see* K-L observed efficiency
$\hat{L}(k)$, defined, 309
$L_1$ distance, defined, 7, 294, 320
$L_1$ regression, 294, 299
quasi-likelihood, 320, 323
$L_1$ observed efficiency, defined, 294, 298, 321
$L_1$ regression, 299, 300
quasi-likelihood, 323
$L_1$ regression, 293, 304, 307
L1AICc, 310, 326
defined, 297
$L_1$ regression, 297
$L_2$ distance, defined, 6, 18, 91, 121, 144, 201, 294, 320
$L_1$ regression, 294
quasi-likelihood, 320
univariate autoregressive models, 117
univariate regression, 17, 27, 48, 280
$L_2$ expected distance, misspecified MA(1) models, 123, 124
univariate autoregressive models, 111

univariate regression, 49, 52
$L_2$ expected efficiency, defined, 8, 52, 111, 123
misspecified MA(1) models, 124
univariate autoregressive models, 111
univariate regression, 54
$L_2$ observed efficiency, defined, 7, 61, 117, 124, 278, 285, 298, 321
$L_1$ regression, 300
misspecified MA(1) models, 125, 388
quasi-likelihood, 323
univariate autoregressive models, 287, 384
univariate regression, 281, 369, 373
$L_2$ *see also* tr$\{L_2\}$ and det$(L_2)$
Least absolutes regrerssion *see* $L_1$ regression
Local polynomial estimator, 330
Location–scale regression models, 312
Log-Beta distribution, 48, 158, 214
Log-$\chi^2$ distribution, 47
Logistic regression, 306, 316
MA(1) model, true model, 120, 387
Mallows Cp *see* Cp
MASE, 331, 340
Mean average squared error *see* MASE
Mean integrated squared error *see* MISE
MeCp, 367
defined, 147
multivariate regression, 393
vector autoregressive models, 406
MISE, 332, 359
Model selection criterion *see*, AIC, AIC$\alpha$, AICc, AICc$_0$, AICc$_1$, AICcm, AICcR, AICcR*, AICm, AICR, AICR*, AICu, BFPE, BIC, BP, BR, Cp, Cp*, CV(1), CV($d$), DCVB, FIC, FPE, FPE4, FPE$\alpha$, FPEm, FPEu, GCV, GM, HQ, HQc, ICOMP, $\hat{L}(k)$, L1AICc, NB, PRESS, $R^2_{adj}$, RCp, Rp, RTp, SIC, Sp, Tp
MPEP, 257, 270, 275
defined, 257
MSEP, 251
defined, 252
Multistep autoregressive model, 127
Multivariate normal distribution, 142, 144, 200, 202

Multivariate regression model, general model, 142
    overfitted model, 143
    true model, 142, 392
    underfitted model, 143
Naive bootstrap, defined, 267, 270, 273, 275
    multivariate regression, 273
    univariate autoregressive models, 270
    univariate regression, 267
    vector autoregressive models, 275
Nason's cross-validation, wavelets, 362
NB, 279
Noncentral Beta distribution, 47
Noncentral $\chi^2$ distribution, 46
Noncentral log-Beta distribution, 47
Noncentral log-$\chi^2$ distribution, 47
Nonparametric regression, true model, 330
Normal distribution, 16, 18, 89, 92
Observed efficiency, 2
Overfitting, 8
Poisson regression, 316
PRESS *see* CV(1)
Quasi-likelihood, 317
Quasi-likelihood *see also* extended quasi-likelihood
$R^2_{adj}$, 280, 290, 367, 411
    defined, 31
    univariate regression, 25, 43, 370
RCp, 304
    defined, 305
Real data examples, highway data, 377
    housing data, 409
    tobacco leaf data, 399
    Wolf yearly sunspot numbers, 386
Refined bootstrap, defined, 269, 272, 275
    univariate autoregressive models, 268
    vector autoregressive models, 275
Refined bootstrapped, defined, 265
    multivariate regression, 272
    univariate regression, 265
Robust regression, 293

Rp, 366
  defined, 366
  misspecified MA(1) models, 390
  univariate autoregressive models, 380
  univariate regression, 373
RTp, 306
  defined, 307
  $L_1$ regression, 297
  quasi-likelihood, 320
Schwarz Information Criterion *see* SIC
Semiparametric regression, 348
  candidate model, 349
  true model, 349
SIC, 22, 255, 280, 366
  defined, 23, 96, 149, 206, 320
  $L_1$ regression, 297
  misspecified MA(1) models, 124, 390
  multivariate regression, 149, 157, 168, 176, 397
  quasi-likelihood, 322
  univariate autoregressive models, 96, 97, 101, 118, 386
  univariate regression, 29, 40, 45, 373
  vector autoregressive models, 205, 207, 216, 227, 232, 402
  wavelets, 357
Signal-to-noise ratio, 24
Sp, 366
SPE, 144
  defined, 258
SSE, 17
Stepwise regression, univariate regression, 427
Superlinear, 26
Tp, 306
TrBFPE, 367
  defined, 274, 276
  vector autoregressive models, 403
TrCp, 367
  defined, 146
  multivariate regression, 399
  vector autoregressive models, 406
TrCV, 367

defined, 258, 260, 261
vector autoregressive models, 403
TrCV($d$), defined, 260, 261
TrDCVB, 367
defined, 274, 276
vector autoregressive models, 403
Tr$\{L_2\}$ distance, defined, 144, 201
multivariate regression, 179
vector autoregressive models, 233
Tr$\{L_2\}$ expected distance, multivariate regression, 169, 171
vector autoregressive models, 223
Tr$\{L_2\}$ expected efficiency, defined, 168, 223
multivariate regression, 169
vector autoregressive models, 224
Tr$\{L_2\}$ observed efficiency, defined, 176, 230
multivariate regression, 177, 393
vector autoregressive models, 234, 405
True model, extended quasi-likelihood, 317
$L_1$ regression, 294
multivariate regression, 142
nonparametric regression, 330
semiparametric regression, 349
univariate autoregressive models, 91
univariate moving average MA(1) model, 120
univariate regression, 16
vector autoregressive models, 200
U distribution, 158, 213
Underfitting, 8
Univariate autoregressive model, general model, 89
true model, 91, 284, 379
Univariate regression model, general model, 16
overfitted model, 17
true model, 16, 277, 369
underfitted model, 17
VAR model *see* vector autoregressive model
Vector autoregressive model, general model, 199
true model, 401
Wavelets, 352, 364
Wishart distribution, 235